Principles of Crop Production

Principles of Crop Production

Edited by **Corey Aiken**

SYRAWOOD
PUBLISHING HOUSE

New York

Published by Syrawood Publishing House,
750 Third Avenue, 9th Floor,
New York, NY 10017, USA
www.syrawoodpublishinghouse.com

Principles of Crop Production
Edited by Corey Aiken

© 2016 Syrawood Publishing House

International Standard Book Number: 978-1-68286-241-4 (Hardback)

Printed in the United States of America.

Contents

Permissions

List of Contributors

Preface

I am honored to present to you this unique book which encompasses the most up-to-date data in the field. I was extremely pleased to get this opportunity of editing the work of experts from across the globe. I have also written papers in this field and researched the various aspects revolving around the progress of the discipline. I have tried to unify my knowledge along with that of stalwarts from every corner of the world, to produce a text which not only benefits the readers but also facilitates the growth of the field.

The book focuses on various principles of crop production by elucidating the theoretical and conceptual applications. It aims to outline methods to improve the yield of crop plants through selective plant breeding and hybridization. The book discusses practical applications, methods and practices in plant breeding, biotechnology in crop improvement and seed production of field crops. It will provide valuable knowledge to the interested readers, experts and agriculturists. It is designed to outline future developments and challenges in this field.

Finally, I would like to thank all the contributing authors for their valuable time and contributions. This book would not have been possible without their efforts. I would also like to thank my friends and family for their constant support.

Editor

EFFECTS OF LEONARDITE APPLICATIONS ON YIELD AND SOME QUALITY PARAMETERS OF POTATOES (*Solanum tuberosum* L.)

Arif SANLI[1]*, Tahsin KARADOGAN[1], Muhammet TONGUC[2]

[1]*Süleyman Demirel University, Faculty of Agriculture, Field Crops Department, Isparta, TURKEY*
[2]*Süleyman Demirel University, Faculty of Agriculture, Agricultural Biotechnology Department, Isparta, TURKEY*
*Corresponding author: arifsanli@sdu.edu.tr

ABSTRACT

The present study was carried out at Suleyman Demirel University research farm in Isparta during 2008 and 2009 crop seasons to determine the effects of leonardite (54.5% organic matter) on yield and quality of potato. Four leonardite doses (0, 200, 400, 600 kg ha^{-1}) and four potato cultivars (Van Gogh, Milva, Lady Olympia, Agata) were used in the study. A total of seven traits (plant height, tuber number per plant, marketable tuber yield, total tuber yield, protein content, vitamin C content and specific gravity) were measured. Significant differences were detected between control and leonardite applications for all the traits studied. Statistically significant differences were not found between the leonardite doses for plant height and specific gravity. There was no significant difference between 400 and 600 kg ha^{-1} leornardite application for tuber number per plant, marketable tuber and total tuber yield, protein and vitamin C contents. Leonardite applications increased number of tubers per plant by 22%, marketable tuber yield by 38% and total tuber yield by 15% compared with the control. These results suggest that 400 kg ha^{-1} leonardite combination to the standard fertilization will be sufficient to obtain adequate yield and quality tubers in potatoes.

Keywords: Fertilization, leonardite, potato, yield

INTRODUCTION

Potato plants need significant amounts of fertilizers to produce marketable tubers and high yield. In order to provide necessary nutrients, both chemical and natural fertilizers are used in potato production. Intensive use of chemical fertilizers could cause environmental pollution (Ece et al., 2007). Therefore there is a growing interest to use natural alternatives of chemical fertilizers in agricultural production.

Leonardite is a concentrated form of humic and fulvic acids, and has appearance of coal but it does not reach the compactness of coal (Erkoç, 2009). Leonardite is rich in organic matter (50-75%) and its humic acid content could change between 30-80% (Akinremi et al., 2000). Humic acid and other humic compounds stimulate root and shoot development, increase both available plant nutrients and nutrient uptake from soil, enhance plants resistance to biotic and abiotic stress factors (Akinremi et al., 2000; Dursun et al., 2002; Serenella et al., 2002; Cimrin and Yilmaz, 2005; Ünlü et al., 2010). Effects of humic acid containing fertilizers on plant yield and nutrient uptake depend on humic acid source, concentration, application type, plant species and cultivars (Chen and Aviad, 1990).

Leonardite both provides nutrients and increases absorption of available nutrients in different soil types (Chen and Aviad, 1990). It also reduces soil pH around roots and helps converting unavailable nutrients to plant accessible forms (Vaughan and Donald, 1976). Leonardite could be used as a soil amendment because it releases nutrients slowly to soil thus preventing nutrient loss due to leaching and evaporation (Sibanda and Young, 1989). Humic acid and leonardite increase cation exchange capacity of soil and cell permeability of capillary roots, thus they serve to increase nutrient availability and act as plant growth promoters and stimulates plant development (O'Donnell, 1973; Malcolm and Vaughan, 1979). Humic substances increase water holding capacity of soil and provide plants the opportunity to develop better (Suganya and Sivasamy, 2006; Selim et al., 2009a; Hassanpanah and Khodadadi, 2009). In addition, it was stated that the coal-humic fertilizers activated the biochemical processes in plants (respiration, photosynthesis and chlorophyll content) and increased the quality and yield of potatoes.

Humic acid applications led to significant increase in soil organic matter content and improve plant growth and crop yield (Erik et al., 2000; Hartwigson and Evans, 2000; Hafez, 2003; El-Desuki, 2004). Mahmoud and Hafez

(2010) reported increased yield and quality of potato with increased level of humic acid application. Selim et al. (2011) reported beneficial effects of humic acid application on to plant growth parameters, tuber production and some biochemical indicators of potato crop. Both liquid and solid forms of humic acid applications are effective to increase yield of potato plants. Humic acid applications between 15 to 30 L ha^{-1} increased potato total yield by 2250 kg ha^{-1} and marketable tuber yield by 2750 kg ha^{-1} (Hopkins and Stark, 2003). Verlinden et al. (2009) reported potato tuber production increased 13% by foliar application and 17% by soil fertilization. Soil application of humic acid had a significant increase in plant growth characters, photosynthetic pigments, total and marketable yield and tuber root quality of sweet potato. In addition, humic acid application significantly increased chemical composition of tuber roots and reduced the weight loss and decay percentages (El Sayed Hameda et al., 2011).

New leonardite mines have been discovered and started to the production of this organic material in Turkey. The aim of the present study was to evaluate the effects of leonardite fertilization on yield and some quality parameters of potatoes grown in Isparta province.

MATERIALS AND METHODS

Study was conducted at the research farms of Suleyman Demirel University, Isparta (37° 45' N, 30° 33' E, altitude 1035 m), during 2008 and 2009 crop seasons. Soil type of the experimental area was loam with a pH of

8.2. Nutrient content of the experimental area was determined for the entire location and it was divided two parts and each part was used to grow potatoes during first and second year of the study separately. Leonardite doses used at the research were determined after soil chemical analysis to supplement nutrient content of the soil. Total nitrogen content of the entire location was 0.21% (macro Kjelhdal method), extractable phosphorus and potassium contents were 18.5 mg kg^{-1} (Olsen method) and 166 mg kg^{-1}, respectively. Organic matter content of soil was 1.3% (Walcley-Black method). Total precipitation between April and September was 127 mm and 178 mm for the first and second years of the experiment, respectively and for the same period the long term average rain fall was 188 mm. Average daily temperature for the same period was 19.7 °C and 18.6 °C, for 2008 and 2009 crop seasons, respectively. Long term average daily temperature between April and September was 18.3 °C.

The experiment was established as factorial design with two factors in a randomized complete block design with three replications. Four doses of leonardite (0, 200, 400, 600 kg ha^{-1}) and four potato cultivars (Agata, very early; Milva mid-early; Van Gogh, mid-late; Lady Olympia, late) were used. Commercial leonardite was purchased from Mineral Tarım (Fethiye, Muğla) and potato tubers were purchased from the Potato Producers Organization (Sandıklı, Afyon). Some chemical and physical properties of leonardite used in the study were given in Table 1.

Table 1. Some physical and chemical properties of leonardite used in the study.

Organic matter (%)	Humic + Fulvic acids (%)	N (%)	P (mg kg^{-1})	K (mg kg^{-1})	pH	Humidity (%)	Heavy metal content
54.5	50.5	1.79	24	135	6.80	16	Below threshold values

Tubers were planted at the first week of May at each growing season and leonardite was broadcast at appropriate doses during the planting. Each plot was 6 m in length and consisted of 4 rows. Planting space between rows were 70 cm and tubers were planted 30 cm apart within rows. Nitrogen (200 kg ha^{-1}) and phosphorus (100 kg ha^{-1}) fertilization were uniformly applied as ammonium sulphate and diamonium phosphate to plots during planting (Karadoğan, 1994).

Weed control was done with hand two to three times depending on weed density. Irrigation was performed with drip irrigation system when available soil moisture dropped below 50% in soil. During the blooming period, plant height was recorded from 20 plants from middle two rows of each plot. Tubers were harvested at the second half of September for each growing year.

Plant height and number of tubers per plant were

determined from middle row of each plot on 20 plants. Rest of the plants within the same two rows were harvested with a machine to determine marketable tuber yield (>45 mm) and total tuber yield.

Protein contents of tubers were determined using macro Kjehldal method (Kaçar and İnal, 2008). Vitamin C content was determined according to AOAC (1975) method and results were expressed as mg 100 g fresh weight (fw)$^{-1}$. In order to calculate specific gravity, approximately 5 kg of tubers were weighted both in air and in water, and specific gravity was calculated according to following formula (Esendal, 1990): weight of tubers in air / (weight in air) – (weight in water).

Data was subjected to the analysis of variance (ANOVA) procedure with SAS statistical program (SAS, 1998). Means were separated using Duncan's multiple range tests at the 0.05 significance level.

RESULTS

According to ANOVA results, cultivar (C) and dose (D) effects were significant for all examined traits at 1% level of significance (Table 2). No statistically significant interactions were detected between Y x D and Y x D x C interactions. Y x C interaction was important for marketable tuber yield (1% level) and total tuber yield (5% level). C x D interaction was important for marketable tuber yield (1% level) and protein content (5% level).

Table 2. Results of Analyses Variance (ANOVA) for the traits measured in the study.

Source of variation	Df	Plant height (cm)	Tuber number per plant	Marketable tuber yield (t ha^{-1})	Total Tuber yield (t ha^{-1})	Protein content (%)	Vitamin C content (mg 100 g fw^{-1})	Specific gravity (g/cm^3)
Year (Y)	1	**	**	**	**	ns	ns	ns
Blok (Year)	4	ns	*	*	**	ns	*	**
Cultivar (C)	3	**	**	**	**	**	**	**
Y x C	3	ns	ns	**	*	ns	ns	ns
Doses (D)	3	**	**	**	**	**	**	**
Y x D	3	ns	ns	ns	ns	ns	ns	ns
C x D	9	ns	ns	**	ns	*	ns	ns
Y x C x D	9	ns	ns	ns	ns	ns	ns	ns
Error	60							
CV		4.8	11.1	10.1	7.7	3.3	8.3	0.3

df, degrees of freedom ; ns, non significant; * P<0.05; ** P<0.01

Plant Height

Leonardite applications increased plant height significantly compared to the control (70.7 cm). Statistically significant differences were not observed between leonardite doses (73.8-74.4 cm) used in the study. Plant height of cultivar Van Gogh was found to be lower than the other 3 cultivars used in the study (Table 3).

Tuber number per plant

The lowest number of tubers per plant was recorded (5.3 tubers) in control. The highest number of tubers per plant was obtained from 600 kg ha^{-1} leonardite application (6.5 tubers), but no statistically significant differences were detected between 400 and 600 kg ha^{-1} leonardite applications (Table 3). Among the tested cultivars, Milva had the highest number of tubers per plant (7.7 tubers) and Agata had the lowest number of tubers per plant (4.4 tubers).

Marketable tuber yield

The lowest marketable tuber yield was found in the control (17.5 t ha^{-1}) and marketable tuber yield increased with increased leonardite doses (Table 3). The yield increase between the control and 200 kg ha^{-1} leonardite application was 18%. Compared to the control, marketable tuber yield increased by 31% and 38% for 400 and 600 kg ha^{-1} leonardite applications, respectively. But a statistically significant difference was not found between these two doses for marketable tuber yield (Table 3). Differences were observed for marketable tuber yield between the cultivars. Marketable tuber yield of cultivars Van Gogh, Milva and Lady Olympia was over 20 t ha^{-1} whereas yield of Agata was 12.2 t ha^{-1}(Table 3). There were no statistically significant differences between 400 and 600 kg ha^{-1} leonardite applications for marketable tuber yield in Van Gogh, Milva and Lady Olympia. Marketable tuber yield increased by 4.6 t ha^{-1}, 7.7 t ha^{-1} and 9.1 t ha^{-1} for Van Gogh, Milva and Lady Olympia at 400 kg ha^{-1} leonardite doses compared with the control, but marketable tuber yield did not increase significantly for Agata among the control and leonardite doses.

Total tuber yield

Leonardite applications increased tuber yield and yield increase associated with increased leonardite doses. Average tuber yield was 29.2 t ha^{-1} in control plots where only standard fertilization was practiced whereas tuber yields of 200, 400 and 600 kg ha^{-1} leonardite applications were 31.5, 33.2 and 33.5 t ha^{-1}; respectively. While 400 and 600 kg ha^{-1} leonardite doses gave the highest total tuber yield, there was no statistical difference between these two doses (Table 3). The highest tuber yields were obtained from Milva (37.0 t ha^{-1}) and Lady Olmpia (36.5 t ha^{-1}) followed by Van Gogh (32.7 t ha^{-1}) and Agata (21.3 t ha^{-1}).

Protein content

Leonardite applications increased protein content of all cultivars compared with control. The highest increase was observed for 400 (2.08%) and 600 kg ha^{-1} (2.12%) leonardite applications, but difference detected between these doses was not significant (Table 4). The highest amount of protein was found in Lady Olympia (2.08%) and Van Gogh (2.07%). With increased leonardite doses protein content usually increased in tested cultivars except in Van Gogh, in which protein content did not increased after 200 kg ha^{-1} leonardite application (Table 4).

Vitamin C content

Leonardite applications increased Vitamin C content. Average vitamin C content was 16.3 mg 100 g fw^{-1} in control whereas average vitamin C content was 17.5, 18.6 and 19.1 mg 100 g fw^{-1} in 200, 400 and 600 kg ha^{-1}

Table 3. Effects of different doses of leonardite applications on plant height, tuber number, marketable tuber yield and total tuber yield of potato cultivars.

	2008-2009					2009-2010					2-year average				
	Leonardite Doses (kg ha^{-1})														
Cultivars	0	200	400	600	Mean	0	200	400	600	Mean	0	200	400	600	Mean
	Plant Height (cm)														
Van Gogh	68.3	70.0	70.2	71.7	70.1	70.5	72.7	73.5	71.7	72.1	69.4	71.3	71.9	71.7	71.1b
Milva	69.6	75.0	75.1	75.1	73.7	72.3	77.3	77.8	77.8	76.3	71.0	76.2	76.5	76.5	75.0a
Lady Olympia	67.5	73.6	71.9	75.3	72.1	72.0	75.7	77.6	76.1	75.4	69.8	74.6	74.8	75.7	73.7a
Agata	71.7	71.6	74.2	73.3	72.7	73.9	74.2	73.5	74.1	73.9	72.8	72.9	73.8	73.7	73.3a
Mean	69.3	72.5	72.9	73.9	**72.1**	72.0	75.0	75.6	74.9	**74.4**	70.7b	73.8a	74.2a	74.4a	
	Tuber Number Per Plant (number plant $^{-1}$)														
Van Gogh	4.9	6.1	6.1	6.5	5.9	5.4	5.5	6.3	6.5	5.9	5.1	5.8	6.2	6.5	5.9b
Milva	6.4	7.2	7.7	7.9	7.3	7.6	7.8	8.3	8.6	8.1	7.0	7.5	8.0	8.3	7.7a
Lady Olympia	5.1	5.7	6.3	6.4	5.9	5.4	6.7	6.9	6.9	6.5	5.3	6.2	6.6	6.6	6.2b
Agata	3.7	4.3	4.3	4.1	4.1	3.7	5.1	5.0	5.1	4.7	3.7	4.7	4.6	4.6	4.4c
Mean	5.0	5.8	6.1	6.2	**5.9**	5.5	6.3	6.6	6.8	**6.3**	5.3c	6.1b	6.4ab	6.5a	
	Marketable Tuber Yield (t ha^{-1})														
Van Gogh	18.3	21.6	22.3	23.6	21.5	20.4	23.2	25.3	26.4	23.8	19.4	22.4	23.8	25.0	22.6c
Milva	17.2	22.9	25.8	26.7	23.2	23.7	28.3	30.4	31.9	28.6	20.4	25.6	28.1	29.3	25.9a
Lady Olympia	16.1	20.9	25.0	25.2	21.8	21.7	24.0	31.0	32.6	27.3	18.9	22.5	28.0	28.9	24.6b
Agata	11.1	12.4	12.7	13.4	12.4	11.6	11.8	11.8	13.1	12.1	11.8	12.1	12.3	13.2	12.2d
Mean	15.7	19.5	21.4	22.2	**19.7**	19.4	21.8	24.6	26.0	**23.0**	17.5c	20.6b	23.0a	24.1a	
Lsd cultivar x dose : 2.49 Lsd year x cultivar : 1.76															
	Total Tuber Yield (t ha^{-1})														
Van Gogh	29.1	32.0	32.6	33.4	31.8	32.5	32.7	34.5	35.0	33.7	30.8	32.3	33.6	34.2	32.7b
Milva	30.3	36.2	38.2	38.5	35.8	35.5	37.6	39.6	39.5	38.1	33.0	36.9	38.9	39.0	37.0a
Lady Olympia	32.6	35.0	36.2	36.2	35.0	33.2	36.3	41.6	41.2	38.1	32.9	35.7	38.9	38.7	36.5a
Agata	20.5	2.17	22.2	22.7	21.8	20.3	20.4	20.9	21.5	20.8	20.4	21.1	21.6	22.1	21.3c
Mean	28.1	31.2	32.3	32.7	**31.1**	30.4	31.8	34.2	34.3	**32.7**	29.2c	31.5b	33.2a	33.5a	
Lsd year x cultivar : 1.99															

Means followed by the same letters in rows and columns are not significantly different at P < 0.05 level

leonardite applications, respectively. There was no significant difference between 400 and 600 kg ha^{-1} leonardite application for vitamin C content (Table 4). Among the tested cultivars, Agata (20.2 mg 100 g fw^{-1}) and Van Gogh (13.7 mg 100 g fw^{-1}) had the highest and the lowest vitamin C concentrations; respectively.

Specific gravity

Leonardite applications increased the specific gravity compared to control plots, but there was no statistically significant differences observed between different leonardite doses for specific gravity (Table 4). However significant differences were detected between potato cultivars for their specific gravity. The highest specific gravity was observed in Lady Olympia (1.091 g cm^{3} $^{-1}$) followed by Van Gogh (1.089 g cm^{3} $^{-1}$), Milva (1.080 g cm^{3} $^{-1}$) and Agata (1.078 g cm^{3} $^{-1}$).

DISCUSSION

Even though plant height is not an important agronomic or quality parameter for potato it could be used to have an idea about vegetative growth and plant development, thus an estimate of yield (Maity and Chatterjee, 1977). Statistically significant differences were not detected between 200, 400 and 600 kg ha^{-1} leonardite doses for plant height, but leonardite applications significantly increased plant height compared with control. Increased plant height may result from increased soil fertility and available plant nutrient elements due to leonardite applications. Good plant development due to increased soil moisture content and nutrient availability increases stolon number and length per tuber (Mahmoud and Hafez, 2010). Similar results were reported by Erik et al. (2000) and Sajid et al. (2012) for onion who reported that humic acid applications led to significant increase in soil organic matter, thus improving plant growth, neck and plant height and crop production.

Leonardite applications increased number of tubers reported in the present study. Leonardite applications also increased both total tuber yield and marketable tuber yield as compared with control, and the best results for yield increase were obtained from 400 and 600 kg ha^{-1} (Table 3). As was mentioned before humic acid increase soil fertility, soil moisture and provides essential nutrients for better growth of plants and hence increase yield of the crops.

Table 4. Protein and vitamin C contents and specific gravity of leonardite applied potato cultivars.

	2008-2009					2009-2010					2-year average				
	Leonardite Doses (kg ha[-1])														
Cultivars	0	200	400	600	Mean	0	200	400	600	Mean	0	200	400	600	Mean
Protein Content (%)															
Van Gogh	1.89	2.01	2.14	2.16	2.05	1.90	2.03	2.21	2.20	2.09	1.90	2.02	2.18	2.18	2.07a
Milva	1.86	1.89	1.97	2.04	1.94	1.92	1.94	2.00	2.10	1.99	1.89	1.92	1.99	2.07	1.97c
Lady Olympia	2.00	2.03	2.10	2.14	2.07	1.97	2.10	2.14	2.15	2.09	1.99	2.07	2.12	2.15	2.08a
Agata	1.96	2.03	2.05	2.06	2.03	1.99	2.00	2.03	2.08	2.02	1.97	2.02	2.04	2.07	2.02b
Mean	1.93	1.99	2.07	2.10	**2.02**	1.95	2.02	2.10	2.14	**2.05**	1.94c	2.00b	2.08a	2.12a	
Lsd cultivar x dose : 0.108															
Vitamin C Content (mg100g fw[-1])															
Van Gogh	13.3	13.5	13.8	15.8	14.1	12.9	13.0	13.3	13.9	13.3	13.1	13.3	13.5	14.8	13.7d
Milva	15.5	16.1	19.6	20.4	17.9	16.1	17.8	18.9	20.2	18.2	15.8	16.9	19.3	20.3	18.1c
Lady Olympia	18.2	18.7	20.5	19.4	19.2	18.0	19.5	20.0	20.8	19.6	18.1	19.1	20.3	20.1	19.4b
Agata	18.0	20.9	20.4	20.8	20.0	18.2	20.3	22.0	21.3	20.5	18.1	20.6	21.2	21.1	20.2a
Mean	16.2	17.3	18.6	19.1	**17.8**	16.3	17.7	18.6	19.1	**17.9**	16.3c	17.5b	18.6a	19.1a	
Specific Gravity (g cm[3 -1])															
Van Gogh	1.086	1.087	1.093	1.091	1.089	1.085	1.087	1.091	1.090	1.088	1.085	1.087	1.092	1.090	1.089b
Milva	1.077	1.082	1.081	1.080	1.080	1.079	1.081	1.081	1.082	1.081	1.078	1.081	1.081	1.081	1.080c
Lady Olympia	1.087	1.089	1.091	1.092	1.090	1.092	1.092	1.093	1.094	1.093	1.089	1.091	1.092	1.093	1.091a
Agata	1.077	1.078	1.078	1.077	1.078	1.076	1.078	1.078	1.078	1.078	1.077	1.078	1.078	1.078	1.078d
Mean	1.081	1.084	1.086	1.085	**1.084**	1.083	1.085	1.086	1.087	**1.085**	1.082b	1.084a	1.085a	1.085a	

Means followed by the same letters in rows and columns are not significantly different at P < 0.05 level

Mahmoud and Hafez (2010) reported that the vegetative growth parameters, yield, tuber size, weight, quality and nutritive value of potato tubers were significantly increased with increased levels of humic acid applications. Hopkins and Stark (2003) showed humic acid applications increased total and marketable yield of potato crop. It was also reported that humic acid application increased tuber yield, number of tubers and marketable tuber yield compared to standard fertilization as was observed in our study (Ezzat et al., 2009; Mahmoud and Hafez, 2010; Selim et al., 2009b). Seyedbagheri (2010) found increased leonardite doses significantly increased marketable tuber yield from 11.4% to 22.3%. Yield increase resulting from use of humic substances were also reported for other crops such as; cucumber (El-Shabrawy et al., 2010), maize (Ayuso et al., 1996), wheat (Delfine et al., 2005) and cabbage (Syabryai et al., 1965).

Protein content is an important quality parameter and depends on the amount of assimilated N by the plants (Selim et al., 2009a). Leonardite used in the study contained 1.8% N (Table 1). Leonardite also serves to increase available N, P and other plant nutrients, such as K; in the soil which in turn could be taken up by plants (David et al., 1994). The lowest rates for protein and vitamin C contents were found in control (Table 4). For both traits, the best results were obtained from 400 and 600 kg ha[-1] leonardite applications. It was also noted that cultivars differed from each other for protein and vitamin C contents; such as, while Van Gogh had the highest level of protein, the same cultivar had the lowest level of vitamin C content (Table 4). Differences observed among cultivars for protein and Vitamin C contents may depend on genetic background of the cultivars. Potassium fertilization and humic acid applications affect

concentration of vitamin C and amount of antioxidative compounds in solanaceous crops (Ünlü et al., 2010; Wuzhong, 2002).

Specific gravity depends on dry matter content of tubers affected by starch accumulation and could be used as a quality trait (Hassanpanah et al., 2011). Even though fertilization is necessary for high yield, it prolongs necessary time for tuber maturity and excessive fertilization could cause tubers to accumulate excess water. Water accumulation reduces dry matter content of tubers and in turn specific gravity and quality of tubers (Herman et al., 1960). In the present study, specific gravity was higher in tubers harvested from leonardite applied plots than the tubers harvested from control. Similarly, it was reported that N containing organic fertilizers increase dry matter contents of potato (Jarvan and Edesi, 2009) and maize kernels (Ouggiotti et al., 2004).

Based on the research findings and related discussion cited in this study it was concluded that leonardite applications increase number of tubers per plant (22%), marketable tuber yield (38%) and total tuber yield (15%). In addition, leonardite applications improve protein and vitamin C contents, and specific gravity of tubers which are important quality traits for both industrial and fresh market consumption. Differences between 400 and 600 kg ha[-1] leonardite doses were insignificant for marketable and total tuber yield, protein and vitamin C contents, and specific gravity suggesting that 400 kg ha[-1] leonardite application to standard fertilization would be sufficient to obtain good yield with high quality tubers.

LITERATURE CITED

Akinremi, O.O., R.L. Janzen, R.L. Lemke, F.J. Larney, 2000. Response of canola, wheat and green beans to leonardite additions. Can. J. Soil Sci. 80:437-443.

AOAC, 1975. Association of Official Analytical Chemists. 12th ed. Washington DC.

Ayuso, M., T. Hernandez, C. Garcia, J. Pascual, 1996. Stimulation of barley growth and nutrient absorption by humic substances originating from various organic materials., Bio-source Tech. 57:251-257.

Chen, Y., T. Aviad, 1990. Effects of humic substances on plant growth. In: Humic Substances in Soil and Crop Sciences Amer. J. Soil Sci. 34:161-186.

Cimrin, K.M., I. Yılmaz, 2005. Humic acid applications to lettuce do not improve yield but do improve phosphorus availability. Acta Agri. Scand. 55:58-63.

David, P.P., P.V. Nelson, D.C. Sanders, 1994. Humic acid improves growth of tomato seedling in solution culture. J. Plant Nutr. 17:173-184.

Delfine, S., R. Tognetti, E. Desiderio, A. Alvino, 2005. Effect of foliar application of N and humic acids on growth and yield of durum wheat. Agro. Sus. Dev. 25:183-191.

Dursun, A., I. Guvenc, M. Turan. 2002. Effects of different levels of humic acid on seedling growth and macro and micronutrient contents of tomato and eggplant. Acta Agrobotanica. 56:81-88.

Ece, A., K. Saltali, N. Eryigit, F. Uysal, 2007. The effects of leonardite applications on climbing bean (Phaseolus vulgaris L,) yield and the some soil properties. J. Agronomy. 6:480-483.

EL-Desuki, M., 2004. Response of onion plants to humic acid and mineral fertilizers application. Annl. Agric. Sci. 42:1955-1964.

El Sayed Hameda, E.A., A. Saif El Dean, S. Ezzat, A.H.A. El Morsy, 2011. Responses of productivity and quality of sweet potato to phosphorus fertilizer rates and application methods of the humic acid. Int. Res. J. Agric. Sci. Soil Sci. 1:383-393.

El-Shabrawy R.A., A.Y. Ramadan, Sh M. EI-Kady, 2010. Use of humic acid and some biofertilizers to reduce nitrogen rates on cucumber (Cucumis sativus L.) in relation to vegetative growth, yield and chemical composition. J. Plant Produc. Mansoura Univ. 1(8):1041-1051.

Erik, B., G. Feibert, C.C. Shock and L.D. Saundres, 2000. Evaluation of humic acid and other non conventional fertilizer additives for onion productivity. Malheur Experiment Station, Oregon State University Ontario.

Erkoc, I., 2009. Effects of Sulphur and Leonardit on Phosphor Efficiency in Greenhouse Grown Tomato. MSc. Thesis. Department of Horticulture Institute of Natural and Applied Sciences Universty of Cukurova,, p. 127.

Esendal, E., 1990. Starch Sugar Crops Breeding, Vol. I: Potato. OMU, Agriculture Faculty Publication, No:101.

Ezzat A.S., U.M. Saif Eldeen, A.M. Abd EI-Hameed, 2009. Effect of irrigation water quantity, antitranspirant and humic acid on growth, yield, nutrients content and water use efficiency of potato (Solanum tuberosum L.). J. Agric. Sci. Mansoura Univ., 34(12):11585-11603.

Hafez, M. Magda, 2003. Effect of some sources of nitrogen fertilizer and concentration of humic acid on the productivity of squash plant. Egypt. J. Appl. Sci. 19:293-309.

Hartwigson, J.A. and M.R. Evans, 2000. Humic acid seed and substrate treatments promote seedling root development. HortScience 35:1231-1233.

Hassanpah, D., M. Khodadadi, 2009. Evaluation of potassium humate effects on germination, yield and yield components

of HPS-II/67 hybrid true potato seeds under in vitro and in vivo conditions. Am. J. Plant Physiol. 4: 52-57.

Hassanpah, D., H. Hassanabadi, S.H. Chakherchaman, 2011. Evaluation of cooking quality characteristics of advanced clones and potato cultivars. Afr. J. Food Tech. 6:72-79.

Herman-Timm, L.D., T. Doneen, J.C. Lyons, V. Bishop, H. Schweers, J.R. Stockton, 1960. Potato quality lowered in field tests with high nitrogen fertilization. California Agriculture. 13p.

Hopkins, B., J. Stark, 2003. Humic acid effects on potato response to phosphorus. Idaho Potato Conference January 22-23, p87-92.

Jarvan, M., L. Edesi, 2009. The effect of cultivation methods on the yield and biological quality of potato. Agro. Res. 7:289-299.

Kacar, B., A. Inal, 2008. Plant Analysis, Ankara. Nobel Publication, No:1241, 892pp.

Karadogan, T., 1994. Effects of different nitrogen sources and applications times on potato yield and yield components. Turk. J. Agric. 19:417-421.

Mahmoud Asmaa, R., M. Hafez Magda, 2010. Increasing productivity of potato plants (Solanum tubersoum L.) by using potassium fertilizer and humic acid application. Int. J. Acad. Res. 2:83-88.

Malcolm, R.E., D. Vaughan, 1979. Effects of humic acid on invertase activities in plant tissues and their interaction with an invertase inhibitor. Soil Biol. Biochem. 11:65-72.

Maity, S., B.N. Chatterjee, 1977. Growth attributes of potato and their inter-relationship with yield. Potato Res. 20:337-341.

Neilsen, G.H., D. Neilsen, L.C. Herbert, E.J. Hogue, 2004. Response of apple to fertigation of N and K under conditions susceptible to the development of K deficiency. HortScience. 129:26-31.

O'Donnell, R.W., 1973. The auxin-like effects of humic preparations from leonardite. Soil Sci. 116:106-112.

Quaggiotti, S., B. Ruperti, D. Pizzeeghello, O. Francisco, V. Tugnoli, N. Serenella, 2004. Effect of low molecular size humic substances on nitrate uptake and expression of genes involved in nitrate transport in maize (Zea mays L.). J. Exp. Bot. 55:803–813.

Paramasivam, S., A.K. Alva, A. Fares, K.S. Sajwan, 2001. Estimation of nitrate leaching in an entisol under optimum citrus production. Soil Sci. Soc. Am. J. 65:914-921.

Sajid, M., A. Rab, S.T. Shah, I. Jan, I. Haq, B. Haleema, M. Zamin, R. Alam, H. Zada, 2012. Humic acids affect the bulb production of onion cultivars. Afr. J. Microbiol. Res. 6:5769-5776.

SAS Institute 1998. INC SAS/STAT users guide release 7.0, Cary, NC, USA.

Selim, E.M., A.A. Mosa, A.M. El-Ghamry, 2009a. Evaluation of humic substances fertigation through surface and subsurface drip irrigation systems on potato grown under Egyptian sandy soil conditions. Agr. Water Manage. 96:1218-1222.

Selim, E.M., A.S. El-Neklawy, S.M. El-Ashry, 2009b. Beneficial effects of humic substances ferrtigation on soil fertility to potato grown on sandy soil. Aust. J. Basic Appl. Sci. 3:4351-4358.

Selim, E.M., S.I. Shedeed, F.F. Asaad, A.S. El-Neklawy, 2011. Interactive effects of humic acid and water stress on chlorophyll and mineral nutrient contents of potato plants. J. Appl. Sci. Res. 7:531-537.

Serenella, N., D. Pizzeghelloa, A. Muscolob, A. Vianello, 2002. Physiological effects of humic substances on higher plants. Soil Biol. Biochem. 34:1527-1536.

Seyedbagheri, M., J.M. Torell, 2001. Effects of humic acids and nitrogen mineralization on crop production in field trials: Humic substances: structures, models and functions:

Proceedings of the Fifth Humic Substances, Boston, Massachusetts, USA, 21-23 March 2001, pp.355-359.

Seyedbagheri, Mir-M., 2010. Influence of humic products on soil health and potato production. Potato Res. 53:341-349.

Sibanda, H.M., S.D. Young, 1989. Competitive adsorption of humus acids and P on goethite, gibbsite and two tropical soils. J. Soil Sci. 37:197-204.

Suganya, S., R. Sivasamy, 2006. Moisture retention and cation exchange capacity of sandy soil as influenced by soil additives. J. Appl. Sci. Res. 2:949-951.

Syabryai V.T., V.A. Reutov, L.M. Vigdergauz, 1965. Preparation of humic fertilizers from brown coal. Geol. Zh. Akad. Nauk Ukr. 25:39-47.

Unlu, H., H. Ozdamar-Unlu, Y. Karakurt, 2010. Influence of humic acid on the antioxidant compounds in pepper fruit. J. Food Agric. Environ. 8:434-438.

Vaughan, D., I.R. MacDonald, 1976. Some effects of humic acid on cation uptake by parenchyma tissue. Soil Biol. Biochem. 8:415-421.

Verlinden, G., B. Pycke, J. Mertens, F. Debersaques, K. Verheyen, G. Baert, J. Bries, G. Haesaert, 2009. Application of humic substances results in consistent increases in crop yield and nutrient uptake. J. Plant Nutr. 32:1407-1426

Wuzhong, N., 2002. Yield and quality of fruits of Solanaceous crops as affected by potassium fertilization. Better Crop Int. 16:6-8.

COMPARISON OF ORGANIC AND TRADITIONAL PRODUCTION SYSTEMS IN CHICKPEA (*Cicer arietinum* L.)

Sevgi CALISKAN[1]*, Cahit ERDOGAN[2], Mehmet ARSLAN[2], Mehmet Emin CALISKAN[1]

[1]*Nigde University, Faculty of Agricultural Sciences and Technologies, Nigde, TURKEY*
[2]*Mustafa Kemal University, Faculty of Agriculture, Field Crops Department, Hatay, TURKEY*
*Corresponding author: sevcaliskan@gmail.com

ABSTRACT

Yield and yield components, and protein content of chickpea (*Cicer arietinum* L.) were compared in traditional and organic production systems since organic farming has many advantages on environment, animal and human health over traditional production systems. In the study, organic production system, green manure, farmyard manure and effective microorganisms and their combinations were tested as six treatments. Seed yield and some yield components were found higher in the traditional production system than those of organic production systems. The highest protein content was obtained in green manure and farmyard manure applications. Among the organic production systems, green manure and farmyard manure applications could be strongly recommended for organic chickpea producer since the highest seed yields with 2729 in 2007 and 3838 kg ha^{-1} in 2008 were found in green manure + farmyard manure treatment among the organic production systems.

Keywords: chickpea; farmyard manure; green manure; organic production; traditional production; yield

INTRODUCTION

In modern industrial agriculture, high productivity capacity in plants depends on new productive cultivars, mineral fertilizers, pesticides, growth regulating materials, and some other agronomical practices. Although these inputs have resulted in high yield and production increase, unconscious and uncontrolled usage of them have also caused serious problems concerning human and environmental health such as contamination of the soil and groundwater. Especially pesticides, chemical fertilizers, and hormones disturb the natural balance and human health. Researchers have developed various production techniques to prevent these adverse effects. Several alternative production systems have been developed and organic farming system has been approved and accepted as the a reliable farming model among these systems in many countries (Bettiol et al., 2004). With organic farming, it is possible to produce food of a better quality, healthier, and more nutritious than those of produced with the conventional counterparts (Warman and Haward, 1998). Therefore, in recent years, use of organic food has increased in parallel with the increased interest in the environmental protection and safe foods in the developed countries (Caliskan, 2007).

Generally, yield of crops-grown organically are lower than those of the conventionally grown crops (Warman

and Haward, 1998) due to malnutrition of plants and effects of diseases and pests. Accordingly, use of organic manures including green manure, farm yard manure, effective microorganisms (EM) composts derived from various wastes are important factors to maintain and improve soil fertility for a sustainable crop production (Ashiona et al., 2006; Javaid, 2006; Talgre et al., 2012).

Legume crops are grown not only as part of the human diet but also for improving soil fertility. As a legume, chickpea (*Cicer arietinum* L.) is considered as one of the most important grain legumes in the world (Namvar et al., 2011). It is also an important crop in Turkey according to sowing area and production (Sepetoglu et al., 2008; Ozalkan et al., 2010; Cagirgan et al., 2011; Toker et al., 2012). On the basis of FAO statistical database, chickpea production area occupied 446.218 ha and produced 530.634 tons in Turkey in 2010 (FAOSTAT, 2012). Chickpea is a major food legume and an important source of protein in many countries. It is used in crop rotation (Sahin and Gecit, 2006) and has atmospheric nitrogen fixation ability (Jain et al., 1999) and play an important role in the maintenance of soil fertility (Gunes et al. 2006) and also widely used as green manure (Mohammadi et al., 2010a). In addition, it especially could play a key role in organic cropping systems (Mohammadi et al., 2010a). Cereals and grain legumes share a 15.3% of organic agriculture production in Turkey (Caliskan, 2007).

Organic chickpea with production of 3614 t is the second most important legume crop after the lentil in Turkey (Anonymous, 2012a).

The purpose of this study was to evaluate the individual or mixture characteristics of different organic manures such as green manure, farmyard manure and effective microorganism (EM) in organically grown chickpea and also to compare them with traditional grown chickpea.

MATERIALS AND METHODS

Field study and experimental arrangements

A field experiment was conducted at the Mustafa Kemal University, in Hatay (36° 39' N, 36° 40' E; 83 m above sea level) located in the Eastern Mediterranean region of Turkey in the 2006-2007 and the 2007-2008 growing seasons. In the experiment, six organic production systems, which were consisted of the sole and combined usage of green manure (GM), farmyard manure (FYM) and Effective Microorganisms (EM), were compared with a traditional production system in chickpea production. According to organic production system in Turkey (Anonymous, 2012b), the experimental area had not been cultivated for five years before starting this experiment. The FYM was obtained from the experimental farm of the Faculty of Agriculture, Animal Science Department and the application rate of the FYM was 30 tons ha^{-1} which was applied a week before sowing. As green manure, common vetch (Vicia sativa L.) was sown and incorporated with soil at the flowering stage in each year. EM solution consisted of mixed culture of beneficial microorganism including a predominant population of lactic acid bacteria (Lactobacillus sp.), yeast (Saccharomyces sp.) and a small proportion of

photosynthetic bacteria (Rhodopseudomonas sp.). EM is available in a dormant state and requires activation before application. Activation involved the preparation of a solution containing one part basic EM + 1 part of sugar solution + 20 parts water by volume. The activation was kept away from direct sunlight for 3 days. The activated EM solution was dissolved in water in a ratio of 1:1000 and applied to soil seven days before sowing. During the course of the experiment, foliar applications of activated EM were carried out at V6 and R4 growth stages at a dilution of 1:300. The mineral fertilizers urea and diammonium phosphate (DAP) were used as source of nitrogen (30 kg N ha^{-1}) and phosphorus (60 kg P ha^{-1}), respectively. Half of dose of N and all of the P were distributed and incorporated thoroughly into the soil at sowing. The remaining N was applied at two equal doses at the 6-8 leave and flowering stages. The soil (0-40) was Clayey and some chemical and physical properties of experimental area and FYM were given in Table 1. The field trial was arranged in a Randomized Complete Block Design with three replications. Each plot has 99 m^2 areas (9.9 m x 10 m) and consisted of 66 rows 0.15 m apart. The traditional growing plots were formed nearly 200 m away from organic production plots. Chickpea seeds (cv Cevdetbey 98) were sown on December 19 in 2007 and on December 31 in 2008. The pre-sowing herbicide trifluralin (α,α,α-trifliofo-2,6-dinitro-N,N-dipropyl-p-toluidine) was applied to soil at the rate of 2000 cc ha^{-1} and standard local cultivation practices were applied in the traditional farming system. Synthetic herbicides or insecticides were not used in the organic plots; weeds were controlled by hand during the growing period. The daily climatic data were recorded. Air temperature, humidity and precipitation data of the experimental site were presented in Table 2.

Table 1. Chemical characteristics of soil and farmyard manure used in the experiment

	E.C (mS cm^{-1})	pH	CaCO$_3$ (%)	OM (%)	N (%)	P (mg kg^{-1})	K (mg kg^{-1})
Soil	1.0	7.8	21.7	17.3	0.13	322.8	287.3
FYM	38.5	8.7	39.0	55.0	2.35	5800.0	26000.0

FYM: Farmyard manure; OM: organic matter

Table 2. Monthly rainfall, temperature and relative humidity during the growing period of 2007 and 2008

	Mean Temperature (°C)		Rainfall (mm)		Relative humidity (%)	
	2007	2008	2007	2008	2007	2008
December	8.4	8.6	6.5	210.1	70.8	67.1
January	7.6	5.6	109.4	126.1	70.1	55.8
February	10.6	9.5	198.8	127.9	77.0	59.9
March	13.9	16.5	130.4	109.5	66.4	62.2
April	15.9	14.6	124.2	27.3	64.6	65.2
May	22.6	20.9	32.7	87.2	68.5	68.2
June	25.6	25.9	4.4	21.3	63.5	63.1

Plots were harvested on 27 June in 2007, and on 1 July in 2008. The morphological characteristics such as plant height (cm), first pod height (cm), number of branch per plant, number of pod per plant and harvest index (%) were recorded on twenty plants, selected randomly from each plot before harvest. Then plots were harvested by hand and threshed by using a threshing machine. After harvest, seed yield (kg ha^{-1}) and 100-seed weight (g) were determined for each treatment. Seed protein content was determined using a micro-Kjeldahl digest procedure (AOAC, 2005).

Statistical Analysis

Data obtained for the traits were statistically analyzed using the ANOVA technique for Randomized Complete Block Design according to the general linear model procedure described in the Statistical Analysis System (SAS Institude 1996). The means were compared by the Duncan's multiple range test (Steel and Torrie, 1997).

RESULTS AND DISCUSSION

Yield and yield components

Yield and yield components of chickpea plants grown under conventional and organic production systems based on the two experimental years were given in Tables 3 and 4. The ANOVA results indicated that the year x treatment interactions were significant for all evaluated traits except protein content. Hence the results for each year were presented and discussed separately. The results obtained from the variance analysis indicated that treatments had statistically significant effects on all of studied traits except for plant height, number of branches per plant in 2007 and 100-seed weight in 2008.

Plant height is an important morphological trait for chickpea and it was significantly influenced by the applications of various organic treatments in 2008 (Table 3). The highest plant height was recorded in traditional production system in both years (71.5 and 62.1 cm in 2007 and 2008). Organic production system without any supplement had high plant height (71.2 cm) in 2007, whereas green manure + farmyard manure application and EM application had high plant height mean (55.1 cm) in 2008 (Table 3).

Table 3. The mean values and ANOVA results for plant and yield characteristics of chickpea grown under the traditional and organic production systems in 2007 and 2008

Treatment	Plant height (cm)		First pod height (cm)		Number of branches per plant		Number of pods per plant	
	2007	2008	2007	2008	2007	2008	2007	2008
Traditional	71.5 a	62.1 a	36.0 c	26.7 b	3.5 ab	5.1 a	39.6 b	99.0 a
Organic	71.2 a	52.6 c	38.5 ab	27.2 b	3.3 bc	1.2 c	38.9 b	55.5 e
FYM	69.4 ab	53.9 bc	39.4 a	28.1 ab	3.4 ab	1.3 c	44.7 a	62.2 cd
EM	69.7 ab	55.1 b	35.7 c	29.5 a	3.3 abc	1.6 c	35.5 d	57.9 de
GM	66.7 b	54.4 b	37.4 bc	23.9 c	3.3 bc	4.7 a	36.2 cd	62.5 cd
GM + EM	69.3 ab	54.3 bc	37.4 bc	24.7 c	3.1 c	4.1 b	34.4 d	66.9 c
GM + FYM	67.9 ab	55.1 b	38.1 ab	24.5 c	3.6 a	4.9 a	37.8 bc	89.6 b
Mean	69.4	55.4	37.5	26.4	3.4	3.3	38.2	70.5
LSD (%5)	4.3	1.8	1.7	2.0	0.3	0.6	2.1	5.2
ANOVA								
Treatment MS	8.8ns	29.0**	5.2**	12.9**	0.06ns	9.74**	34.9**	853.4**
T x Year MS	12.60**		10.37**		4.72**		427.71**	
CV (%)	3.5	1.8	2.6	4.2	4.3	10.4	3.1	4.2

FYM: Farmyard manure, GM: Green manure, EM: Effective microorganism, MS: Mean square
*, **: F-test significant at $p < 0.05$, and $p < 0.01$, respectively. ns: not significant.

Mean first pod height was affected by the production systems. However, number of branches per plant was not affected by the production systems in 2007. Farmyard manure application had the highest first pod height with 39.4 cm in 2007, and effective microorganism (EM) application had the highest first pod height with 29.5 cm in 2008. In the 2008 growing season, plant development and branch formation were poor and the average of first pod height was 26.4 cm which was lower than the first pod height (37.5 cm) of the previous year. The highest number of branches per plant was obtained in the traditional production system. Regarding to organic applications, the green manure + farmyard manure combination had the best results for first pod height in both years (3.6 and 4.9 in 2007 and 2008). On the average, number of branches per plant obtained in 2007 was less than that of obtained in 2008. Pod number per plant could be the responsible trait for yield change (Toker and Cagirgan, 2004) under the conventional and organic production systems in the two years of study. In 2008, pod number per plant was higher than that of in 2007 in the traditional system. Pod number per plant seemed to be highly influenced by change in cultural and environmental conditions.

The 100-seed weight and the harvest index of chickpea grown under the traditional and organic production systems are shown in Table 4. One hundred seed was not affected by the production systems in 2008. The average of 100-seed weights were different between 2007 and 2008 growing seasons such as 47.1 g and 46.3 g in 2007 and 2008, respectively. The 100-seed weight was higher for the traditionally grown chickpea than those of organic production systems in 2008. The highest 100-seed weight was obtained from the green manure application in 2007, contrary to the traditional production system in 2008 (Table 4). There were major harvest index differences between production systems. The highest harvest index was obtained in green manure treatments in both years. The highest harvest index value (40.4%) was found in the green manure + farmyard manure combination in the first year. In the second year, the highest harvest index value (41.2%) was obtained from the green manure + farmyard manure application. The GM + EM combination (40.8%) and the traditional model (39.3 %) have followed it. The organic production system without any supplement and sole EM application resulted in the lowest harvest index and followed by sole FYM treatment in both years (Table 4).

Table 4. The mean values and ANOVA results for yield, yield components and protein content of chickpea grown under the traditional and the organic production systems in 2007 and 2008.

Treatment	100-seed weight (g)		Harvest index (%)		Protein content (%)		Seed yield (kg ha^{-1})	
	2007	2008	2007	2008	2007	2008	2007	2008
Traditional	46.8 b	48.1 a	38.3 ab	39.3 b	20.4 d	20.4 cd	2972 a	4002 a
Organic	45.2 c	46.0 ab	28.2 d	29.3 d	20.1 d	20.1 d	1538 f	1830 e
FYM	47.9 ab	46.0 ab	30.0 cd	29.4 d	21.4 b	20.8 bcd	2468 d	2133 d
EM	47.1 ab	46.9 ab	28.6 d	29.3 d	20.6 cd	20.6 bcd	1891 e	1926 e
GM	48.3 a	45.8 b	32.1 c	36.4 c	21.1 bc	21.1 bc	2462 d	2430 c
GM + EM	47.1 ab	45.3 b	37.4 b	40.8 ab	21.5 b	21.4 b	2590 c	2843 b
GM + FYM	47.7 ab	46.0 ab	40.4 a	41.6 a	22.3 a	22.6 a	2729 b	3838 a
Mean	47.1	46.3	33.6	35.1	21.1	21.0	2379	2715
LSD (%5)	1.4	2.2	2.3	2.0	0.7	0.8	12.2	18.2
ANOVA								
Treatment MS	3.1**	2.5 ns	76.3**	97.0**	1.8**	2.1**	7396.9**	23792.5**
T x Year MS	3.28*		4.17*		0.09		440772.10**	
CV (%)	1.6	2.7	3.9	3.2	1.8	2.3	2.9	3.8

FYM: Farmyard manure, GM: Green manure, EM: Effective microorganism, MS: Mean square
*,**: F-test significant at p <0.05, and p <0.01, respectively. ns: not significant.

Seed yields of chickpea grown under the traditional and organic production systems are shown in Table 4. The means of seed yield for the production systems were significantly different. The mean seed yields were 2379 kg ha^{-1} and 2715 kg ha^{-1} in 2007 and 2008, respectively. Generally, seed yields in 2008 were higher than those of in 2007 due to favorable climatic conditions during the second year. In 2007, germination and crop growth were better at the start, but high temperatures were adversely affected during the flowering and pod initiation stages. Furthermore, total rainfall during second year (709.3 mm) was higher than the first year (606.4 mm). Islam (2012) also observed that growth and yield of chickpea were better during the cropping season of higher rainfall due to abundant supply of moisture when crop was grown under rainfed conditions. Chickpea seed yields were higher for the traditional system in both years. The results also showed that the means of seed yield was 2972 and 4002 kg ha^{-1} in the traditional system in 2007 and 2008, respectively. Chemical fertilizers used in the conventional production system caused more yield and yield component in both years. Chemical fertilizers are essential in the traditional agriculture, since they provide essential plant nutrients. Essential plant nutrient elements are considered a prerequisite for the enhanced plant development (Sahin and Gecit, 2006; Rosculete et al., 2010; Namvar et al., 2011). The lowest seed yield was obtained from the organic production system without organic and inorganic fertilizer application in both years (Table 4). In general, chickpea seed yield and yield components were lower in the organic production systems. These results are in agreement with the results of Vadavia et al. (1991) and Rosculete et al. (2010). They reported that seed yield and yield components of chickpea were increased with application of mineral fertilization. The low yields of organic production approach were generally attributed to the differences between organic and traditional production practices, particularly due to crop rotations, methods of controlling weeds and pathogens and fertilization practices. Green and farm yard manures were found to be more effective on crop growth, seed yield and yield components of chickpea. Green manure improves organic materials of soil, enriches soil nitrogen content, and transforms nutrients into the absorbable form, so the seed yield and yield components increases (Fageria, 2007; Mohammadi et al., 2010b). Furthermore, potential benefits of green manure are reduced in nitrate leaching risk and lower fertilizer usage in the succeeding crops. However, influence of green manure might depend on soil, crop, environment to environment, type of green manure crop, and management techniques (Fageria, 2007). Besides, green manure crops are also known to increase soil N and availability of P for the following crop. Astier et al. (2006) studied the effects of two different green manure plants (vetch and oat) on yield of maize. They reported that green manure had a significant effect on maize grain yield, N and P uptakes. Astier et al. (2006) concluded that vetch provided more yield than oat. Furthermore, the application of farm yard manure improves physical conditions of the soil, provides energy for activity of microorganisms, increases nutrient supply and improves the efficiency of macro elements (Ashiono et al., 2006; Khan et al., 2009). Yaduvanshi et al. (2003) reported that the application of N, P, K and their combination with green manure and farmyard manure significantly increased the rice yield. They also resulted in increase of soil N, P, K and organic C, and reduced soil pH. In our study, farmyard manure application was also significantly affected plant growth and seed yield, but the green manure + farmyard manure applications were found to be more effective seed yield and quality in organic chickpea production. Also, farmyard manure applications increased the efficiency of green manure. Similar results were reported by Vadavia et al. (1991) and Tolanur and Badanur (2003) in chickpea and Astier et al. (2006) in maize.

In this study, application of effective microorganisms (EM) alone did not increase the seed yield significantly (Table 4). On the contrary, the applications of EM combined with green manure were found to be more effective on plant growth and seed yield. In earlier studies, there were conflict reports in regard to the effect of EM application on plant growth, seed yield and yield parameters. Some researchers have reported that the application of the EM had increased plant growth and yield (Javaid, 2006). Similarly, Daly and Stewart (1999) reported that application of EM to onion, pea and sweet corn increased yields by 29%, 31% and 23%, respectively. In contrast, some other researchers have reported that the effect of EM on plant growth and yield was negative or no effects (Daiss et al., 2008). However, some studies also demonstrated that the negative effects of EM could be overcome by repeated applications of the EM (Javaid, 2010).

Protein content

The means of protein content for different treatments were given in Table 4. The seed protein content ranged from 20.1 to 22.6%, and differences between protein content of organic and traditional chickpea treatments were significant in both years (Table 4). In general, high seed protein contents were obtained in the organic production systems as compared to the traditional production system. The highest protein content was also obtained from the green manure + farmyard manure application, followed by the green manure + effective microorganisms (GM + EM) in both years (Table 4). The lowest protein content (20.1 %) in two years was obtained from organic production without any fertilizer application. It was also determined that seed protein contents were higher in the organic production systems combined with green manure than those of the other production systems. Similar findings regarding positive effects of green manuring on seed protein content of chickpea were also reported by Williams and Singh (1986) and Mohammadi et al. (2010a).

CONCLUSION

Results of this study clearly indicated the effects of the traditional and organic production systems on growth, yield and yield components of chickpea cv. Cevdetbey 98. The traditional production system had the highest seed yield in 2007 (2972 kg ha^{-1}) and in 2008 (4002 kg ha^{-1}). In the organic production systems, the highest seed yields (2729 and 3838 kg ha^{-1} in corresponding years, respectively) were obtained from green manure + farmyard manure treatment. Therefore, the green manure and farmyard manure applications could be recommended in a organic chickpea production.

ACKNOWLEDGEMENT

This study was supported by The State Planning Organization of Turkey (DPT) under the research project number 03 K 120860. The authors wish to thank to the DPT for financial support.

LITERATURE CITED

Anonymous,2012a.http://organik.tarim.gov.tr/sayfam.asp?sid=4 1&pid=41&ld=Organik%20Tar%FDm%20%DCretim%20V erileri.

Anonymous,2012b.http://organik.tarim.gov.tr/Kitaplar/ot_yonet melik/index.html#/2

A.O.A.C. (2005): Official Methods of Analysis 18th edition, Association of Official Analytical Chemist, Washington D. C. U.S.A

Ashiono, G.B., J.P. Ouma and S.W. Gatwiku, 2006. Farmyard manure as alternative nutrient source in production of cold tolerant sorghum in the dry highlands of Kenya. J. of Agron. 5 (2): 201-204.

Astier, M., J.M. Maass, J.D. Etchevers-Barra, J.J. Pena, F. Leon Gonzalez, 2006. Short-term green manure and tillage management effects on maize yield and soil quality in an Andisol. Soil and Tillage Res., 88: 153-159.

Bettiol, W., R. Ghini, J.A.H. Galvão, R.C. Siloto, 2004. Organic and conventional tomato cropping systems. Sci. Agric., 61 (3): 253-259.

Cagirgan, M.I., C. Toker, M. Karhan, M. Aksu, S. Ulger, H. Canci, 2011. Assessment of endogenous organic acid levels in ascochyta blight [*Ascochyta rabiei* (Pass.) Labr.] susceptible and resistant chickpeas (*Cicer arietinum* L.). Turkish J. of Field Crops, 16, 121-124.

Caliskan, S. 2007. Organic farming and organic field crops production in Turkey. Journal of Agricultural Faculty, Mustafa Kemal Univ., 12(1-2): 37-46.

Daly, M.J., D.P.C. Stewart, 1999. Influence of "effective microorganism" (EM) on vegetative production and carbon mineralization a preliminary investigation. J. Sustainable Agric., 14: 15-25.

Fageri, N.K., 2007. Green manuring in crop production. J. of Plant Nutrition, 30: 691-719.

FAOSTAT, 2012. Food and Agriculture Organization Statistical Database. http://faostat.fao.org/site/339/default.aspx. (Accessed in November, 2012).

Gunes, A., N. Cicek, A. Inal, M. Alpaslan, F. Eraslan, E. Guneri, T. Guzelordu, 2006. Genotypic response of chickpea (*Cicer arietinum* L.) cultivars to drought stress implemented at pre- and post- anthesis stages and its relations with nutrient uptake and efficiency. Plant & Soil Environ., 52 (8): 368 – 376.

Islam, M., 2012. The effect of different rates and forms of sulfur on seed yield and micronutrient uptake by chickpea. Plant and Soil Environ. 58 (9): 399-404.

Jain, P.C., P.S. Kushawaha, U.S. Dhakal, H. Khan, and S.M. Trivedi, 1999. Response of chickpea (*Cicer arietinum* L.) to phosphorus and biofertilizer. Legume Res., 22: 241-244.

Javaid, A., 2006. Foliar application of effective microorganism on pea as an alternative fertilizer. Agron. for Sustainable Development, 26: 257-262.

Khan, A., M.T. Jan, K.B. Marwat, and M. Arif, 2009. Organic and inorganic nitrogen treatments effects on plant and yield attributes of maize in a different tillage systems. Pakistan J. Botany, 41 (1): 99-108.

Mohammadi, K., A. Ghalavand, M. Aghaalikhani, 2010a. Effect of organic matter and biofertilizer on chickpea quality and biological nitrogen fixation. World Academy of Sci., Engineering and Tech., 68: 1144-1149.

Mohammadi, K., A. Ghalavand, M. Aghaalikhani, 2010b. Study the efficacies of green manure application as chickpea pre plant. World Academy of Sci., Engineering and Tech. 70: 233-236.

Namvar, A., R.S. Sharifi, M. Sedghi, R.A. Zakaria, T. Khandan, and B. Eskandarpour, 2011. Study on the Effects of Organic and Inorganic Nitrogen Fertilizer on Yield, Yield Components, and Nodulation State of Chickpea (*Cicer arietinum* L.). Commun. in Soil Sci. and Plant Analysis, 42: 1097-1109.

Ozalkan, C., H.T. Sepetoglu, I. Daur, O.F. Sen, 2010. Relationship between some plant growth parameters and grain yield of chickpea (*Cicer arietinum* L.) during different growth stages. Turkish J. of Field Crops, 15 (1): 79-83.

Rosculete, C.A., E. Rosculete, G. Matei, L. Dadulescu, 2010. The influence of compost and mineral fertilization on the chickpea production from the barren gangue from husnicioara quarry in Mehedinti county. Res. J. of Agric. Sci., 42 (3): 280-284.

SAS, 1998. SAS/STAT: user's guide version 7.0//SAS Institute. Carry, USA.

Sahin, N., and H.H. Gecit, 2006. The effects of different fertilizing methods on yield and yield components in chickpea (*Cicer arietinum* L.). J. of Agric. Sci., 12 (3): 252-258.

Sepetoglu, H., M. Altinbas, I. Daur, 2008. Uptake of same essential nutrients in chickpea during different growth stages in relation to biomass yield. Turkish J. of Field Crops, 13 (1): 1-11.

Steel, R.G.D., and J.H. Torrie, 1997. Principles and procedures of statistics: A biometrical
approach. 3rd ed. McGraw-Hill, New York.

Talgre, L., E. Lauringson, H. Roostalu, A. Astover, 2009. The effects of green manures on yields and yield quality of spring wheat. Agron. Res. 7(1): 125-132.

Talgre, L., E. Lauringson, H. Roostalu, A. Astover, A. Makke 2012. Green manure as a nutrient source for succeeding crops. Plants Soil and Environ., 58 (6): 275-281.

Toker, C., M.I. Cagirgan, 2004. The use of phenotypic correlation and factor analysis in determining characters for grain yield in chickpea (*Cicer arietinum* L.). Hereditas, 140 (3): 226-228.

Toker, C., F.O. Ceylan, N.E. Inci, T. Yildirim, M.I. Cagirgan, 2012. Inheritance of leaf shape cultivated chickpea (*Cicer arietinum* L.). Turkish J. of Field Crops, 17 (1):16-18.

Tolanur, S.I. and V.P. Badanur, 2003. Effect of integrated use of organic manure, green manure and fertilizer nitrogen on sustaining productivity of *rabi* sorghum-chickpea system and fertility of a Vertisol. J. of the Indian Soci. of Soil Sci., 51 (1): 41-44.

Vadavia, A.T., K.K. Kalaria, J.C. Patel, and N.M. Baldha, 1991. Influence of organic, inorganic and biofertilizers on growth, yield and nodulationof chickpea. Indian J. of Agron., 36(2): 263-264.

Warman, P.R. and K.A. Havard, 1998. Yield, vitamin and mineral contents of organically and conventially grown potatoes and sweet corn. Agric., Ecosystems and Environ., 68: 207-216.

Williams, P.C. and U. Singh, 1986. Nutritional Quality and the Evaluation of Quality in Breeding Programmes. The Checkpea. (Ed. M.C. Saxena and K.B.Sing), C.A.B. International, The International Center For Agricultural Research in the Dry Areas. Wallingford, Oxon, UK, p. 321-356.

Yaduvanshi, N.P.S., 2003. Substitution of inorganic fertilizers by organic manures and the effect on soil fertility in a rice-wheat rotation on reclaimed sodic soil in India. J. of Agric. Sci., 140: 161-168.

3

EVALUATION OF YIELD AND SEED REQUIREMENTS STABILITY
OF BREAD WHEAT(*Triticum aestivum* L.) *VIA* AMMI MODEL

Velimir MLADENOV[1]*, Borislav BANJAC[1], Mirjana MILOŠEVIĆ[2]

[1] *University of Novi Sad, Faculty of Agriculture, Novi Sad, REPUBLIC SERBIA*
[2] *Institute of Field and Vegetable Crops, Novi Sad, REPUBLIC SERBIA*
* Corresponding author: vmladenov@polj.uns.ac.rs

ABSTRACT

High quality seed of wheat is the key to successful agriculture. Improvement and evaluation of agronomic traits have been the primary objective of breeders for many years under variable environments. The objective of the present research was to determine influence of genotype, environment and their interaction on yield and randman of seed as a seed quality represent and to evaluate stability through AMMI model. Grain samples were obtained from ten winter wheat cultivars grown in 2009/10 and 2010/11 at three locations in Serbia: Novi Sad, Sremska Mitrovica and Pančevo. Yield and randman of seed were investigated and statistically analized via AMMI model which shows significant differences between genotypes at various locations. Best performer was Simonida with average yield 8.22 t·ha^{-1}. Analyses of randman of seed indicate at differences in the main effect and interaction.

Key words: AMMI model, randman of seed, stability, wheat (*Triticum aestivum* L.), yield.

INTRODUCTION

Wheat is a major crop contributing importantly to the nutrient supply of the global population and also a very versatile crop; it shows wide adaptation to diverse agro-ecological conditions and cropping technologies (Pena, 2007). High quality seed is the key to successful agriculture (Banu, 2004). Quality of seed is most often defined as a unifed sum of seed features that after sowing lead to a rapid and uniform germination, forming of a strong and healthy seedlings which will give the neccesary number of plants in favorable and unfavorable environmental conditions. The seed quality is also reflected in the final growth, maturation of plants, their uniformity and stability of yield (Molnar et al., 2005; Chloupek et al., 2003). It has long been recognized that wheat productivity and quality of seed vary considerably as a result of genotype, environment and their interaction (GEI). While wheat growers consider yield as a major issue, millers and bakers emphasize variability in the functional properties of flour as their biggest concern (Denčić et al., 2011). Precisely because of this situation it is necessary to include as many parameters into consideration. Improvement and evaluation of agronomic traits has been the primary objective of breeders for many years under variable environments. Breeders have also measured and selected for grain yield and most related traits such as thousand grain weight, test weight, randman of seed and other related traits (Rubio et al., 2004; Dimitrijević, 2011). All these traits are affected by the growing environment as well as by genetic factors, and numerous studies have described the GEI (Doehlert and

McMullen, 2000; Doehlert et al., 2001). However, evaluation of genotypes across diverse environments and over several years is needed in order to identify spatially and temporally stable genotypes that could be recommended for release as new cultivars and/or for use in the breeding programs (Sharma et al., 2010). Large differential genotypic responses occur under varying environmental conditions (Mkumbira et al., 2003). This phenomenon is referred to as the GEI and is important in plant breeding programs (Mohammadi, 2010). An understanding of the cause of the GEI can help to identify superior genotypes based on traits. Usually a number of genotypes are tested over a number of sites and years and multiple traits are recorded and it is often difficult to determine the pattern of genotypic performance across environments. Numerous methods have been used to understand the causes of the interactions, although strategies may differ in overall appropriateness.

The objectives of the present research were two-fold (i) to determine influence of genotype, environment and their interaction on yield and randman of seed as a seed quality represent; (ii) to evaluate stability throught AMMI model.

MATERIALS AND METHODS

Samples

Grain samples were obtained from ten winter wheat cultivars grown in 2009/10 and 2010/11 at three locations in Serbia: Novi Sad, Sremska Mitrovica and Pančevo. The ten winter wheat cultivars used in this study were: Evropa 90, NSR 5, Pobeda, Renesansa, Ljiljana, Cipovka,

Dragana, Simonida, NS 40 S and Zvezdana (Table 1). All of these cultivars were designed in Institute of Field and Vegetable Crops, Novi Sad so the fact that all cultivars were agronomically suitable for production in these locations was not questionable. The wheat cultivars

were planted in a randomized complete block design with three replications. Plots of 5 m^2 with 10 rows spaced 12.5 cm apart were seeded at a rate of ≈230 kg·ha^{-1}. Sowing in both growing seasons was completed by the end of October, while harvest was ended in first ten-day period of July.

Table 1. Genotype, pedigree and year of release of 10 winter wheat cultivars and environments description

Genotype	Pedigree	Year of release	Environment		
			Code	Location-Veg. season	
Evropa 90	Talent x NS rana 2	1990			
NS rana 5	NS rana 1/Tisa x Partizanka/3/Mačvanka 1	1991			
Pobeda	Sremica x Balkan	1990	E1	Rimski Šančevi-2009/10	
Renesansa	Jugoslavija x NS 55-25	1994	E2	Sremska Mitrovica-2009/10	
Ljiljana	NS3287-3 x Rodna	2000	E3	Pančevo-2009/10	
Cipovka	NS rana 5 x Rodna	2002	E4	Rimski Šančevi-2010/11	
Dragana	Sremka 2 x Francuska	2002	E5	Sremska Mitrovica-2010/11	
Simonida	NS 63-25/Rodna x NS-3288	2003	E6	Pančevo-2010/11	
NS 40 S	NS 694 x NSA 88-3141	2005			
Zvezdana	NS 63-27/Stamena x NS rana 5	2006			

Sample Analyses

Yield YLD (t·ha^{-1}) was determined in field. Randman of seed (RND) is defined as a ratio between natural and pure seed without any admixtures. RND was determined when the 4·100 g of natural seed was sifted through rectangular aperture size 2.2 mm. The rest of seeds on the sieve was measured and expressed as a percentage (%). Tests were performed on the harvested seed of each cultivar for each replication.

Statistical Analyses

Minimum, maximum, mean values, standard deviation and variance were calculated as indicators of trait variability. These statistical calculations were done using StatSoft, Inc. (2011), STATISTICA (data analysis software system), version 10 (www.statsoft.com). The sustainability index (SI) was estimated by the following formula (Babarmanzoor et al., 2009).

SI = [(Y-σ)/ Y$_{max}$] × 100

where Y - average performance of a genotype, σ - standard deviation and Y$_{max}$ – maximum value of a genotype in any year. The values of sustainability index were divided arbitrarily into 5 groups viz. very low (upto 20%), low (21-40%), moderate (41-60%), high (61-80%) and very high (above 80%). Genotype by environment interaction (GEI) was tested using AMMI (Additive Main Effects and Multiplicative Interaction) analysis by Zobel et al. (1998). Data processing was performed in GenStat 9th Edition VSN International Ltd (www. vsn-intl.com).

Growing Season Conditions

Reactions of genotypes were observed in two growing seasons (data not shown). Generally speaking, the weather conditions were very favourable for wheat production, the temperature on all locations across two years were higher than average, with more rainfall than usual and insulation much less than average. (Republic Hydro-meteorological Service of Serbia, 2011).

RESULTS AND DISCUSSION

The AMMI method is used for three main purposes. The first is model diagnoses, AMMI is more appropriate in the initial statistical analysis of yield trials, because it provides an analytical tool of diagnosing other models as sub cases when these are better for particular data sets (Gauch, 1988). Secondly, AMMI clarifies the G × E interaction and it summarizes patterns and relationships of genotypes and environments (Zobel et al., 1998; Crossa et al., 1990). The third use is to improve the accuracy of yield estimates (Ilker et al., 2011).

Table 2. AMMI analysis for the yield of examined wheat cultivars

Source [1]	df	SS	MS	F	p	The share of the total variation %
Total	179	485,6	2,713	-	-	100
Treatments	59	448,3	7,599	32,88	0,000**	92,32
Genotypes	9	29,1	3,236	14,00	0,000**	5,99
Environments	5	374,1	74,812	72,61	0,000**	77,04
Block	12	12,4	1,030	4,46	0,000**	2,55
Interactions	45	45,1	1,003	4,34	0,000**	9,29
IPCA$_1$	13	25,9	1,991	8,61	0,000**	57,43
IPCA$_2$	11	11	1,000	4,33	0,000**	24,39
IPCA$_3$	9	5,6	0,626	2,71	0,007**	12,42
IPCA$_4$	7	1,8	0,257	1,11	0,360^{ns}	3,99
Residuals	5	0,8	0,163	0,71	0,621^{ns}	-
Error	108	25	0,231	-	-	-

[1] All sources were tested in relation to the error

AMMI analysis of variance for yield showed that the total sum of squares attributed to the impact of environments 77.04%, GEI was represented with 9.29% while 5.99% was the effect of genotype. Differences in soil and surface conditions of the site have caused a large sum of squares environments in total variation and precisely this fact was reflected with the axiom that environments was the most responsible for the variation in yield. The genetic constitution of cultivars is a precondition for expression of yield, nevertheless wheat is grown in the open field and the yield is quantitative trait so the environmental factors are crucial determinant of yield expression. There were significant differences between genotypes at various locations. It can be seen from the ratio of the sum of squares among interaction and genotype since the interaction was two times higher than the share of genotype (Table 2) .

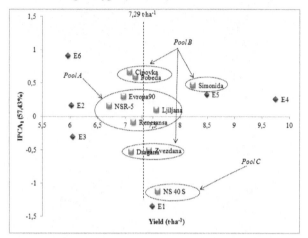

Figure 1. Biplot of the AMMI model for yield for 10 examined wheat varieties grown across six environments

In further course of analyses from the interaction sum of squares four main components were separated, of which the first one (IPCA$_1$) explains 57.43% of the

structure. Due to the high proportion of first principal component graphical display of AMMI analyses is provided in form of AMMI 1 biplot (Figure 1).

Biplot shows that the most genotypes achived yield that slighly deviate from the grand mean of the experiment (7.29 t·ha^{-1}). Best performer was Simonida with average yield 8.22 t·ha^{-1} and also achived good interaction with the E5 environment. Largest contribution to the GEI gave E1 and E6, since they were most distant from the axis of stability. In relation to the value of interaction, genotypes were grouped in three pools according to stability:

pool A-stable genotypes: Renesansa, Ljiljana, NSR5 and Evropa 90

pool B-medium stable genotypes: Simonida, Zvezdana, Dragana, Pobeda and Cipovka

pool C- minimum stable genotypes: NS 40S.

For achiving highest yield most suitable environments were E4 and E5, while the lowest values od yield were recorded at environments E6, E2 and E3.

High sustainability index (%) was estimated in the cases of Simonida (69.03) and Zvezdana (64.69), while the rest of genotypes showed moderate sustainability index (41-60), Table 3. These results prove that SI is not a suitable stability index for discriminating stable genotypes with high grain yield.

Table 3. Basic statistics parameters for yield (t·ha^{-1}) of 10 winter wheat genotypes across two seasons: minimum, maximum and mean values, standard deviation, variance with sustainability index in six environments

Genotype	Minimum value	Maximum value	Mean value	Standard deviation	Variance	Sustainability index (SI) %
Evropa 90	5.47	9.48	6.96	1.74	3.03	55.01
NS rana 5	4.89	9.19	6.70	1.56	2.43	55.98
Pobeda	5.08	10.38	7.18	1.97	3.89	50.14
Renesansa	5.68	9.79	7.14	1.64	2.67	56.21
Ljiljana	5.97	10.34	7.56	1.66	2.74	57.10
Cipovka	5.27	9.76	7.07	1.70	2.87	55.01
Dragana	5.03	9.24	7.10	1.57	2.45	59.93
Simonida	6.83	10.01	8.22	1.31	1.72	69.03
NS 40 S	5.03	10.09	7.57	2.01	4.04	55.17
Zvezdana	5.99	9.25	7.43	1.44	2.08	64.69

AMMI analyses showed that the variation of randman was mainly determined by the influence of environment (60.39%), GEI (20.43%) and finally genotype (15.34%), Table 4.

Minor proportion of genotype indicates that randman of seed is trait to which the genotype effect is under expression of morphological properties of grain (size of a

seed) while everything else is under the influence of living conditions through different phenological stages but mostly during grain filling. Due to the very high share of the first principal component of the total variation of interaction (72.55%) stability of genotypes and environment was presented in the form of AMMI1 biplot (Figure 2).

Table 4. AMMI analysis for the randman of seed in examined wheat cultivars

Source [1]	df	SS	MS	F	p	The share of the Total Variation %
Total	179	2115,4	11,82	-	-	100
Treatments	59	2034,2	34,48	62,39	0,000[**]	96,16
Genotypes	9	324,6	36,06	65,26	0,000[**]	15,34
Environments	5	1277,5	255,50	142,24	0,000[**]	60,39
Block	12	21,6	1,80	3,25	0,001[**]	1,02
Interactions	45	432,1	9,60	17,38	0,000[**]	20,43
IPCA$_1$	13	313,5	24,11	43,63	0,000[**]	72,55
IPCA$_2$	11	45,4	4,13	7,47	0,000[**]	10,51
IPCA$_3$	9	34,7	3,86	6,98	0,000[**]	8,03
IPCA$_4$	7	32,3	4,62	8,36	0,000[**]	7,48
Residuals	5	6,2	1,24	2,24	0,060[ns]	-
Error	108	59,7	0,55	-	-	-

[1] All sources were tested in relation to the error

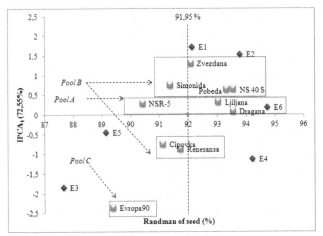

Figure 2. Biplot of the AMMI model for randman of seed for 10 examined wheat varieties grown across six environments

The position of environment points indicates at differences among main effect and interaction. Cultivars recorded largest randman of seed in the E6, which was at the same time the most stable environment for the expression of this trait. For the achievement of stable randman of seed after E6, E5 was most suitable. Nevertheless, this environment has not showned to be suitable for high randman of seed values, considering that

the lowest main has been achieved here (89.15%). Environments E3 and E1 had the highest scores of interaction and that fact distinguished them as the unfavourable for stable randman of seed. Similar to yield and for randman of seed three groups in relation to stability were allocated:

pool A- stable genotypes: Ljiljana, NSR5 and Dragana

pool B-medium stable genotypes: Pobeda, NS40S, Simonida, Cipovka, Renesansa and Zvezdana.

pool C- minimum stable genotypes: Evropa 90.

Zvezdana has been classifed in *pool B* although its interaction deviates from the values of other genotypes in that group. But the deviation is less than the least stable genotype from the *pool C*. Zvezdana achieved the value of randman of seed at the average level of two-year experiment and also had a positive interaction with E2 and E1 and negative with E4. These facts indicates that its stability is conditioned by the terms of year and not by locality.

Very high sustainability index (above 90%) was estimated in the all cases (Table 5). These results prove that SI was not a suitable stability index for discriminating stable genotypes with high randman of seed.

Table 5. Basic statistics parameters for randman of seed (%) of 10 winter wheat genotypes across two seasons: minimum, maximum and mean values, standard deviation, variance with sustainability index in six environments

Genotype	Minimum value	Maximum value	Mean value	Standard deviation	Variance	Sustainability index (SI) %
Evropa 90	85.90	95.27	89.39	3.61	13.01	90.04
NS rana 5	85.03	93.27	90.42	2.99	8.96	93.74
Pobeda	86.80	96.60	93.30	3.77	14.18	92.69
Renesansa	89.50	94.73	91.71	2.10	4.41	94.59
Ljiljana	89.05	96.47	93.04	3.00	9.00	93.34
Cipovka	88.13	93.90	91.11	2.98	8.88	93.85
Dragana	89.93	95.87	93.53	2.75	7.55	94.69
Simonida	86.77	94.23	91.36	3.13	9.81	93.63
NS 40 S	86.80	97.03	93.54	3.98	15.87	92.30
Zvezdana	85.33	96.57	92.08	4.70	22.10	90.49

In the presented results, all characteristics of wheat have changed under the influence of locality – this was reflected through significant GEI. Calculated values of GEI showed that there were differences in a stability of genotype across both investigated traits. Genotypes were grouped differently according to stability, expressed throught interaction scores (AMMI model) and SI index. Nevertheless, AMMI model is more comprehensive and results of this analysis are considered as plausible. Since that AMMI analysis fully explained sum of squares it provides a good estimation of genotypes on the observed locations.

LITERATURE CITED

Babarmanzoor A., M.S. Tariq, A. Ghulam and A. Muhammad, 2009. Genotype × environment interaction for seed yield in Kabuli Chickpea (*Cicer arietinum* L.) genotypes developed through mutation breeding. Pak J Bot 41(4): 1883-1890.

Banu S., J.M. Duxbury, J.G. Lauren, C. Meisner, R. Islam, 2004. Wheat seed quality – A study on farmers seed. Paper presented at 4th International Crop Science Congress, Brisbane, Queensland, Australia, 26 September – 01 October 2004.

Chloupek O., P. Hrstkova, D. Jurecka, 2003. Tolerance of barley seed germination to cold- and drought-stress expressed as seed vigour. Plant Breeding 122:199-203.

Crossa J., H.G. Gauch and R.W. Zobel, 1990. Additive main effect and multiplicative interaction analysis of two international maize cultivar trials. Crop Sci 30: 493–500.

Denčić S., N. Mladenov, B. Kobiljski, 2011. Effects of genotype and environment on breadmaking quality in wheat. Inter. J. Plant Prod. 5(1): 71-82.

Dimitrijević M., D. Knežević, S. Petrović, V. Zečević, J. Bošković, M. Belić, B. Pejić, B. Banjac, 2011. Stability of yield components in wheat (*Triticum aestivum* L). Genetika, Vol 43, No.1, 29 -39.

Doehlert D.C., M.S. McMullen, 2000. Genotypic and environmental effects on oat milling characteristics and groat hardness. Cereal Chem. 77: 148-154.

Doehlert D.C., M.S. McMullen, J.J. Hammond, 2001. Genotypic and environmental effects on grain yield and quality of oat grown in North Dakota. Crop Sci. 41: 1066-1072.

Gauch H.G., 1988. Model selection and validation for yield trials with interaction. Biometrics 44: 705-715.

GenStat 9th Edition VSN International Ltd (www. vsn-intl.com). 2009.

Ilker E., H. Geren, R. Unsal, I. Sevin, F. Aykut Tonk, M. Tosun, 2011. AMMI-biplot analysis of yield performances of bread wheat cultivars grown at different locations. Turkish J. of Field Crops 16(1):64-68.

Mkumbira J., N.M. Mahungu, U. Gullberg, 2003. Grouping locations for efficient cassava evaluation in Malawi. Exp. Agric. 39:167-179.

Mohammadi R., M. Armion, D. Kahrizi, A. Amri, 2010. Efficiency of screening techniques for evaluating durum wheat genotypes under mild drought conditions. Inter. J. Plant Prod. 4(1): 11-24.

Molnar I., M. Vujaković, M. Milošević, R. Kastori, D. Milošev, S. Šeremešić, 2005. Uticaj đubrenja i plodoreda na životnu sposobnost semena ozime pšenice. Field Veg. Crop Res. 41:257-267.

Pena R.J., 2007. Current and future trends of wheat quality needs. In: Buck, HT, Nisi JE, Salomon N (Eds.).Wheat production in stressed environments. Springer, p: 411-424.

Republic Hydro-meteorological service of Serbia 2011 http://www.hidmet.gov.rs/ciril/meteorologija/index.php

Rubio J., J.I. Cubero, L.M. Martin, M.J. Suso, F. Flores, 2004. Biplot analysis of trait relations of white lupin in Spain. Euphytica 135: 217-224.

Sharma R.C., A.I. Morgounov, H.J. Braun, B. Akin, M. Keser, D. Bedoshvili, A. Bagci, C. Martius, M. van Ginkel, 2010. Identifying high yielding stable winter wheat genotypes for irrigated environments in Central and West Asia. Euphytica 171: 53-64.

StatSoft, Inc. 2011. STATISTICA (data analysis software system), version 10 (www.statsoft.com).

Zobel R.W., M.J. Wright, H.G. Gauch, 1998. Statistical analysis of yield trial. Agron. J. 80:388 393.

4

EFFECT OF IRRIGATION AMOUNTS APPLIED WITH DRIP IRRIGATION ON MAIZE EVAPOTRANSPIRATION, YIELD, WATER USE EFFICIENCY, AND NET RETURN IN A SUB–HUMID CLIMATE

Hayrettin KUSCU[1]*, Abdullah KARASU[1], Mehmet OZ[1], Ali Osman DEMIR[2], İlhan TURGUT[3]

[1]*Uludag University, Mustafakemalpasa Vocational School, Department of Crop and Animal Production, Mustafakemalpasa, Bursa, TURKEY.*
[2]*Uludag University, Faculty of Agriculture, Department of Biosystems Engineering, Gorukle, Bursa, TURKEY.*
[3]*Uludag University, Faculty of Agriculture, Department of Field Crops, Gorukle, Bursa, TURKEY.*
*Corresponding author: kuscu@uludag.edu.tr

ABSTRACT

The purpose of this study was to determine the effect of irrigation amount applied with drip irrigation on field maize (*Zea mays* L.) evapotranspiration (ET), yield, water use efficiency, yield response factor (ky) and net return in a sub–humid environment of Turkey. Irrigation management treatments were created as 125%, 100%, 75%, 50%, 25% and 0% replenishment of water depleted in the 90 cm root zone from 100% replenishment treatment in every seven days. Irrigation amounts ranged from 76 to 1120 mm in 2007 and from 91 to 997 mm in 2008. The treatments resulted in seasonal ET of 311–1078 mm and 298–1061 mm in 2007 and 2008, respectively. The average grain yields varied from 5570 to 16535 kg ha^{-1}. In both seasons, irrigation significantly affected yields, which increased with irrigation up to a level (1100 mm of irrigation water amount), but additional amounts of irrigation did not increase it any further. Yields increased linearly with seasonal ET. The yield response factor (ky) averaged 0.89 over the two seasons. Maximum water use efficiency (WUE) and irrigation water use efficiency (IWUE) values were obtained for the treatment of 25% deficit irrigation. A further increase in water amount from reference irrigation (T–100) increased grain yield but reduced both the WUE and IWUE. The reference irrigation treatment gave the highest net return of $3212 ha^{-1}. The results revealed that the full irrigation is the best choice for higher yield and net income. The results also suggest that 25% deficit irrigation approach may be a good strategy for increase water use efficiencies when full irrigation is not possible.

Keywords: Maize, net return, water–yield relationships, water use efficiency, yield response factor

INTRODUCTION

Maize is one of the most important cereal crops in Turkey (İlker, 2011) and the cultivated area is about 0.60 million hectares (FAO stat, 2009). Maize production in Turkey is about 4.25 million tons and it covers about 95% of corn consumption in the country (FAO stat, 2009). Today, the government of Turkey through its financial assistance programme is trying to encourage farmers to grow maize. It is grown almost all over the country under varied soil and climatic conditions. Most of the maize in Turkey is irrigated and is grown under low rainfall and heat stress conditions. In these conditions, irrigation is the major factor determining yield. It is consequently essential to determine the water regimes leading to highest yield. Maize has been reported in the literature to have high irrigation requirements (Stone et al., 1996; Karam et al.,

2003). Maize grain yield increased significantly by irrigation water amount and irrigation frequency (Yazar et al., 1999; Kara and Biber, 2008; Farré and Faci, 2009). However, water availability is usually the most important natural factor limiting expansion and development of agriculture in Marmara region of Turkey. Competition for water from other sectors such as industry and domestic use will force irrigation to operate under water scarcity.

When water supplies are limiting, the farmer's goal should be to maximize net income per unit water used rather than per land unit. Deficit irrigation, by reducing irrigation water use, can aid in coping with situations where water supply is restricted. In field crops, a well-designed deficit irrigation regime can optimize water productivity over an area when full irrigation is not possible (Fereres and Soriano, 2007). The correct application of deficit irrigation requires thorough understanding of the yield response to water (crop sensitivity to drought stress) and of the economic impact of reductions in harvest (English, 1990). However, maize has been reported to be very sensitive to drought (Otegui et al., 1995). Lamm et al. (1995) stated that it is difficult to plan deficit irrigation for maize without causing yield reduction. Payero et al. (2006a) reported that trying to increase crop water productivity by imposing deficit irrigation for maize might not be a beneficial strategy in a semiarid climate. Karam et al. (2003) found that grain and dry matter yield, and leaf area index was reduced by severity of water stress. Pandey et al. (2000) stated that yield reduction (22.6–26.4%) caused by deficit irrigation was associated with a decrease in kernel number and weight. The effects of deficit irrigation for the same crop may vary with location. Climate and soil type of the location are perhaps the most important factors dictating the influence of deficit irrigation (Igbadun et al., 2008).

Shortage in irrigation water supplies in the Marmara region has motivated farmers to find ways to produce crops, especially maize, with less irrigation water, such as using more efficient irrigation systems and changing from fully–irrigated to deficit irrigated cropping systems. Furrow irrigation is the most common method used for irrigating row crops such as maize in the Marmara region of Turkey. However, the drip irrigation method is becoming more popular because of numerous advantages over other methods (Hanson et al., 1997). Some advantages of drip irrigation over other irrigation methods include improved water and nutrient management, improved saline water management, potential for improved yields and crop quality, reducing the incidence of diseases and weeds in dry row middles, greater control on applied water resulting in less water and nutrient loss through deep percolation, and reduced total water

requirements (Ayars et al., 1999; Dogan and Kirnak., 2010). During the past decade, Turkish government has been financially supporting the farmers who are willing to set up drip irrigation system. Therefore, the use of drip irrigation is increasing substantially each year in the region. However, local information from the Marmara region of Turkey on the response of maize yield with drip irrigation is very limited, especially dealing with the effect of limited water allocations. In Marmara climatic region, little attempt has been made to assess the water–yield relationships and optimum water management programs of maize for recently developed hybrids.

The main aim of this study is to examine the effect of different irrigation amounts applied with drip irrigation on evapotranspiration, grain yield, water use efficiency, total production cost and net return of maize grown in a sub–humid climate of Turkey.

MATERIALS AND METHODS

Field experiments were conducted at the experimental station of the Mustafakemalpasa Vocational School, Uludag University located in Bursa, Turkey (40° 02′ N, 28° 23′ E, 25 m above sea level) for two consecutive summer seasons (2007 and 2008). The climate in this part of the country is classified as sub–humid according to Thornthwaite climate classification system (Feddema, 2005). The climatic parameters during the crop growing seasons are summarized in Table 1. The soil in the experimental field was clay loam. The soil moisture content at field capacity (–33 kPa) and permanent wilting point (–1500 kPa) was 36% and 21% on an oven dry weight loss basis, (moisture content on dry weight basis) respectively. The average bulk density of the soil was 1.41 g cm^{-3}. The plant available soil moisture was 186.1 mm m^{-1}.

Table 1. Some climatic parameters in 2007, 2008 and between 1975 and 2007 at Mustafakemalpasa in Bursa, Turkey.

Months	Climatic Parameters								
	Temperature (°C)			Humidity (%)			Precipitation (mm)		
	2007	2008	Average[a]	2007	2008	Average	2007	2008	Average
May	19.9	18.1	17.2	61	66.7	66.3	12.1	24.8	42.9
June	24.6	23.1	21.6	55	63.2	61.2	47.2	10.8	23.4
July	26.2	24.3	23.6	51	60.9	61.1	13.4	0	13.9
August	26.4	24.1	23.3	53	62.0	61.7	1.0	0	14.9
September	21.4	20.2	19.6	57	76.1	64.8	3.4	87.2	31.2

[a] Average values between 1975 and 2007.

The experiments were conducted using a randomized complete block design with three replications. The area of each plot was 22.75 m^2 (long: 5.00 m, wide: 4.55 m). A

buffer zone spacing of 2 m was provided between the plots. Experimental plots received 180 kg ha^{-1} nitrogen and 120 kg ha^{-1} P$_2$O$_5$. The maize hybrid Ada–523 was planted at a spacing of 0.20 m × 0.65 m at a plant population of 76920 plants per hectare (Çarpıcı et al., 2010). In 2007, maize was planted on May 10, and harvested on October 10. In 2008, maize was planted on May 17 and harvested October 18. Grain yields were determined by hand harvesting the 3.8 m sections of the five adjacent center rows in each plot. The harvest area in each plot was 12.35 m^2. Grain yields were adjusted to a constant moisture basis of 150 g kg^{-1} water.

Each experiment consisted of six irrigation levels, i.e. the amount of water in different treatments was 0 (T–0), 25 (T–25), 50 (T–50), 75 (T–75), 100 (T–100) and 125% (T–125) of water depleted in the 90 cm root zone in every seven days (Panda et al., 2004). Gravimetric method was used in determining the amount of water need to bring the soil in the crop root zone to field capacity. All irrigation treatments were applied on the same day. Crops were irrigated by drip irrigation. Irrigation water was pumped directly from a well to the drip irrigation system. The lateral lines were laid adjacent to each crop row. The laterals had an outer diameter of 16 mm and pressure-compensating emitters spaced every 0.3 m. Each emitter had a nominal flow rate of 1.6 L h^{-1} at a pressure of 100 kPa. The sub-main line was connected to a water meter and a control valve. During the first 2 weeks all the treatments received a daily amount of 5–7 mm irrigation water in order to establish plants.

Soil moisture contents were monitored in 0.3 m depth increments to 1.5 m prior to and after irrigation weekly from the plots of the second replication (block) throughout the growing season. Soil samples were taken at positions immediately under the drippers. Soil moisture was determined by the gravimetric method (oven dry basis).

Actual crop evapotranspiration under the different irrigation treatments was estimated using the following from of the water balance equation (Garrity et al., 1982):

$$ET = I + P \pm \Delta S - R - D$$

(1)

where ET is evapotranspiration (mm), I is the irrigation water (mm), P is the precipitation (mm), ΔS is the change in soil water storage (mm), R is the runoff, and D is the drainage below the root zone. In the equation, I was measured by water meters, P was observed at the meteorological station nearby the experimental area, ΔS was obtained from gravimetric moisture observations in the soil profile to a depth of 90 cm. In this study, surface runoff was assumed to be negligible because the amount of irrigation water was controlled through the drip irrigation.

The water use–yield relationship was determined using the Stewart's model (Stewart et al., 1977). Water use efficiency (WUE, kg m^{-3}) was calculated as grain yield divided by seasonal ET. Irrigation water use efficiency (IWUE, kg m^{-3}) was also determined according to Howell (2001).

Estimates of net return were calculated as

$$R_{net} = R_{gross} \times Y_{grain} - C_{water} - C_{production}$$

where, R$_{net}$ is net return ($ ha^{-1}), R$_{gross}$ is commodity price ($ t^{-1}), Y$_{grain}$ is grain yield (t ha^{-1}), C$_{water}$ is cost of water ($ ha^{-1}) and C$_{production}$ is cost of production ($ ha^{-1}). Gross return was calculated by assuming a unit of the commodity price $320.63 t, the average price in Bursa province of Turkey during the period 2003–2008 (Turk stat, 2010). The cost of water for each irrigation treatment was calculated by multiplying the cost of unit volume of water and the total quantum of irrigation water required for the maize crop (Panda et al., 2004). The main source of the irrigation water was the groundwater. All other production costs including labor (installation, irrigation, planting, weeding, cultivation, fertilizer application, spraying, and harvesting), land preparation, seeds, fertilizers, chemicals (insecticides and pesticides) were assumed constant across all water treatments except I$_0$. The irrigation cost was eliminated from the production cost for I$_0$ treatment. Data required for the production cost were collected from 20 sample farms by field survey, and it was determined as $1550.85 ha^{-1}. The production cost was not included in land rental cost.

Data were subjected to analysis of variance for grain yield using MINITAB (University of Texas at Austin) software. The F–protected least significant difference (LSD) was calculated at the 0.05 probability level according to Steel and Torrie (1980).

RESULTS AND DISCUSSION

Water applied and evapotranspiration (ET)

Seasonal water applied and seasonal ET values for the different treatments are shown in Table 2. The amount of irrigation water applied varied from 76 to 1120 mm in 2007 and from 91 to 997 mm in 2008. Active root depth for maize assumed to be 90 cm, and therefore, deep percolation measurements were made 90–120 cm soil depth. Result indicated that percolation occurred only with T–125 treatment of about 9% calculated from 2 years average. Regression analysis indicated a linear relationship between seasonal ET and seasonal water applied (Fig. 1). In a similar study, Payero et al. (2008) reported the strong quadratic relationships between

Table 2. Seasonal water applied, seasonal maize evapotranspiration (ET), grain yield, water use efficiency (WUE) and irrigation water use efficiency (IWUE) for each irrigation treatment obtained in 2007 and 2008 at Mustafakemalpasa in Bursa, Turkey.

Year	Treatment	Seasonal water applied (mm)	Seasonal ET (mm)	Grain yield (kg ha^{-1})	WUE (kg m^{-3})	IWUE (kg m^{-3})
2007	T–0	76	311	5650 e	1.82	–
	T–25	285	452	6690 d	1.48	0.50
	T–50	494	621	11680 c	1.88	1.44
	T–75	702	809	15610 b	1.93	1.59
	T–100	911	992	15920 b	1.60	1.23
	T–125	1120	1078	16340 a	1.52	1.02
	LSD$_{(P=0.05)}$	–	–	346.1	–	–
2008	T–0	91	298	5490 f	1.84	–
	T–25	272	447	6240 e	1.40	0.41
	T–50	454	652	12110 d	1.86	1.82
	T–75	635	824	15240 c	1.85	1.79
	T–100	816	957	16480 b	1.72	1.52
	T–125	997	1061	16730 a	1.58	1.24
	LSD$_{(P=0.05)}$	–	–	75.0	–	–

seasonal ET and seasonal water applied for maize under sprinkler irrigation. On the other hand, Mengu and Ozgürel (2008) determined a linear relationship for these two variations.

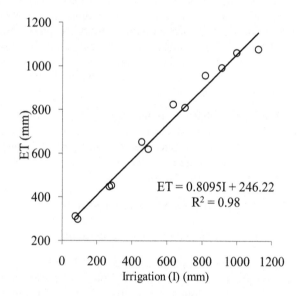

Figure 1. Relationship between irrigation and seasonal evapotranspiration (ET) for maize obtained at Bursa, Turkey, for combined years (2007 and 2008)

The seasonal values of ET per treatment ranged from 311 to 1078 mm in 2007 and from 298 to 1061 mm in 2008. As expected, the highest seasonal ET occurred in the T–125 treatment and the lowest ET occurred in the non–irrigated treatment (T–0). Yildirim and Kodal (1998) reported that seasonal ET in maize varied between 300 and 1024 mm in Ankara, Turkey. Under furrow irrigation applications, seasonal ET of maize obtained by Gencoglan and Yazar (1999) was 1026 mm for full irrigation treatment and 410 mm for non–irrigated treatment in the Cukurova region of Turkey. Oktem et al. (2003) found that seasonal ET for maize by using drip irrigation method in Sanliurfa conditions of Turkey varied between 1040–

701 mm depending on irrigation scheduling. The values of seasonal ET obtained in this study are in agreement to those values reported in the previous literature for maize.

Grain yield and water–yield relationships

Irrigation treatments also resulted in differences in grain yield as shown in Table 2. This ranged from 5650 to 16340 kg ha^{-1} in 2007 and from 5490 to 16730 kg ha^{-1} in 2008 for the different irrigation regimes. Increased water amounts resulted in a relatively higher yield, since water deficit was the main yield–limiting factor in both years. The maximum yield was obtained at T–125 and the minimum yield at T–0 in both 2007 and 2008. However, in 2007, there was no significant difference between the treatments T–100 and T–75 i.e. irrigated with a 25 percent deficit. The results in this study are in agreement with previous studies. For instance, Bozkurt et al. (2011) reported that the highest grain yield was found in 120% of evaporation from a Class A Pan under the Eastern Mediterranean climatic conditions in Turkey. Yazar et al. (2002) reported also that the highest average maize grain yield obtained from full irrigation treatment using drip irrigation method. However, Yildirim and Kodal (1998) stated that applications of excessive irrigation water did not increase grain yields at the important level.

The relation between applied water and grain yield was evaluated for each experimental year (Fig. 2). The relationship between applied water and grain yield was quadratic. Small irrigation amounts increased yield, more or less linearly up to a level where the relationship was curvilinear because part of the water applied is not used in ET. At a point of 1100 mm of irrigation water amount, yield reached its maximum value (16730 kg ha^{-1}). Moreover, the regression equation shows that additional amounts of irrigation did not increase it any further (Fig. 2). Nonlinear relationships have also been reported by Gencoglan and Yazar (1999), Kipkorir et al. (2002), Bozkurt et al. (2006), and Farré and Faci (2009). However, Payero et al. (2006b) reported that there was linear relationship between grain yield and seasonal irrigation water amount.

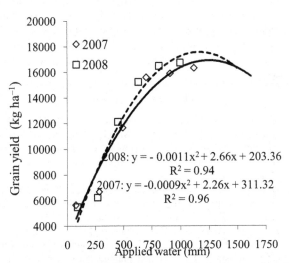

Figure 2. Relationship between applied water and grain yield for maize

A linear relationship was found between seasonal ET and grain yield in both years (Fig. 3). Grain yield responded linearly to crop water consumption. The linearity between grain yield and ET has also been reported by Cakir (2004), Oktem (2006) and Igbadun et al. (2007).

A good linear relationship between relative evapotranspiration deficit and relative yield decrease was observed from combining data over the two years (Fig. 4). The slope of the line in Fig. 4 represents that the yield response factor (ky) is 0.89 (R^2=0.89). This value implies that the rate of yield decrease is proportionally slightly lower (ky<1) than the relative evapotranspiration deficits. The ky value obtained in this study is similar to the literature data (Yildirim et al., 1996; Yazar et al., 2002; Karam et al., 2003; Mengu and Ozgurel, 2008). On the other hand, our result was lower than the values of ky obtained by Kipkorir et al. (2002) as 1.21, by Cakir (2004) as 1.29, by Dagdelen et al. (2006) as 1.04, and by Oktem (2008) as 1.23.

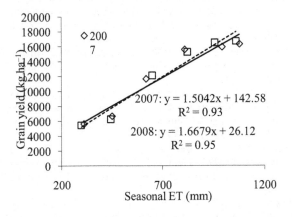

Figure 3. Relationship between seasonal ET and grain yield for maize

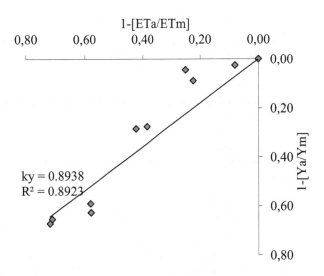

Figure 4. Relationship between relative grain yield decrease and relative ET deficit for maize throughout the total growing season during 2 years (2007–2008)

WUE and IWUE were different based on the treatments and years (Table 2). WUE values ranged between 1.40 and 1.93 kg m^{-3} for the both years. In the first year, the maximum values of both WUE and IWUE were obtained from the T–75 treatment as 1.93 and 1.59 kg m^{-3}, respectively. In the second year, higher water use efficiencies were found for the T–75 and T–50 treatments. However, the efficiencies decreased in the other deficit irrigation treatments (T–25 and T–0) and excessive irrigation treatment (T–125). On the contrary, Oktem (2006) stated that WUE increased as the amount of irrigation water increased. The ranges of WUE and IWUE obtained in this study were close to those reported in the previous literature for maize. Howell et al. (1995) reported WUE range of 0.89–1.45 kg m^{-3}, Yazar et al. (1999) reported WUE ranges of 0.87–1.42 kg m^{-3}, Oktem et al. (2003) reported WUE range of 1.04–1.36 kg m^{-3} and Oktem (2006) reported IWUE range of 1.07-1.43 kg m^{-3}. However, the range of WUE and IWUE obtained in this study were higher than those reported by Igbadun et al. (2008) and Pandey et al. (2000). Generally, WUE and IWUE are influenced by crop yield potential, irrigation method, estimation and measurement of ET, crop environment, and climatic characteristics of the region. The results related to the efficiencies shows that when irrigation water is limited, 25% deficit irrigation can be applied for increase the water use efficiencies. Mansouri–Far et al. (2010) reported that irrigation water can be conserved and yields maintained in maize plant (as sensitive crop to drought stress) under water limited conditions through improved fertilizer managements and selecting more tolerant hybrids. On the other hand, the feasibility of increasing either the WUE or IWUE is a decision that needs to be based not only on the biophysical response of the crop but also on economic factors. Often the objective of producers is not to increase yields but to increase profits (Payero et al., 2008). Determining the level of irrigation needed to optimize profits can be complex and depends on both biophysical and economic factors (Norton et al., 2000; English et al., 2002; Payero et al., 2008).

Net return

Total cost, gross return and net return of maize at different irrigation levels are presented in Table 3. The total cost of production increased with increase in irrigation levels. The net return increased sharply from T–25 to T–75 treatment due to the sharp increase in grain yield. The net return from T–75 to T–125 did not increase considerably because of the insignificant improvement in grain yield.

Table 3. Economic analysis of drip-irrigated maize under different irrigation schedules (average data of 2 years)

Treatment	Seasonal water applied (mm)	Total production cost ($ ha^{-1})	Gross return ($ ha^{-1})	Net return ($ ha^{-1})
T-0	76	1596	1785	189
T-25	285	1700	2167	467
T-50	494	1805	3813	2008
T-75	702	1910	4945	3035
T-100	911	2015	5227	3212
T-125	1120	2119	5301	3182

The reference irrigation treatment (T–100) gave the highest net return of $3212 ha^{-1}. The results revealed that the full irrigation is the best choice for higher yield and net income under drip irrigation. In a similar study, Panda et al. (2004) reported that the highest net return was obtained when irrigation was scheduled at 45% depletion of available soil water. But, net return values (170–206 $ ha^{-1}) determined in their study, were also considerably lower than the values (189–3212 $ ha^{-1}) determined in this study. The reasons for this difference may be lower local production cost, higher local commodity price and higher grain yield at the study area.

CONCLUSION

This study evaluated the effect of different seasonal irrigation amounts on maize evapotranspiration, grain yield, water use efficiency, and net return in a subhumid climate during 2007 and 2008. Increased water amounts resulted in a relatively higher yield, since water deficit was the main yield-limiting factor. This finding supported the hypothesis that less water stress would produce higher yield. In both years, seasonal water applied and grain yield of maize exhibited strong quadratic relationships. The average yield response factor (ky), which indicates the effect of water stress on reducing crop yield, averaged 1.07 over the 2 years. The value of ky obtained for this study could be used for the purposes of irrigation management and water allocation scheduling over irrigation schemes under limited irrigation water supply. In this study, higher values of both WUE and IWUE were obtained when irrigation was scheduled at 75 percent available soil moisture depletion (T–75). On the other hand, full irrigation (T–100) gave the highest net return. Finally, the overall results clearly revealed that in order to obtain higher yield and net income of maize in a sub-

humid climate under drip irrigation, crops during the summer season should be irrigated at 100% soil water depletion every week. The results also suggest that 25% deficit irrigation approach may be a good strategy for increase water use efficiencies when full irrigation is not possible.

LITERATURE CITED

Ayars, J.E., C.J. Phene, R.B. Hutmacher, K.R. Davis, R.A. Schoneman, S.S. Vail, R.M. Mead, 1999. Subsurface drip irrigation of row crops: a review of 15 years of research at the Water Management Research Laboratuary. Agric. Water Manage. 42: 1-27.

Bozkurt, Y., A. Yazar, B. Gencel, M.S. Sezen, 2006. Optimum lateral spacing for drip–irrigated corn in the Mediterranean Region of Turkey. Agric. Water Manage. 85: 113–120.

Bozkurt, S., A. Yazar, G.S. Mansuroglu, 2011. Effects of different drip irrigation levels on yield and some agronomic characteristics of raised bed planted corn. Afric. J. Agric. Res. 6(23): 5291–5300.

Cakir, R. 2004. Effect of water stress at different development stages on vegetative and reproductive growth of maize. Field Crops Res. 89: 1–16.

Çarpıcı, E.B., N. Çelik, G. Bayram, 2010. Yield and quality of forage maize as influenced by plant density and nitrogen rate. Turk. J. Field Crops. 15(2): 128-132.

Dagdelen, N., E. Yilmaz, F. Sezgin, T. Gurbuz, 2006. Water–yield relation and water use efficiency of cotton (*Gossypium Hirsutum* L.) and second crop corn (*Zea mays* L.) in western Turkey. Agric. Water Manage. 82 (1–2): 63–85.

Dogan, E., H. Kirnak, 2010. Water temperature and system pressure effect on drip lateral properties. Irrig. Sci. 407-419.

English, M., 1990. Deficit irrigation I: Analytical framework. J. Irrig. Drain. Eng. 116(3): 399–410.

English, M.J., K.H. Solomon, G.J. Hoffman, 2002. A paradigm shift in irrigation management. J. Irrig. Drain. Eng. 128: 267–277.

FAO stat, 2009. http://faostat.fao.org (accessed 11.21.11).

Farré, I., J.M. Faci, 2009. Deficit Irrigation in maize for reducing agricultural water use in a Mediterranean environment. Agric. Water Manage. 96: 383–394.

Feddema, J.J., 2005. A revised Thornthwaite-type global climate classification. Physical Geography. 26(6): 442–466.

Fereres, E., M.A. Soriano, 2007. Deficit irrigation for reducing agricultural water use. J. Exp. Botany. 58(2): 147–159.

Garrity, P.D., D.G. Watts, C.Y. Sullivan, J.R. Gilley, 1982. Moisture deficits and grain sorghum performance, evapotranspiration yield relationships. Agron. J. 74: 815–820.

Gencoglan, C., A. Yazar, 1999. The effects of deficit irrigations on corn yield and water use efficiency. Turk. J. Agric. For. 23: 233–241.

Hanson, B.R., L.J. Schwankl, K.F. Schulback, G.S. Pettygrove, 1997. A comparision of furrow, surface drip and subsurface drip irrigation on lettuce yield and applied water. Agric. Water Manage. 33: 139–157.

Howell, T.A., A. Yazar, A.D. Schneider, D.A. Dusek, K.S. Copeland, 1995. Yield and water use efficiency of corn in response to LEPA irrigation. Trans. ASAE 38(6): 1737–1747.

Howell, T.A. 2001. Enhancing water use efficiency in irrigated agriculture. Agron. J. 93(2): 281–289.

Igbadun, H.E., A.K.P.R. Tarimo, B.A. Salim, H.F. Mahoo, 2007. Evaluation of selected crop water production functions for an irrigated maize crop. Agric. Water Manage. 94: 1–10.

Igbadun, H.E., B.A. Salim, A.K.P.R. Tarimo, H.F. Mahoo, 2008. Effects of deficit irrigation scheduling on yields and soil water balance of irrigated maize. Irrig. Sci. 27: 11–23.

İlker, E., 2011. Correlation and path coefficient analyses in sweet corn. Turk. J. Field Crops 16(2): 105–107.

Kara, T., C. Biber, 2008. Irrigation frequencies and corn (*Zea mays* L.) yield relation in Northern Turkey. Pakistan J. Bio. Sci. 11(1): 123–126.

Karam, F., J. Breidy, C. Stephan, J. Rouphael, 2003. Evapotranspiration, yield and water use efficiency of drip irrigated corn in the Bekaa Valley of Lebanon. Agric. Water Manage. 63(2): 125–137.

Kipkorir, E.C., D. Raes, B. Massawe, 2002. Seasonal water production functions and yield response factors for maize and onion in Perkerra, Kenya. Agric. Water Manage. 56: 229–240.

Lamm, F.R., H.L. Manges, L.R. Stone, A.H. Khan, D.H. Rogers, 1995. Water requirement of subsurface drip–irrigated corn in northwest Kansas. Trans. ASAE. 38(2): 441–448.

Mansouri–Far, C., S.A.M.M. Sanavy, S.F. Saberali, 2010. Maize yield response to deficit irrigation during low–sensitive growth stages and nitrogen rate under semi–arid conditions. Agric. Water Manage. 97: 12–22.

Mengu, G.P., M. Ozgürel, 2008. An evaluation water–yield relations in maize (*Zea mays* L.) in Turkey. Pakistan J. Bio. Sci. 11(4): 517–524.

Norton, N.A., R.T. Clark, J.P. Schneekloth, 2000. Effects of alternative irrigation allocations on water use, net returns, and marginal user costs. In: Western Agricultural Economics Association Annual Meetings, Vancouver, British Columbia, p.13.

Oktem, A., M. Simsek, A.G. Oktem, 2003. Deficit irrigation effects on sweet corn (*Zea mays* saccharata sturt) with drip irrigation system in a semi–arid region I. Water–yield relationship. Agric. Water Manage. 61: 63–74.

Oktem, A., 2006. Effect of different irrigation intervals to drip irrigated dent corn (*Zea mays* L. indentata) water–yield relationship. Pakistan J. Bio. Sci. 9(8): 1476–1481.

Oktem, A., 2008. Effect of water shortage on yield, and protein and mineral compositions of drip–irrigated sweet corn in sustainable agricultural systems. Agric. Water Manage. 95: 1003–1010.

Otegui, M.E., F.H. Andrade, E.E. Suero, 1995. Growth, water use, and kernel abortion of maize subjected to drought at silking. Field Crops Res. 40(2): 87–94.

Panda, R.K, S.K. Behera, P.S. Kashyap, 2004. Effective management of irrigation water for maize under stressed conditions. Agric. Water Manage. 66: 181–203.

Pandey, R.K., J.W. Maranville, A. Admou, 2000. Deficit irrigation and nitrogen effects on maize in a Sahelian enviroment. I. Grain yield and yield components. Agric. Water Manage. 46(1): 1–13.

Payero, J.O., N.L. Klocke, J.P. Schneekloth, D.R. Davison, 2006a. Comparison of irrigation strategies for surface–irrigated corn in west central Nebraska. Irrig. Sci. 24: 257–265.

Payero, J.O., M. Steven, S. Irmak, D.D. Tarkalson, 2006b. Yield response of corn to deficit irrigation in a semiarid climate. Agric. Water Manage. 84: 895–908.

Payero, J.O., D.D. Tarkalson, S. Irmak, D. Davison, J.L. Petersen, 2008. Effect of irrigation amounts applied with subsurface drip irrigation on corn evapotranspiration, yield, water use efficiency, and dry matter production in a semiarid climate. Agric. Water Manage. 95: 895–908.

Steel, R.G.D., J.H. Torrie, 1980. Principles and Procedures of Statistics. A Biometrical Approach. 2. ed. New York: McGraw–Hill, 631p.

Stewart, J.I., R.H. Cuenca, W.O. Pruit, R.M. Hagan, J. Tosso, 1977. Determination and utilization of water production functions for principal California crops. W–67 California Contribution Project Report. Davis, USA: University of California.

Stone, L.R., A.J. Schlegel, R.E. Gwin, A.H. Khan, 1996. Response of corn, grain sorghum, and sunflower to irrigation in the High Plains of Kansas. Agric. Water Manage. 30: 251–259.

Turk stat, 2010. Turkish Statistical Institute. http://www.turkstat.gov.tr/ (Accessed 05.11.11).

Yazar, A., T.A. Howell, D.A. Dusek, K.S. Copeland, 1999. Evaluation of crop water stress index for LEPA irrigated corn. Irrig. Sci. 18, 171–180.

Yazar, A., S.M. Sezen, B. Gencel, 2002. Drip irrigation of corn in the Southeast Anatolia Project (GAP) area in Turkey. Irrigation and Drainage. 51: 293–300.

Yildirim O, S. Kodal, F. Selenay, Y.E. Yildirim, A. Ozturk, 1996. Corn grain yield response to adequate and deficit irrigation. Turk. J. Agric. For. 20(4): 283–288.

Yildirim, E., S. Kodal, 1998. Effect of irrigation water on corn grain yield in Ankara conditions. Turk. J. Agric. For. 22: 65–70.

5

NITROGEN CONCENTRATIONS AND NITROGEN YIELDS OF ABOVE-GROUND DRY MATTER OF CHICKPEA DURING CROP GROWTH COMPARED TO PEA, BARLEY AND OAT IN CENTRAL EUROPE

R.W. Neugschwandtner[1]*, H. Wagentristl[2], H.-P. Kaul[1]

[1]*BOKU - University of Natural Resources and Life Sciences, Vienna, Department of Crop Sciences, Division of Agronomy, Tulln, AUSTRIA*
[2] *BOKU - University of Natural Resources and Life Sciences, Vienna, Department of Crop Sciences, Experimental Farm Groß-Enzersdorf, Groß-Enzersdorf, AUSTRIA*
**Corresponding author: reinhard.neugschwandtner@boku.ac.at*

ABSTRACT

Alternative crops like chickpea could become of interest under Pannonian climate conditions in Central Europe due to forecasted changes in climate. Therefore a two-year trial was conducted to evaluate concentrations, uptake and yields of nitrogen (N) during crop growth of chickpea (*Cicer arietinum* L.) compared to pea (*Pisum sativum* L.), barley (*Hordeum vulgare* L.) and oat (*Avena sativa* L.) as affected by N fertilization with either calcium ammonium nitrate (CAN) or the depot fertilizer Basacote® Plus 6M (DF) in eastern Austria. Chickpea had the lowest above-ground dry matter (AGDM) and N yields among the four crops in 2006; however, it could gain higher AGDM and N yields than those of barley and oat under drought conditions in 2007. N concentrations and N yields throughout crop growth were increased by increasing rates of N fertilization (with CAN showing generally higher values than DF). Chickpea had a high crop N uptake rate and a high relative N uptake rate even under drought conditions. Thus, results indicated that chickpea could be an alternative crop in dry environments for achieving reasonably N yields in Central European growing conditions.

Keywords: Chickpea, *Cicer arietinum*, nitrogen, uptake rate, Central Europe

INTRODUCTION

Expected changes in agro-climatic conditions and a substantial deficit of protein sources for livestock are two challenges for European agriculture. The changes in agro-climatic conditions in Central Europe are expected to go along with an increase in air temperature, changes in the amount and distribution of precipitation, and prolonged growing seasons. This may lead to a lower productivity of rainfed spring crops due to a higher risk of drought (Trnka et al., 2011). Promising opportunities may arise under these conditions for adopting crops with a pronounced warm-season growth habit such as chickpea (*Cicer arietinum* L.) in comparatively cool, northern latitude areas (Gan et al., 2009). Introducing a new grain legume to Central European agricultural systems would be also beneficial for reducing the substantial deficit of protein sources in the European Union where around two-thirds of soybean meal and soybeans [*Glycine max* (L.) Merr.] for livestock feed is imported (Henseler et al., 2013).

Chickpea is mainly produced in arid or semiarid environments (Canci and Toker, 2009a, b). Due to several morphological and physiological advantages, the crop can effectively cope with drought conditions (Serraj et al., 2004; Cutforth et al., 2009; Zaman-Allah et al., 2011). Chickpea is of high importance in human diets in many areas of the world. Additionally, chickpea grains can be used as energy and protein-rich feed in animal diets and chickpea straw as forage for ruminants (Bampidis and Christodoulou, 2011). Chickpea yields, yield components and protein contents are affected by production system and fertilization regime (Caliskan et al, 2013).

Although chickpea is not a common crop in Central Europe, it could provide an alternative for food and feed protein production in the face of climate change. Recently, the plant has been introduced to semiarid regions in Australia (Siddique and Sykes, 1997), to the Northern Great Plains in North America (Miller et al., 2002) and in western Canada (Anbessa et al., 2007). In eastern Austria, chickpea can achieve higher grain and straw yields than spring sown barley (*Hordeum vulgare* L.) and oat (*Avena sativa* L.) under drought conditions (Neugschwandtner et al., 2013). Furthermore, the adoption of chickpea in Central Europe could lead to crop diversification and improved productivity of sustainable agricultural systems as legumes satisfy a bulk of their N demand from atmospheric N through symbiosis with N fixing rhizobia. Thereby the demand for N fertilizer inputs within crop rotations is minimized (van Kessel and Hartley, 2000), and positive yield effects are caused through the transfer of biologically fixed N via crop residues to subsequent non-legume crops (Kaul, 2004).

Currently, little information exists on the agronomy and the performance of chickpea grown in northern latitudes (Gan et al., 2009). Therefore, the objective of the presented work was to evaluate N concentrations, N yields and N uptake of chickpea during crop growth under Central European growing conditions as compared to pea, barley and oat and as affected by fertilizer form and fertilizer rate to gain information for a possible introduction of this crop to Central Europe.

MATERIAL AND METHODS

Experimental site and weather conditions

The experiment was carried out in Raasdorf (48° 14' N, 16° 33' E; altitude: 153 m a.s.l.) in eastern Austria at the experimental farm of BOKU University. The soil is classified as a chernosem of alluvial origin and rich in calcareous sediments (pH 7.6). The texture is silty loam; the content of organic substance is at 2.2-2.3%.

The mean annual temperature is 10.6°C, the mean annual precipitation is 538 mm (1980-2009). Table 1 shows the long-term average monthly temperatures and precipitation from February to July and the deviations

Table 1. Long-term average monthly temperature and precipitation (1980-2009) and deviations during the 2006 and 2007 growing seasons

	Temperature (°C)			Precipitation (mm)		
	Long term mean (1980-2009)	2006 (±)	2007 (±)	Long-term mean (1980-2009)	2006 (±)	2007 (±)
February	1.7	-1.9	3.8	26.4	-7.7	17.7
March	5.8	-2.1	2.3	38.5	7.7	28.0
April	10.7	1.3	2.1	35.3	30.3	-34.4
May	15.6	-0.5	1.6	56.1	16.7	-9.8
June	18.5	0.6	2.8	72.3	-9.9	-3.9
July	20.8	2.8	1.9	59.1	-52.3	-6.2

during the 2006 and 2007 growing seasons. The temperature was considerably higher in 2007 than in 2006 (except for July). Monthly precipitation was well above average in March and April in 2006. Contrary to that, the growing season 2007 was characterized by a severe spring drought without rainfall from the end of March to the beginning of May.

Experimental factors

Two chickpea genotypes were tested under different N fertilization in comparison to common varieties of pea and the non-legume crops barley and oat with similar vegetation periods. The experiment was set up in a randomized complete block design with two replications. The chickpea variety Kompolti and commercial seeds of a chickpea genotype of unknown origin obtained from a trade company were planted (both are Kabuli type chickpeas). The seeds had been multiplied on-farm. Pea cv. Attika and Rosalie, barley cv. Xanadu and oat cv. Jumbo were used as standards of comparison. The nitrogen fertilizer calcium ammonium nitrate (CAN) (27% N, 10% Ca) and the depot fertilizer Basacote® Plus 6M (DF) (16% N, 3.5% P, 10% K, 1.2% Mg, 5% S and micronutrients) were applied right after sowing at two N fertilization levels (10 and 20 g N m^{-2}) (10 CAN, 10 DF, 20 CAN, 20 DF) supplemented by an unfertilized control. Maximum temperature from sowing to harvest was 35.5°C (2006) and 37.9°C (2007); minimum temperature was 4.9°C (2006) and 3.1°C (2007).

Crop management and measurements

Seeds were sown with an Oyjard plot drill (row distance: 12 cm; plots size: 30 m^2). Chickpea nodulates with symbiotic bacteria Mesorhizobium cicieri and M. mediterraneum (Toker et al., 2007), and therefore, seeds were inoculated with Mesorhizobium ciceri (Jost GmbH), seeds of pea with Rhizobium leguminosarum (Radicin No4, Jost GmbH) according to product specifications before sowing. Inoculation was performed as eastern Austrian soils may not contain the specific rhizobia for chickpea to ensure an effective plant-microbe association for nitrogen fixation. Inoculation of chickpea seeds has been shown to increase yield and protein content of seeds (El Hadi and Elsheikh, 1999; Farzaneh et al., 2009). Sowing was performed on 14 April 2006 and on 11 April 2007, respectively, with a sowing rate of 90 seeds m^{-2} for chickpea and pea and 300 seed m^{-2} for barley and oat. Weed control was performed mechanically. Above-ground dry matter (AGDM) development was determined by harvesting (0.24 m^2 per plot) at intervals of about 14 days from May until the end of June (2006: 5 May, 22 May, 9 June, 27 June; 2007: 14 May, 31 May, 14 June, 26 June). The final harvest was performed at full ripeness of the plants on 0.96 m^2 per plot (chickpea: 1 August 2006 and 23 July 2007; pea: 20 July 2006 and 9 July 2007; barley: 18 July 2006 and 23 July 2007; oat: 24 July 2006 and 23 July 2007). Plant samples were dried at 100°C for 24 h.

Nitrogen determination and calculations

For nitrogen determination, plant samples were first ground to pass through a 1 mm sieve. Nitrogen concentrations were determined as an average of duplicate samples per plot of about 500 mg each with a combustion technique using a LECO-2000CN auto analyzer (LECO, 1994).

Crop nitrogen uptake rate (CUR_N) and relative nitrogen uptake rate (RUR_N) were calculated for each period between subsequent harvest dates according to Hunt (2002) as follows:

$$(1) \quad CUR_N \; (g \; m^{-2} \; d^{-1}) = \frac{N_2 - N_1}{t_2 - t_1}$$

$$(2) \quad RUR_N \; (mg \; g^{-2} \; d^{-1}) = \frac{\ln N_2 - \ln N_1}{t_2 - t_1}$$

where N_2 and N_1 indicate the final and initial nitrogen yield of the AGDM and t_2 and t_1 indicate the end and the start day of each period.

Statistics

Statistical analyses were performed using software SAS version 9.2. Analyses of variance (PROC GLM) with subsequent multiple comparisons of means were performed. Means were separated by least significant differences (LSD), when the F-test indicated factorial effects on the significance level of $p < 0.05$. Genotype differences within chickpea and pea, respectively, were not significant, so data were pooled for analysis.

RESULTS AND DISCUSSION

Data are presented for N fertilization (main effect) and interaction of crop × year based on analysis of variance results.

Above-ground dry matter production (AGDM)

Growth and yield analysis of chickpea compared to pea, barley and oat have been already described by Neugschwandtner et al. (2013). In both years until end of June (HD 4), the AGDM of chickpea was significantly lower than those of pea, barley and oat. Final AGDM (HD 5) of chickpea in 2006 was significantly lower than those of the other crops. With dry conditions in 2007, chickpea's final AGDM was less impaired than those of the other crops resulting in similar levels of AGDM in all four crops as chickpea is well adapted to drought stress (Serraj et al., 2004; Cutforth et al., 2009; Zaman-Allah et al., 2011) (Figs. 1a-e). Fertilization with readily available CAN enhanced early AGDM production (HD 2) of all crops compared to the unfertilized controls. The DF treatments resulted in intermediate AGDM. At HD 3 and 4 all N fertilization regimes had higher AGDM than the controls. Also at final harvest (HD 5), fertilizer regimes 20 CAN, 10 CAN and 20 DF had higher AGDM than control (with 10 DF lying in between) (Table 2). Early N application may cause sufficient plant and root

development and thereby enables a better adaptation to the post anthesis drought stress (Gevrek and Atasoy, 2012).

Table 2. Above-ground dry matter (AGDM), N concentration and N yield during crop development as affected by fertilization (in g fertilizer N m^{-2}) (means across crops and years)

	Harvest dates				
	1	2	3	4	5
AGDM yield (g m^{-2})					
0	38^a	127^b	360^b	632^b	954^b
10 CAN	37^a	154^a	433^a	805^a	1115^a
10 DF	37^a	142^{ab}	437^a	756^a	1058^{ab}
20 CAN	37^a	156^a	454^a	820^a	1156^a
20 DF	37^a	147^{ab}	440^a	828^a	1130^a
N concentration (%)					
0	4.46^d	3.40^e	2.87^c	2.28^b	1.85^b
10 CAN	5.05^{ab}	4.65^b	3.27^b	2.39^b	1.94^{ab}
10 DF	4.83^c	4.14^d	3.02^c	2.36^b	1.88^b
20 CAN	5.15^a	4.95^a	3.59^a	2.67^a	2.11^a
20 DF	4.98^b	4.37^c	3.25^b	2.45^b	1.96^{ab}
N yield of AGDM (g N m^{-2})					
0	1.78^a	4.9^d	10.1^c	14.3^b	16.9^b
10 CAN	1.94^a	6.9^{ab}	13.2^b	18.1^a	19.9^{ab}
10 DF	1.88^a	5.8^c	12.7^b	17.4^a	18.5^b
20 CAN	1.97^a	7.6^a	15.7^a	19.7^a	22.5^a
20 DF	1.90^a	6.3^{bc}	13.6^b	19.0^a	19.9^{ab}

CAN: calcium ammonium nitrate; DF: depot fertilizer Basacote® Plus 6M. Different letters indicate significant differences between means (p < 0.05).

N concentrations and N yields of above-ground dry matter

The N concentrations (%) of the AGDM generally decreased with plant growth. Starting with HD 3 the N concentrations of all crops were higher in the dry year of 2007 compared to 2006. The N concentrations of chickpea in 2006 were lower than those of pea at early growth (HD 1 and 2) and at final harvest (HD 5); but they were higher than those of barley and oat at HD 3 and 4. In 2007, chickpea's N concentrations were always higher than those of barley and pea (except for HD 1); at HD 2 they were similar to, at HD 3 and 4 higher and at HD 5 lower than those of pea (Figs. 1f-j). The N concentrations of AGDM were increased by N fertilization at HD 2 in the following order: 20 CAN > 10 CAN > 20 DF > 10 DF > control; thus, easily soluble N fertilizer CAN increase N concentrations more strongly at early growth compared to DF. At final harvest, the N concentrations were significantly higher in the 20 CAN treatment than in the 10 DF and control treatments (with 20 DF and 10 CAN lying in between) (Table 2). Differences between N concentrations during crop growth between years and fertilizer levels have already been reported for oat by Maral et al. (2012). Turpin et al. (2002) reported higher N concentrations in the AGDM of chickpea due to N fertilization during crop growth until flowering stage in experiments conducted in Australia. Contrary to that, higher N rates in our experiment resulted in higher N concentrations until maturity.

The N yield is a function of AGDM production and N concentration. Up to about the middle of June (HD 3), N yields were generally higher in 2007 than in 2006 (except for oat), mainly due to higher N concentrations and with pea and barley also due to higher AGDM. At the end of June (HD 4) chickpea, barley and oat still achieved higher N yields in 2007 compared to 2006 despite drought stress due to significantly higher N concentrations. Strongly impaired final AGDM of pea, barley and oat in 2007 resulted in a lower N yield compared to 2006 (although N concentrations of AGDM of these crops were higher in 2007). In contrast, chickpea was able to compensate slightly lower AGDM in 2007 by increased N

concentrations, resulting in a higher N yield than in 2006 (Figs. 1k-o). N fertilization clearly affected N yields during crop growth. At early growth (HD 2), N yields were high especially in treatments fertilized with CAN. At final harvest, the N yield of AGDM was significantly higher in the 20 CAN treatment than in the 10 DF and control treatments (with 20 DF and 10 CAN lying in between) (Table 2). Soltani et al. (2006) and Koutroubas et al. (2009) reported that variations of the N yield of chickpea were mainly linked to corresponding AGDM variations. Contrary to that, our results show that variations of both AGDM production and N concentrations affected the N yields of the crops. Furthermore, Caliskan et al. (2013) reported that fertilization also increases harvest indices and protein yields of chickpeas.

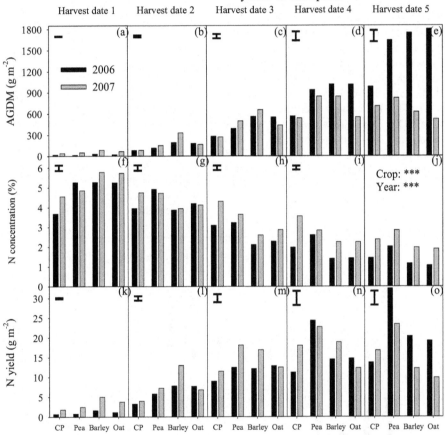

Figure 1. (a-e) Above-ground dry matter (AGDM), (f-j) N concentration and (k-o) N yield on harvest dates 1-5 as affected by crop and year. Error bars are LSD. CP = chickpea.

Crop nitrogen uptake rate (CUR$_N$) and relative nitrogen uptake rate (RUR$_N$)

The CUR$_N$ of chickpea was lowest in the first observation period (HD 1-2) among the tested crops (except for oat in 2007). Between HD 2-3 chickpea's CUR$_N$ was lower than that of pea but higher than those of the cereals. In the more humid year of 2006 the CUR$_N$ of chickpea was the lowest among the four crops starting from middle of June (HD 3) until harvest; however, with dry conditions in 2007, chickpea's CUR$_N$ was clearly higher than those of the other crops between HD 3-4. The

CUR$_N$ was significantly lower in the dry year of 2007 than in 2006 in the last sampling period (HD 4-5) with negative values for chickpea, barley and oat and just a slightly positive value for pea (Figs. 2a-d). Drought stress significantly reduces the CUR$_N$ as described for wheat (*Triticum aestivum* L.) by Abreu et al. (1993). A negative CUR$_N$ during seed development until maturity has also been reported for oilseed rape (*Brassica napus* L.) by Barlóg and Grzebisz (2004). A high variability of the CUR$_N$ between crops, years and during crop development is in accordance with Gastal and Lemaire (2002). N fertilization affected CUR$_N$ until middle of June. The amount and seasonal distribution of N uptake was affected by N rate and N form as reported by McTaggart and Smith (1995). There was a significant fertilization × year interaction on CUR_N between HD 1-2 insofar that N fertilization increased the CUR_N with a higher increase after application of CAN than DF in 2006 whereas just the highest CAN treatment significantly increased CUR_N in 2007 (data not shown); thus, easily soluble N fertilizer CAN increase CUR_N more strongly than the DF at early growth. Between HD 2-3 the CUR_N were ranked as follows: 20 CAN ≥ 20 DF, 10 CAN, 10 DF > control. No differences in the CUR_N after middle of June due to fertilization were observed (Table 3).

The RUR_N of chickpea was between HD 2-3 in both years slightly higher than that of peas and significantly higher than those of barley and oat; in the dry year of 2007 chickpea's RUR_N highly surpassed those of the other crops between HD 3-4. From end of June (HD 4) until final harvest (HD 5) all crops had a positive RUR_N in 2006 whereas in 2007 dry conditions caused a negative RUR_N for the crops (with chickpea's RUR_N lying between pea's low negative RUR_N and the cereals' higher negative RUR_N) (Figs. 2e-h). The RUR_N was increased with N fertilization between HD 1-2 in 2006 whereas no fertilizer effect was observed in 2007 (data not shown). No statistical differences in the RUR_N occurred between end of May (HD 2) and end of June (HD 4). The RUR_N between the last two harvest dates (HD 4-5) was impaired by N fertilization with 20 DF compared with the control

with all other N fertilization treatments having a slightly lower RUR_N than the control (Table 3).

Table 3. Crop nitrogen uptake rate (CUR_N) and relative nitrogen uptake rate (RUR_N) during crop development as affected by fertilization (in g fertilizer N m^{-2}) (means across crops and years)

	Harvest dates			
	1-2	2-3	3-4	4-5
CUR_N (g m^{-2} d^{-1})				
0	0.18^d	0.34^b	0.29^a	0.11^a
10 CAN	0.29^{ab}	0.41^{ab}	0.27^a	0.16^a
10 DF	0.23^{cd}	0.45^{ab}	0.31^a	0.05^a
20 CAN	0.33^a	0.51^a	0.26^a	0.14^a
20 DF	0.26^{bc}	0.47^{ab}	0.30^a	0.09^a
RUR_N (mg m^{-2} d^{-1})				
0	63^c	48^a	23^a	7^a
10 CAN	83^a	46^a	22^a	2^{ab}
10 DF	74^b	54^a	18^a	3^{ab}
20 CAN	85^a	51^a	15^a	3^{ab}
20 DF	81^{ab}	52^a	24^a	-1^b

CAN: calcium ammonium nitrate; DF: depot fertilizer Basacote® Plus 6M. Different letters indicate significant differences between means (p < 0.05).

Low and negative CUR_N and RUR_N before harvest may be due to ammonia (NH_3) volatilization from aboveground plant parts as observed by Bahrani et al. (2011) for wheat in the period from anthesis to maturity or due to leaf drop.

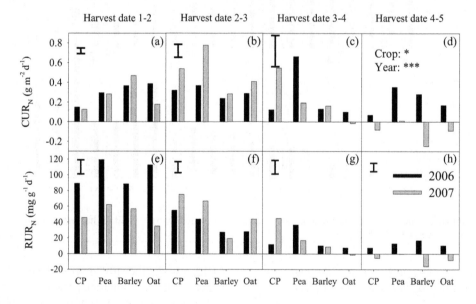

Figure 2. (a-d) Crop nitrogen uptake rate (CUR_N) and (e-h) Relative nitrogen uptake rate (RUR_N) between the harvest dates as affected by crop and year. Error bars are LSD. CP = chickpea.

CONCLUSION

Chickpea had a lower AGDM in 2006 than pea, barley and oat but exceeded pea and oat in the dry year of 2007 due to its adaptability to drought stress. The N concentrations of all crops were higher in the dry year; however, chickpea was the only crop that could achieve a

higher N yield as its AGDM was only slightly reduced under drought conditions. N fertilization clearly affected N concentrations and N yields of the crops during crop growth with CAN showing slightly higher N concentrations and N yields a final harvest than the DF. Chickpea showed a high CUR_N and a high RUR_N under drought conditions. Thus, results indicated that chickpea could be an alternative crop for achieving a reasonable N yield in dry Central European growing conditions.

LITERATURE CITED

Abreu, J.D.M., Flores, I., Abreu, F.D. and M. Madeira. 1993. Nitrogen uptake in relation to water availability in wheat. Plant Soil 154:89-96.

Anbessa, Y., Warkentin, T., Bueckert, R., Vandenberg, A. and Y. Gan. 2007. Post-flowering dry matter accumulation and partitioning and timing of crop maturity in chickpea in western Canada. Can J Plant Sci 87:233-240.

Bahrani, A., Abad, H.H.S. and A. Aynehband. 2011. Nitrogen remobilization in wheat as influenced by nitrogen application and post-anthesis water deficit during grain filling. Afr J Biotechnol 10:10585-10594.

Bampidis, V.A. and V. Christodoulou. 2011. Chickpeas (*Cicer arietinum* L.) in animal nutrition: A review. Anim Feed Sci Technol 168:1-20.

Barłóg, P. and W. Grzebisz. 2004. Effect of timing and nitrogen fertilizer application on winter oilseed rape (*Brassica napus* L.). II. Nitrogen uptake dynamics and fertilizer efficiency. J Agron Crop Sci 190:314-323.

Caliskan, S., Erdogan, C., Arslan, M. and M.E. Caliskan. 2013. Comparison of organic and traditional production systems in chickpea (*Cicer arietinium* L.). Turk J Field Crops 18:34-29.

Canci, H. and C. Toker. 2009a. Evaluation of annual wild *Cicer* species for drought and heat resistance under field conditions. Genet Resour Crop Ev 56:1-6.

Canci, H. and C. Toker. 2009b. Evaluation of yield criteria for drought and heat resistance in chickpea (*Cicer arietinum* L.). J Agron Crop Sci 195:47-54.

Cutforth, H.W., Angadi, S.V., McConkey, B.G., Entz, M.H., Ulrich, D. Volkmar, K.M., Miller, P.R. and S.A. Brandt. 2009. Comparing plant water relations for wheat with alternative pulse and oilseed crops grown in the semiarid Canadian prairie. Can J Plant Sci 89:823-835.

El Hadi, E.A. and E.A.E. Elsheikh. 1999. Effect of *Rhizobium* inoculation and nitrogen fertilization on yield and protein content of six chickpea (*Cicer arietinum* L.) cultivars in marginal soils under irrigation. Nutr Cycl Agroecosys 54:57-63.

Farzaneh, M., Wichmann, S., Vierheilig, H. and H.-P. Kaul. 2009. The effects of arbuscular mycorrhiza and nitrogen nutrition on growth of chickpea and barley. Pflanzenbauwiss 13:15-22.

Gan, Y.T., Warkentin, T.D., McDonald, C.L., Zentner, R.P. and A. Vandenberg. 2009. Seed yield and yield stability of chickpea in response to cropping systems and soil fertility in northern latitudes. Agron J 101:1113-1122.

Gastal, F. and G. Lemaire. 2002. N uptake and distribution in crops: An agronomical and ecophysiological perspective. J Exp Bot 53:789-799.

Gevrek, M.N. and G.D. Atasoy. 2013. Effect of post anthesis drought on certain agronomical characteristics of wheat under two different nitrogen application conditions. Turk J Field Crops 17:19-23.

Henseler, M., Piot-Lepetit, I., Ferrari, E., Mellado, A.G., Banse, M., Grethe, H., Parisi, C. and S. Hélaine. 2013. On the asynchronous approvals of GM crops: Potential market impacts of a trade disruption of EU soy imports. Food Policy 41:166-176.

Hunt, R., Causton, D.R., Shipley, B. and A.P. Askew, 2002. A modern tool for classical plant growth analysis. Ann Bot 90:485-488.

Kaul, H.-P. 2004. Pre-crop effects of grain legumes and linseed on soil mineral N and productivity of subsequent winter rape and winter wheat crops. Bodenkultur 55:95-102.

Koutroubas, S.D., Papageorgiou, M. and S. Fotiadis. 2009. Growth and nitrogen dynamics of spring chickpea genotypes in a mediterranean-type climate. J Agr Sci Camb 147:445-458.

LECO Corporation. 1994. CN-2000 Carbon/Protein/Nitrogen Elemental Analyzer—Instruction Manual, Version 4. LECO Corporation, St. Joseph, MI, pp. 146.

McTaggart, I.P. and K.A. Smith. 1995. The effect of rate, form and timing of fertilizer N on nitrogen uptake and grain N content in spring malting barley. J Agric Sci Camb 125:341-353.

Maral, H., Dumlupinar, Z., Dokuyucu, T. and A. Akkaya. 2012. Impact of genotype and nitrogen fertilization rate on yield and nitrogen use by oat (*Avena sativa* L.) in Turkey. Turk J Field Crops 17:177-184.

Miller, P.R., McConkey, B.G., Clayton, G.W., Brandt, S.A., Staricka, J.A., Johnston, A.M., Lafond, G.P., Schatz, B.G., Baltensperger, D.D. and K.E. Neill. 2002. Pulse crop adaptation in the northern Great Plains. Agron J 94:261-272.

Neugschwandtner, R.W., Wichmann, S., Gimplinger, D.M., Wagentristl, H. and H.-P. Kaul. 2013. Chickpea performance compared to pea, barley and oat in Central Europe: Growth analysis and yield. Turk J Field Crops 18:179-184.

Serraj, R., Krishnamurthy, L., Kashiwagi, J., Kumar, J., Chandra, S. and J.H. Crouch. 2004. Variation in root traits of chickpea (*Cicer arietinum* L.) grown under terminal drought. Field Crops Res 88:115-127.

Siddique, K.H.M. and J. Sykes. 1997. Pulse production in Australia: Past, present and future. Aust J Exp Agr 37:103-111.

Soltani, A., Robertson, M.J., Rahemi-Karizaki, A., Poorreza, J. and H. Zarei. 2006. Modelling biomass accumulation and partitioning in chickpea (*Cicer arietinum* L.). J Agron Crop Sci 192:379-389.

Toker, C., Lluch, C., Tejera, N.A., Serraj, R. and K.H.M. Siddique, 2007. Abiotic stresses. In: Chickpea Breeding and Management (eds S.S. Yadav, B. Redden, W. Chen & B. Sharma), CAB Int. Wallingford, UK, pp: 474-496.

Trnka, M., Eitzinger, J., Semerádová, D., Hlavinka, P., Balek, J., Dubrovský, M., Kubu, G., Štěpánek, P., Thaler, S., Možný, M. and Z. Žalud. 2011. Expected changes in agroclimatic conditions in Central Europe. Climatic Change 108:261-289.

Turpin, J.E., Herridge, D.F. and M.J. Robertson. 2002. Nitrogen fixation and soil nitrate interactions in field-grown chickpea (*Cicer arietinum*) and fababean (*Vicia faba*). Aust J Agric Res 53:599-608.

van Kessel, C. and C. Hartley. 2000. Agricultural management of grain legumes: has it led to an increase in nitrogen fixation? Field Crops Res 65:165-181.

Zaman-Allah, M., Jenkinson, D.M. and V. Vadez. 2011. A conservative pattern of water use, rather than deep or profuse rooting, is critical for the terminal drought tolerance of chickpea. J Exp Bot 62:4239-4252.

6

PERFORMANCE OF SOME LOCAL NIGERIAN TURFGRASSES IN SOLE AND MIXED STANDS

Stephen OYEDEJI[1]*, Augustine Onwuegbukiwe ISICHEI[2], Adekunle OGUNFIDODO[3]

[1]*Department of Plant Biology, University of Ilorin, Ilorin, Nigeria.*
[2]*Department of Botany, Obafemi Awolowo University, Ile-Ife, Nigeria.*
[3]*Department of Mathematics, Obafemi Awolowo University, Ile-Ife, Nigeria.*
*Corresponding author: oyedeji.s@unilorin.edu.ng

ABSTRACT

The study assessed the performance of *Axonopus compressus, Chysopogon aciculatus, Sporobolus pyramidalis, Eleusine indica* and *Dactyloctenium aegyptium* in turf establishment. The five grass species planted in sole and mixed stands were varied with the legume - *Desmodium triflorum*. Ground cover differed significantly among grass species and their mixtures from 4 - 11 weeks after planting (WAP) but ground cover in the legume and no legume subplots were not significantly different from 4 – 6 WAP. *Eleusine, Axonopus* and *Dactyloctenium* and their mixtures had higher ground cover than those of *Sporobolus* and *Chrysopogon*. There were significant differences in ground cover among grasses and mixtures at 3 and 6 weeks after clipping (WAC), and grass-legume subplots and subplots without the legume were different at 3 and 4 WAC. Recovery weeks after trampling was faster in sole stands and mixtures with *Axonopus* and *Eleusine*, indicating they are better adapted to trampling.

Key words: ground cover, growth rate, legume, mixed stands, trampling-tolerant, warm-season turfgrasses.

INTRODUCTION

Turfgrasses serve aesthetic, recreational and environmental purposes to most landscapes; providing recreational areas, erosion control and other ecological benefits (Kir et al., 2010). The durability of recreational turfs which is the function of adaptation to the abiotic and biotic environments and the suitability to specificities of usage like trampling intensities have been given little attention in Africa. Consequently, most sports fields tend to be excessively used and proper maintenance is often lacking.

Research studies have demonstrated rapid germination and growth of turfgrass to be critical for successful establishment of athletic field turf (Murphy, 2004; Murphy and Park, 2005). Establishment and management of turf without adequate consideration of the adaptation of the sown grass to the environmental context often cause failure of the turf. Successful turf management often begins with proper selection of species that are adapted to the climatic conditions to be experienced (Busey, 2003).

There are several studies on the performance of turfgrasses under Mediterranean (Volterrani et al., 2001; Geren et al., 2009; Demiroglu et al., 2010; Kir et al., 2010; Salman et al., 2011) and temperate climates (Gaussoin, 1994; Steir and Koeritz, 2008) but far less is known about the performance and response of tropical turfgrasses to trampling. Although trampling-tolerant turfgrasses have been the focus of many research efforts over the past three decades (Shearman and Beard, 1975; Hacker, 1987; Taivalmaa et al., 1998; Park et al., 2005) and the performance of some tropical species and cultivars have been studied (Wood and Law, 1972; Evans, 1988; Minner et al., 1993), surprisingly none of these studies have investigated the performances of warm season turfgrasses and their response to trampling. Even in well-established turf, intensive use is a major factor which reduces turf density, particularly use in rainy weather in finer-textured soils with poor drainage (Murphy, 2004).

Differences in trampling tolerance among plant species have been viewed by some authors from the morphological point of view (Bates, 1938; Tachibana, 1976; Sun and Liddle, 1993). Species in trampled habitats are believed to have prostrate growth in which most of their growing point is in contact with the ground surface. The strength of aerial organs in some grass species has been reported to favour trampling tolerance (Kobayashi and Hori, 2000).

In this study, the performance of five grasses; three creepers and two bunch- type (tussock) grasses, at different combinations under the low-mowing (clipping) and the trampling conditions in a tropical climate was assessed.

MATERIALS AND METHODS

The study area

The field study was carried out at the Teaching and Research Farm (T&RF), Obafemi Awolowo University Campus, Ile-Ife, Nigeria: latitude 7°23' N and longitude 4° 37' E. The vegetation of the area falls within the rain forest belt where there are two prominent seasons, the rainy season and the dry season. The dry season is short, usually from November to March while the rainy season lasts the remaining seven months. The rainfall pattern is bimodal with peaks in July and September/October, and total annual precipitation is between 1200 mm and 1500 mm with over 1800 mm in some very rainy years. Daily temperatures do not fluctuate considerably with a maximum between 27 °C and 35 °C and a minimum between 18.9 °C and 23.3 °C.

Soil characteristics of the experimental site

The study area is associated with rolling topography with slope range of 3-6 % and lies within the basement complex with the underlying rock consisting of granites, gneisses, and undifferentiated schists that were mapped and grouped as Iwo Association by Smyth and Montgomery (1962), and classified as ferruginous tropical soils.

The experimental site has a slightly acidic (pH = 6.4), well-drained sandy-loam soil with the sand, silt and clay in the proportions 72.8: 15.1: 12.1. Soil organic carbon, total N and available P were 19.8 g/kg, 0.18 g/kg and 3.35 mg/kg, respectively. The soil exchangeable cations Na^+, K^+, Ca^{2+} and Mg^{2+} were 0.09, 0.29, 1.65 and 0.11 cmol.kg^{-1} respectively.

Experimental design and protocol

The five perennial turfgrasses, *Axonopus compressus* (Sw.) P. Beauv., *Chrysopogon aciculatus* (Retz.) Trin., *Dactyloctenium aegyptium* (L.) Willd., *Eleusine indica* (L.) Gaertn., and *Sporobolus pyramidalis* (P. Beauv.), with their mixtures tested, made fifteen entries. The former three grass species are creeping while the latter two species are tussocks. The entries were tested with and without the association with *Desmodium triflorum* (Linn.) DC. The fifteen entries of grass species and their mixtures are the whole-plot levels, whereas the legume versus no legume factor is the sub-plot factor. The set-up consisted of 90 subplots of 0.5 m × 0.5 m each separated by a 0.2 m alley and arranged in a 15 by 2 factorial treatment fitted in a split-plot design and was replicated three times. The grass species and *Desmodium* in all the treatments were established on 15th June by planting tillers. Grass ground cover was measured weekly, starting from 14th July using a 0.04 m² quadrat with a regular 4 × 4 grid and placed at two different points selected using random numbers, each point corresponds to positions along x and y axis for each subplot. The grid nodes of the quadrats touching grass species in each sub-plot were counted from the two selected points weekly at the same points in all the subplots. Percentage ground cover was calculated by dividing the number of points touching grass species by thirty-two (16 nodes x 2) and multiplied by 100. Eleven weeks after planting (11 WAP), the grasses in each subplot were clipped 2 cm above ground and left to re-grow for another 3 weeks after which ground cover measurement resumed. Starting from seven weeks after clipping (7 WAC), the grasses were trampled daily by a 66 kg man wearing soccer boots for two weeks. The trampling consisted of pressing the foot 30 times on top of the plants at different positions within each subplot. Weekly assessment of ground cover of green turfs in each subplot weeks after trampling (WAT) was done for 6 weeks.

Data processing and statistical analyses

The effects of grass species or mixture and presence/absence of legume on percentage ground cover were examined using three-way ANOVA model performed with the Proc ANOVA in SAS for Windows (SAS, 2003). Means were separated using least significant difference (LSD) at α = 0.05. The linear model for the ANOVA procedure is represented by:

$$Y_{ijk} = \mu + R_i + G_j + RG_{(ij)} + L_k + (GL)_{jk} + e_{(ijk)}$$

Where $i = 1, 2, ..., 3; j = 1, 2, ..., 15; k = 1, 2$

R_i is the replicated block factor (random); G_j is the grass species/combinations effect or whole-plots factor (fixed); L_k is the legume effect or subplot factor (fixed); $RG_{(ij)}$ is the whole-plot error (random); $(GL)_{jk}$ is the grass species-legume interaction effect (fixed); and $e_{(ijk)}$ is the sub-plot error (random).

RESULTS

During the period of establishment of the turf (4 – 11 WAP), there were significant differences in ground cover of the grass species and their mixtures (Table 1a). Sole stand of *Eleusine* had the highest ground cover except at 9 WAP (96.35%) and 10 WAP (98.44%). Most grass species and their mixtures had 100% ground cover at 11 WAP except the sole stands of *Chrysopogon* (88.54%), *Sporobolus* (95.31%), and grass mixture of *Chrysopogon + Sporobolus* (93.75%) (Table 1b). There were significant differences due to the presence of the legume (*Desmodium*) from 7 - 11 WAP, but not from 4 - 6 WAP (Table 1a). The interaction between grass species/mixtures and *Desmodium* was significant from 7 - 11 WAP.

Weeks after clipping (WAC), there were significant differences in ground cover in grass species and their mixtures at 3 WAC and 6 WAC (Table 2). Although, the sole stands *Axonopus* and *Dactyloctenium*, with *Axonopus + Dactyloctenium* had 100% ground cover at 4 WAC, the ground cover in these sole stands were not consistent from our observations at 5 WAC. Ground-cover in subplots with *Chrysopogon + Sporobolus* and *Chrysopogon + Dactyloctenium* reduced by about 6.5% at 5 WAC. All grass species with their mixtures had 100% ground cover at 7 WAC, except *Chrysopogon + Sporobolus* with 99.48% ground cover (Table 2). There was significant difference in the ground cover of grass species and their mixtures between subplots with *Desmodium* and those without the legume at 3 WAC and 4 WAC and average ground cover was 100% in subplots with legume at 7 WAC (Table 3). The interaction between the grass species/mixtures and legume was significant only at 6 WAC (Table 2).

Table 1a. F-statistics to determine the effect of treatment and their interactions on ground cover (%) in grasses and their mixtures weeks after planting (WAP)

Source	df	4 WAP		5 WAP		6 WAP		7 WAP	
		F	P	F	P	F	P	F	P
Grass	14	2.25	0.0333	5.38	<.0001	6.18	<.0001	13.58	<.0001
Grass × Rep.	28								
Legume	1	2.22	0.1470	0.39	0.5361	2.63	0.1152	6.95	0.0131
Grass × Legume	14	1.48	0.1789	0.86	0.6073	1.36	0.2336	2.05	0.0489
Grass × Legume × Rep.	30								
Total	89								
		8 WAP		9 WAP		10 WAP		11 WAP	
Grass	14	15.21	<.0001	9.70	<.0001	6.77	<.0001	10.36	<.0001
Grass × Rep.	28								
Legume	1	22.15	<.0001	20.22	<.0001	10.67	0.0027	20.51	<.0001
Grass × Legume	14	5.16	<.0001	4.71	0.0002	3.78	0.0011	8.23	<.0001
Grass × Legume × Rep.	30								
Total	89								

Table 1b. Percentage ground cover in grass species and their mixtures weeks after planting (WAP) in trial plots. LSD$_{0.05}$ is least significant difference at $\alpha=0.05$ and means with the same letter(s) down the column are not significant.

	Grass species/Grass combination	4WAP (%)	5WAP (%)	6WAP (%)	7WAP (%)	8WAP (%)	9WAP (%)	10WAP (%)	11WAP (%)
1	*Axonopus compressus*	15.63[bcde]	23.44[cdef]	53.65[bcd]	76.56[bcde]	87.50[bcd]	90.62[abc]	100.00[a]	100.00[a]
2	*Chrysopogon aciculatus*	11.46[bcde]	13.54[ef]	23.44[f]	38.54[f]	54.17[e]	68.23[d]	88.02[cd]	88.54[c]
3	*Sporobolus pyramidalis*	5.21[e]	15.10[ef]	26.04[ef]	50.52[f]	61.46[e]	65.10[d]	93.23[bc]	95.31[b]
4	*Eleusine indica*	29.69[a]	53.65[a]	75.00[a]	93.23[a]	98.44[a]	96.35[ab]	98.44[ab]	100.00[a]
5	*Dactyloctenium aegyptium*	13.02[bcde]	29.17[bcde]	58.85[abc]	84.90[abcd]	94.27[abc]	95.83[abc]	100.00[a]	100.00[a]
6	*Axonopus + Chrysopogon*	10.42[bcde]	19.27[def]	38.54[cdef]	72.40[de]	84.37[cd]	94.79[abc]	100.00[a]	100.00[a]
7	*Axonopus + Sporobolus*	8.85[cde]	15.10[ef]	34.90[def]	66.15[e]	82.29[d]	84.90[c]	98.96[a]	100.00[a]
8	*Axonopus + Eleusine*	20.31[abc]	44.79[ab]	67.52[ab]	89.58[ab]	92.19[abcd]	96.87[a]	100.00[a]	100.00[a]
9	*Axonopus + Dactyloctenium*	9.90[bcde]	22.92[cdef]	45.31[cde]	75.00[cde]	87.00[bcd]	87.50[abc]	100.00[a]	100.00[a]
10	*Chrysopogon + Sporobolus*	12.50[bcde]	12.50[f]	20.83[f]	43.23[f]	58.33[e]	60.94[d]	84.90[d]	93.75[b]
11	*Chrysopogon + Eleusine*	21.87[ab]	38.54[abc]	56.25[abc]	82.29[abcd]	90.62[abcd]	93.23[abc]	96.87[ab]	99.48[a]
12	*Chrysopogon + Dactyloctenium*	9.37[cde]	22.40[def]	40.10[cdef]	65.62[e]	84.37[cd]	89.06[abc]	100.00[a]	99.48[a]
13	*Sporobolus + Eleusine*	17.71[abcd]	41.15[ab]	67.71[ab]	85.94[abc]	92.71[abcd]	93.23[abc]	100.00[a]	100.00[a]
14	*Sporobolus + Dactyloctenium*	7.81[de]	18.23[def]	44.79[cde]	66.67[e]	86.46[cd]	85.42[bc]	100.00[a]	100.00[a]
15	*Eleusine + Dactyloctenium*	18.23[abcd]	33.33[bcd]	75.52[a]	91.15[a]	97.40[ab]	97.92[a]	100.00[a]	100.00[a]
	LSD$_{0.05}$	12.49	15.89	21.19	13.51	10.43	11.27	5.38	3.00

There were significant differences in ground cover among the grass species and their mixtures in all the weeks after trampling (1 – 6 WAT). Subplots with *Desmodium* had significantly higher ground cover of grass species/mixtures except at 1 WAT. There were no significant interactions between grass species/mixtures and presence or absence of *Desmodium* at 1, 2 and 6 WAT (Table 4a). Sole stand of *Dactyloctenium* and *Chrysopogon + Dactyloctenium* had below 50% ground cover after trampling. Regrowth was fast in stands of *Axonopus* (sole and mixed) but only the sole stand had 100% cover at 6 WAT (Table 4b). Ground cover of grasses and their mixtures in both subplots with *Desmodium* and those without were below 100% at 6 WAT (Table 2).

DISCUSSION

The consistent dominance of *Eleusine* during the establishment period could be ascribed to its high tiller density, combined with higher leaf conductance and greater net photosynthesis especially in areas with abundant rainfall (Kobayashi and Hori, 2000). Increased tillering will not only favour ground cover but also enhances the visual quality and the overall performance of the turf (Park et al., 2005). This is achieved as increase in aboveground parts cover up the bare ground rapidly. Murphy and Park (2005) emphasized fast germination and rapid establishment in turfgrass species, as necessary to provide a dense turf within the shortest period. The other tussock species (*Sporobolus*) as well as the three creeping species (*Chrysopogon*, *Axonopus* and *Dactyloctenium*) had low ground cover. The tall slender stems, narrower leaves and fewer tillers in *Sporobolus* probably accounted for the lower ground cover. This is in conformity with the report by Sharma (1984). The low ground cover in *Chrysopogon* was due to its uneven spread resulting from the continued apical dominance. By contrast, the other creepers (*Axonopus* and *Dactyloctenium*) lost apical

dominance sooner and their axillary buds spread out and covered the ground fast. The lower ground cover in the *Chrysopogon* and *Sporobolus* also affected the ground cover of their mixed stands were poorest among the mixed grass species. This corroborates Austin's (1982) report that performance of species in mixed stands can be modelled from their individual performance in monoculture.

Table 2. F-statistics to determine the effect of treatment and their interactions on ground cover (%) in grasses and their mixtures weeks after clipping (WAC). Bottom part table shows ground cover (%) in grasses and their mixtures from 3 – 7 WAC in trial plots. LSD$_{0.05}$ is least significant difference at α=0.05 and means with the same letter(s) down the column are not significant.

		3 WAC		4 WAC		5 WAC		6 WAC		7 WAC	
Source	df	F	P	F	P	F	P	F	P	F	P
Grass	14	2.70	0.0123	1.64	0.1275	1.63	0.1308	2.10	0.0459	1.00	0.4793
Grass × Rep.	28										
Legume	1	10.81	0.0026	6.99	0.0129	0.84	0.3674	1.47	0.2347	1.00	0.3253
Grass × Legume	14	0.92	0.5535	1.15	0.3627	1.18	0.3415	2.95	0.0063	1.00	0.4777
Grass × Legume × Rep.	30										
Total	89										
						Ground cover (%)					
Grasses and mixtures											
Axonopus compressus		98.96[a]		100.00[a]		98.96[a]		100.00[a]		100.00[a]	
Chrysopogon aciculatus		86.98[de]		83.85[c]		88.54[ab]		97.40[a]		100.00[a]	
Sporobolus pyramidalis		85.94[e]		89.06[bc]		90.10[ab]		95.83[ab]		100.00[a]	
Eleusine indica		89.06[bcde]		97.40[ab]		99.48[a]		98.44[a]		100.00[a]	
Dactyloctenium aegyptium		95.83[abc]		100.00[a]		99.48[a]		100.00[a]		100.00[a]	
Axonopus + *Chrysopogon*		94.79[abcd]		97.92[ab]		99.48[a]		100.00[a]		100.00[a]	
Axonopus + *Sporobolus*		98.44[a]		96.87[ab]		91.15[ab]		100.00[a]		100.00[a]	
Axonopus + *Eleusine*		98.44[a]		98.96[ab]		95.83[a]		98.96[a]		100.00[a]	
Axonopus + *Dactyloctenium*		97.92[a]		100.00[a]		100.00[a]		100.00[a]		100.00[a]	
Chrysopogon + *Sporobolus*		92.71[abcde]		90.10[abc]		83.85[b]		92.19[b]		99.48[b]	
Chrysopogon + *Eleusine*		88.90[bcde]		96.87[ab]		98.95[a]		98.44[a]		100.00[a]	
Chrysopogon + *Dactyloctenium*		96.35[ab]		95.83[ab]		89.06[ab]		96.35[ab]		100.00[a]	
Sporobolus + *Eleusine*		92.19[abcde]		92.19[abc]		96.35[a]		97.92[a]		100.00[a]	
Sporobolus + *Dactyloctenium*		96.35[ab]		97.92[ab]		97.92[a]		100.00[a]		100.00[a]	
Eleusine + *Dactyloctenium*		88.02[cde]		96.35[ab]		97.92[a]		99.48[a]		100.00[a]	
LSD$_{0.05}$		8.09		10.62		11.69		4.38		0.39	

Table 3. Ground cover (%) of grasses and their mixtures in subplots with *Desmodium* (Legume) and without the legume (No Legume) weeks after planting, clipping and trampling. LSD$_{0.05}$ is least significant difference at α=0.05 and means with the same letter(s) down the column at WAP, WAC and WAT are not significant.

			Ground Cover of Grass (%)					
Subplot	4 WAP	5 WAP	6 WAP	7 WAP	8 WAP	9 WAP	10 WAP	11 WAP
Legume	15.07 ± 1.67[a]	26.18 ± 2.42[a]	50.95 ± 3.34[a]	75.00 ± 2.66[a]	87.15 ± 1.72[a]	90.07 ± 1.50[a]	98.75 ± 0.47[a]	99.44 ± 0.23[a]
No Legume	13.19 ± 1.59[a]	27.57 ± 2.56[a]	46.18 ± 3.65[a]	69.24 ± 3.28[b]	79.72 ± 3.07[b]	83.26 ± 2.79[b]	95.97 ± 1.33[b]	97.43 ± 0.93[b]
LSD$_{0.05}$	4.56	5.80	7.74	4.93	3.81	4.11	1.96	1.09
	3 WAC	4 WAC	5 WAC	6 WAC	7 WAC			
Legume	96.89 ± 0.70[a]	97.36 ± 0.96[a]	95.76 ± 1.36[a]	98.68 ± 0.58[a]	100.00 ± 0.00[a]			
No Legume	89.91 ± 1.84[b]	93.75 ± 1.51[a]	94.44 ± 1.36[a]	97.99 ± 0.57[a]	99.93 ± 0.07[a]			
LSD$_{0.05}$	8.09	3.88	4.27	1.60	0.14			
	1 WAT	2 WAT	3 WAT	4 WAT	5 WAT	6 WAT		
Legume	61.11 ± 2.87[a]	85.62 ± 1.25[a]	88.18 ± 1.02[a]	93.20 ± 0.70[a]	96.46 ± 0.59[a]	95.90 ± 0.79[a]		
No Legume	60.62 ± 3.21[a]	81.62 ± 1.81[b]	85.21 ± 1.61[b]	88.82 ± 1.40[b]	92.15 ± 1.45[b]	93.40 ± 1.19[b]		
LSD$_{0.05}$	2.05	3.40	2.82	1.25	1.58	1.64		

The mixed stand of *Axonopus* with either *Eleusine* or *Sporobolus* had the highest ground cover (98.44%) at 3WAC despite the low value in the sole stand of *Eleusine* and *Sporobolus* (Table 3), because clipping did not have much effect on ground cover of *Axonopus* unlike tussock species which were quite affected by clipping. The long sub-crown internodes and coarse nature of the leaves in *Eleusine* and *Sporobolus* contributed to the losses of ground cover after clipping. This reduction in

aboveground growth over time with regular clippings has previously been documented by other defoliation studies (Savelle and Heady, 1970; Heady, 1975) which is in agreement with Hawes' (1980) report, that creeping grasses are better adapted to low mowing. This explains why combinations of either *Axonopus* or *Dactyloctenium* with *Sporobolus* resulted in a high ground cover. Although *Chrysopogon* possesses some features such as fine short leaf blades and short internodes that enhance adaptability to low mowing, its slow growth and uneven spread of the aerial parts accounted for the fluctuations in ground cover at 5 WAC in mixtures with *Sporobolus* and *Dactyloctenium*, and the low performance during this experiment. Unlike sole *Axonopus* and *Dactyloctenium* which reached 100% ground cover at 4 WAC, the turf of sole *Chrysopogon* require about 7 weeks to reach full ground-cover if mowed low to about 2 cm to the ground.

Table 4a. F-statistics to determine the effect of treatment and their interactions on ground cover (%) in grasses and their mixtures weeks after trampling (WAT).

| Source | df | 1 WAT | | 2 WAT | | 3 WAT | |
		F	P	F	P	F	P
Grass	14	9.88	<0.0001	5.84	<0.0001	6.21	<0.0001
Grass × Rep.	28						
Legume	1	0.04	0.8376	6.36	0.0172	6.47	0.0163
Grass × Legume	14	1.56	0.1490	1.19	0.3302	2.91	0.0068
Grass × Legume × Rep.	30						
Total	89						
		4 WAT		**5 WAT**		**6 WAT**	
Grass	14	25.67	<0.0001	17.09	<0.0001	14.56	<0.0001
Grass × Rep.	28						
Legume	1	24.85	<0.0001	30.27	<0.0001	8.53	0.0066
Grass × Legume	14	4.22	0.0005	4.30	0.0004	0.72	0.7419
Grass × Legume × Rep.	30						
Total	89						

Table 4b Ground cover (%) in grass species and their mixtures weeks after trampling (WAT) in trial plots. LSD$_{0.05}$ is least significant difference at $\alpha = 0.05$ and means with the same letter(s) down the column are not significant.

	Grasses and mixtures	1WAT (%)	2WAT (%)	3WAT (%)	4WAT (%)	5WAT (%)	6WAT (%)
1	*Axonopus compressus*	76.04[ab]	89.58[ab]	94.79[a]	97.92[a]	100.00[a]	100.00[a]
2	*Chrysopogon aciculatus*	51.56[d]	82.29[bcd]	84.37[cde]	88.02[fg]	93.75[bcd]	94.27[cdef]
3	*Sporobolus pyramidalis*	70.31[abc]	87.58[ab]	91.67[abc]	94.79[abc]	97.92[ab]	95.83[abcde]
4	*Eleusine indica*	66.15[abcd]	77.77[cd]	89.06[abcd]	93.23[bcd]	98.44[a]	96.87[abcde]
5	*Dactyloctenium aegyptium*	13.02[f]	61.46[e]	67.62[f]	72.92[h]	76.56[f]	75.00[g]
6	*Axonopus + Chrysopogon*	64.06[bcd]	90.10[ab]	89.58[abcd]	94.79[abc]	97.40[ab]	96.87[abcde]
7	*Axonopus + Sporobolus*	80.21[a]	93.23[a]	88.54[abcd]	93.75[bc]	96.87[ab]	96.87[abcde]
8	*Axonopus + Eleusine*	75.00[ab]	87.00[abc]	89.58[abcd]	95.33[ab]	98.96[a]	98.96[ab]
9	*Axonopus + Dactyloctenium*	67.19[abcd]	85.94[abc]	87.50[abcd]	90.10[def]	91.15[d]	92.71[ef]
10	*Chrysopogon + Sporobolus*	68.75[abc]	82.29[bcd]	82.29[de]	88.54[efg]	92.19[cd]	93.75[def]
11	*Chrysopogon + Eleusine*	57.29[cd]	85.42[abc]	89.06[abcd]	94.27[bc]	98.96[a]	98.44[ab]
12	*Chrysopogon + Dactyloctenium*	34.37[e]	73.96[d]	78.65[e]	85.42[g]	86.46[e]	91.15[f]
13	*Sporobolus + Eleusine*	65.10[abcd]	84.38[abc]	86.98[bcd]	91.68[def]	96.35[abc]	96.87[abcde]
14	*Sporobolus + Dactyloctenium*	66.67[abcd]	86.98[abc]	87.50[abcd]	90.10[def]	91.67[d]	94.79[bcdef]
15	*Eleusine + Dactyloctenium*	57.29[cd]	85.94[abc]	93.23[ab]	94.27[bc]	97.92[ab]	97.40[abcd]
	LSD$_{0.05}$	15.94	2.05	7.73	1.25	4.34	4.51

The percentage ground cover one week after trampling (1 WAT) was an indication of resistance of the turfgrass species to trampling pressure as suggested by Minner and Valverde (2005) for cool-season turfgrasses. Subsequent weeks were used to assess the grass' inherent recuperative potential. The greatest reduction in cover due to trampling was in the sole stand of *Dactyloctenium* due to its soft tender stems and leaves (i.e. low tensile strength). There was loss of verdure in the other grasses but their recovery rate (recuperative potential) was high. Despite its increase in ground cover from 13.02% to 61.46% at 2 WAT (Table 4b), *Dactyloctenium* would not be recommended for use in athletic turf or heavily trampled sites since it will result to bare ground if intensive trampling, such as in this experiment, would last for three weeks or more.

Apart from wear and divots removal on the grasses, other impacts of trampling include soil compaction and soil displacement (Park et al., 2010). Wear injury, defined by the immediate result of the crushing, tearing, and shearing actions of foot and vehicular traffic (Park et al., 2007), affects aboveground plant parts (stems, leaves, inflorescences). Only the sole *Axonopus* had 100% cover at 6 WAT but the performance of other grasses and their mixtures was also high except for sole stand of *Dactyloctenium* (Table 4b). The performance of mixed stands of *Dactyloctenium* with *Eleusine* and *Sporobolus* was better possibly due to positive interactions of these tussock species with *Dactyloctenium*. The tall and flexible stems of the tussocks, could enable cushioning of *Dactyloctenium*, potentially reducing direct trampling impacts such as bruising and shearing of the vegetative parts in the mixed stands.

The lack of significant effect of the legume species (*Desmodium*) on ground cover of the grasses and their mixtures during the early stages (4 - 6 WAP) could be due to abundance of soil nutrients, such as nitrogen and phosphorus, required for carbon uptake in both grass and legume during this period. However at 7 WAP when most of the grasses and their mixtures have covered above 50% of the ground, the demand for mineral nutrients to sustain this vegetative cover increases with enhanced competition for soil N up to 11 WAP. It is possible that the grass-legume subplots had significantly higher ground cover than the no legume grass subplots from 7 - 11 WAP due to the contribution from the legume. Most species of *Desmodium* have been reported to fix nitrogen (Dzowela, 1986; Amudavi et al., 2007) thus reducing the dependence of grass fields on inorganic fertilizers (Vendramini et al., 2010; Caddel et al., 2012). Biological nitrogen fixation is a symbiotic process in which rhizobium bacteria change inert atmospheric N_2 to NH_3 biologically useful to the legume species. Legume nitrogen fixation starts with the formation of nodule, which can be seen 2 - 3 weeks after planting, depending on the legume species and germination conditions (Lindermann and Glover, 2011). Some legume species are inoculated with *Rhizobium* to stimulate nodulation. But an uninoculated legume such as *Leucaena leucocephala* is able to form as much as 36 nodules per plant at 12 WAP (Khonje, 2012). The number of nodules per plant varies among species and their shapes depend on the life cycle of the legume (Caddel et al., 2012; Lindermann and Glover, 2012).

The clipping in the grasses resulted in active meristematic growth which induces an increase in nutrient demand. The grass-legume subplots will benefit from soil nitrogen enrichment from the legume and reduced evaporative loss of water due to closed mat formed by the legume underneath the grasses at early stage after clipping. Conversely, in pure grass stands, clipping will uncover more soil surface. Since growth is sigmoidal in nature, this difference in ground cover at later stages (4 – 7 WAC) between the subplots with legume and those without the legume were insignificant. Response of plants to defoliation, even within a single species, is a variable that is sensitive to timing, nutrient availability, and plant associations (Maschinski and Whitham, 1989).

The presence of *Desmodium* did not confer additional strength on the grasses thus grasses grown with or without the legume were recuperating from the same effect at 1 WAT and ground cover were not significantly different. This observation in ground cover at 1 WAT could also result from that fact that the non-exclusive nature of the trampling affected both grasses and the legume in the grass-legume subplots, thereby reducing their activity. Lindermann and Glover (2011) asserted that any stress that reduces legume activity will ultimately reduce nitrogen fixation. This is because nitrogen fixation is a symbiotic process that requires the plant to contribute some amount of photosynthates and other nutritional factors to the bacteria and this contribution is often compromised when the plant (legume) is stressed. However, when such stress is corrected the legume responds directly and normal physiological process is resumed (Lindermann and Glover, 2011) as observed from 2 – 6 WAT in our result.

CONCLUSION

Fast establishment and high ground cover in turf grasses does not necessarily express resistance to trampling. Tough aerial organs with increased growth have proved a better tool to adjudging resistance to trampling than increased ground cover alone. The individual performance of the grasses in the sole stands correlated with their performances in the mixed stands. Fast recovery after trampling, a function of the growth of the grass species, was facilitated by the presence of *Desmodium*. We recommend the use of grass mixes in turfs composed of species that should have been previously identified to perform well in sole stands (e.g. *Axonopus* and *Eleusine*). Features like rapid leaf elongation, fast growth and leaf toughness in *Axonopus* and *Eleusine* should be used to breed trampling-resistant turfgrass cultivars. Inconspicuous legumes such as *Desmodium* should be planted in turfs to reduce dependence on organic fertilizers and to improve technical quality of the turf swards.

LITERATURE CITED

Amudavi, D., Z. Khan and J. Pickett, 2007. Enhancing the push-pull strategy. *LEISA Magazine*, 23(4): 8-10

Austin, M.P., 1982. Use of relative physiological performance value in the prediction of performance of multispecies mixtures from monoculture performance. *J. Ecol.*, 70: 559-570

Bates, G.H., 1938. Life forms of pasture plants in relation to treading. *J. Ecol.*, 26: 452-454

Busey, P., 2003. Cultural management of weeds in turfgrass: a review. *Crop Sci.*, 43: 1899-1911

Caddel, J., D. Redfearn, H. Zhang, J. Edwards and S. Deng, 2012. *Forage Legumes and Nitrogen Production*. Oklahoma Cooperative Extension Fact Sheets PSS-2590. Division of Agricultural Sciences and Natural Resources, Oklahoma State University. Available at http://osufacts.okstate.edu

Demiroglu, G., H. Geren, B. Kir and R. Avcioglu, 2010. Performances of some cool season turfgrass cultivars in Mediterranean environment: II. *Festuca arundinacea* Screb., *Festuca ovina* L., *Festuca rubra* spp. *rubra* L., *Festuca rubra* spp. *trichophylla* Gaud and *Festuca rubra* spp. *commutate* Guad. *Turk J Field Crops.*, 15: 180-187

Dzowela, B.H., 1986. Value of a forage legume component in summer beef fattening systems in Malawi. *In*: Haque, I., S. Jutzi, and P.J.H. Neate (eds.), *Potentials of Forage Resumes in Farming Systems of Sub-Saharan Africa*. Proceedings of a Workshop held at ILCA, Addis Ababa, Ethiopia, 16 -19 September 1985. ILCA, Addis Ababa, Ethiopia.

Evans, G.E., 1988. Tolerance of selected Bluegrass and Fescue taxa to simulated human foot traffic. *J. Environ. Hort.*, 6: 10-14

Gaussoin, R.E., 1994. Choosing traffic-tolerant turfgrass varieties. *Sports Turf*, 10: 25-26

Geren, H., R. Avcioglu and M. Curaoglu, 2009. Performance of some warm-season turfgrasses under Mediterranean conditions. *Afr. J. Biotech.*, 8: 4469-4474

Hacker, J.W., 1987. Wear tolerance in amenity and sports turf: A review 1980-1985. *Acta Hort.*, 195: 35-41

Hawes, D.T., 1980. Response of warm- and cool-season turfgrass polystands to nitrogen and topdressing. *In*: *Proceedings of the Third International Turfgrass Research Conference*, Munich, Germany. July 11-13, 1977. International Turfgrass Society, ASA, CSSA and SSSA, Madison, WI, USA.

Heady, H.F., 1975. Physical effects of grazing animals. *In*: Heady, H.F. (Ed.), *Rangeland Management*. McGraw-Hill, USA.

Khonje, D.J., 2012. Adoption of the rhizobium inoculation technology for pasture improvement in sub-Saharan Africa. *In*: *FAO Corporate Document Repository* available at http://www.fao.org/wairdocs/ILRI/x5536E/x5536e1g.htm

Kir, B., R. Avcioglu, G. Demiroglu and A. Simic, 2010. Performances of some cool season turfgrass species in Mediterranean environment: I. *Lolium perenne* L., *Festuca arundinacea* Schreb., *Poa pratensis* L., and *Agrostis tenuis* Sibth. *Turkish J. Field Crop.*, 15: 174-179

Kobayashi, T. and Y. Hori, 2000. Photosynthesis and water-relation traits of the summer annual C_4 grasses, *Eleusine indica* and *Digitaria adscendens*, with contrasting trampling tolerance. *Ecol. Res.* 15: 165-174

Lindermann, W.C. and C.R. Glover, 2011. Nitrogen Fixation by Legumes. New Mexico University Extension Guide A-129. New Mexico State University, New Mexico. Available at http://aces.nmsu.edu/pubs/_a/A-129/welcome.html

Maschinski, J. and T.G. Whitham, 1989. The continuum of plant response to herbivory: The influence of plant-association, nutrient availability, and timing. *Amer. Nat.*, 134: 1-19

Minner, D.D. and F.J. Valverde, 2005. Performance of established cool-season grass species under simulated traffic. *Int. Turfgrass Soc. Res. J.*, 10: 393-397

Minner, D.D., J.H. Dunn, S.S. Burghrara and B.S. Fresenburg, 1993. Traffic tolerance among cultivars of Kentucky bluegrass, tall fescue, and perennial ryegrass. *Int. Turfgrass Soc. Res. J.*, 7: 687-694

Murphy, J.A. and B.S. Park, 2005. Turfgrass establishment procedures for sports fields. Rutgers Cooperative Research and Extension, New Jersey Agricultural Experiment Station, Bulletin E300.

Murphy, J.A. 2004. Maintaining athletic fields. Rutgers Cooperative Research and Extension, New Jersey Agricultural Experimental Station, Fact Sheet FS105. The State University of New Jersey, New Brunswick.

Park, B.S., T.J. Lawson, H. Samaranayake, and J.A. Murphy, 2010. Tolerance and recovery of Kentucky bluegrass subjected to seasonal wear. *Crop Sci.*, 50: 1526-1536

Park, B.S., J.A. Murphy, T.J. Lawson, W.K. Dickson and B. Clark, 2007. Did Kentucky bluegrass and tall fescue cultivars and selections differ in response to traffic stress in 2007? *In*: *2007 Rutgers Turfgrass Proceedings of the New Jersey Turfgrass Association*, December 4-6, 2007.

Park, B.S., J.A. Murphy, W.A. Meyer, S.A. Bonos, J. den Haan, D.A. Smith and T.J. Lawson, 2005. Performance of Kentucky bluegrass within phenotypic classifications as affected by traffic. *Int. Turfgrass Soc. Res. J.*, 10: 618-626

Salman, A., R. Avcioglu, H. Oztarhan, A.C. Cevheri and H. Okkaoglu, 2011. Performances of different cool season turf grasses and some mixtures under Mediterranean environmental condition. *Int. J. Agric. Biol.*, 13: 529-534

SAS Institute, 2003. *SAS/STAT 9 Users Guide*. SAS Institute Inc., Cary, NC, USA.

Savelle, G.D. and H.F. Heady, 1970. Mediterranean annual species: Their response to defoliation. Process XI International Grassland Congress.

Sharma, B.M., 1984. Ecophysiological studies of *Eleusine indica* (L.) Gaertn.and *Sporobolus pyramidalis* P. Beauv. at Ibadan, Nigeria. *J. Range Managmt.*, 37: 275-276

Shearman, R.C. and J.B. Beard, 1975. Turfgrass wear tolerance mechanisms: I. Wear tolerance of seven turfgrass species and quantitative methods for determining wear tolerance. *Agron. J.* 67: 208-211

Smyth, A.J. and R.F. Montgomery, 1962. *Soils and Land Use in Central Western Nigeria*. Ibadan Government Printers.

Steir, J.C. and E.J. Koeritz, 2008. Seeding dates for tall fescue (*Festuca arundinacea* Schreb.) athletic fields established in a temperate continental climate. *Acta Hort.*, 783: 49-56

Sun, D. and M.J. Liddle, 1993. The morphological responses of some Australian tussock grasses and the importance of tiller number in their resistance to trampling. *Biol. Cons.*, 65: 43-49.

Tachibana, H., 1976. Changes and revegetation in sphagnum moors destroyed by human treading. *Ecol. Rev.*, 18: 133-210.

Taivalmaa, S.L., H. Talvitie, L. Jauhiainen and O. Niemelainen, 1998. Influence of wear-stress on turfgrass species and cultivars in Finland. *J. Turf. Sci.* 74: 52-62

Vendramini, J., M.B. Adjei, and A.E. Kretschmer Jr., 2010. Florida Carpon Desmodium. *Florida Forage Handbook Document SS-AGR-11*, Institute of Food and Agricultural Sciences (IFAS), University of Florida. Available at http://edis.ifas.ufl.edu/pdf/pdffiles/AG/AG24900.pdf

Volterrani, M., S. Miele, S. Magni, M. Gaetani and G. Pardini, 2001. Bermudagrass and seashore paspalum overseeded with seven cool-season turfgrasses. *Int. Turfgrass Soc. Res. J.*, 9: 957-961

Wood, G.M. and A.G. Law, 1972. Evaluating bluegrass cultivars for wear resistance. In: *1972 Agronomy Abstracts*. ASA, Madison, WI. pp.65.

7

IMPACT OF GENOTYPE AND NITROGEN FERTILIZER RATE ON YIELD AND NITROGEN USE BY OAT (*Avena sativa* L.) IN TURKEY

Hasan MARAL[1], Ziya DUMLUPINAR[2], Tevrican DOKUYUCU[3] and Aydin AKKAYA[3]*

[1]*Karamanoglu Mehmetbey University, Ermenek Vocational School, Karaman, TURKEY.*
[2]*Kahramanmaras Sutcu Imam University, Agricultural Faculty, Department of Agricultural Biotechnology, Kahramanmaras, TURKEY.*
[3]*Kahramanmaras Sutcu Imam University, Agricultural Faculty, Department of Agronomy, Kahramanmaras, TURKEY.*
*Corresponding author: zdumlupinar@ksu.edu.tr

ABSTRACT

Efficient nitrogen (N) use is one of the most crucial issues for crop management in developing countries. Since Turkish agricultural practices tends to use lower inputs of N fertilizers, a field trial was carried out in 2007-08 and 2008-09 cropping years to determine response of six oat genotypes (Seydisehir, Apak, Yesilkoy-330, Amasya, Checota and Yesilkoy-1779) to three N rates (0, 100 and 200 kg ha^{-1}). According to the results, differences between the years were significantly important for all investigated traits except soil nitrogen content at planting (NCP), nitrogen content at maturity (NCM) and nitrogen accumulation in grain (NAG). Genotypes significantly varied for all traits except NCP, nitrogen in grain (NG) and NAG. However, differences among N rates were significant for all traits except NCP. Among the genotypes Checota had the highest nitrogen use efficiency (NUE) and grain yield (GY) (20.6 kg kg^{-1} N and 2590 kg ha^{-1}, respectively).

Keywords: *Avena sativa* L., grain yield, nitrogen use efficiency, oats.

INTRODUCTION

Hexaploid oat (*Avena sativa* L. and *Avena byzantina* Coch.) is widely used for human food and animal feed. Despite oat production decreases gradually in the world as well as in Turkey, demand in oat for human consumption increased due to dietary benefits of whole grain and β-glucan (soluble fiber) (Achleitner et al., 2008).

In cereals N fertilization accounts for significant proportion of the total input costs and may affect plant growth, development, yield and quality (Mohr et al., 2005). Reduced use of N fertilizer is likely to decrease both production costs and pollution, but could also result in reduced yields (Cassman et al., 2003). However, increasing N concentration does not always increase grain yield due to diminishing returns, and the excessive use of N raises potential adverse environmental and health concerns (Bohlool et al., 1992) and incidence of foliar pathogens and lodging of the plant (Samonte et al., 2006).

Nitrogen use efficiency is a complex trait (Muurinen et al., 2007) that comprises N uptake efficiency and N utilization (Moll et al., 1982; Ortiz-Monasterio et al., 1997). Studies have shown that genetic variation in cereals for nitrogen uptake efficiency (Kelly et al., 1995; Singh and Arora, 2001) and nitrogen utilization efficiency (Woodend et al., 1986; Papakosta, 1994; Singh and Arora, 2001). Przulj and Momcilovic (2003) reported that grain N content in wheat mainly represents N accumulated in

vegetative parts until anthesis and is translocated to the grain during the reproductive phase. Nitrogen in the form of protein and amino acids is a component of the pre-anthesis portion that is potentially available for grain filling (Schnyder, 1993). In barley, 10 to 100% of grain N is taken up during vegetative growth and translocated during the grain filling period (Carreck and Christian, 1991; Bulman and Smith, 1994). Sanford and MacKown (1987) found that variation in final spike N may be associated with variation in total N uptake. Bulman and Smith (1994) reported that post-heading N uptake in barley was generally not related to N concentration and N per plant at heading instead, it was highly correlated with total dry matter accumulation after anthesis and total plant N and grain at harvest. Environmental conditions during the pre- and post-anthesis periods are likely to have different effects on N accumulation. N uptake is influenced by available water (Clarke et al., 1990), the supply of nitrate (Cox et al., 1985; Papakosta and Gagianas, 1991), genotype requirements and nitrogen use efficiency, and other properties of the genotype and growing conditions (Przulj and Momcilovic, 2003).

Slafer and Peltonen-Sainio (2000) put targets on plant breeders to develop cultivars with increased yield potential associated with higher NUE-improved ability to absorb N more efficiently from the soil and partition the greater part of the absorbed N into the grain. Limited researches were devoted to determine NUE of oat genotypes in literature. Therefore, the aims of the current study were to evaluate nitrogen fertilization response of six oat genotypes released between 1963 and 2004 for nitrogen use efficiency, its components and grain yield.

MATERIALS AND METHODS

Plant materials

Oat genotypes Apak (registered in 1963), Yesilkoy-1779 (registered in 1964), Yesilkoy-330 (registered in 1975), Checota (registered in 1986), Seydisehir (registered in 2004) and Amasya (landrace) were evaluated for nitrogen fertilizers based on nitrogen use efficiency, its components and grain yield.

Field trials

Field experiments were carried out in 2007-08 and 2008-09 cropping years in Kahramanmaras province of Turkey (East–Mediterranean Region of Turkey, located between 37° 53′ N, 36° 58′ E and 507 m above sea level). The experiment was arranged in a Randomized Complete Block Design with three replications. Climate of the region is typical of Mediterranean climate and some

climatic data are shown in Table 1. Wheat plant was planted in previous years without fertilization to consume nitrogen amount in the experiment lands. Some chemical and physical traits of experiment soils sampled from 0-30 and 30-60 cm topsoil and analyzed. According to analysis experiment soils were loamy and alkaline, high in lime, adequate in phosphate and potassium and low in organic matter. Soil pH varied from 7.54 to 7.64. Nitrogen content of the soil from each plots at planting (NCP) were also measured and are given in Table 2. The experiments are planted on the dates of 18 November 2007 and 2008. Plot sizes were arranged as 6 x1.2 m and there were six plant rows in each plot. The sowing density was 350 seeds m^{-2}.Besides different amount and application of N fertilization (0, 100 and 200 kg ha^{-1}), a certain amount (80 kg ha^{-1}) of phosphorus (P) was applied at planting. Half of the nitrogen was applied at planting, while the rest was applied as top dressing on the dates of 10 March 2008 in the first year and 19 February 2009 in the second year at tillering. Herbicide (Tribenuron–Methyl 75%) was used for weed control. At the maturity stage four rows in the middle of plots were harvested.

Table 1 Some average climatic data belong to experiment (2007-2009) and long-term years (1930–2009) of Kahramanmaras province.

Months	Rainfall (mm)			Temperature (°C)			Relative Humidity (%)		
	2007-08	2008-09	Long-term	2007-08	2008-09	Long-term	2007-08	2008-09	Long-term
November	105.9	105.9	90.2	13.2	13.2	11.4	64.1	64.0	64.0
December	96.2	96.2	128.1	6.1	6.1	6.6	65.5	66.0	71.0
January	78.6	107.5	122.6	3.3	4.5	4.9	55.0	69.0	70.0
February	121.5	221.2	110.1	5.5	7.2	6.3	61.4	78.8	65.0
March	69.5	158.0	95.0	14.4	9.4	10.4	59.6	67.2	60.0
April	54.7	82.5	76.3	18.1	15.1	15.3	55.5	59.4	58.0
May	23.7	43.4	39.9	20.2	20.5	20.4	56.5	51.9	54.0
June	0.0	3.7	6.2	27.3	26.8	25.1	49.8	48.2	50.0
Total	550.1	818.4	668.4						
Mean				13.5	12.8	12.6	58.4	63.0	61.5

Investigated traits

Experiment soil was sampled from 0-30 cm depth from each plots and nitrogen content in soil at planting (NCP) was determined by Dumas procedure (Barbottin et al., 2005). Nitrogen content at flowering (NCF) and nitrogen content at maturity (NCM) were determined by Dumas procedure (Pan et al., 2006). Grain samples after harvest were dried at 65 °C for 48 hours and hulls removed by hand. Grain samples ground and screened through 0.5 mm sieve then nitrogen in grain (NG) was determined by Dumas procedure (Barbottin et al., 2005). Nitrogen accumulation in grain (NAG) was calculated by the formula; total grain N content (NG) / grain filling period (GFP) (Pan et al., 2006). Nitrogen utilization efficiency (UTE) was calculated by the formula: grain yield / total above ground N (Muurinen et al., 2006; Dawson et al., 2008). Grain filling period was determined as days between anthesis and maturity. Nitrogen use efficiency (NUE) was determined as: grain yield / total available N (NCP plus N amount applied to the plots) (Muurinen et al., 2006). Nitrogen remobilization

efficiency (NRE) was calculated as: amount of remobilized N (grain N content) / N uptake at flowering (Barbottin et al., 2005) and nitrogen harvest index (NHI) was calculated by the equation: grain N yield / total above ground N (Muurinen et al., 2006; Peng and Bouman, 2007). Grain yield (GY) was also determined by weighing of grain products obtained from the plot, harvested after excluding side effects of the plots.

Data Analysis

Factorial analyses of variance were conducted to determine the significance of the effects of year, genotype and nitrogen on NCP, NCF, NCM, NG, NAG, UTE, NUE, NRE, NHI and GY. The means of significant traits were analyzed and ranked by using the least significant difference (LSD) mean comparison test. Correlation analyses were used to determine the relationships among the investigated traits (SAS, 1999). In addition, linear regressions were used to calculate genetic improvements for grain yield and nitrogen use efficiency and to estimate relationship between grain yield and N rates.

RESULTS

Average air temperature of the experiment years and long-term were similar. The amount of rainfall differed between years. In 2007-08, precipitation was lower than long-term. In 2008-09, precipitation was higher than in the first year and long-term average. The precipitation was mostly in spring (Table 1).

Table 2. Average data belong to soil nitrogen content at planting (NCP, mg kg $^{-1}$), nitrogen content at flowering (NCF, g N plant $^{-1}$), nitrogen content at maturity (NCM, g N plant $^{-1}$), nitrogen in grain (NG, g N grain $^{-1}$), nitrogen accumulation in grain (NAG, mg N grain day $^{-1}$) and grain filling period (GFP day).

		NCP	NCF	NCM	NG	NAG	GFP
		ns	**	ns	**	ns	**
Years	2007-08	0.19	4.19 a	2.52	1.66 b	0.060	27.3 b
	2008-09	0.18	3.30 b	2.56	2.15 a	0.064	34.6 a
	LSD	0.01	0.17	0.12	0.09	0.003	1.21
		ns	**	**	ns	ns	*
	Seydisehir	0.18	4.15 a	2.68 a	1.90	0.059	32.7 a
	Apak	0.20	3.88 ab	2.66 a	1.88	0.063	29.7 c
	Yesilkoy-330	0.18	3.72 bc	2.33 b	1.96	0.063	31.2 abc
Genotypes	Amasya	0.19	3.40 d	2.30 b	1.88	0.062	30.2 bc
	Checota	0.19	3.76 bc	2.59 a	1.93	0.060	32.1 ab
	Yesilkoy-1779	0.19	3.54 cd	2.69 a	1.89	0.065	29.7 c
	LSD	0.02	0.30	0.20	0.16	0.006	2.10
		ns	**	**	**	**	**
	0	0.19	3.32 c	2.39 b	1.75 c	0.055 c	32.1 a
N Rates	100	0.19	3.76 b	2.52 b	1.93 b	0.060 b	31.6 a
	200	0.19	4.15 a	2.72 a	2.05 a	0.071 a	29.2 b
	LSD	0.01	0.21	0.14	0.12	0.004	1.48
	Mean	0.18	3.74	2.54	1.90	0.062	
	CV (%)	19.33	12.36	12.31	13.36	15.93	10.19
Year x Genotype		ns	**	**	ns	**	**
Year x N Rate		ns	**	**	ns	ns	Ns
Genotype x N Rate		ns	ns	ns	ns	ns	Ns
Year x Genotype x N Rate		ns	ns	**	ns	ns	Ns

** Significant at 1%, * significant at 5% and ns: not significant

Mean values of NCP, NCF, NCM, NG, NAG and GFP are given in Table 2, mean values of UTE, NUE, NRE, NHI and GY in Table 3 and Pearson correlation coefficients of the investigated traits in Table 4. The plant breeding achievements of oat genotypes released between 1963 and 2004, in respect to nitrogen use efficiency and grain yield are shown in Figure 1 and 2.

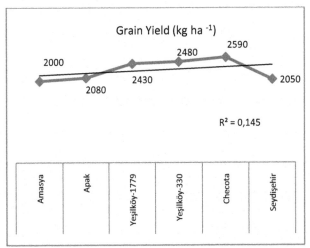

Figure 2. Genetic improvements for GY of six oat genotypes released between 1963 and 2004 evaluated with linear regression. Changes in grain yield with release year of six oat genotypes over two years data are illustrated.

Nitrogen content at planting was not different for years, genotypes and N rates. There was no fertilizer application in previous years and wheat was planted to consume nitrogen amount in the experiment soils. Therefore, the experiment soils N content was almost zero before planting. No interaction occurred among years, cultivars and N rates.

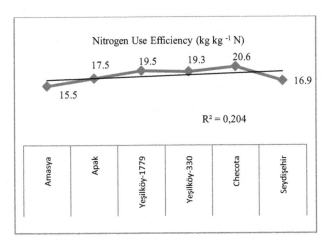

Figure 1. The plant breeding achievements of oat genotypes released between 1963 and 2004, in respect to nitrogen use efficiency. Changes in nitrogen use efficiency with release year of six oat genotypes observed for two years.

Significant differences ($P<0.01$) were recorded for NCF between years, genotypes and N rates. Year x genotype and year x N rate interactions were also significant ($P<0.01$). Seydisehir genotype had the highest NCF amount (4.15 g N plant^{-1}) and Amasya had the lowest NCF amount (3.40 g N plant^{-1}). Nitrogen content at flowering in the first year (4.19 g N plant^{-1}) was higher than the second year (3.30 g N plant^{-1}). At 200 kg ha^{-1} N rate, NCF was the highest with 4.15 g N plant^{-1} (Table 2).

Differences in NCM was significant for genotypes and N rates ($P<0.01$), whereas it was not significant for years (Table 2). In addition, year x genotype, year x N rate and year x genotype x N rate interactions were also significant for NCM ($P<0.01$). There was no significant differences between year one and year two, whereas 200 kg ha^{-1} N treatment had the highest NCM (2.72 g N plant^{-1}) (Table 2). Two groups occurred for genotypes, Yesilkoy-330 and Amasya genotypes were in the same group with lower NCM with 2.33 and 2.30 g N plant^{-1}, while Yesilkoy-1779, Seydisehir, Apak and Checota with higher NCM with 2.69, 2.68, 2.66 and 2.59 g N plant^{-1}, respectively(Table 2). Genotypes were not different for NG, while years and N rates varied significantly

($P<0.01$). In addition, no interaction occurred among year, genotype and N rates for NG (Table 2). In the first experiment year NG (1.66 g N grain^{-1}) was lower than second year (2.15 g N grain^{-1}). Nitrogen in grain was the lowest in control treatment (1.75 g N grain^{-1}), whereas it was higher in 100 and 200 kg ha^{-1} N treatments with 1.93 and 2.05 g N grain^{-1},respectively. The differences in NAG was significant for N rates ($P<0.01$), while did not vary over years and genotypes (Table 2). There was a year x genotype interaction for NAG ($P<0.01$). At 200 kg ha^{-1} N treatment 0.071 mg N grain day^{-1} was determined, while the other N rates had lower NAG (Table 2). Genotypes ($P<0.05$), years ($P<0.01$), nitrogen rates ($P<0.01$) and year x genotype interaction ($P<0.01$) were different for GFP (Table 2). Seydisehir genotype had the highest grain filling with 32.7 days, while Apak and Yesilkoy-1779 genotypes were the earliest. In the first experiment year GFP was 27.3 days, while it was 37.6 days in the second experiment year. The nitrogen rates had significant effect on GFP. The earliest nitrogen rate was 200 kg ha^{-1} with 29.2 days, while control treatment was the latest with 32.1 days (Table 2).

Table 3 Average data for nitrogen utilization efficiency (UTE, kg kg^{-1} N), nitrogen use efficiency (NUE, kg kg^{-1} N), nitrogen remobilization efficiency (NRE, kg N kg^{-1} N), nitrogen harvest index (NHI, kg N kg^{-1} N) and grain yield (GY, kg ha^{-1}).

		UTE	NUE	NRE	NHI	GY
		**	**	**	**	**
Years	2007-08	62.2 a	12.8 b	0.43 b	0.64 b	1623.2 b
	2008-09	32.3 b	23.9 a	0.75 a	0.72 a	2920.1 a
	LSD	3.10	1.62	0.02	0.02	130.78
		**	**	**	**	**
	Seydisehir	41.3 b	16.9 bc	0.55 c	0.68 bc	2050 b
	Apak	38.5 b	17.5 bc	0.58 bc	0.67 c	2080 b
Genotypes	Yesilkoy-330	51.8 a	19.3 ab	0.60 ab	0.76 a	2480 a
	Amasya	48.5 a	15.5 c	0.64 a	0.71 b	2000 b
	Checota	51.4 a	20.6 a	0.57 bc	0.61 d	2590 a
	Yesilkoy-1779	52.2 a	19.5 ab	0.60 ab	0.64 cd	2430 a
	LSD	5.37	2.76	0.04	0.04	224.67
		**	**	**	**	**
N Rates	0	37.1 c	#	0.54 b	0.61 c	1704.9 c
	100	50.0 b	22.6 a	0.61 a	0.69 b	2309.9 b
	200	54.7 a	13.8 b	0.62 a	0.74 a	2800 a
	LSD	3.79	1.62	0.02	0.02	158.65
	Mean	47.2	18.35	0.59	0.68	2271.6
	CV (%)	15.57	18.5	10.46	9.29	14.94
Year x Genotype		**	ns	ns	**	**
Year x N Rate		**	**	ns	ns	**
Genotype x N Rate		ns	ns	ns	ns	ns
Year x Genotype x N Rate		ns	ns	ns	ns	ns

** Significant at 1%, ns: not significant and, # no measured data

Nitrogen utilization efficiency was significant over years, genotypes and N rates ($P<0.01$). Year x genotype and year x N rate interactions were also significant ($P<0.01$) for UTE (Table 3). Nitrogen utilization efficiency was higher in the first year (62.2 kg kg^{-1} N) than second year (32.3 kg kg^{-1} N). In addition, UTE was affected by increasing of N rates. The highest UTE with 54.7 kg kg^{-1} N was obtained from 200 kg ha^{-1} N treatment, while control treatment had the lowest (37.1 kg kg^{-1} N). Within the genotypes Yesilkoy-1779, Yesilkoy-330,

Checota and Amasya had the higher UTE with 52.2, 51.8, 51.4and 48.5kg kg^{-1} N, respectively, while Apak and Seydisehir had lower UTE with 38.5 and 41.3 kg kg^{-1} N, respectively. Significant differences were recorded for NUE over years, genotypes and N rates ($P<0.01$). Interaction between year x N rate was significant for NUE (Table 3). In the first experiment year NUE was lower with 12.8 kg kg^{-1} N, while it was higher in the second year with 23.9 kg kg^{-1} N (Table 3). No data measured for control treatment due to nonexistent available N (very small amount N content at planting). On the other hand, NUE of 100 kg ha^{-1} N treatment (22.6 kg kg^{-1} N) was higher than 200 kg ha^{-1} N treatment (13.8 kg kg^{-1} N) (Table3). Nitrogen use efficiency differed markedly among the genotypes. Checota genotype had the highest NUE (20.6 kg kg^{-1} N), whereas Amasya genotype had the lowest (15.5 kg kg^{-1} N). Nitrogen remobilization efficiency varied for years, genotypes and N rates (*P*<0.01). However, there was no significant interaction for NRE (Table 3). Amasya genotype had the highest NRE value (0.64 kg N kg^{-1} N), while Seydisehir genotype was the lowest (0.55 kg N kg^{-1} N). Based on the LSD, there was no difference between Amasya, Yesilkoy-330, and Yesilkoy-1779 genotypes (Table 3).Nitrogen remobilization efficiency of the second year (0.75 kg N kg^{-1} N) was higher than the first year (0.43 kg N kg^{-1} N). Control treatment had lower NRE value (0.54 kg N kg^{-1} N), whereas 100 and 200 kg ha^{-1} N rates had similar NRE values (0.61 and 0.62 kg N kg^{-1} N) (Table 3). Nitrogen harvest index was significantly different for years, genotypes and N rates (*P*<0.01). Also, year x genotype interaction of NHI was significant (*P*<0.01). Yesilkoy-330 genotype had the highest NHI with 0.76 kg N kg^{-1} N and, Amasya and Seydisehir genotypes followed this genotype with 0.71 and 0.68 kg N kg^{-1}N, respectively, whereas Checota had the lowest NHI with 0.61 kg N kg^{-1} N (Table 3). Based on the LSD Checota and Yesilkoy-1779 genotypes were similar.

In the first year NHI value was 0.64 kg N kg^{-1} N, while 0.72 kg N kg^{-1} N in the second year. Nitrogen harvest index differed for N rates, 200 kg ha^{-1} N treatment was the highest (0.74 kg N kg^{-1} N), while control treatment was the lowest (0.61 kg N kg^{-1} N). Grain yield varied for years, genotypes and N rates (*P*<0.01). Year x genotype and year x N rate interactions were also significant for GY (*P*<0.01). Checota genotype had the highest GY (2590 kg ha^{-1}) and, Yesilkoy-330 and Yesilkoy-1779 genotypes followed this genotype (2480 and 2430 kg ha^{-1}). However,

Amasya, Seydisehir and Apak genotypes had lower GY with 2000, 2050 and 2080 kg ha^{-1}, respectively (Table 3). Grain yield was lower in year one (1623.23 kg ha^{-1}), while it was higher in year two (2920.1 kg ha^{-1}). Nitrogen rates had significant effect on GY, the lowest GY was obtained from control treatment (1704.9 kg ha^{-1}), whereas the highest GY was obtained from 200 kg ha^{-1} N treatment (2800 kg ha^{-1}).

Genetic improvements for NUE and GY of six oat genotypes released between 1963 and 2004 evaluated with linear regression (Figure 1 and Figure 2). There was a moderate relationship (R^2= 0.204) between NUE and year of genotype releases except 2004. NUE was calculated 15.5 and 17.5 kg kg^{-1} N for Amasya landrace and Apak cultivar which was released in 1963. Nitrogen use efficiency was improved to 20.6 kg kg^{-1} N in Checota which was registered in 1986 (Figure 1). The grain yield improvement of oat genotypes was intermediate (R^2= 0.145) between that lacks of improvement in Seydisehir which is released latest (Figure 2). It is noteworthy that Seydisehir genotype (2050 kg ha^{-1}) released in 2004 intercepts between Amasya genotype (2000 kg ha^{-1}) and Apak genotype (2080 kg ha^{-1}) released in 1963. On the other hand, grain yield improvement of Checota genotype released in 1986 was the highest with 2590 kg ha^{-1} (Figure 2).

According to Pearson correlation coefficients over the mean values of investigated traits, some of the traits were found highly correlated to each other (Table 4). Grain yield was highly correlated with NG, NAG, GFP, UTE, NUE, NRE and NHI (r= 0.61**, r= 0.32**, r= 0.33**, r= 0.91**, r= 0.61**, r= 0.73** and r= 0.39**, respectively), while it was negatively correlated with NCF (r= -0.23*) (Table 4). On the other hand, NUE was highly correlated with NG, GFP, UTE, NRE and GY (r= 0.29*, r= 0.56**, r= 0.71**, r= 0.58** and r= 0.61**), while a negative relationship was determined with NCF (r= -0.56**).

Table 4. Pearson Correlation Coefficients of investigated traits

	1	2	3	4	5	6	7	8	9	10	11
1. NCP	---										
2. NCF	-0.01	---									
3. NCM	0.0001	0.41**	---								
4. NG	-0.03	-0.16	0.16	---							
5. NAG	0.07	0.14	0.23*	0.64**	---						
6. GFP	-0.09	-0.04**	-0.05	0.38**	-0.37**	---					
7. UTE	-0.11	0.52**	-0.14	0.48**	0.16	0.39**	---				
8. NUE	-0.09	-0.56**	-0.02	0.29*	-0.17	0.56**	0.71**	---			
9. NRE	-0.12	-0.52**	0.02	0.69**	0.25**	0.55**	0.75**	0.58**	---		
10. NHI	-0.02	-0.03	-0.10	0.53**	0.38**	0.20*	0.33**	-0.06	0.46**	---	
11. GY	-0.10	-0.23*	0.17	0.61**	0.32**	0.33**	0.91**	0.61**	0.73**	0.39**	---

DISCUSSION

The two year experiment results indicate that oat genotypes varied in NCF due to differences in their vegetative periods. Differences in genotypes were influenced by genetics. Przulj and Momcilovic (2003) also reported differences among barley cultivars for NCF.

Nitrogen content at flowering was also affected by N rates. Higher NCF was obtained from the highest N application. In addition, environmental influence was also significant for NCF. Lower rainfall in year one resulted in higher NCF when compared to higher rainfall in year two. Seasonal differences such as rainfall probably accounted for year x genotype and year x N rate interactions. Similar results were also reported by Muurinen et al. (2007). Genotypes were found different for NCM because of genetic influence. Genotypes responded to higher N rates with higher NCM. In previous work, Muurinen et al. (2007) stated that oat had higher mean total plant N content at maturity. Experiment years were not found different, while year x genotype, year x N rate and year x genotype x N rate interactions were found significant. This situation may be due to different climatic conditions between year one and two. Muurinen et al. (2007) reported large variation over years for NCM, which was in contrast with our findings. This situation may be due to regional conditions. In addition, Przulj and Momcilovic (2003) stated N losses from anthesis to maturity in the above ground plant parts might be caused by loss of some plant parts during the post-anthesis period for instance old leaves. Nitrogen in grain was not significantly different for genotypes. However, genotypes responded to higher N rates with higher NG. In addition, experiment years were found different due to environmental influence for NG. In our study oat genotypes were not significantly different for NAG while Muurinen et al. (2007) reported differences among species for NAG. However, genotypes responded higher N rates with higher NAG. In addition, NAG was not affected by years while year x genotype interaction was significant, which may be caused by significant changes in GFP across the years due to environmental effects and which NAG was calculated by the formula (NG/GFP). Przulj and Momcilovic (2003) also reported that poor growing conditions suppressed N accumulation during pre-anthesis, diminished translocated N amount, and increased post-anthesis N uptake. Genotypes, years and nitrogen rates showed variation for grain filling period. This may be due to genetic influence and climatic conditions. Wych et al. (1982) and Peltonen-Sainio and Rajala (2007) reported genetic influence for GFP.

Oat genotypes varied for nitrogen utilization efficiency. Isfan (1993) defined UTE as an essential physiological parameter contributing to improved NUE, which is in agreement with our findings. Higher nitrogen rates also promoted the higher UTE. Experiment years differed for UTE and, year x genotype and year x N rate interactions were significant for UTE. This situation may be due to environmental influence. Przulj and Momcilovic (2003) indicated variation between years. Ortiz et al. (1998) reported that improvement in UTE was achieved through reduced plant height and lodging, which is resulted in higher grain yield due to improved harvest index in wheat. Differences in UTE for cereals were also reported in previous works (Woodend et al., 1986; Papakosta, 1994; Singh and Arora, 2001). There were differences in NUE among genotypes. Muurinen et al. (2006) reported differences among cultivars of oat, barley and wheat. Calderini et al. (1995) reported differences in NUE for wheat. Nitrogen use efficiency was the highest at 100 kg ha^{-1} N rate (22.6 kg kg^{-1} N). Wetselaar and Farquhar (1980) and Papakosta and Gagianas (1991) reported N losses between anthesis to maturity under field conditions, which may explain the reason for lower N rate

was the most efficient than the highest level of N rate. In contrast with our findings, Muurinen et al. (2006) indicated that cultivars responded to higher N rates with higher NUE. Differences in years were significant for NUE and year x N rate interaction was significant. In previous works, for barley and oat, authors reported that agronomic efficiency of N use rather than NUE and varied widely depending on growing conditions (Isfan, 1993; Delogu et al., 1998; Sinebo et al., 2003; Muurinen et al., 2006). Oat genotypes differed for nitrogen remobilization efficiency, which demonstrates the ability of the crop to remove N from vegetative tissue. Nitrogen remobilization efficiency also increased with high level of N rates. In addition, experiment years varied for NRE. Nitrogen movement during grain filling in the soil might be limited by lower rainfall in year one as resulted in lower N content in grain. Muurinen et al. (2007) explained the differences with dry weather conditions during grain filling period, which is in line with our results. However, also reported that NRE might not be controlled by environment, which is not in agreement with our findings.

Nitrogen harvest index varied for genotypes. Nitrogen harvest index ratios of the genotypes were 61 to 76% which is similar with the previous works (Rattunde and Frey, 1986; Welch and Leggett, 1997; McMullan et al.,1988). Nitrogen harvest index responded to high level of N rates. In addition, NHI was affected by years and year x genotype interaction was significant. This situation might be due to precipitation differences between years. Oat genotypes differed for grain yield. Grain yield was ranked between 2000 and 2590 kg ha^{-1}. Grain yield of the genotypes was influenced by genetics. Checota genotype had the highest grain yield with the highest UTE and NUE. Genotypes responded higher grain yield with high level of N rates. In addition, grain yield varied in experiment years and year x genotype and year x N rate interactions were important.

According to changes in NUE with year of six oat genotype releases, NUE improvement of Checota genotype was the highest with year of release of 1986. Genetic improvement for NUE was moderate that lacks of improvement in Seydisehir which had lower NUE than Apak, Yesilkoy-1779, Yesilkoy-330 and Checota even though released latest (Figure 1). On the other hand, changes in grain yield with the release year of six oat genotypes were intermediate. Seydisehir genotype lacks of improvement even though released latest while, grain yield improvement of Checota genotype was the highest that released in 1986 (Figure 2).

Nitrogen in grain, NAG, UTE and NUE were positively correlated with GY, while NCF was negatively correlated. N accumulation and the ratio of translocated N in grain affected grain yield positively. In the second experiment year, growing conditions were favorable and highest N content came from pre-anthesis at the high level of nitrogen treatments, whereas in year one half of the N amount in grain accumulated during grain filling period. Also, there was a positive relationship between NG and NAG. Przulj and Momcilovic (2003) stated that low temperature and water deficit, lead to the inhibition of N absorption even N is available in the soil. Higher efficiency of N utilization resulted in higher grain yield (Przulj and Momcilovic, 2003). There was also a positive correlation between UTE and NUE. Muurinen et al. (2006) reported positive correlation between UTE and NUE and also stated that NUE would be improved through UTE which is in line with our results. Nitrogen remobilization efficiency and NHI were also positively correlated with GY, while positively correlated to each other as well.

CONCLUSIONS

The current study determines the response of six oat genotypes to nitrogen fertilization based on NUE, its components and GY. Among the genotypes Checota had the highest NUE and GY. In addition, genetic improvement of Checota genotype for NUE and GY was the highest. The cultivars with high NCF, NRE and NHI could be used in the development of cultivars with the desired N balance. Oat breeders should take advantage of the variation among oat cultivars in NUE, NRE, UTE and NHI so that in the future selected genotypes are both high yields and efficient utilizes of N.

LITERATURE CITED

Achleitner, A., N.A. Tinker, E. Zechner, H. Buerstmayr, 2008. Genetic diversity among oat varieties of worldwide origin and associations of AFLP markers with quantitative. Theor. Appl. Genet. 117: 1041-1053.

Bohlool, B.B., J.K. Ladha, D.P. Garrity, T. George, 1992. Biological nitrogen fixation for sustainable agriculture. A perspective. Plant Soil. 14: 1-11.

Barbottin, A., C. Lecomte, C. Bouchard, M.H. Jeuffroy, 2005. Nitrogen remobilization during grain filling in wheat: genotypic and environmental effects. Crop Sci. 45: 1141-1150.

Bulman, P., D.L. Smith, 1994. Post-heading uptake, retranslocation, and partitioning in spring barley. Crop Sci. 34: 977-984.

Calderini, D.F., S. Torres-Leon, G.A. Slafer, 1995. Consequences of wheat breeding on nitrogen and phosphorus yield, grain nitrogen and phosphorus concentration and associated traits. Ann. Bot.76: 315-322.

Carreck, N.L., D.G. Christian, 1991. Studies on the pattern of nitrogen uptake and translocation to grain of winter barley intended for malting. Ann. Appl. Biol. 119: 549-559.

Cassman, K.G., A. Dobermann, D.T. Walters, H. Yang, 2003. Meeting cereal demand while protecting natural resources and improving environmental quality. Ann. Rev. Environ. Resour. 28: 315-358.

Clarke, J.M., C.A. Campbell, H.W. Cutforth, R.M. De Pauw, G.E. Winkleman, 1990. Nitrogen and phosphorus uptake, translocation, and utilization efficiency of wheat in relation to environmental and cultivar yield and protein levels. Can. J. Plant Sci. 70: 965-977.

Cox, M.C., C.Q. Qualset, D.W. Rains, 1985. Genetic variation for nitrogen assimilation and translocation in wheat. II. Nitrogen assimilation in relation to grain yield and protein. Crop Sci. 25: 435-440.

Dawson, J.C., D.R. Huggins, S.S. Jones, 2008. Characterizing nitrogen use efficiency in natural and agricultural ecosystems to improve the performance of cereal crops in low-input and organic agricultural systems. Field Crops Res. 107: 89-101.

Delogu, G., L. Cattivelli, N. Pecchioni, D. DeFalcis, T. Maggiore, A.M. Stanca, 1998. Uptake and agronomic efficiency of nitrogen in winter barley and winter wheat. Eur. J. Agron. 9: 11-20.

Isfan, D., 1993. Genotypic variability for physiological efficiency index of nitrogen in oats. Plant Soil. 154: 53-59.

Kelly, J.T., R.K. Bacon, B.R. Wells, 1995. Genetic variability in nitrogen utilization at four growth stages in soft red winter wheat. J. Plant Nutr. 18: 969-982.

McMullan, P.M., P.B.E McVetty, A.A. Urquhart, 1988. Dry matter and nitrogen accumulation and redistribution and their relationship to grain yield and grain protein in oats. Can. J. Plant Sci. 68: 983-993.

Mohr, R.M., C.A. Grant, W.E. May, 2005. N, P and K: Fertilizer management for oats. Top Crop Manager. 5: 30.

Moll, R.H., E.J. Kamprath, W.A. Jackson, 1982. Analysis and interpretation of factors which contribute to efficiency of nitrogen utilization. Agron. J. 74: 562-564.

Muurinen, S., G.A. Slafer, P. Peltonen-Sainio, 2006. Breeding effects on nitrogen use efficiency of spring cereals under northern conditions. Crop Sci. 46: 561-568.

Muurinen, S., J. Kleemola, P. Peltonen-Sainio, 2007. Accumulation and translocation of nitrogen in spring cereal cultivars differing in nitrogen use efficiency. Agron. J. 99: 441-449.

Ortiz-Monasterio, J.I., K.D. Sayre, S. Rajaram, M. McMahon, 1997. Genetic progress in wheat yield and nitrogen use efficiency under four nitrogen rates. Crop Sci. 37: 898-904.

Ortiz, R., S. Madsen, S.B. Anderson, 1998. Diversity in Nordic spring wheat cultivars (1901-93). Acta Agri. Scand. Sect. B. Soil Plant Sci. 48: 229-238.

Pan, J., Y. Zhu, D. Jiang, T. Dai, Y. Li, W. Cao, 2006. Modeling plant nitrogen uptake and grain nitrogen accumulation in wheat. Field Crops Res. 97: 322-336.

Papakosta, D.K., A.A. Gagianas, 1991. Nitrogen and dry matter accumulation, remobilization, and losses for Mediterranean wheat during grain filling. Agron. J. 83: 864-870.

Papakosta, D.K., 1994. Analysis of wheat cultivar in grain yield, grain nitrogen yield and nitrogen utilization efficiency. J. Agron. Crop Sci. 172: 305-316.

Peltonen-Sainio, P., A. Rajala, 2007. Duration of vegetative and generative development phases in oat cultivars released since 1921. Field Crops Res. 101: 72-79.

Peng, S., B.A.M. Bouman, 2007. Prospects for genetic improvement to increase lowland rice yields with less water and nitrogen. In:Spiertz, J.H.J., Struikand, P.C., Van Laar, H.H. (eds), Scale and Complexity in Plant Systems Research. Gene-Plant-Crop Relations. Springer, pp. 251-256.

Przulj, N., V. Momcilovic, 2003. Dry matter and nitrogen accumulation and use in spring barley. Plant Soil Environ. 49(1): 36-47.

Rattunde, H.F., K.J. Frey, 1986. Nitrogen harvest index in oats: Its repeatability and association with adaptation. Crop Sci. 26: 606-610.

Samonte, S.O.P.B., L.T. Wilson, J.C. Medley, S.R.M. Pinson, A.M. McClung, J.S. Lales, 2006. Nitrogen utilization efficiency: Relationships with grain yield, grain protein, and yield-related traits in rice. Agron. J. 98: 168-176.

Sanford, D.A., C.T. MacKown, 1987. Cultivar differences in nitrogen remobilization during grain fill in soft red winter wheat. Crop Sci. 27: 295-300.

SAS Institute, 1999. SAS system for windows release 8.01. SAS Inst., Cary, NC.

Schnyder, H., 1993. The role of carbohydrate storage and redistribution in the source-sink relations of wheat and barley during grain filling-a review. New Phytol. 123: 233-245.

Singh, V.P., A. Arora, 2001. Intraspecific variation in nitrogen uptake and nitrogen utilization efficiency in wheat (*Triticum aestivum* L.). J. Agron. Crop Sci. 186: 239-244.

Sinebo, W., R. Gretzmacher, A. Edelbauer, 2003. Genotypic variation for nitrogen use efficiency in Ethiopian barley. Field Crops Res. 85: 43-60.

Slafer, G.A., P. Peltonen-Sainio, 2000. Yield trends of temperate cereals in high latitude countries from 1940 to1998. Agric. Food Sci. Finl. 10: 121-131.

Welch, R.W., J.M. Leggett, 1997. Nitrogen content, oil content and oil composition of oat cultivars (*A. sativa*) and wild *Avena* species in relation to nitrogen fertility, yield and partitioning of assimilates. J. Cereal Sci. 26: 105-120.

Wetselaar, R., G.D. Farquhar, 1980. Nitrogen losses from tops of plants. Adv. Agron. 33: 263-302.

Woodend, J.J., A.D.M. Glass, C.O. Person, 1986. Intraspecific variation for nitrate uptake and nitrogen utilization in wheat (*T. aestivum* L.) grown under nitrogen stress. J. Plant Nutr. 9: 1213-1225.

Wych, R.D., R.L. McGraw, D.D. Stuthman, 1982. Genotype x year interaction for length and rate of grain filling in oats. Crop Sci. 22: 1025-1028.

8

EFFECTS OF PLANT GROWTH-PROMOTING RHIZOBACTERIA ON SOME MORPHOLOGIC CHARACTERISTICS, YIELD AND QUALITY CONTENTS OF HUNGARIAN VETCH

Halil YOLCU[1]*, Adem GUNES[2], M. Kerim GULLAP[3], Ramazan CAKMAKCI[4]

[1]*Kelkit Aydın Dogan Vocational Training School, Gumushane University, Gumushane, TURKEY.*
[2]*Department of Soil Science, Faculty of Agriculture, Ataturk University, Erzurum, TURKEY.*
[3]*Narman Vocational Training School, Ataturk University, Narman, Erzurum, TURKEY.*
[4]*Department of Agronomy, Faculty of Agriculture, Ataturk University, TURKEY.*
*Corresponding author: halil-yolcu@hotmail.com

ABSTRACT

This research study was designed to determine the effects of plant growth-promoting rhizobacteria (PGPR) on some morphological characteristics, yield and quality contents of Hungarian vetch (*Vicia pannonica* Crantz.) in the Kelkit Aydın Dogan Vocational Training School Research Area during 2008-2009 and 2009-2010 plant growing season. The research consisted of a control (without plant growth-promoting rhizobacteria) and 12 different plant growth-promoting rhizobacteria (*Pseudomonas putida* PPB310, *Bacillus cereus* BCB51, *Pantoea agglomerans* PAB58, *Pseudomonas fluorescens* PFC82, *Pseudomonas fluorescens* PF84, *Arthrobacter mysorens* AM235, *Paenibacillus polymyxa* PP315, *Pantoea agglomerans* PAA362, *Bacillus atrophaeus* BA361, *Bacillus megaterium* BMA424, *Bacillus megaterium* BMA479 and *Bacillus subtilis* BS521) seed inoculations and three replicates. Some morphological characteristics (stem diameter, plant height and leaf number), dry matter yield, crude protein, crude protein yield, ADF, NDF, macro and micro-elements (B, Ca, Na, K, Mg, P, S, Cu, Fe, Mn and Zn) were tested in the research study. According to the results of the research study, some of the PGPR treatments had positive effect on morphological characteristics (especially PFC82, BA361 and PAB58), dry matter yield (very little PAA362), crude protein (very little PFC82), ADF and NDF (especially BA361) and macro and micro-elements (especially PAA362) of Hungarian vetch.

Keywords: Hungarian vetch, plant growth-promoting rhizobacteria, dry matter yield, crude protein, macro and micro-elements

INTRODUCTION

Organic animal production activities have begun to spread throughout the world in recent years. As a result of these phenomena, the requirement of organic forage crops production increases for organic livestock. Organic forage crops production in the fertilization especially shows important differences compared to conventional production. Crop rotation systems, intercropping mixtures, green manures, solid and liquid farmyard manures, poultry manures, compost, zeolite and biological fertilizers are used by farmers in the organic agriculture instead of chemical fertilizers (Yolcu 2010).

Uses of plant growth-promoting rhizobacteria isolated from various plants have started to spread in organic agriculture areas by the aim of plant nutrition. Micro-organisms promote the circulation of plant nutrients and reduce the need for chemical fertilizers (Cakmakci et al., 2007a). N_2-fixing and P-solubilizing bacteria may be important for plant nutrition by increasing N and P uptake of the plants (Cakmakci et al., 2006). Improvement of plant growth can be achieved by the direct application of plant growth-promoting rhizobacteria to seeds (Cakmakci et al., 2007b). Plant growth-promoting rhizobacteria

(PGPR) are beneficial bacteria that colonize plant roots and enhance plant growth using a wide variety of mechanisms (Ashrafuzzaman et al., 2009).

Vetches are commonly used in organic animal feeding throughout the world. Hungarian vetch seeded in the autumn in cold climates doesn't generally suffer from the effects of harsh winters. Hungarian vetch growing faster by benefiting from snow water and early spring rains competes better against weeds. As a result of this, Hungarian vetch provides plenty quality hay production for organic animal feeding.

Numerous studies have been conducted on different crops on the topic of the effects of PGPR (Javaid 2009; Yolcu et al. 2011; Khanna and Sharma 2011; Krey et al. 2011). However, there are currently no adequate number of studies on the topic of the effects of PGPR on the morphological characteristics, yield and quality contents of Hungarian vetch. In this regard, the aim of the present study was to investigate the effectiveness of twelve PGPR on some morphological characteristics, yield and quality contents of Hungarian vetch.

MATERIALS AND METHODS

Characterisation and isolation of bacterial strains

Twelve plant growth-promoting rhizobacteria, 2 fix N_2 (*Arthrobacter mysorens* AM235 and *Bacillus subtilis* BS521) and 10 solubilise P and fix N_2 (*Pseudomonas putida* PPB310, *Bacillus cereus* BCB51, *Pantoea agglomerans* PAB58, *Pseudomonas fluorescens* PFC82, *Pseudomonas fluorescens* PF84, *Paenibacillus polymyxa* PP315, *Pantoea agglomerans* PAA362, *Bacillus atrophaeus* BA361, *Bacillus megaterium* BMA424 and *Bacillus megaterium* BMA479) were used in this study. All strains used in the present study were previously isolated from the acidic rhizosphere of tea growing areas (Cakmakci et al., 2010).

Field experiment and growth conditions

A research study was carried out at the Kelkit Aydin Dogan Vocational Training School Research Area of Gumushane University in the Northeast of Turkey (1412 m elevation, 40° 08' N, 39° 25' E). Hungarian vetch was sown on October 14, 2008 and October 17, 2009 at a seeding rate of 80 kg ha^{-1} (Acikgoz 2001). The study design was a randomised complete block with thirteen treatments replicated three times. Each plot size was 3.0 m x 1.68 m, with a 24 cm row spacing and plots were separated by a 2.5 m buffer zone. The treatments were: control (without plant growth-promoting rhizobacteria), 2 fix N_2 (AM235 and BS521) and 10 solubilise P and fix N_2 (PPB310, BCB51, PAB58, PFC82, PF84, PP315, PAA362, BA361, BMA424 and BMA479). The plots were irrigated twice with 15 day intervals after the rains stopped (Serin and Tan, 2001) and were harvested at the embodiment period of bottom fruits (Acikgoz 2001) in 2009 and 2010. Climatic data during the years of study and long term means were shown in Table 1.

Table 1. Climatic data of the location in 2008, 2009, 2010 and the long-term average (1986-2006) at Kelkit, Turkey.

	J	F	M	A	M	J	J	A	S	O	N	D	
Years						Total Precipitation (mm) (Monthly)							**Total**
2008	40.8	23.3	38.4	51.4	28.4	35.8	2.6	20.0	30.3	35.2	21.1	34.4	361.7
2009	21.3	45.6	57.9	96.3	63.6	25.3	37.4	0.0	71.0	35.1	127.2	33.0	613.7
2010	73.2	24.1	58.2	49.2	57.5	93.5	12.8	0.0	8.2	87.1	1.2	14.2	479.2
1986-2006	33.1	35.5	38.3	57.7	68.3	45.1	14.8	13.8	26.3	50.6	45.1	37.6	466.2
					Mean air temperature (°C) (Monthly)								**Mean**
2008	-6.1	-4.4	8.1	11.6	11.7	16.6	20.1	21.5	17.1	11.9	6.6	-0.7	9.5
2009	-0.2	3.2	3.8	7.8	12.9	18.1	19.6	18.0	14.8	13.0	4.7	4.2	10.0
2010	2.0	4.3	6.3	9.0	14.4	19.4	22.3	23.5	19.6	11.9	7.9	5.3	12.2
1986-2006	-1.8	-1.0	3.1	9.4	13.3	16.8	20.2	20.1	16.3	11.3	4.4	0.5	9.4
					Mean relative humidity (%) (Monthly)								**Mean**
2008	70.7	71.4	63.0	65.0	68.3	69.6	68.5	69.4	68.3	73.0	72.9	73.2	69.4
2009	71.2	68.7	67.9	63.5	65.5	66.4	67.6	65.8	72.8	67.7	77.2	73.7	69.4
2010	73.0	68.1	65.7	69.1	67.2	66.9	65.5	59.5	65.2	74.6	65.2	64.5	67.0

Plant analysis

Stem diameter, plant height and leaf number were determined as the mean of six plants. Plant samples were oven-dried at 68^{0}C for 48 hours and weighted to find the dry matter yield for each year. Dried Hungarian vetch samples were ground with a Wiley mill to pass a 1-mm screen and then analyzed for chemical characteristics. The total nitrogen was found by using the Kjeldahl method and the crude protein was calculated by multiplying the N content by 6.25 (Bremner 1996). The ADF and NDF concentrations of samples were determined according to Van Soest (1963). B, Ca, Na, K, Mg, P, S, Cu, Fe, Mn and Zn concentrations of Hungarian vetches were found as a result of wet digestion of dried and ground sub-samples using a HNO_3-H_2O_2 acid mixture (2:3 v/v) in three steps (first step: 145°C, 75%RF, 5 minutes; second step: 180°C, 90%RF, 10 min and the third step: 100°C, 40 %RF, 10 minutes) in a microwave

(Bergof Speedwave Microwave Digestion Equipment MWS-2) (Mertens 2005a), and inductively Couple plasma spectrophotometer (Perkin-Elmer, Optima 2100 DV, ICP/OES, Shelton, CT 06484-4794, USA) (Mertens 2005b). Chemical characteristics of research soils (0-45 cm) in 2008 and 2009 were shown in Table 2.

Table 2. Chemical characteristics of research soils (0-45 cm) in 2008 and 2009.

Soil Properties	Units	2008	2009
Cation exchangeable capacity	cmol$_c$ kg^{-1}	23.2	20.1
Total N	mg kg^{-1}	5.3	13.7
pH	(1:2 soil:water)	7.9	7.4
Organic matter	g kg^{-1}	1.1	2.7
CaCO$_3$	g kg^{-1}	25.3	28.7
Available P	mg kg^{-1}	14.2	12.8
Exchangeable Ca	cmol$_c$ kg^{-1}	15.0	17.0
Exchangeable Mg	cmol$_c$ kg^{-1}	1.51	1.71
Exchangeable K	cmol$_c$ kg^{-1}	2.2	2.5
Exchangeable Na	cmol$_c$ kg^{-1}	0.7	0.5
Available Fe	mg kg^{-1}	5.4	5.0
Available Mn	mg kg^{-1}	6.1	7.7
Available Zn	mg kg^{-1}	2.2	2.2
Available Cu	mg kg^{-1}	2.3	2.6

Statistical analysis

The results were subjected to analysis of variance (ANOVA) and significant means were compared with Duncan's multiple range test method using the SPSS 13.0 statistical package program (SPSS Inc., 2004).

RESULTS AND DISCUSSION

Morphological characteristics

Most of PGPR inoculants provided higher stem diameter than that of the control as the mean of 2009 and 2010 (Table 3). The greatest stem diameter of Hungarian vetch was found in BA361 (1.89 mm) PGPR inoculant. This inoculation was followed by BMA479 (1.86 mm) and PFC82 (1.86 mm) plant growth-promoting rhizobacteria inoculants. Many of the PGPR inoculants had a higher plant height than that of the control in the mean of 2009 and 2010 (Table 3). The

highest plant height was obtained in PFC82 (51.0 cm) followed by PP315 (49.9 cm), PF84 (49.4 cm) and PAB58 (48.4 cm) rhizobacteria inoculants. Similarly, Cakmakci et al. (2007a) in barley and Javaid (2009) in blackgram [*Vigna mungo* (L.) Hepper] reported differences in terms of plant height with bacteria inoculations. PAB58 (19.4), PFC82 (18.7) and BA361 (18.5) rhizobacteria inoculants had higher leaf numbers than the control (17.0) in the mean of 2009 and 2010. The other rhizobacteria inoculants caused similar or lower leaf numbers compared to the control (Table 3). In addition, other researchers reported that plant growth-promoting rhizobacteria affected the leaf numbers of soybean (Dashti et al., 1997) and fodder maize (Hamidi 2006). Effects (p<0.01) of plant growth-promoting rhizobacteria on all morphological characteristics in 2009 and 2010 were shown in Figure 1.

Table 3. Effects of plant growth-promoting rhizobacteria (PGPR) on some morphological characteristics, dry matter yield and forage quality in Hungarian vetch (throughout two years)

Treatments	Stem Diameter (mm)	Plant Height (cm)	Leaf Number (leaf plant^{-1})	Dry Matter Yield (kg ha^{-1})	Crude Protein (%)	C. Protein Yield (kg ha^{-1})	ADF (%)	NDF (%)
Control	1.71ef	43.9e	17.0cd	4593ab	16.8ab	771a	31.6c-e	51.3ab
PPB310	1.80bc	45.0de	17.6b-d	4170c	16.5a-c	688b	30.1e	47.7de
BCB51	1.80bc	45.6c-e	16.4de	4191bc	16.0b-d	671b	32.9bc	44.9fg
PAB58	1.74c-e	48.4a-c	19.4a	2524f	15.9b-e	401e	30.8e	50.6a-c
PFC82	1.86ab	51.0a	18.7ab	2932e	17.4a	510c	31.7c-e	48.7cd
PF84	1.79cd	49.4ab	16.5de	4303a-c	16.1b-d	693b	34.5a	49.1b-d
AM235	1.66f	45.3de	18.3a-c	3682d	14.8e-g	545c	33.9ab	42.9g
PP315	1.66f	49.9ab	15.6e	4538a-c	14.4fg	653b	32.6b-d	43.7g
PAA362	1.71ef	43.9e	16.4de	4653a	13.9g	647b	30.6e	46.6d-f
BA361	1.89a	45.0de	18.5ab	4450a-c	15.4d-f	685b	28.7f	44.6fg
BMA424	1.73de	42.6e	17.7b-d	4354a-c	14.9e-g	649b	31.4de	52.5a
BMA479	1.86ab	47.4b-d	17.5b-d	3116e	15.6c-e	486cd	31.5c-e	46.1ef
BS521	1.78cd	47.2b-d	16.9de	2897e	15.2d-f	440de	30.7e	48.3c-e
Mean	1.77	46.5	17.43	3877	15.6	603	31.6	47.5
2009	1.72b	41.3b	16.04b	2961b	15.9a	470b	30.8b	47.0b
2010	1.81a	51.7a	18.82a	4794a	15.3b	736a	32.4a	48.0a
T	**	**	**	**	**	**	**	**
Y	**	**	**	**	**	**	**	*
T*Y	**	**	**	**	**	**	**	**

T : Treatments, Y: Year, *Significant at %5 level. **significant at %1 level

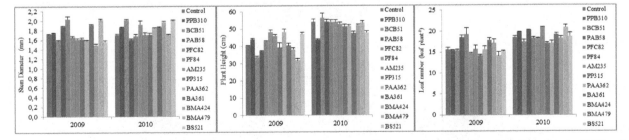

Figure 1. Effects of plant growth-promoting rhizobacteria (PGPR) on stem diameter, plant height and leaf number in 2009 and 2010.

Yield and Quality

Inoculation of Hungarian vetch seeds with plant growth-promoting rhizobacteria caused significant differences in the dry matter yield of Hungarian vetch

in the mean of 2009 and 2010 (Table 3). The highest dry matter yield was determined in PAA362 (4653 kg ha^{-1}) rhizobacteria inoculation. However, PAA362 (4653 kg ha^{-1}), the control (4593 kg ha^{-1}), PP315 (4538 kg ha^{-1}), BA361 (4450 kg ha^{-1}), BMA424 (4354 kg ha^{-}

[1]) and PF84 (4303 kg ha[-1]) rhizobacteria inoculations had statistically similar yields in terms of dry matter. The other inoculation applications yielded lower dry matter than that of the control (Table 3). Also in other studies, Dashti et al. (1997) in soybean, Cakmakci et al. (2001) and Canbolat et al. (2006) in barley, Rugheim and Abdelgani (2009) in faba bean reported an increase of yield with bacteria inoculations. However, Yolcu et al. (2011) determined that one of five PGPR decreased the dry matter yield of Italian ryegrass. Furthermore, Berggren et al. (2005) reported that the *P. putida* strain A313 reduced pea dry matter production. Conversely, other researchers stated that microorganism strains had no effect on the growth or yield of pea (Chanway et al., 1989) and soybean (Javaid and Mahmood, 2010).

The highest crude protein concentration in Hungarian vetch was found in PFC82 (17.4 %) rhizobacteria inoculant in the mean of 2009 and 2010 (Table 3). However, the PFC82 rhizobacteria inoculant was similar to the control and PPB310 rhizobacteria inoculants. The other rhizobacteria inoculants had similar or lower crude protein concentration than that of the control (16.8 %). Similarly, Estevez et al. (2009) stated that the symbiotic N2 fixation of legume plants does not always produce advantageous results with PGPR and rhizobia co-inoculation. Furthermore, Zaidi et al. (2003) reported that the soil nitrogen content did not show appreciable differences as a result of the inoculation in chickpea. In addition, Berggren et al. (2005) determined that the Pseudomonas putida strain A313 reduced the pea nitrogen content. The highest crude protein yield was determined in the control (771 kg ha[-1]) in the mean of 2009 and 2010. All treatments had a significantly lower crude protein yield than that of the control (Table 3). Also in another study, Yolcu et al. (2011) reported that one of five PGPR inoculants decreased the crude protein yield in Italian ryegrass. ADF and NDF concentrations were affected by different PGPR inoculations in the mean of 2009 and 2010. BA361 (28.7%) rhizobacteria inoculants gave lower ADF concentration than that of the control (31.6 %). Most of the PGPR inoculations had lower NDF concentrations than that of the control (Table 3). Similarly, Mishra et al. (2008) stated that Dual inoculation, i.e., *Azospirillum* with indigenous AM consortia decreased ADF and NDF concentrations of *Panicum maximum*. Effects of plant growth-promoting rhizobacteria (PGPR) on dry matter yields, crude protein concentrations, crude protein yield, ADF and NDF in 2009 and 2010 were shown in Figure 2.

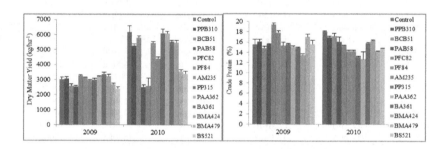

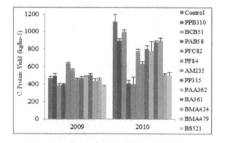

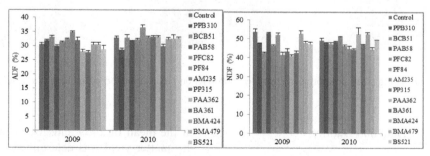

Figure 2. Effects of plant growth-promoting rhizobacteria (PGPR) on dry matter yield, crude protein concetration, crude protein yield, ADF and NDF in 2009 and 2010.

Table 4. Effect of plant growth-promoting rhizobacteria (PGPR) on macro- and micro-nutrient concentrations (mg kg^{-1}) in Hungarian vetch (throughout two years)

Treatments	B	Ca	Na	K	Mg	P	S	Cu	Fe	Mn	Zn
Control	43cd	24781c	2665a	8671f	2814c	1733d	2730e	18.5bc	999d	63b	113f
PPB310	46b	25225c	2639a	9368e	2439e	1710d	3015c	17.7bc	941d	57c	125de
BCB51	42de	23904d	2597a	9763d	2301fg	1625e	2662e	17.6 c	679g	56cd	124de
PAB58	45bc	23026ef	2682a	8374g	2339ef	1720d	2569f	20.9a	723fg	57c	147b
PFC82	40e	23368e	2678a	7953h	2606d	1511f	2875d	19.1 b	1452a	66ab	130cd
PF84	46b	24222d	2611a	9755d	2743c	1717d	2883d	17.5 c	764ef	53de	128cde
AM235	43cd	26672b	2639a	11943b	2762c	2288b	2463g	18.9bc	583h	51e	125de
PP315	42de	25108c	2739a	10095c	3085b	1736d	2738e	17.8bc	1158b	67a	134c
PAA362	50a	27854a	2590a	12979a	3279a	2190c	3754a	20.9a	818e	67a	124de
BA361	38f	20822g	2684a	8700f	2227g	2339b	2697e	18.1bc	931d	65ab	124de
BMA424	42de	22537f	2447b	4985i	1366h	2954a	2229h	18.3bc	1073c	53de	115f
BMA479	43cd	18570h	2638a	13019a	2827c	1745d	3241b	18.9 b	771ef	56c	156a
BS521	45bc	24942c	2646a	9892cd	2717c	1676de	3181b	17.5 c	757ef	57c	120ef
Mean	43	23925	2635	9654	2577	1919	2849	18.6	896	59	128
2009	43	26833a	2615	10489a	3010a	1647b	2825b	18.8	1208a	69a	142a
2010	43	21018b	2655	8819b	2145b	2190a	2873a	18.4	585b	49b	114b
T	**	**	*	**	**	**	**	**	**	**	**
Y	ns	**	ns	**	**	**	**	ns	**	**	**
T*Y	**	**	**	**	**	**	**	**	**	**	**

T : Treatments, Y: Year, *Significant at %5 level. **significant at %1 level

Macro- and micro-nutrient concentrations

All PGPR inoculants caused variation on macro and micro – nutrient concentrations of Hungarian vetch in the mean of 2009 and 2010. Inoculants of PAA362, PPB310 and PF84 in B, PAA362 and AM235 in Ca, BMA479, PAA362, AM235, PP315, BS521, BCB51, PF84 and PPB310 in K, PAA362 and PP315 in Mg and BMA424, BA361, AM235 and PAA362 in P had higher concentrations than that of the control in the mean of 2009 and 2010 (Table 4). The other PGPR inoculants had similar or lower B, Ca, K, Mg and P concentrations than that of the control. Similarly, Yolcu et al. (2011) reported that some PGPR treatments increased concentrations of K, Ca, Mg and B, and also the other PGPR treatments caused similar or lower concentrations than that of the control in Italian ryegrass. Elkoca et al. (2010) determined an increase of P, K, Ca, B and Mg concentration in common bean with PGPR treatments. Furthermore, Cakmakci et al. (2009) found that PGPR treatments caused a similar or higher concentration of P, K, Ca and Mg than that of the control in spinach leaves. Conversely, it was reported that PGPR treatments had no effect on the K and Mg concentration of barley (Cakmakci et al., 2007a).

Inoculants of PAA362, BMA479, BS521, PPB310, PF84 and PFC82 in S, PAA362 and PAB58 in Cu, PFC82, PP315 and BMA424 in Fe, PAA362 and PP315 in Mn and BMA479, PAB58, PP315, PFC82, PF84, PPB310, AM235, BCB51, PAA362 and BA361 PGPR treatments in Zn had higher concentrations than that of the control in the mean of 2009 and 2010 (Table 4). The other PGPR inoculants had similar or lower S, Cu, Fe, Mn and Zn concentrations than those of the control. Similarly, Yolcu et al. (2011) reported that some PGPR treatments increased the concentrations of S, Cu and Zn, and also other PGPR treatments had a similar or lower concentration than that of the control in Italian ryegrass. Furthermore, Cakmakci et al. (2007a) stated that treatments of PGPR had a similar or higher Fe, Mn, Zn and Cu concentration than that of the control in barley. Elkoca et al. (2010) found significant increases of Fe, Mn, Cu and Zn concentration in the common bean with PGPR treatments. In addition, Shirmardi et al. (2010) reported that rhizobacteria inoculants produce higher Zn and Cu, similar Fe and similar or lower Mn concentration than that of the control. Conversely, it was determined that PGPR treatments had no effect on the Fe, Mn and Zn concentration of spinach leaves (Cakmakci et al., 2009). Effects of plant growth-promoting rhizobacteria (PGPR) on B, Ca, Na, K, Mg, P, S, Cu, Fe, Mn and Zn concentration in 2009 and 2010 were shown in Figure 3, 4, 5 and 6.

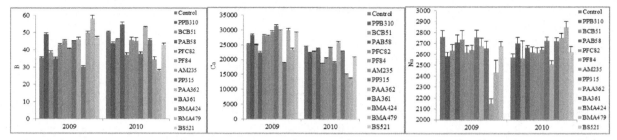

Figure 3. Effects of plant growth-promoting rhizobacteria (PGPR) on B, Ca and Na concentrations (mg kg^{-1}) in 2009 and 2010.

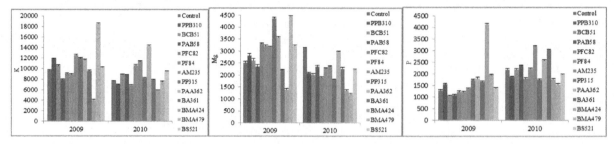

Figure 4. Effects of plant growth-promoting rhizobacteria (PGPR) on K, Mg and P concentrations (mg kg^{-1}) in 2009 and 2010.

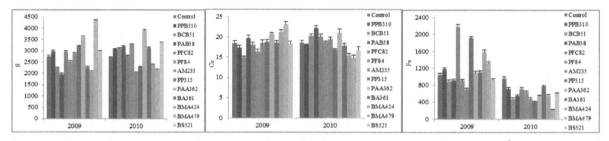

Figure 5. Effects of plant growth-promoting rhizobacteria (PGPR) on S, Cu and Fe concentrations (mg kg^{-1}) in 2009 and 2010.

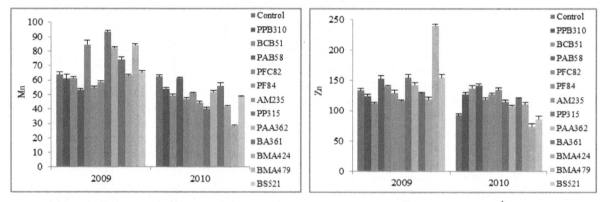

Figure 6. Effects of plant growth-promoting rhizobacteria (PGPR) on Mn and Zn concentrations (mg kg^{-1}) in 2009 and 2010.

CONCLUSIONS

Overall, some of the PGPR treatments (especially PFC82, BA361 and PAB58) had a positive effect on the morphological characteristics of Hungarian vetch. PAA362 in dry matter yield and PFC82 in crude protein concentration produced a very slight increase. *Rhizobium* spp are naturally present in the root zone of legumes. However, the addition of PGPR may cause competition or incompatibility between native *Rhizobium* spp and PGPR. Therefore, this competition and incompatibility may lead to the reduced efficiency of both bacteria in terms of dry matter yield, crude protein concentration and crude protein yield. Some of the rhizobacteria (especially BA361) was effective in terms of ADF and NDF. Some rhizobacteria (especially PAA362) increased the macro and micro-element concentrations (B, Ca, K, Mg, P, S, Cu, Mn and Zn) of Hungarian vetch.

The results of our research study show that inoculation of a legume plant seed with the plant growth-promoting rhizobacteria does not cause a significant increase in dry matter yield and hay quality except macro and micro-element concentrations. However, our results should be tested by studies to be conducted with various PGPR in different legume forage crops.

LITERATURE CITED

Acikgoz, E., 2001. Forage Crops. University of Uludag, Publication No: 182, Bursa,584p.

Ashrafuzzaman, M., F.A. Hossen, M.R. Ismail, M.A. Hoque, M.Z. Islam, S.M. Shahidullah, S. Meon, 2009. Efficiency of plant growth-promoting rhizobacteria (PGPR) for the enhancement of rice growth. Afr J Biotechnol. 8: 1247-1252.

Berggren, I., S. Alstrom, J.W.L. van Vuurde, A.M. Martenson, 2005. Rhizoplane colonisation of peas by *Rhizobium leguminosarum* bv. *viceae* and a deleterious *Pseudomonas putida*. FEMS Microbiol Ecol. 52:71-78.

Bremner, J.M., 1996. Nitrogen-total In: Methods of soil analysis Part III (Bartels, J.M., and Bigham, J.M., eds.). ASA SSSA Publisher *Agron* No: 5 Madison WI, USA, pp 1085–1121.

Cakmakci, R., F. Kantar, F. Sahin, 2001. Effect of N_2-fixing bacterial inoculations on yield of sugar beet and barley. J. Plant Nutr. Soil Sci. 164: 527-531.

Cakmakci, R., F. Dönmez, A. Aydın, F. Sahin, 2006. Growth promotion of plants by plant growth-promoting rhizobacteria under greenhause and two different field soil conditions. Soil Biol Biochem. 38: 1482-1487.

Cakmakci, R., M. F. Dönmez, Ü. Erdoğan, 2007a. The effect of plant growth promoting rhizobacteria on barley seedling growth, nutrient uptake, some soil properties, and bacterial counts. Turk J. Agric. For. 31: 189-199.

Cakmakci, R., M. Erat, Ü. Erdoğan, M.F. Dönmez, 2007b. The influence of plant growth-promoting rhizobacteria on growth and enzyme activities in wheat and spinach plants. J. Plant Nutr. Soil Sci. 170: 288-295.

Cakmakci, R., M. Erat, B. Oral, Ü. Erdoğan, F. Sahin, 2009. Enzyme activities and growth promotion of spinach by indole-3-acetic acid-producing rhizobacteria. J Hortic Sci Biotech. 84: 375-380.

Cakmakci, R., M.F. Dönmez, Y. Ertürk, M. Erat, A. Haznedar, R. Sekban, 2010. Diversity and metabolic potential of culturable bacteria from the rhizosphere of Turkish tea grown in acidic soils. Plant Soil. 332: 299-318.

Canbolat, M.Y., S. Bilen, R. Cakmakcı, F. Sahin, A. Aydın, 2006. Effect of plant growth-promoting bacteria and soil compaction on barley seedling growth, nutrient uptake, soil properties and rhizosphere microflora. Biol Fertil Soils. 42: 350-357.

Chanway, C.P., R.K. Hynes, L.M. Nelson, 1989. Plant growth-promoting rhizobacteria: Effects on growth and nitrogen fixation of lentil (*Lens esculenta* moench) and pea (*Pisum sativum* L.). Soil Biol Biochem. 21: 511-517.

Dashti, N., F. Zhang, R. Hynes, D. L. Smith, 1997. Application of plant growth-promoting rhizobacteria to soybean (Glycine max [L] Merr.) increases protein and dry matter yield under short-season conditions. Plant Soil. 188: 33-41.

Elkoca, E., M. Turan, M.F. Dönmez, 2010. Effects of single, dual and triple inoculations with bacillus subtilus , bacillus megaterium and rhizobium leguminosarum bv. phaseoli on nodulation, nutrient uptake, yield and yield parameters of common bean (Phaseolus vulgaris L. cv. 'Elkoca-05'). J Plant Nutr. 33: 2104-2119.

Estevez, J., M.S. Dardanelli, M. Megias, D.N. Rodriguez-Navarro, 2009. Symbiotic performance of common bean and soybean co-inoculated with rhizobia and *Chryseobacterium balustinum* Aur9 under moderate saline conditions. Symbiosis. 49: 29-36.

Hamidi, A., 2006. The effects of application of plant growth promoting rhizobacteria (PGPR) on the yield of fodder maize (Zea Mays L.). Pajouhesh and Sazandegi. 19: 16-22.

Javaid, A., 2009. Growth, nodulation and yield of blackgram [Vigna mungo (L.) Hepper] as influenced by biofertilizers and soil amendments. Afr J Biotechnol. 8: 5711-5717.

Javaid, A., N. Mahmood, 2010. Growth, nodulation and yield response of soyaben to biofertilizers and organic manures. Pak J Bot. 42: 863-871

Khanna, V., P. Sharma, 2011. Potential for enhancing lentil (Lens culinaris) productivity by co-inoculation with PSB, plant growth-promoting rhizobacteria and rhizobium. Indian J Agr Sci. 81: 932-934.

Krey, T., M. Caus, C. Baum, S. Ruppel, B. Eichler-Lobermann, 2011. Interactive effects of plant growth-promoting rhizobacteria and organic fertilization on P nutrition of Zea mays L. and Brassica napus L. J. Plant Nutr. Soil Sci. 174: 602-613.

Mertens, D., 2005a. AOAC Official Method 922,02, Plants Preparation of Laboratuary Sample, Official Methods of Analysis, 18th edn, Horwitz, W,, and G,W, Latimer, (Eds), Chapter 3, pp1-2, AOAC-International Suite 500, 481, North Frederick Avenue, Gaitherburg, Maryland 20877-2417, USA.

Mertens, D., 2005b. AOAC Official Method 975,03, Metal in Plants and Pet Foods, Official Methods of Analysis, 18th edn, Horwitz, W,, and G,W, Latimer, (Eds), Chapter 3, pp 3-4, AOAC-International Suite 500, 481, North Frederick Avenue, Gaitherburg, Maryland 20877-2417, USA.

Mishra, S., S. Sharma, P. Vasudevan, 2008. Comparative effect of biofertilizers on fodder production and quality in guinea grass (*Panicum maximum* Jacq.). J Sci Food Agr. 88: 1667-1673.

Rugheim, A.M.E., M. E. Abdelgani, 2009. Substituting Chemical Fertilizers with Microbial Fertilizers For Increasing Productivity of Faba Bean (*Vicia faba L.*)in Arid Lands. (in Li S.C., Wang Y.J., Cao F.X., Huang P., Zhang Y. Eds.) Progress in environmental science and technology, Vol II, PTS A AND B: 1910-1918.

Serin, Y., M. Tan, 2001. Forage Legumes. Ataturk University, Agriculture Faculty Publication No: 190, Erzurum, Turkey, 177 p.

Shirmardi, M., G.R. Savaghebi, K. Khavazi, A. Akbarzadeh, M. Farahbakhsh, F. Rejali, A. Sadat, 2010. Effect of Microbial Inoculants on Uptake of Nutrient Elements in Two Cultivars of Sunflower (*Helianthus annuus* L.) in Saline Soils. Notulae Scientia Biologicae. 2: 57-66.

SPSS., 2004. SPSS for Windows, Release 13.0. SPSS Inc. Chicago, IL, USA.

Van Soest, P.J., 1963. The use of detergents in the analysis of fibrous feeds. II. A. rapid method for determination of the fiber and lignin. JAOAC. 46: 829-835.

Yolcu, H., 2010. Effect of biologic fertilizers on some morphologic properties and yield of Hungarian vetch. Turkey I. Organic Livestock Congress. Pp. 263-267. 1-4 July. Kelkit/Gümüşhane, Turkey.

Yolcu, H., M. Turan, A. Lithourgidis, R. Çakmakçı, A. Koç, 2011. Effects of plant growth-promoting rhizobacteria and manure on yield and quality characteristics of Italian ryegrass under semi arid conditions. AJCS. 5: 1730-1736.

Zaidi, A., Md. S. Khan, Md. Amil, 2003. Interactive effect of rhizotrophic microorganisms on yield and nutrient uptake of chickpea (Cicer arietinum L.,). Eur J Agron. 19: 15-21.

9

DRY MATTER YIELD AND SILAGE QUALITY OF SOME WINTER CEREALS HARVESTED AT DIFFERENT STAGES UNDER MEDITERRANEAN CLIMATE CONDITIONS

Hakan GEREN

Ege University, Faculty of Agriculture, Department of Field Crops, Izmir, TURKEY
Corresponding author: hakan.geren@ege.edu.tr

ABSTRACT

Winter cereals can provide feed earlier than annual grasses since they are generally more adaptable to early sowing due to their higher tolerance of dry conditions. Cereals are also better suited to single-cut silage-making, whereas annual grasses require multiple cuts or grazings to be fully utilised. A field and laboratory experiments were conducted to evaluate the effect of different harvest stages on the dry matter yield and silage quality of some winter cereals, during 2009-2011 growing season. Effects of three different harvest stages (early heading, milky stage, mid-dough stage) on five cereals (*Hordeum vulgare*, *Triticum aestivum*, *Secale cereale*, *Triticosecale*, *Avena sativa*) were tested. The experiment was arranged in split block with four replications. Results indicated that, it was possible to produce an average of 10.9 t ha^{-1} dry matter yield and an average of 9.2% crude protein content at mid-dough stage in regions with Mediterranean-type climates. It was also concluded that *Avena sativa* should be preferred for high biomass yield and should be cut at the beginning of mid-dough maturity stages for higher quality silage.

Key words: winter cereals, harvest stages, dry matter yield, silage quality, CP, ADF

INTRODUCTION

Producers have some problems in meeting the required feed for livestock in some regions of the world. Winter cereals, to overcome these obstacles, supply an important alternative. In addition to using the grain feed, the entire plant can be used as ensiled roughage when harvested at the milk-dough stage (Delogu et al., 2002). Harvesting small grain cereals for forage has several advantages: i) land might be double cropped, ii) the risk of crop loss from rain, wind or hail is decreased, iii) circumstances sometimes make it desirable, even, necessary, to use these crops for forage even though they were planted for another purpose, e.g., weather stressed wheat with a low level of grain production might be more profitable if harvested as forage.

Such silage is best exploited in environments with low rainfall. On the other hand, harvesting the entire crop at the milk-dough stage in an environment marked by high fertility and good water supply makes it possible to maximize yields per unit area via double-cropping such as winter cereal + summer crop (Siefers and Bolsen, 1997).

Many of studies that have focused on the yield and quality potential of various cereal species ear-marked for whole-plant silage production (Bocchi et al., 1996) have found that their high variability in yield and nutritive value are linked to year and environment. Indeed, it is a known fact that the nutritional value of forages is dependent largely on seasonal temperature, light and rainfall trends, soil type, energy inputs applied over the growing cycle and, to a lesser extent, to the cultivar within the species.

If winter cereal crops are to be harvested for silage, management recommendations are needed to obtain the best compromise between forage yield and quality for a given farm situation or weather conditions. On the other hand, stage of maturity at harvest has a major effect on yield and quality of cereals (Cherney and Martin, 1982; Bergen et al., 1991). Although Schneider et al. (1991) found little effect of stage of maturity at harvest on triticale silage quality, Ben-Ghedalia et al. (1995) found that the quality of silage made from cereals declined with maturity at harvest. Yield increases and quality declines as the crop matures, although in cereals, quality may keep improve as grain development takes place (Khorasani et al., 1997; Budaklı Çarpıcı et al. 2010). The optimal stage of harvest for barley and oat to maximize yield and quality traits is the soft-dough stage (Bergen et al., 1991; Juskiw et al., 2000); while for triticale and rye it ranges from the boot to early milk stages (Twidwell et al., 1987; Schneider et al., 1991). Research in other areas has shown little difference in forage yield among oats (*Avena sativa*), barley (*Hordeum vulgare*) and triticale (*x Triticosecale*) when harvested for silage (McCartney and Vaage, 1994).

Barley was shown to have better forage quality as compared to oats due to its greater proportion of highly-digestible inflorescence during development (Cherney and Martin, 1982). Brink and Martin (1986) also found higher digestibility in barley as compared to oats, but this did not translate to higher yields of digestible dry matter. In this sense it is important to know the relations of production and quality as the crop proceeds phenology to determine the right time to cut, especially for animals with high requirements for winter cereals.

The objective of this experiment was to evaluate the agronomic performance and optimum cutting time for silage making in different winter cereals under Mediterranean climatic conditions.

MATERIALS AND METHODS

Location of Experiment

The experiment was carried out during 2009 and 2011 growing season (Table 1) on a silty-clay loam soil with 7.8 pH (Table 2) at Bornova experimental area (38°27.236 N, 27°13.576 E) in Agricultural Faculty of Ege University, Izmir, Turkey, at about 20 m a.s.l. with typical Mediterranean climate characteristics.

Table 1: Monthly average temperatures and total precipitations recorded at Bornova-Turkey location during the 2009 and 2011 growing seasons

| | ------- Temperature (°C) ------- | | | ------ Total precipitation (mm) ------ | | |
	2009-10	2010-11	LA	2009-10	2010-11	LA
November	14.6	18.1	13.2	160.3	32.4	80.3
December	13.1	13.3	9.9	151.8	155.7	122.3
January	10.6	9.0	8.1	143.2	100.9	109.7
February	12.6	10.3	8.6	301.3	107.3	89.8
March	13.3	12.0	10.8	16.1	18.8	72.3
April	17.4	14.5	15.0	20.4	65.3	48.9
May	21.8	20.1	20.2	27.1	29.0	32.2
Mean or total	**14.8**	**13.9**	**12.3**	**820.2**	**509.4**	**555.5**

LA: Long-term average

Table 2: Soil characteristics of the experiment area

Soil (0-30 cm)			
Sand (%)	32.7	pH	7.8
Clay (%)	30.6	OM (%)	1.1
Silt (%)	36.7	N (%)	0.1
CaCO₃ (%)	18.6	P (ppm)	0.4
Salt (%)	0.07	K (ppm)	400

Field applications and experimental design

Five winter cereals (*Hordeum vulgare* [Akhisar-98], *Triticum aestivum* [Cumhuriyet-75], *Secale cereale* [Aslim-95], *Triticale* [Ege yıldızı], *Avena sativa* [Faikbey]) and three harvest stages (*i*: early heading, *ii*: milky stage, *iii*: mid-dough) were tested. The experimental design was a split plot arrangement of a randomized complete block with four replications. The harvesting stages were main plots, and cereals were sub-plots. Each sub-plot was consisted of 10 rows with 20 cm apart and 5 m length (10 m²).

All cultivars were sown by hand on 15th November in both years (2009-2010), at a density of 350 viable seeds m⁻², in to the field which pioneer crop was forage turnip. The basic pre-sowing fertilization rates for all plots were 30 kg N ha⁻¹, 35 kg P ha⁻¹, 35 kg K ha⁻¹; a top dressing of 120 kg N ha⁻¹ was applied as follows; 50% of total N at the 3-4 leaf stage and the remainder at the early stem elongation stage.

Measurements, silage making and chemical analysis

Plots were harvested at 3 different stages of cereals, cutting mid 6 rows of plots in order to avoid border effects (net 4.8 m²), by cutting the plants leaving a 3-5 cm stubble height. Harvested fresh forage were weighed and dried to a constant weight at 65°C during 48 h. In each plot 5-6 kg of fresh mixture samples were taken at the stage and were chopped mechanically then wilted for 24 h. The samples without additives were pressed using a special apparatus (Petterson, 1988) into glass jars of 3 litres capacity. The jars were then tightly sealed and kept in storage without light for app. 70 days for fermentation.

pH value of matured silage samples was also determined (Alcicek and Ozkan, 1996). Matured silage samples of each component were dried at 65°C for 48 h. The dried samples were reassembled and ground in a mill passed through a 1 mm screen. The crude protein (CP) was calculated by multiplying the Kjeldahl N concentration by 6.25. The neutral detergent fibre (NDF) and acid detergent fibre (ADF) concentrations were measured to Ankom Technology.

Statistical analysis

All data were statistically analyzed using analysis of variance (ANOVA) with the Statistical Analysis System (SAS, 1998). Probabilities equal to or less than 0.05 were considered significant. If ANOVA indicated differences between treatment means a LSD test was performed to separate them (Stell et al., 1997).

RESULTS AND DISCUSSION

Experimental area is located in the Mediterranean zone of the country with quite mild winters and hot summers. Field studies were started in late autumn with low air temperature and satisfactory moisture levels were experienced in the germination and emergence period of relatively small seeds. Therefore, stands were excellent in both years. The results are summarized in Table 3 and 4. The year effect was the main source of variation in all characters tested, while the Year x Cereal x Harvest Stages (YxCxHS) interactions for all traits were not significant but for dry biomass yield and CP content.

Table 3: Effect of different harvest stages on the plant height, dry matter yield and silage pH values of some winter cereals.

Cereals	--------- 2010 ---------				--------- 2011 ---------				--------- 2 yrs average ---------			
	I	II	III	Mean	I	II	III	Mean	I	II	III	Mean
--- Plant height (cm) --												
Barley	96.4	100.4	101.1	99.3	89.6	92.1	90.1	90.6	93.0	96.3	95.6	95.0
Wheat	86.8	97.9	100.5	95.1	92.4	97.8	91.8	94.0	89.6	97.8	96.2	94.5
Rye	86.8	102.3	104.7	97.9	74.7	85.4	88.3	82.8	80.7	93.9	96.5	90.3
Triticale	120.4	124.4	126.1	123.6	117.9	120.3	140.1	126.1	119.2	122.4	133.1	124.9
Oat	134.5	182.1	185.3	167.3	106.2	111.7	122.0	113.3	120.4	146.9	153.7	140.3
Mean	105.0	121.4	123.6	116.7	96.2	101.5	106.5	101.4	100.6	111.4	115.0	109.0
LSD (.05)	Year (Y):4.6 Cereal (C):7.3 Harvest Stage (HS):5.6											
	YxC:10.3 YxHS:ns CxHS:12.6 YxCxHS:ns CV(%):10.1											
--- Dry matter yield (kg ha^{-1}) ---												
Barley	6062	8167	10588	8272	5430	6897	10097	7474	5746	7532	10342	7873
Wheat	5931	7208	9494	7544	5993	6622	9153	7256	5962	6915	9323	7400
Rye	5244	6774	9541	7186	3214	5767	8131	5704	4229	6270	8836	6445
Triticale	7480	9147	12517	9715	7913	9795	13707	10472	7697	9471	13112	10093
Oat	7667	9102	13096	9955	6858	9989	12935	9927	7263	9546	13016	9941
Mean	6477	8079	11047	8535	5882	7814	10804	8167	6179	7947	10926	8351
LSD (.05)	Y:134 C:211 HS:164 YxC:298 YxHS:ns CxHS:366 YxCxHS:517 CV(%):3.82											
--- Silage pH ---												
Barley	4.68	4.41	4.01	4.37	4.47	4.33	4.27	4.36	4.58	4.37	4.14	4.36
Wheat	5.53	4.87	4.69	5.03	4.35	4.24	4.05	4.21	4.94	4.56	4.37	4.62
Rye	4.48	4.45	4.34	4.42	5.02	4.42	4.14	4.53	4.75	4.44	4.24	4.48
Triticale	4.28	4.26	4.16	4.23	4.46	4.14	3.97	4.19	4.37	4.20	4.06	4.21
Oat	4.41	4.38	3.98	4.26	4.87	4.16	3.83	4.29	4.64	4.27	3.91	4.27
Mean	4.68	4.47	4.23	4.46	4.64	4.26	4.05	4.32	4.66	4.37	4.14	4.39
LSD (.05)	Y:0.080 C:0.126 HS:0.096 YxC:0.176 YxHS:ns CxHS:0.215 YxCxHS:0.308 CV(%): 4.32											

(I: early heading stage, II: milky stage, III: mid-dough stage)

Plant height

The plant height was affected by YxC interactions. The highest plant height (167.3 cm) was obtained from oat in the first year, whereas the lowest was 95.1 cm for wheat in the first year. Year effect was also significant and average cereal height of first year (116.7 cm) was higher than the second year (101.4 cm) due to the average temperature and total precipitation of the first year which was clearly higher than second year (Table 2). The plant height was also affected by CxHS interaction that means the maximum plant height was recorded at mid-dough harvest stage in the first year (123.6 cm) and oat crop had the highest. The plant height of cereals increased with progressing harvesting stage in both years. There is diversity of information's about plant height of cereals that maturation is an important factor to increase plant height (McCartney and Vaage, 1994; De Ruiter et al., 2002).

Dry Matter Yield

The highest dry matter yield (13707 kg ha^{-1}) was obtained from triticale at mid-dough harvest stage in the second year, whereas the lowest yield (3241 kg ha^{-1}) was in rye at early heading harvest stage in the second year.

Year effect was also significant and average dry dry matter yield of first year (8535 kg ha^{-1}) was higher than the second year (8167 kg ha^{-1}) because of favourable climatic parameters of the first year (Table 2). Two year average monitored that average dry matter yields across cultivars were 6179, 7947 and 10926 kg ha^{-1}, respectively (Table 3). Dry matter production increased with delaying harvesting stage and, triticale and oat were the most productive (~10 t ha^{-1}) cereals tested in the study. In the study, the dry matter yields of triticale and oat for three harvest stages were high; with an average 170% and 180% increase from early heading to milk dough, as expected and according to the data reported by Brink and Martin (1986) and Cherney and Martin (1982), respectively. This finding can be attributed to the completion of growing cycle by the crop and to the storage of newly formed photosynthetic activity in the grain. The accumulated dry matter by early heading is the determinant factor for the yield capacity of the subsequent growth stage (mid dough) and it is also largely dependent on the climatic conditions of the experimental environments (De Ruiter and Hanson, 2004; De Ruiter et al., 2002). Barley also showed the acceptable dry matter yield performance at both milky and mid dough harvests. In effect, the crop combines a high earliness with good cold tolerance (Pugnaire and Chapin, 1992).

Silage pH

Y x C x HS interaction was significant on silage pH. The favourable silage pH value (3.83) was obtained from oat harvested at mid-dough stage in the second year, whereas the highest pH (5.53) was in wheat harvested at early heading stage in the first year. Year effect was also significant on pH values and average pH of first year (4.46) was higher than the second year (4.32). Two year average showed that average pH values among harvest stages were 4.66, 4.37 and 4.14, respectively (Table 3). The most important physicochemical parameter for the evaluation of silage quality is a pH below 5, which was observed for all the silages tested. All indicators were characteristic of good silage conservation whatever the treatments were. The silage quality was especially confirmed by the proportion of fermentation products at the end of the storage period (Kristensen et al., 2010). Wheat silage at early heading stage had undergone secondary fermentations, which were characterized by the presence of some amounts of butyric acid and ammonia N (Siefers and Bolsen, 1997). All milky stage and mid-dough stage silages were satisfactorily preserved, as evidence by a pH range of 4.14 to 4.37. pH values of cereal silages decreased with delaying harvesting stage as expected because of the crops contained high level moisture and low level of carbohydrates in the first cut. Mid-dough triticale and oat silages performed slightly better than the other cereal silages and harvest stages in terms of pH values in the study.

Crude protein (CP) content

There were significant Y x C and Y x HS interactions for CP content of silages. The highest CP content recorded in oat in both years (11.9-11.8%) whereas the lowest CP was 9.4% for rye in 2010 (Table 4). CP content was also affected by YxHS interactions that mean the highest CP level was determined at early heading stage (12.3%) in 2011 whereas the lowest CP was at mid-dough stage (9.0%) in 2010. Between-year differences in CP invariably had greater influence than the cultivar and maturation effects. For example, mean CP content was significantly higher in 2011 (10.8%) than in 2010 (10.5%). The two year average has shown that average CP contents declined by increasing crop maturity. CP levels showed considerable variation in the silages of cereals. Generally, CP was lower for rye but was somewhat higher for oat and triticale. The largest variation in CP content occurred with the triticale (12.4 vs. 9.5%) depending on maturity. As with grass or legume silage production, cereal forage production is a compromise between forage quality and nutrient yield. As the crop developing from the boot to soft dough stage, CP content decreases and fiber concentration increases (Brand et al., 2003). On the other hand, both dry matter and CP yield increase with advancing maturity (Pereira-Crespo et al., 2010) and appear to be maximized at the soft dough stage. The difference in stage of maturity among the forages in the present experiment may have influenced both composition and yield. However, the effect was likely minimal since changes in nutrient density between the milk and soft dough stages appear to be relatively small (McCartney and Vaage, 1994).

Table 4: Effect of different harvest stages on the content of CP, NDF and ADF of some winter cereals silage.

Cereals	2010				2011				2 yrs average			
	I	II	III	Mean	I	II	III	Mean	I	II	III	Mean
	Crude protein (%)											
Barley	11.7	10.1	8.6	10.2	11.7	10.0	9.0	10.2	11.7	10.1	8.8	10.2
Wheat	11.7	10.0	8.6	10.1	12.6	10.9	10.0	11.1	12.1	10.5	9.3	10.6
Rye	11.0	9.3	7.9	9.4	11.6	9.5	8.7	9.9	11.3	9.4	8.3	9.7
Triticale	12.6	11.0	9.4	11.0	12.3	10.8	9.5	10.8	12.4	10.9	9.5	10.9
Oat	13.6	11.7	10.1	11.8	13.5	11.7	10.7	11.9	13.5	11.7	10.4	11.9
Mean	12.1	10.4	9.0	10.5	12.3	10.6	9.5	10.8	12.2	10.5	9.2	10.7
LSD (.05)	Y:0.099	C:0.156	HS:0.121	YxC:0.220	YxHS:0.171	CxHS:ns	YxCxHS:ns	CV(%):2.21				
	NDF (%)											
Barley	55.3	57.4	60.0	57.6	52.1	55.2	58.5	55.3	53.7	56.3	59.3	56.4
Wheat	57.2	59.8	63.2	61.4	54.0	57.5	60.4	57.3	55.6	58.7	61.2	58.5
Rye	59.2	61.8	63.2	61.4	56.4	59.3	61.6	59.1	57.8	60.6	62.4	60.3
Triticale	58.0	60.5	61.5	60.0	54.8	58.1	60.3	57.7	56.4	59.3	60.9	58.9
Oat	53.1	54.6	56.2	54.6	49.4	51.5	53.7	51.5	51.3	53.1	54.9	53.1
Mean	56.6	58.8	60.6	58.7	53.4	56.3	58.9	56.2	55.0	57.6	59.7	57.4
LSD (.05)	Y:0.268	C:0.424	HS:0.329	YxC:ns	YxHS:0.465	CxHS:0.735	YxCxHS:ns	CV(%):1.11				
	ADF (%)											
Barley	35.4	37.4	38.9	37.2	32.3	35.3	38.0	35.2	33.9	36.3	38.5	36.2
Wheat	39.4	41.2	42.8	41.1	35.2	38.8	42.3	38.7	37.3	40.0	42.6	39.9
Rye	40.0	42.4	45.1	42.5	37.2	41.3	44.7	41.1	38.6	41.9	44.9	41.8
Triticale	34.0	35.9	38.5	36.2	31.2	33.2	37.1	33.8	32.6	34.6	37.8	35.0
Oat	29.9	32.1	35.4	32.5	27.2	30.5	34.6	30.8	28.6	31.3	35.0	31.6
Mean	35.7	37.8	40.2	37.9	32.6	35.8	39.3	35.9	34.2	36.8	39.7	36.9
LSD (.05)	Y:0.180	C:0.284	HS:0.220	YxC:0.402	YxHS:0.311	CxHS:0.492	YxCxHS:ns	CV(%):1.16				

Neutral Detergent Fibre (NDF) content

There were significant Y x HS and C x HS interactions on NDF content of cereal silages. The favourable average NDF content (53.4%) was recorded at early heading stage in the second year, whereas the highest content (60.6%) was in the first year at mid-dough stage. On the other hand, the lowest average NDF content (51.3%) was determined in oat harvested at early heading stage, while the maximum NDF (62.4%) was in rye cut at mid-dough stage. Year effect was also significant (Table 4) and average NDF content of first year (58.7%) was higher than the second year (56.2%). The two year average demonstrated that average NDF content among harvest stages were 55.0%, 57.6% and 59.7%, respectively (Table 4). NDF contents of cereal silages increased with delaying harvesting stage as expected because of the progressing maturation of cereals. At large, mid-dough oat silage performed better than the other cereal silages with regard to NDF content in the study. The rye silage, however, had greater NDF concentrations than the other silages. It also had a greater ADF concentration than the other silages. Many research workers emphasized that rye silage is not a good option to feed livestock due to the lower quality and biomass yield (Neal et al., 2010,).

Acid Detergent Fibre (ADF) content

Excepting Y x C x HS interaction, all factors and their interactions were significant on ADF content of silages. Average ADF content of cereal silage in 2010 (37.9%) was higher than in 2011 (35.9%). Two year average shown that average ADF content among harvest stages (early heading, milky stage and mid-dough) were 34.2%, 36.8 and 39.7%, respectively (Table 4), while average ADF contents among cereal silages were 36.2%, 39.9%, 41.8%, 35.0% and 31.6% in barley, wheat, rye, triticale and oat, respectively. Generally, oat silage had performance better than the other cereal silages with regard to ADF content in the study. ADF contents of cereal silages increased from early heading stage to mid-dough stage that means maturation of cereals. Both the oat and triticale had lower ADF concentrations than the rye and wheat, despite their harvests at earlier stages of growth. These results were comparable to values reported by Pereira-Crespo et al. (2010) and Brand et al. (2003) for barley and oat harvested as forage at the milk to soft dough stage. Brand et al. (2003) also reported higher ADF levels for triticale compared with barley and oat. ADF levels in the barley silages, although higher than the wheat or rye silage, were very similar to values reported by many researchers (McCartney and Vaage, 1994; De Ruiter and Hanson, 2004).

CONCLUSION

Winter cereals can provide feed earlier than annual grasses like *Lolium italicum*, etc because they are generally more adaptable to early sowing due to higher tolerance of dry conditions. Cereals are also better suited to single-cut silage-making, whereas annual grasses require multiple cuts or grazings to be fully utilised. The results of our two-year study testing a total of five winter cereals (*Hordeum vulgare Triticum aestivum, Secale cereale, Triticale, Avena sativa*) showed that it was possible to produce an average of 10.9 t ha^{-1} dry matter yield with an average of 9.2% crude protein content at mid-dough stage in regions with Mediterranean-type climates. It was also concluded that *Avena sativa* should be preferred for high biomass yield and should be cut at the beginning of mid-dough maturity stages for higher quality silage.

LITERATURE CITED

Alcicek,A. and K.Ozkan. 1996. Zur quantitativen Bestimmung von Milch-, Essig- und Buttersäuren in Silage mit Hilfe eines Destillationsverfahren, J.of Agr.Fac.of Ege Univ, 33(2-3):191-198.

Ben-Ghedalia,D., A.Kabala, J.Miron and E.Yosef. 1995. Silage fermentation and in vitro degradation of monosaccharide constituents of wheat harvested at two stages of maturity. J. Agric. Food Chem. 43:2428–2431.

Bergen,W.G., T.M.Byrem and A.L.Grant. 1991. Ensiling characteristics of whole-crop small grains harvested at milk and dough stages. J. Anim. Sci. 69:1766–1774.

Bocchi,S., G.Lazzaroni, N.Berardo and T.Maggiore. 1996. Evaluation of triticale as a forage plant through the analysis of the kinetics of qualitative from stem elongation to maturity. Kluwer Acad. Publishers, 827-834.

Brand,T.S., C.W.Cruywagen, D.A.Brandt, M.Viljoen and W.W.Burger. 2003. Variation in the chemical composition, physical characteristics and energy values of cereal grains produced in the Western Cape area of South Africa, South African Journal of Animal Science 33(2):117-126.

Brink,G.E. and G.C.Martin. 1986. Barley vs oat companion crops. I. Forage yield and quality response during alfalfa establishment. Crop Sci. 26:1060-1067.

Budaklı Çarpıcı, E., N.Çelik and G.Bayram, 2010. Yield and quality of forage maize as influenced by plant density and nitrogen rate, Turk. J. Field Crops, 15(2):128-132.

Cherney,J.H. and G.C.Martin. 1982. Small grain crop forage potential: I.Biological and chemical determinants of quality and yield, Crop Sci. 22:227-231.

De Ruiter, J.M, and R.Hanson. 2004. Whole crop cereal silage-production and use in dairy, beef, sheep and deer farming. Christchurch, NZ: NZ Institute for Crop and Food Research Ltd.

De Ruiter,J.M., R.Hanson, A.S.Hay, K.W.Armstrong and R.D.Harrison-Kirk. 2002. Whole-crop cereals for grazing and silage: balancing quality and quantity, Proceedings of the New Zealand Grassland Association 64: 181–189.

Delogu,G., N.Faccini, P.Faccioli, F.Reggiani, M.Lendini, N.Berardo and M.Odoardi. 2002. Dry matter yield and quality evaluation at two phenological stages of forage triticale grown in the Po Valley and Sardinia Italy, Field Crops Research 74:207-215.

Juskiw,P.E., J.H.Helma and D.F.Salmona. 2000. Forage yield and quality for monocrops and mixtures of small grain cereals, Crop Sci. 40:138–147.

Kaiser,A.G., J.W.Piltz, H.M.Burns and N.E.Griffiths. 2004. Successful Silage, Edition 2.Dairy Australia and NSW Department of Primary industries, NSW DPI, Orange.

Khorasani,G.R., P.E.Jedel, J.H.Helm and J.J.Kennelly. 1997. Influence of stage of maturity on yield components and chemical composition of cereal grain silages. Can. J. Anim. Sci. 77:259–267.

Kristensen,N.B., K.H.Sloth, O.Højberg, N.H.Spliid, C.Jensen and R.Thøgersen. 2010. Effects of microbial inoculants on corn silage fermentation, microbial contents, aerobic stability, and milk production under field conditions. *J. Dairy Sci.* 93:3764–3774.

McCartney,D.H. and A.S.Vaage. 1994. Comparative yield and feeding value of barley, oat and triticale silages, *Canadian Journal of Animal Science 74*: 91-96.

Neal,J.S., W.J.Fulkerson and L.C.Campbell. 2010. Differences in yield among annual forages used by the dairyindustry under optimal and deficit irrigation. Crop and Pasture Science 61(8):625–638.

Pereira-Crespo,S., B.Fernández-Lorenzo, J.Valladares, A.González-Arráez and G.Flores. 2010. Effects of seeding rates and harvest date on forage yield and nutritive value of pea-triticale intercropping, Options Méditerranéennes, A no. 92,-The contributions of grasslands to the conservation of Mediterranean biodiversity, 215-218.

Petterson,K. 1988. Ensiling of Forages: Factors affecting silage fermentation and quality, Swedish University of Agricultural Sciences, Department of Animal Nutrition and Management, Uppsala, 46p.

Pugnaire,F.I. and F.S.Chapin. 1992. Environmental and physiological factors governing nutrient resorption efficiency in barley, Oecologia 90:120-126.

SAS Institute. 1998. INC SAS/STAT user's guide release 7.0, Cary, NC, USA.

Schneider,S., R.Vogel and U.Wyss. 1991. Die eignung von triticale zur bereitung von ganzpflanzensilage. Landwirtschaft Schweiz Band 4:407–411.

Siefers,M.K. and K.K.Bolsen. 1997. Agronomic and silage quality traits of winter cereals. 1st Silage Congress in Turkey, Bursa, 201-203.

Stell,R.G.D., J.A.Torrie and D.A.Dickey. 1997. Principles and Procedures of Statistics. A.Biometrical Approach 3rd Edi. Mc Graw Hill Book. INC. NY.

Twidwell,E.K., K.D.Johnson, J.H.Cherney and H.W.Ohm. 1987. Forage yield and quality of soft red winter wheats and a winter triticale. Appl. Agric. Res. 2:84–88.

10

EFFECTS OF MIXTURE RATIO AND ROW SPACING IN HUNGARIAN VETCH (*Vicia pannonica* Crantz.) AND ANNUAL RYEGRASS (*Lolium multiflorum* Lam.) INTERCROPPING SYSTEM ON YIELD AND QUALITY UNDER SEMIARID CLIMATE CONDITIONS

Alpaslan KUSVURAN[1]*, Mahmut KAPLAN[2], Recep Irfan NAZLI[3]

[1]*Kizilirmak Vocational High School, Cankiri Karatekin University, Cankiri, TURKEY*
[2]*Department of Field Crops, Faculty of Agriculture, Erciyes University, Kayseri, TURKEY*
[3]*Department of Field Crops, Faculty of Agriculture, Cukurova University, Adana, TURKEY*
Corresponding author: akusvuran@gmail.com

ABSTRACT

This study was conducted at the Middle Kizilirmak basin of Turkey (40°20′N, 33°58′E, elevation 550 m), during the 2011–2012 and 2012–2013 growing seasons, to determine the effects of different mixture ratios (sole Hungarian vetch (HV), annual ryegrass (AR); 80%HV+20%AR, 60%HV+40%AR, 40%HV+60%AR, and 20%HV+80%AR) and row spacings (20, 30, and 40 cm) on the forage yield and quality of the HV and AR intercropping system. The experiment was planned in a randomized complete block design, where a split-plot arrangement of mixture ratios was considered as the main plot with the row spacings considered as subplots. According to the mean of 2 years, the different mixture ratios and row spacings had a statistically significant effect on all of the properties. At the end of the research, among the different mixture ratios and row spacing interactions, the highest green herbage yield (33.4 t ha^{-1}), hay yield (7.5 t ha^{-1}), lowest neutral detergent fiber (52.2%), and crude ash (7.8%) rates were obtained from the 60%HV+40%AR mixture and 30-cm row spacing interaction. The highest crude protein (CP) rate (17%), CP yield (1156 kg ha^{-1}) and ADF (39.5%) ratios were obtained from the 80%HV+20%AR mixture and 30-cm row spacing interaction. The highest RFV value (107) was obtained from the 80%HV+20%AR mixture and 20-cm row spacing interaction. These results show that increasing the HV ratio in the mixture has positive effects on the yield and quality, whereas raising the row spacing has adverse effects on the forage values. Therefore, 80%HV+20%AR or 60%HV+40%AR with a 30-cm row spacing interaction can be suggested for forage production in an HV and AR intercropping system.

Key words: Acid detergent fiber, Animal feeding, Crude protein, Forages, Neutral detergent fiber, Relative feed value

INTRODUCTION

Intercropping had been neglected in research on plant production systems in Europe, possibly due to the complexity of these systems (Hauggard-Nielsen et al. 2009), but afterwards, in forage crop production, many intercropping systems were used for different purposes (Acar et al. 2006). This system allows lower inputs through reduced fertilizer and pesticide requirements, and it contributes to a greater uptake of water and nutrients, increased soil conservation, and high productivity and profitability (Lithourgidis et al. 2011; Akman et al. 2013) compared to monocrop systems.

Forage grasses or cereals are commonly grown with legumes in a mixture because of their ability to increase the herbage yield and to produce forage with more balanced nutrition for livestock feeding (Koc et al. 2013). Legumes are a particularly good source of protein (Eskandari et al. 2009) and incorporating them into an

intercropping system could be of paramount importance for the nutritive value of forage (Nadeem et al. 2010). Another advantage of grass-legume intercropping is that nitrogen (N) can be transferred from the legume into the soil; hence, grasses can use it during their growth (Lauk and Lauk 2009; Mariotti et al. 2009). Accordingly, this system has risen in popularity lately worldwide and it has been a common cropping method in rain-fed areas, particularly in Mediterranean countries such as Turkey (Lithourgidis et al. 2006; Dhima et al. 2007). It is most important that mixtures must contain at least 1 grass and legumes (Acar et al. 2006).

Although there is a large diversity of *Vicia* species, in the Mediterranean Basin, including Turkey, vetches, particularly common vetch (*Vicia sativa* L.) and Hungarian vetch (HV) (*Vicia pannonica* Crantz.) are the most common annual forage crops cultivated (Acikgoz 2001; Uzun et al. 2004). HV is a winter-hardy and drought-resistant legume species, which is widely used in regions with cool winter growing conditions (Uzun et al. 2004; Albayrak et al. 2011). It has satisfactory forage yields with tiny plentiful and palatable leaves, and good quality hay with high crude protein (CP) (Tuna and Orak 2007; Unal et al. 2011). Moreover, that is generally recommended in dry regions (Uzun et al. 2004).

HV has a semierect growth habitus with a leaning tendency in pure stands, especially during rainy years. Intensive early spring rains lead to the decay of plant parts close to the ground to, as a consequence of high humidity. This results in a decrease in the rate of forage yield and quality (Iptas 2002). Cereals can provide support for climbing vetches, improve light interception, and thus facilitate mechanical harvesting (Nadeem et al. 2010). For instance, HV can be grown with forage grasses or cereals under Mediterranean conditions, during which the fallow+cereal system is performed.

One of the most important forage crops is annual ryegrass (AR) (*Lolium multiflorum* Lam.), which is a cool-season grass that is suitable for quality herbage production on account of its rich protein, minerals, and water-soluble carbohydrate content (Kusvuran and Tansi 2005). It is generally a highly nutritious grass that may be presented as forage for beef cattle through grazing, dried out and fed as hay, or ensiled and fed as silage (Acikgoz 2001; Kusvuran and Tansi 2011; Durst et al. 2013), and desirably eaten by livestock, especially in milk production (Kusvuran 2011). In recent years, one of the cultivars of AR, 'Caramba', has quite well adapted to Turkey's climate and soil conditions (Ozkul 2012), which has been recognized as potential roughage for ruminant animals (Van Niekerk et al. 2008; Catanese et al. 2009). Furthermore, (Ozkul et al. 2012) reported that it will rise crucially in the use and importance of AR in future and it will be preferred by livestock as a sole crop when compared to cereals such as barley and triticosecale (Van Niekerk et al. 2008; Catanese et al. 2009). Accordingly, HV+AR herbage can be used for feeding directly to livestock as fresh material or used as hay or silage.

Some factors affect the growth of the species used in intercropping, including cultivar selection, seeding ratios, mixture ratios, row spacing and competition between the mixture components (Dhima et al. 2007; Akman et al. 2013), extra work in preparing and planting seed, and crop management practices (Lithourgidis et al. 2011).

In vetch mixtures with cereals or grasses, it is necessary to know the ratios of the vetch and cereal/grass species (Albayrak et al 2004; Balabanli and Turk 2006), because it affects the growth rate of the individual species in the mixtures as well as the forage yield and quality (Lithourgidis et al. 2006; Lauk and Lauk 2009). In the mixtures, for example, an increased proportion of the cereal in its mixture with vetch significantly decreases the stand lodging, and has a positive influence on the forage yield, but the forage nutrient is of a poorer quality (Karagic et al 2012), because a high cereal ratio in the botanic composition of legume+cereal mixtures causes low protein. Mariotti et al. (2009) reported that cereals had

a higher belowground competitive ability than legumes and legumes had a higher aboveground competitive ability than cereals in their mixtures, and that the competitive ability of the plants showed differences among the species. Between plant species, there may be aboveground competition for light and space, and belowground competition for water and nutrients (Mariotte et al. 2012). Thus, these competition conditions have important influences on the mixtures and these factors must be considered in this system.

The effects of different mixture ratios in the intercropping system have been evaluated in many studies. In these studies (Albayrak et al. 2004; Lithourgidis et al. 2006; Pinar 2007; Ozel 2010), increasing the ratio of those vetches whose forage quality was higher in the mixtures, increased the both forage yield and nutrient content, while some researchers reported the opposite findings; that increasing rate of the cereal/grass whose dry matter content was higher, resulted in a higher forage yield, but lower forage quality (Orak and Uygun 1996; Balabanli and Turk 2006; Tuna and Orak 2007; Dhima et al. 2007; Gunduz 2010; Bedir 2010). Moreover, the optimal forage yield and CP contents were obtained when the legume and cereal ratio was equal in the mixture (Basbag et al. 1999).

In addition to the mixture ratios, row spacing is another important factor for higher yield realization through light penetration in the crop canopy. An advantage of narrow row spacing is more equidistant plant spacing, which leads to increased canopy leaf area development and greater light interception earlier in the season (Shibles and Weber 1966). These changes in the canopy formation increase the crop growth rate and dry matter accumulation (De Bruin and Pedersen 2008; Albayrak et al. 2011).

HV and AR have been grown at different row spacings for forage in several studies. Generally, these species were grown at a 20-cm row spacing (Geren et al. 2003; Pinar 2007; Darvishi et al. 2009; Gunduz 2010; Bedir 2010). Moreover, while some researchers planted at a 25-cm row spacing (Gultekin 2008; Unal et al. 2011; Yolcu et al. 2012), both species were grown at a 30-cm row spacing by Kilavuz (2006), Balabanli and Turk (2006), and Nadeem et al. (2010) in different researches.

Some researchers reported the most suitable row spacing in AR as 30 cm (Orak and Uygun 1996; Kusvuran and Tansi 2011; Kara 2013) and they reported that increasing the row spacing had a negative impact on the forage yield and quality. The highest forage yields and nutrient values were obtained at 17-, 20-, and 25-cm row spacings in HV by Uca et al. (2007), Nizam et al. (2007), and Bagci (2010), respectively. These researchers reported that while the herbage quality increased gradually depending on the rising row spacing, a significant decreasing was determined from the forage yield at narrower row spacings. Contrary the these researches, Orak and Tuna (1994) and Akkopru et al. (2007) reported that optimal values were obtained at 35- and 40-cm row spacings, and they stated that increasing the row spacing had a positive effect on the forage yield and quality. On the other hand, Albayrak et al. (2011) reported that there was no difference in the forage quality with row spacings of between 17 and 35 cm.

To date, many new researches have been conducted where HV was grown with different cereals in an intercropping system, whereas few associated with AR have been carried in recent years.

The objectives of this study were: a) to evaluate the forage production capacities of HV+AR mixtures, b) to assess the effects of different mixture ratios and row spacings on the yield and quality of the mixtures, and c) to determine the most convenient mixture ratio and row spacing interaction in this intercropping mixture system.

MATERIALS AND METHODS

Research Site

The study was carried out at the Research and Implementation Area of the Kizilirmak Vocational High School of Cankiri Karatekin University (40°20′N, 33°58′E, elevation 550 m), which is located on the Middle Kizilirmak Basin of Turkey, during the 2011–2012 and 2012–2013 growing seasons.

The region is under semiarid and central Anatolian climate conditions. It has a longterm annual rainfall of 407.8 mm (TSMS, 2013), and while the average temperature of the first growing season (4.6 °C) was lower than the longterm average value (8.3 °C), it was closer to this value in second growing season (8.5 °C). The total precipitation was 329.1 mm and 309.5 mm during the 2011–2012 and 2012–2013 growing seasons, respectively (Table 1.). Hence, the fallow-cereal system was applied to half of this cultivated area based on the restricted precipitation (TSI 2013).

The major soil characteristics are given in Table 2 for the experimental area. The soil at the experimental site had a clay or clay-loam texture and moderate lime content.

Table 2. Initial chemical and physical properties of the soil.

Property	Depth (0-30 cm)	Depth (30-60 cm)
Sand (%)	13.5	21.3
Silt (%)	31.1	42.1
Clay (%)	55.4	36.6
Soil Texture	Clay	Clay-Loam
pH (1:2.5)	8.03	7.86
Salt (mmhos cm $^{-1}$)	0.50	1.19
Lime (%)	16.0	17.9
P_2O_5 (kg ha^{-1})	6.5	65.4

Table 1. Climate data of the location in 2011-2012-2013 years and long term average (1960-2012) at Cankiri, Turkey.

Months	Mean Temperature (°C)			Total precipitation (mm)		
	2011-2012	2012-2013	Long-term	2011-2012	2012-2013	Long-term
October	9.6	15.5	11.9	45.5	23.2	27.7
November	1.3	7.5	5.6	2.5	28.3	26.8
December	0.8	2.7	1.6	31.9	63.9	47.9
January	-2.1	1.2	-0.6	100.4	40.4	42.3
February	-5.0	4.4	0.9	52.5	22.5	34.5
March	2.2	6.8	5.6	21.0	39.8	36.8
April	13.4	12.1	11.0	6.5	61.6	46.4
May	16.2	18.1	15.7	68.8	29.8	54.4
Total	-	-	-	329.1	309.5	316.8
Average	4.6	8.5	8.3	-	-	-

Plant materials and treatments

The HV 'Tarm Beyazi-98' cultivar and AR 'Caramba' cultivar were used as the plant material of the study, and both species were planted as sole crops and mixtures. A formulation of 4 different HV (*Vicia pannonica* Crantz.) and AR (*Lolium multiflorum* Lam.) mixtures (80%HV+20%AR, 60%HV+40%AR, 40%HV+60%AR, 20%HV+80%AR), as the sole of these species, and 3 different row spacings (20, 30, and 40 cm) were used in this study.

After seeding, 27 kg ha^{-1} of N and 69 kg ha^{-1} of phosphorus were applied as a starter fertilizer using diammonium phosphate. Seeds were planted in October of

both the first and second year, at a rate of 120 kg ha^{-1} for the HV (Acikgoz 2001; Balabanli and Turk 2006; Balabanli 2009) and 45 kg ha^{-1} for the AR (Kusvuran 2011; Kusvuran and Tansi 2011) in each plot. The HV and AR seeds were sown in the same rows. The size of each plot was 12 m^2 (2.4 × 5 m). During both the first and second year, the experiments were carried out without supplementary irrigation.

The harvest time was determined by taking the physiological periods of the HV into consideration. Hence, plots were harvested at the full pod set stage for the HV (Balabanli 2009) and at the beginning of flowering for the ryegrasses (at the end of May) (Kusvuran and Tansi 2005) at the same time for the forage yield.

Following the harvest, the HV ratios in the mixtures were determined in the fresh material.

Experiment design and statistical analyses

The experiments were conducted in a randomized complete block design with a split-plot arrangement (the mixtures were placed in the main plots and the row spacings were placed in subplots) with 18 treatments in 3 replications. Analysis of variance of the experimental results was performed using MSTAT-C statistical software (Freed 1991) and $P < 0.05$ and 0.01 were considered as statistically significant, and means were compared using Duncan's multiple range test at a significance of $P < 0.05$.

Chemical analyses

A total of 500 g of fresh sample was taken from the harvested plants and dried at 70 °C for 48 h. Next, dry matter ratios and hay yields were determined. Afterwards, hay samples were ground in a hand mill with a 1-mm sieve. The crude ash contents of the samples were determined by burning the samples at 550 °C for 8 h. The Kjeldahl method was used to determine the N contents of the dry samples (Kacar and Inal, 2008). CP ratios (CPRs) were calculated using the equation of $N \times 6.25$ and CP yields (CPYs) were determined using the equation of CPR $\times$ hay yield. The neutral detergent fiber (NDF) (Van Soest and Wine, 1967) and acid detergent fiber (ADF) (Van Soest 1963) contents were analyzed with an ANKOM 200 fiber analyzer (ANKOM Technology Corp. Fairport, NY, USA). The dry matter digestibility (DMD), dry matter intake (DMI), and relative feed value (RFV) were calculated using the following equations (Rohweder et al. 1978):

To calculate the RFV, initially, the DMD was calculated from the ADF value by:

DM% = 88.9 – (0.779 × ADF%).

The DMI, based on animal live-weight, was calculated from the NDF value by:

DMI % of BW = 120 / NDF%.

Next, the RFV was calculated from the DMD and DMI by:

RFV = DDM% × DMI% × 1.29.

RFVs were evaluated using the values provided in the standards for dry hays given in Table 3.

Table 3. Relative feed value standards[a]

Quality Standards	CP	ADF, % (DM)	NDF, % (DM)	RFV
The best quality	>19	<31	<40	>151
1	17-19	31-40	40-46	151-125
2	14-16	36-40	47-53	124-103
3	11-13	41-42	54-60	102-87
4	8-10	43-45	61-65	86-75
5	8,00	>45	>65	<75

[a] Relative feed value is assumed to be 100 when the ADF is 41% and NDF is 53% (Rohweder et al 1978).

RESULTS

All of the parameters were significantly influenced by the different mixture ratios, row spacings, and mixture ratio × row spacing interactions (Table 4).

With regard to the different row spacings, the highest plant height (90.4 cm), green herbage (30.0 t ha^{-1}), hay (6.7 t ha^{-1}) and CPY (996 kg ha^{-1}), and CP (15.2%) and ADF (38.2%) rates were obtained from 30-cm row spacing; the highest HV rate (78.0%) and RFV value (104) were observed at 20-cm row spacing; and the highest NDF (58.1%) and crude ash (8.5%) rates were obtained from 40-cm row spacing.

Among the different mixture ratios, the highest plant height (88.5 cm), HV (86.2%), CP rate (15.8%), CPY (1042 kg ha^{-1}), and RFV value (103) were observed from the 80%HV+20%AR mixture; the highest green herbage (30.9 t ha^{-1}) and hay (6.9 t ha^{-1}) yields were obtained from the 60%HV+40%AR mixture; the highest NDF (55.9%) and ADF (38.5%) rates were observed from the 20%HV+80%AR mixture; and the highest crude ash rate (8.2%) was obtained from the 40%HV+60%AR mixture.

For the different mixture ratio × row spacing interactions, the highest plant height (92.3 cm), CP (17.0%) and ADF (39.5%) rates, and CPY (1156 kg ha^{-1}) were obtained from the 80%HV+20%AR × 30-cm interaction; the highest HV rate (90.4%) and RFV (108) values were observed from the 80%HV+20%AR × 20-cm interaction; the highest green herbage (33.4 and 32.9 t ha^{-1}) and hay (7.5 and 7.3 t ha^{-1}) yields were obtained from the 60%HV+40%AR × 30-cm and 40%HV+60%AR × 30-cm interactions, which were not statistically significant, as they were in the same group; the highest NDF rate (59.2%) was observed from the 20%HV+80%AR × 40-cm interaction; and the highest crude ash rate (8.5%) was obtained from the 40%HV+60%AR × 40-cm interaction.

Table 4. Summary of ANOVA and mean squares of the traits determined based on the combined analysis over two-years.

Source of Variation	DF	PH	HVR	GHY	HY	CPR
Replication (R)	2	6.708ns	1.392ns	1,030ns	0,334ns	0,019ns
Year (Y)	1	8214.938**	28.727ns	225,333**	6,700*	50,841**
Error 1	2	4.293	5.952	0,853	0,340	0,069
Mixture ratios (MR)	5	1389.700	13501,332**	92,062**	5,343**	122,893**
YXMR	5	166.235**	41,714**	12,895**	2,304**	13,253**
Row Spacing (RS)	2	88.915**	308,154**	167,031**	5,814**	9,551**
YXRS	2	8.781ns	76,683**	18,986**	0,651**	0,812**
MRXRS	10	36.831**	56,658**	17,611**	1,178**	8,737**
YXMRXRS	10	67.801	16,365**	7,581**	0,191**	3,431**
Error 2	68	4.789	2,918	0,400	0,049	0,052
CV (%)		2.47	2,28	2,29	3,54	1,60
Source of Variation	**DF**	**CPY**	**NDF**	**ADF**	**CAR**	**RFV**
Replication (R)	2	6967,148ns	0.017ns	0.154ns	0.148ns	0.857ns
Year (Y)	1	753838.231**	1.100*	18.008**	38.401**	6.356*
Error 1	2	5298.370	0.034	0.019	0.010	0.181
Mixture ratios (MR)	5	348853.320**	139.577**	32.186**	6.178**	732.790**
YXMR	5	18419.231**	16.173**	1.193**	0.177**	46.733**
Row Spacing (RS)	2	233712.954**	179.941**	16.066**	0.935**	661.211**
YXRS	2	11289.343**	5.343**	14.300**	1.039**	97.255**
MRXRS	10	36068.954**	8.817**	16.335**	0.267**	49.220**
YXMRXRS	10	10368.143**	10.797**	2.624**	0.148**	54.660**
Error 2	68	1100.651	0.068	0.038	0.010	0.350
CV (%)		3.77	0.47	0.52	1.18	0.59

Degrees of freedom, CV: Coefficient of variation, NS: Not significant, *P < 0.05, **P < 0.01, PH: Plant height, HVR: Hungarian vetch ratio GHY: Green herbage yield, HY: Hay yield, CPR: Crude protein ratio CPY: Crude protein yield NDF: Neutral detergent fiber ADF: Acid detergent fiber CAR: Crude ash ratio RFV: Relative feed value

DISCUSSION

While sole HV grows semierect and can reach up to 90 cm in height (Acikgoz 2001; Balabanli 2009), AR grows erect and can rise to a height of 130 cm (Baytekin et al. 2009). Some researchers obtained plant heights of between 34 and 103 cm in sole HV (Balabanli and Turk 2006; Gunduz 2010; Nadeem et al. 2010; Unal et al. 2011), while others obtained heights of between 60 and 133 cm in AR (Kusvuran and Tansi 2005, 2011; Gultekin 2008; Kesiktas 2010; Kusvuran 2011). In this study, the highest plant height (92.3 cm) was observed from the 80%HV+20%AR mixture and 30-cm interaction, and its values diminished with more than 30-cm row spacing.

Aside from genetic factors, the plant species, sowing ratio, soil, climate, and environmental conditions also influence plant heights. The average plant heights obtained from sole sowings of the species in the present study were similar to those reported in previous studies. In the mixtures, higher average plant heights were achieved due to higher HV plant heights in relation to the existence of AR in the mixture. Generally, when the ryegrass rate increased in the mixtures, the plants heights also increased. The higher height of the plants in the mixtures compared to sole HV sowing could be caused by the competition between the HV and AR, and this position might have affected the height of the plant positively. Additionally, as is known, intercropping of forage grasses with legumes provides structural support for climbing vetches and improves light interception (Nadeem et al. 2010). Thus, the obtained higher plant heights from the mixtures can arise from these situations when compared to sole HV sowing.

Separately, while Orak and Uygun (1996), Akkopru et al. (2007), and Bagci (2010) reported that plant heights decreased depending on the increasing row spacing, Kusvuran and Tansi (2011) found that there was no difference with different row spacings. These researchers reported that the high plant height probably resulted from light competition between the legume and grass species in the mixtures.

The vetch ratio in the mixture affects the quality of the harvested herbage. Different vetch ratios in mixtures were observed in previous intercropping systems, ranging from 14% to 91.3%. In these studies, the highest vetch ratio was obtained using vetch at a rate of 70% or higher in the harvested herbage (Albayrak et al. 2004; Kilavuz 2006; Pinar 2007; Nizam et al. 2007; Ozel 2010; Tas 2010). In addition to these studies, while Yolcu et al. (2009) found the average vetch rate as 34% in different HV-cereal mixtures; Gunduz (2010), and Orak and Nizam (2012) found rates of 16% and 42%, respectively, in a 50%HV+50%cereal mixture; Geren et al. (2003) found a vetch rate of 85% in a 50%vetch+50%AR mixture; Albayrak et al. (2004) found a rate of 66% in a different mixture ratio of HV+triticosecale; and Bedir (2010) observed a rate of 14% in different HV+barley mixtures. Similarly, Lithourgidis et al. (2006) reported different legume ratios in various intercropping systems.

Table 5. Forage yield and yield components of Hungarian vetch and annual ryegrass mixtures at different by mixture ratio and row spacing. Data are the means of the 2 years.

Mixture ratios (MR)	Plant height (cm)				Hungarian vetch ratio (%)			
	Row spacing (RS) (cm)				Row spacing (RS) (cm)			
	20	30	40	Mean	20	30	40	Mean
Hungarian Vetch (HV)	79.6 g	78.3 g	75.5 h	77.8 d	-	-	-	-
Annual Ryegrass (AR)	99.8 b	107.6 a	106.9 a	104.8 a	-	-	-	-
80%HV+20%AR	87.8 d	92.3 c	85.1 ef	88.5 b	90.4 a	82.0 c	86.2 b	86.2 a
60%HV+40%AR	87.6 de	88.3 d	84.6 f	86.8 c	84.3 b	69.8 e	74.1 d	76.1 b
40%HV+60%AR	86.2 df	87.6 de	87.4 de	87.1 bc	60.2 f	56.9 g	55.5 g	57.5 c
20%HV+80%AR	88.2 d	88.0 d	84.2 f	86.8 c	33.2 h	24.6 j	30.7 ı	29.5 d
Mean	88.3 b	90.4 a	87.3 b		78.0 a	72.2 c	74.4 b	
LSD (5%)	MR:1.46 RS:1.03 MRxRS:2.52				MR:1.14 RS:0.80 MRxRS:1.97			

Mixture ratios (MR)	Green herbage yield (t ha⁻¹)				Hay yield (t ha⁻¹)			
	Row spacing (RS) (cm)				Row spacing (RS) (cm)			
	20	30	40	Mean	20	30	40	Mean
Hungarian Vetch (HV)	25.9 f	25.0 gh	22.7 j	24.5 e	5.7 gh	5.2 ı	5.1 ı	5.3 d
Annual Ryegrass (AR)	27.0 e	28.1 d	25.6 fg	26.9 cd	6.0 fg	6.3 de	6.1 ef	6.2 c
80%HV+20%AR	29.5 c	30.9 b	28.4 d	29.6 b	6.8 bc	6.8 b	6.4 de	6.7 b
60%HV+40%AR	30.5 b	33.4 a	28.6 d	30.9 a	7.0 b	7.5 a	6.5 cd	7.0 a
40%HV+60%AR	24.7 h	32.9 a	23.9 ı	27.2 c	5.7 h	7.3 a	5.6 h	6.2 c
20%HV+80%AR	25.3 fh	29.6 c	25.2 fh	26.7 d	5.8 gh	7.0 b	5.6 h	6.1 c
Mean	27.2 b	30.0 a	25.7 c		6.1 b	6.7 a	5.8 c	
LSD (5%)	MR:0.42 RS:0.30 MRxRS:0.73				MR:0.15 RS:0.10 MRxRS:0.26			

Mixture ratios (MR)	Crude protein ratio (%)				Crude protein yield (kg ha⁻¹)			
	Row spacing (RS) (cm)				Row spacing (RS) (cm)			
	20	30	40	Mean	20	30	40	Mean
Hungarian Vetch (HV)	18.4 b	20.6 a	17.6 c	18.6 a	1043 bc	1075 ab	898 e	994 b
Annual Ryegrass (AR)	11.1 l	13.0 ı	10.0 m	11.4 f	660 j	815 f	617 k	697 f
80%HV+20%AR	14.9 e	17.0 d	14.8 e	15.8 b	1010 c	1156 a	947 c	1042 a
60%HV+40%AR	13.8 g	14.4 f	13.4 h	13.9 c	957 d	1055 b	883 e	965 c
40%HV+60%AR	13.4 h	13.6 gh	12.6 j	13.2 d	757 g	1002 c	705 hı	821 d
20%HV+80%AR	11.9 k	12.6 j	13.1 ı	12.6 e	684 ıj	871 e	732 gh	762 e
Mean	13.9 b	15.2 a	13.6 c		852 b	996 a	797 c	
LSD (5%)	MR:0.15 RS:0.11 MRxRS:0.27				MR:22.1 RS:15.6 MRxRS:38.2			

Mixture ratios (MR)	NDF (%)				ADF (%)			
	Row spacing (RS) (cm)				Row spacing (RS) (cm)			
	20	30	40	Mean	20	30	40	Mean
Hungarian Vetch (HV)	49.9 k	53.2 ı	54.6 g	52.5 f	34.1 k	36.0 ı	37.3 f	35.8 e
Annual Ryegrass (AR)	59.2 b	58.7 c	64.0 a	60.6 a	39.1 c	37.4 f	42.4 a	39.7 a
80%HV+20%AR	52.2 j	52.4 j	56.5 e	53.7 e	36.7 h	39.5 b	36.9 g	37.7 c
60%HV+40%AR	55.5 f	52.2 j	57.1 d	54.9 d	35.2 j	38.0 e	36.9 gh	36.7 d
40%HV+60%AR	54.7 g	54.0 h	57.0 d	55.2 c	37.4 f	38.8 d	36.9 gh	37.7 c
20%HV+80%AR	54.6 g	53.8 h	59.2 b	55.9 b	39.0 cd	39.4 b	37.0 g	38.5 b
Mean	54.4 b	54.1 c	58.1 a		36.9 c	38.2 a	37.9 b	
LSD (5%)	MR:0.17 RS:0.12 MRxRS:0.30				MR:0.13 RS:0.09 MRxRS:0.22			

Mixture ratios (MR)	Crude ash ratio (%)				RFV			
	Row spacing (RS) (cm)				Row spacing (RS) (cm)			
	20	30	40	Mean	20	30	40	Mean
Hungarian Vetch (HV)	8.1 f	7.9 g	7.9 g	7.9 e	116 a	108 b	102 e	109 a
Annual Ryegrass (AR)	9.2 c	9.4 b	9.9 a	9.5 a	93 j	94 ı	81 k	89 f
80%HV+20%AR	8.1 f	8.1 f	8.2 f	8.1 cd	107 b	103 d	99 g	103 b
60%HV+40%AR	8.3 e	7.8 g	8.3 e	8.1 c	103 d	106 c	98 h	102 c
40%HV+60%AR	8.3 e	7.8 g	8.5 d	8.2 b	103 d	100 g	98 h	100 d
20%HV+80%AR	8.1 f	7.9 g	8.2 f	8.0 d	100 g	101 f	95 ı	98 e
Mean	8.3 b	8.2 c	8.5 a		104 a	102 b	96 c	
LSD (5%)	MR:0.07 RS:0.05 MRxRS:0.12				MR:0.39 RS:0.28 MRxRS:0.68			

Numbers followed by the same letters are not significantly different at the 5% level of significance.
HV: Hungarian vetch AR: Annual ryegrass

In fresh material, the average across the different mixture ratios and row spacing interactions showed that the HV rate ranged between 25% and 90% in the present study. Our findings were similar to those in previous studies. As expected, the HV rate decreased depending on the increased AR ratio in the mixtures. All of the abovementioned researchers stated the same findings in previous studies. The lowest average values determined in 30-cm row spacing averaged across the different row spacings. Researchers commonly use either 20- or 25-cm row spacing for forage production for AR (Geren et al. 2003; Gultekin 2008; Darvishi et al. 2009; Kesiktas 2010). However, 30-cm row spacing was recommended by Orak and Uygun (1996), and Kusvuran and Tansi (2005, 2011). Accordingly, it may be concluded in the present study that the superior growing performance of AR at 30-cm row spacing was effective on decreasing the HV rate in the mixture.

The plant species in mixtures, environmental conditions, competition index, sowing ratios, and harvest times of the species influence the post-harvest ratios of species in mixtures. The competitive ability of the plants show differences among the species (Mariotti et al. 2009), and there may be an aboveground competition for light and space and below grow competition for water and nutrients among the plants species (Mariotte et al. 2012; Koc et al. 2013). The growth of legumes is very slow in winter due to low temperatures, while cereals produce tillers and stems, and cover the small legume seedlings (Twidwell et al. 1987). Therefore, Iptas (2002) and Ozel (2010) reported that the percentage of cereal seeds in the mixtures should not be higher than 40% for successful production.

Green herbage and hay yield values derived from the sole sowing of HV were reported as 6.2–17.0 t ha^{-1} and 1.7–4.0 t ha^{-1}, respectively, in the previous studies (Karadag and Buyukburc 2004; Balabanli and Turk 2006; Unal et al. 2011; Albayrak et al. 2011; Orak and Nizam 2012). In AR, green herbage and hay yield values ranged between 6.2–80.7 t ha^{-1} and 0.7–14.9 t ha^{-1}, respectively, in previous studies (Parlak et al. 2007; Simic et al. 2009; Darvishi et al. 2009; Kesiktas 2010; Kusvuran 2011; Rivera et al. 2013). Green herbage and hay yields of HV and cereal mixtures by earlier researchers in different mixture ratios ranged between 9.8–47.2 t ha^{-1} and 3.4–13.0 t ha^{-}1, respectively.

When compared to previous studies, the green herbage and hay yield values of the present study were quite close to the upper limit values observed in the sole sowing of HV. However, while satisfactory yield values were obtained from the sole sowing of AR and its mixtures with HV, the values were not close to the upper limits specified in some earlier researches. AR can be cut more than once during the vegetation period. The reason the highest yield levels were not reached, was because of single cut of the present study.

While several researchers reported the highest green herbage and hay yields of HV mixtures as 66%–87.5% (Albayrak et al. 2004; Yolcu et al. 2009; Budak et al. 2011; Koc et al. 2013), others reported the highest values for equivalent rate mixtures (50%–50%) (Karadag and Buyukburc 2004; Gunduz 2010; Aksoy and Nursoy 2010; Nadeem et al. 2010). Contrary to other researches, Balabanli and Turk (2006), Tuna and Orak (2007), and Bedir (2010) reported the highest yield values for cereals mixture rates of 75%–80%.

In the present study, while the highest green herbage (33.4 t ha^{-1}) and hay (7.5 t ha^{-1}) yields were observed from the 60%HV+40%AR and 30-cm row spacing interaction, the yields were then decreased based on the decreasing HV ratio. Some researchers (Albayrak et al. 2004; Lithourgidis et al. 2006; Balabanli 2009; Ozel et al. 2010) stated that the HV ratio in the mixtures should not be less than 60% for high yield. They also indicated that increasing the proportion of legumes in mixed forages increases the yields of green herbage and hay per unit area. In contrast to these researchers, Orak and Uygun (1996), Balabanli and Turk (2006), Dhima et al. (2007), Bedir (2010), and Gunduz (2010) reported that the increase of the rate of cereal and its occurrence of above 50% in the mixture produces higher yields, mainly due to its being taller and growing more strongly than vetch species.

In addition to the mixture rate, the highest yields in the present study were determined at 30-cm row spacing, and the yields then decreased based on the increasing row spacing. According to Nizam et al. (2007), Uca et al. (2007), Bagci (2010), and Albayrak et al. (2011), narrow row spacing; for example, 17- and 20-cm row spacing, increased the forage yield compared to wider row spacing. Contrary to these researchers, Orak and Tuna (1994), Orak and Uygun (1996), Akkopru et al. (2007), and Kusvuran and Tansi (2011) obtained top yields in wider row spacings, such as 30, 35, and 40 cm. These differences were probably because of the plant species and various soil and climate conditions at the experiment sites.

Mixed intercropping systems have a significant advantage over sole sowing because of better economics and land use efficiency (Dhima et al. 2007). When the components of a mixture are complementary to each other, a higher yield occurs based on the transfer of symbiotically fixed N grasses (Lauk and Lauk 2009; Akman et al. 2013). Simic et al. (2009) and Kusvuran (2011) reported that increasing the N dose has a positive effect on the yield and quality in AR. Moreover, intercrops can use available environmental resources, such as light, water, and nutrients, more efficiently (Corre-Hellou et al. 2006; Lithourgidis et al. 2011). On the other hand, if the mixture ratio and row spacing are below optimum, then the nutrients, water, and light will not be utilized to their fullest, thus resulting in poor yield (Lone at al. 2010; Albayrak et al. 2011).

Accordingly, grass mixtures with legumes are usually more productive than pure grass or legume sowings and they can support greater animal performance (Hauggard-Nielsen et al. 2009). Additionally, the stand will always be more productive than sole cropping of the component if a mixture establishes a suitable complementary species with the proper sowing design (Mariotti et al. 2009; Koc et al. 2013). The high yielding capacity of a mixture over sole cropping of a component was reported by Nizam et al. (2007), Mariotti et al. (2009), and Albayrak et al. (2004, 2011). However, Pinar (2007) reported that the sole sowing of a species has a higher yield when compared with mixtures.

Parallel with previous researches, this study shows that growing AR with HV in a mixture produced more herbage yield than sole sowing of the species. On the other hand, although the values were considered to be sufficient for livestock feeding, the green herbage and hay yields were lower than the values of HV+cereal mixtures. This might be explained by the higher shooting and tillering capability of cereals compared to AR. Moreover, cereals produce more vegetative components than AR.

In previous researches, the CPR and CPY of sole HV sowings were between 15.4%–24.1% and 561–1461 kg ha-1, respectively (Karadag and Buyukburc 2004; Bedir 2010; Albayrak et al. 2011; Yolcu et al. 2012). While HV has 15%–17% of CP (Balabanli 2009), AR also has CP contents of around the same values (Ozkul et al. 2012). These CPRs and yields of sole AR sowings were reported as between 7.3%–25.4% and 101–5197 kg ha^{-1}, respectively (Parlak et al. 2007; Gultekin 2008; Catanese et al. 2009; Simic et al. 2009; Kusvuran 2011; Durst et al. 2013; Rivera et al. 2013).

The CPR and yields observed in the sole HV sowing of the present study were quite similar to those in previous studies, but the values of sole ryegrass were relatively lower than those in previous studies. When the cutting period is delayed in AR, the CPR and digestibility decrease based on the increased lignification and cellulose concentration. The sole sowing system in AR enables multiple cuts in a growing season and the cuts are carried out during the early stages of the plant. An early cut has a positive influence on the CPR and multiple cuts increase the forage yield. However, in the present study, only 1 cutting was carried out because of the mixture sowings, and the obtained lower CPR can result from this position.

The CPR and yield of the HV-cereal and grass mixtures were between 10.3%–15.3% and 376–1609 kg ha-1, respectively. The highest values were observed at 30-cm row spacing (15.2 and 996 kg ha^{-1}), and as the row spacing was increased to 40 cm, the CP rates and yields decreased gradually. The works of Parlak et al. (2007), Bagci (2010), and Kusvuran and Tansi (2011) supported the results obtained with this study. On the other hand, Akkopru et al. (2007) and Uca et al. (2007) reported increasing CPRs with increasing row spacing, and they found that the best values were at 40-cm row spacing.

Obtaining optimum forage quality is also a very important requirement in forage crop cultivation for animal nutrition, as well as obtaining a high forage yield from the unit area. Plant species, legume ratio in the mixture, and row spacing are the factors directly influencing how the CPR and hay yield affects the CPY. HV, for example, in mixtures, improves the quality of forage by increasing the protein concentration (Vasilakoglou et al. 2008). The mixtures had comparable or better quality values than the grass monocultures. In most cases, the mixtures had higher CP contents than the grass monocultures in all of the mixtures owing to the utilization of symbiotically fixed N and more enhanced interception of light, except for sole HV (Zemenchik et al. 2002; Albayrak et al. 2011). The difference in CP between the grass monocultures and the mixtures can partially be explained by the low fertilizer N rate. The CP content of grasses is known to increase with higher N rates (Buxton 1996; Zemenchik et al. 2002).

The rate of the species in the mixtures ranks first among the factors affecting the quality of the mixture. The highest CPR and yield were determined from the 80%HV+20%AR mixture and 30-cm row spacing interaction in the present study, and previous results are in agreement our findings. In the mixtures, decreased CPYs were observed with decreasing HV ratios. While most researchers found the highest CPRs and yields with 60% or higher HV in the mixture (Albayrak et al. 2004; Karadag and Buyukburc 2004; Balabanli and Turk 2006; Lithourgidis et al. 2006; Bingol et al. 2007; Yolcu et al. 2009; Budak et al. 2011), some others observed the highest yields in equal mixtures (50%–50%) (Basbag et al. 1999; Gunduz 2010; Aksoy and Nursoy 2010) and 60% cereal mixtures (Bedir 2010). HV contains remarkably more CP than grasses and its mixtures with grasses contain more CP than the sole sowing of AR.

A high CPR is expected in mixtures that contain a high level of HV. Additionally, HV naturally includes higher a CPR than grasses. However, its CPY ranks lower in comparison to the 80%HV+20%AR mixture in this study' due to its low biomass productivity among all of the treatments. Similarly, despite the highest hay yield being achieved from the 60%HV+40%AR mixture, a lower CPY was obtained in this mixture because of the decreased CPR, related to the increased AR ratio.

Previous authors reported different results for forage quality parameters such as RFV, NDF, ADF, and crude ash. Albayrak et al. (2011) determined RFV, NDF' and ADF rates for HV, respectively, as 142%, 42%, and 33%. Yolcu et al. (2009) found the NDF and ADF of different mixtures as 57% and 33%, respectively; Aksoy and Nursoy (2010) reported these values of HV-wheat mixture as 43% and 34%, respectively, and Bingol et al. (2007) reported the values in HV-barley mixtures as 57% and 30%, respectively.

Crop species differences in the forage quality between grasses and legumes can be very large. The protein content of legumes is typically much higher than that of grasses, and legume fiber tends to digest faster than grass fiber, allowing the ruminant to eat more of the legume (Eskandari et al. 2009). Along with an increase in the hay production in the mixtures, there is also a change in the chemical composition of the produced. A lower rate of vetch in the mixture, below a definite point, is not desired when the nutrient quality of the mixture is considered (Tas et al. 2007). Grasses have higher cell wall concentrations and a more rapid accumulation of lignin and cellulose than legumes. Thus, this composition leads to a further decline in the digestibility of the mixtures (Buyukburc and Karadag 2002; Tas 2010).

In previous studies, the rates of ADF, NDF, and crude ash in AR were determined. Catenase et al. (2009) reported a NDF rate of 50% and Rivera et al. (2013) reported an ADF rate of 27%. While Durst et al. (2013) found these values as 56% and 31%, Gultekin (2008) reported them as 58% and 37%. Bingol et al. (2007) reported a crude ash rate of 8.9%, while Geren et al. (2003) found this value as 12.9% in the mixture.

CP, NDF, ADF, and RFV are among the most important quality parameters. The total digestible nutrients (TDNs) refer to the nutrients that are available for livestock and are related to the ADF concentration of the forage. As the ADF increases, there is a decline in the TDNs, which means that animals are not able to utilize the nutrients that are present in the forage (Aydin et al. 2010). The RFV is an index used to predict the intake and energy value of the forages and it is derived from the DDM and DMI (Rohweder et al. 1978; Albayrak et al. 2011).

According to the results provided in Table 3, the best NDF (52.2%), ADF (39.5%), and crude ash (7.8%) values were determined with 60% or 80% HV and 30-cm row spacing, and the highest RFV (107) was obtained from the same mixtures, and was sown at 20-cm row spacing. In this study, depending on the increasing HV ratio in mixture, these values were affected positively. Some researchers (Parlak et al. 2007; Simic et al. 2009; Kusvuran 2011) reported that the forage N content increased based on the increasing N fertilization; hence, a higher ADF quality was obtained in the hay (Zhang et al. 1995). In this study, a higher HV ratio in the mixture probably caused this result.

The hay and CPs of this interaction were at satisfactory levels, the CPRs were at high levels, and the crude ash values were at low levels. NDF, ADF, and RFV are significant digestibility indicators. Accordingly, the values observed in the present study were at medium levels.

Considering all of the investigated parameters in the present study, 60% or 80% HV and a 30-cm row spacing interaction can be recommended for high yield and quality forage production.

LITERATURE CITED

Acar, Z., O.O. Asci, I. Ayan, H. Mut and U. Basaran. 2006. Intercropping systems for forage crops. J. of Fac. of Agri. OMU. 21(3):379-386 (In Turkish with English abstract).

Acikgoz, E. 2001. Forages. (Renewed third edition). Amplification Foundation of Uludag University Issue No: 182, Vipas Inc: 58, Bursa. (In Turkish)

Akkopru, E., C.O. Sabanci and M.M. Ertus. 2007. Effects of seeding rate and row spacing on yield and some yield characteristics of Hungarian vetches (*Vicia Pannonica*

Crantz.) Turkey 7th Field Crops Congress of Turkey. 25-27 June, Erzurum, Turkey, 235-238. (In Turkish with English abstracts)

Akman, H, A. Tamkoc and A. Topal. 2013. Effects on yield, yellow berry and black point disease of fertilization applications in Hungarian vetch and durum wheat intercropping system. Digital Proceeding of the ICOEST'2013, Cappadocia, June, 18-21, Nevsehir, Turkey. 839-847.

Aksoy, I. and H. Nursoy. 2010. Determination of the varying of vegetation harvested hungarian vetch and wheat mixture on nutrient content, degradation kinetics, in vitro digestibility and relative feed value. J. of Vet. Fac. of Kafkas Uni.16(6):925-931. (In Turkish with English)

Albayrak, S., M. Guler and M.O. Tongel. 2004. Effects of seed rates on forage production and hay quality of vetch-triticale mixtures. Asian J. of Plant Sci. 3(6):752-756.

Albayrak S, M. Turk and O. Yuksel. 2011. Effect of row spacing and seeding rate on Hungarian vetch yield and quality. Turk J Field Crops. 16(1):54-58.

Bagci, M. 2010. Effects of row spacing and seeding rate of Hungarian vetch *(Vicia pannonica* Crantz cv. Tarmbeyazı-98) on hay yield under central Anatolia conditions. MsC Thesis. Cukurova University, Adana, Turkey. (In Turkish with English abstracts)

Balabanli, C. and M. Turk. 2006. The effects of harvesting periods in some forage crops mixture on herbage yield and quality. J. of Bio. Sci. 6(2):265-268.

Balabanli, C. 2009. Hungarian Vetch (*Vicia pannonica* Crantz.). Forages. Legumes forages. Edited by Avcioglu, R., R. Hatipoglu and Y. Karadag. Republic of Turkey Ministry of Agriculture and Rural Affairs Press, Izmir, Turkey. 2:417-418 (In Turkish).

Basbag, M., I. Gul and V. Saruhan. 1999. The effect of different mixture ratios in some annual legume and cereal mixtures on yield and its components under Diyarbakır conditions. Turkish 3rd. Field Crops Congress (Fodder plant and edible grain legumes) 69-74; 15-20 November 1999. Adana

Baytekin, H., M. Kizilsimsek and G. Demiroglu. 2009. Ryegrasses. Annual Ryegrass (*Lolium multiflorum* Lam.). Forages. Grasses forages. Edited by Avcioglu, R., R. Hatipoglu and Y. Karadag. Republic of Turkey Ministry of Agriculture and Rural Affairs Press, Izmir, Turkey. 2:561-572 (In Turkish).

Bedir, S. 2010. Research on determination of proper seed mixture ratio of Hungarian vetch and barley to be grown under conditions of Karaman. MsC Thesis, Cukurova University, Adana.

Bingol, N.T., M.A. Karsli, I.H., Yilmaz and D. Bolat. 2007. The effects of planting time and combination on the nutrient composition and digestible dry matter yield of four mixtures of vetch varieties intercropped with barley. J. of Vet. Animal Sci. 31:297-302.

Budak, F., T. Tukel and R. Hatipoglu. 2011. Possibilities of growing vetch (*V. pannonica, V.villosa, V. dasycarpa,*) and cereal (barley, oat, triticale) mixtures in fallow fields in Eskişehir Conditions. The J. of Animal and Plant Sci. 21(4):724-729

Buyukburc, U. and Y. Karadag. 2002. The amount of NO_3-N transferred to soil by legumes, forage and seed yield, and the forage quality of annual legume+triticale mixtures. Turk. Agric. For. 26:281-288.

Buxton, D.R. 1996. Quality-related characteristics of forages as influenced by plant environment and agronomic factors. Animal Feed Sci. and Techn. 59(1-3):37-49.

Catanese, F., R.A. Distel and M. Arzadun. 2009. Preferences of lambs offered Italian ryegrass (*Lolium multiflorum* L.) and barley (*Hordeum vulgare* L.) herbage as choices. Grass and Forage Sci. 64:304-309.

Corre-Hellou, G., J. Fustec and Y. Crozat. 2006. Interspecific competition for soil and its interactions with N_2 fixations, leaf expansion and crop growth in pea-barley intercrops. Plant and Soil. 282:195-208.

Darvishi, A., S. Albayrak and S. Altinok. 2009. The morphological characters and forage yields of some annual ryegrass (*Lolium multiflorum* Lam.) Varieties. Turkey 8th Field Crops Congress of Turkey. Hatay, Turkey. 2:861-864. (in Turkish with English abstract)

De Bruin, J.L., and Pedersen, P., 2008. Effect of row spacing and seeding rate on soybean yield. Agron. J. 100:704-710.

Dhima, K.V., A.S. Lithourgidis, I.B. Vasilakoglou and C.A. Dordas. 2007. Competition indices of common vetch and cereal intercrops in two seeding ratio. Field Crops Res. 100:249-256.

Durst, L.V., B.J. Rude and S.H. Ward. 2013. Evaluation of Different Dietary Supplements for Cattle Consuming Annual Ryegrass Baleage. Department Report of the Animal and Dairy Sciences of MSU. 64-69.

Eskandari, H., A. Ghanbari and A. Javanmard. 2009. Intercropping of cereals and legumes for forage production. Notulae Botanicae Horti. Agrobotanici Cluj. 1(1):07-13.

Freed, R.D. 1991. MSTAT-C: A Microcomputer program for the design, Management and Analysis of Agronomic Research Experiments. East Lansing, MI, USA: Michigan State University.

Geren H., H. Soya and R. Avcioglu. 2003. Investigations on the effect of cutting dates on some quality properties of annual Italian ryegrass and hairy vetch mixtures. J. of Agric. Fac. of Ege. 40(2):17-24.

Gultekin, R. 2008. The effects of different forms doses of barnyard manure on seed and forage yield and quality of annual ryegrass (*Lolium multiflorum* Lam.) under Cukurova conditions. MsC Thesis. Cukurova University, Adana, Turkey. (In Turkish with English abstracts)

Gunduz, E.T. 2010. Effects of mixture ratio on hay yield and hay quality of Hungarian vetch+wheat mixture under Diyarbakir conditions. MsC Thesis, Cukurova University, Adana, Turkey. (In Turkish with English abstracts)

Hauggard-Nielsen, H., M. Gooding, P. Ambus, G. Corre-Hellou, Y. Crozat, C. Dahlmann, A. Dibet, P. Von Fragstein, A. Pristeri, M. Monti and E.S. Jensen. 2009. Pea–barley intercropping for efficient symbiotic N2-fixation, soil N acquisition and use of other nutrients in European organic cropping systems. Field Crops Res. 113:64-71.

Iptas, S. 2002. Effects of row spacing, support plant species and support plant mixture ratio on seed yield and yield characteristics of Hungarian vetch (*Vicia pannonica* Crantz). J. Agron. and Crop Sci. 188: 357-362.

Kacar, B. and A. Inal. 2008. Analyses of Plant. Nobel Press. ISBN 978-605-395-036-3, Ankara. (In Turkish)

Kara, I. 2013. Changing of dry matter yield and forage quality of common vetch Hungarian vetch and forage pea harvested in different dates. MsC Thesis, Ataturk University, Erzurum.

Karadag, Y. and U. Buyukburc. 2004. Forage qualities, forage yields and seed yields of some legume-triticale mixtures under rainfed conditions. Acta Agric.Scand., Sect. B, Soil And Plant Sci. 54: 140-148.

Karagic, D. A. Mikic, B. Milosevic, S. Vasiljevic and N. Dusanic, 2012. Common vetch-wheat intercropping: Haylage yield and quality depending on sowing rates. African J. of Biotech. 11:7637–7642.

Kesiktas, M. 2010. Effects of sowing time and nitrogen doses on forage yield of annual ryegrass (*Lolium multiflorum* westerwoldicum-Caramba) in Karaman. MsC Thesis. Cukurova University, Adana, Turkey. (In Turkish with English abstracts)

Kilavuz, D. 2006. Effects of different sowing dates on yield and yield components of some vetch + barley mixtures. MsC Thesis. Van Yuzuncu Yil University.

Kusvuran, A. 2011. The effects of different nitrogen doses on herbage and seed yields of annual ryegrass (*Lolium multiflorum* cv. caramba). African J. of Biotech. 10(60):12916-12924.

Kusvuran, A. and V. Tansi. 2005. The effects of various harvest densities and nitrogen doses on herbage and seed yield of annual ryegrass variety caramba (*Lolium multiflorum* cv. caramba) under the Cukurova Conditions. VI. Field Crops Congress of Turkey, September 5-9th, Antalya, Turkey, 2:797-802. (In Turkish with English abstracts)

Kusvuran, A. and V. Tansi. 2011. The effects of different row spacing on herbage and seed yields of annual ryegrass (*Lolium multiflorum* cv. caramba). Bulgarian J. of Agri. Sci. 17(6):744-754.

Koc, A., S. Erkovan, H.I. Erkovan, U. Oz, M.M. Birben and R. Tunc. 2013. Competitive effects of plant species under different sowing ratios in some annual cereal and legume mixtures. J. of Animal and Vet. Adv. 12(4):509-520

Lauk, R. and N. Lauk. 2009. Dual intercropping of common vetch and wheat or oats, effects on yields and interspecific competition. Agron. Res. 7:21–32.

Lithourgidis, A.S, I.B. Vasilakoglou, K.V. Dhima, C.A. Dordas and M.D. Yiakoulaki. 2006. Forage yield and quality of common vetch mixtures with oat and triticale in two seeding ratios. Field Crops Res. 99:106-113.

Lithourgidis, A.S., C.A. Dordas, C.A. Damalas and D.N. Vlachostergios. 2011. Annual intercrops: an alternative pathway for sustainable agriculture. Australian J. of Crop Sci., 5:396–410.

Lone, B.A., B. Hassan, S. Ansur-ul-haq and M.H. Khan. 2010. Effect of seed rate, row spacing and fertility levels on relative economics of soybean (*Glycine max* L.) under temperate conditions. African J. of Agric. Research. 5:322-324.

Mariotti, M., A. Masoni, L. Ercoli and I. Arduini. 2009. Above- and below-ground competition between barley, wheat, lupin and vetch in cereal and legume intercropping system. Grass Forage Sci. 64:401-412.

Mariotte, P., A. Buttler, D. Johnson, A. Thebault and C. Vandenberghe. 2012. Exclusion of root competition increases competitive abilities of subordinate plant species through root-shoot intecactions. J. Veget. Sci. 23:1148-1158.

Nadeem, M., M. Ansar, A. Anwar, A. Hussain and S. Khan. 2010. Performance of winter cereal-legumes fodder mixtures and their pure stand at different growth stages under rainfed conditions of pothowar. J. of Agri. Res. 48(2):181-192.

Nizam İ, A. Orak, I. Kamburoglu, M.G Cubuk, E. Moralar. 2007. Performance of Barley (*Hordeum vulgare* L.) and Hungarian vetch (*Vicia pannonica* Crantz.) mixtures ratios in different row spacings. 7th Field Crops Congress of Turkey. 25-27 June, 2007, Erzurum, Turkey, 114-118 (in Turkish with English abstract).

Orak, A. and M. Tuna. 1994. A study determining the effect of various row spacing and sowing rates on some yield and yield components of Hungarian vetch (*Vicia pannonica* Crantz). J. of Tekirdag Agric. Fac. 3(1/2):166-170.

Orak, A. and V. Uygun. 1996. Some important morphological characters and green fodder yield of berseem clover (*Trifolium alexandrinum* L.) and Italian ryegrass (*Lolium multiflorum* Lam.) mixtures which have different sowing norms row spacing and mixture rates. 3. Pasture and Forage Crops Meeting of Turkey. 17-19 June, 369-375. (in Turkish with English abstract)

Orak, A. and I. Nizam 2012. Determining the yields of some annual forage crop intercropping suitable for Tekirdag conditions. Research J. of Agri. Sci.-TABAD, 5(2):5-10. (In Turkish with English abstract)

Ozel, A. 2010. Determining the appropriate mixture rations in barley+Hungarian vetch and barley+common vetch mixtures. MsC Thesis. Mustafa Kemal University, Hatay, Turkey. (In Turkish with English abstracts)

Ozkul, H., F. Kirkpinar and K. Tan. 2012. Utilization of caramba (*Lolium multiflorum* cv. caramba) herbage in ruminant nutrition. Animal Production 53(1):21-26. (In Turkish with English abstract)

Parlak, A.O., F. Akgul and A. Gokkus 2007. The effects of different row spacing and nitrogen doses on herbage yield and quality of Annual Ryegrass (*Lolium multiflorum* cv. caramba) in Ankara conditions. 7th Field Crops Congress of Turkey. 25-27 June, Erzurum, Turkey, 139-142. (In Turkish with English abstract)

Pinar, I. 2007. The effect of different rates of mixture on the yield and some yield characteristics of hairy vetch (*Vicia villosa* Roth.)+barley (*Hordeum vulgare* L.) and hungarian vetch (*Vicia pannonica* Crantz.) + barley (*Hordeum vulgare* L.) mixtures. MsC Thesis. Ege University. Izmir, Turkey. (In Turkish with English abstracts)

Rivera, J.D., J.A. White, M.L. Gipson, R. Gipson and R.W. Lemus. 2013. Effects of various nitrogen fertilizer sources on biomass and forage quality of annual ryegrass. Dep. Report of the Animal and Dairy Sci. of MSU. 1:49-53

Rohweder, D.A., R.F. Barnes, and N. Jorgensen. 1978. Proposed hay grading standards based on laboratory analyses for evaluating quality. J. of Animal Sci. 47:747-759 http://jas.fass.org/cgi/reprint/47/3/747.

Shibles, R.M., and C.R. Weber. 1966. Interception of solar radiation and dry matter production by various soybean planting patterns. Crop Sci. 6:55-59.

Simic, A., S. Vuckovic, M. Kresovic, S. Vrbnicanin and D. Bozic. 2009. Changes of crude protein content in Italian ryegrass influenced by spring nitrogen application. Biotechn. in Animal Husbandry. 25(5-6):1171-1179.

Tas, N., A. Kara and Y. Serin. 2007. The effects of mixture rate and cutting time on hay quality in winter and spring sown vetch+wheat mixtures under rainfed conditions. African Crop Sci. Conf. Proceed. 8:173-177.

Tas, N. 2010. Determination of optimum mixture rate and cutting time for vetch+wheat mixtures sown in spring and autumn under irrigated conditions. II. Hay Quality. Anadolu J. of AARI 20 (2):59-69.

TSI. 2013. Turkish Statistical Institute, http://tuikapp.tuik.gov.tr/bitkiselapp/bitkisel.zul, (Accessed December 19, 2013).

TSMS. 2013. Turkish State Meteorological Service, Meteorological District Office of, Record of Climate Data, Adana.

Tuna, C. and A. Orak 2007. The role of intercropping on yield potential of common vetch (*Vicia sativa* L.)/Oat (*Avena sativa* L.) cultivated in pure stand and mixtures. J. of Agri. and Bio. Sci. 2(2):14-19.

Twidwell, E. K., K. D. Johnson, and J. H. Cherney. 1987. Forage potential of soft red winter wheat–hairy vetch mixtures. Appl. Agric. Res. 2. 164-169.

Uca, L., B. Çomaklı and M. Dasci. 2007. The effects of different row spacing and seeding rate on Hungarian vetch (*Vicia pannonica* Crantz.) and hairy vetch (*Vicia villosa* Roth.) of herbage and seed yields. Turkey 7th Field Crops Congress of Turkey. 25-27 June, Erzurum, Turkey, 390-394. (In Turkish with English abstracts)

Unal, S., Z. Mutlu and H.K. Firincioglu. 2011. Performances of some winter Hungarian vetch accessions (*Vicia pannonica* Crantz.) on the highlands of Turkey. Turk J Field Crops. 16(1):1-8.

Uzun A, U. Bilgili, M. Sincik and E. Acikgoz. 2004. Effects of seeding rates on yield and yield components of Hungarian vetch (*Vicia pannonica* Crantz.). Turkish J. of Agri. and Forestry. 28:179-182.

Van Niekerk, W.A., A. Hassen and R.J. Coertze. 2008. Diet quality, intake and growth performance of South African Mutton Merino sheep on *Triticum x Secale* and *Lolium multiflorum* pastures at different grazing pressures. Tropical Grasslands. 42:54-59.

Van Soest, P.J. 1963. The use of detergents in the analysis of fibre feeds. II. A rapid method for the determination of fibre and lignin. J. of the Assoc. of Official Analytical Chemists, 46:829-835.

Van Soest, P.J., and R.H. Wine. 1967. The use of detergents in the analysis of fibrous feeds. IV. Determination of plant cell wall constituents. J. of the Assoc. of Official Anal. Chemists, 50:50-55.

Vasilakoglou, I., K. Dhima, A. Lithourgidis and I. Eleftherohorinos. 2008. Competitive ability of winter cereal-common vetch intercrops against sterile oat. Expl Agric. 44:509–520.

Yolcu H, M. Polat and V. Aksakal. 2009. Morphologic, yield and quality parameters of some annual forages as sole crops and intercropping mixtures in dry conditions for livestock. J. of Food, Agri. and Envi. 7(3-4):594-599.

Yolcu, H., A. Gunes, M.K. Gullap and R. Cakmakci. 2012. Effects of plant growth-promoting rhizobacteria on some morphologic characteristics, yield and quality contents of Hungarian vetch. Turk J Field Crops, 17(2):208-214.

Zemenchik, R.A., K.A. Albrecht and R.D. Shaver. 2002. Improved nutritive value of kura clover- and birdsfoot trefoil-grass mixtures compared with grass monocultures. Agronomy Journal, 94(5): 1131-1138.

Zhang, Y., L.D. Bunting, L.C. Kappel, J.L. Hafley. 1995. Influence of nitrogen fertilization and defoliation frequency on nitrogen constituents and feeding value of annual ryegrass. J. Anim. Sci. 73: 2474-2482.

11

GENOTYPE–ENVIRONMENT INTERACTIONS AND STABILITY ANALYSIS FOR DRY-MATTER YIELD AND SEED YIELD IN HUNGARIAN VETCH (*Vicia pannonica* CRANTZ.)

Mehmet Salih SAYAR[1], Adem Emin ANLARSAL[2], Mehmet BASBAG[3]*

[1] *Crop and Animal Production Department, Bismil Vocational Training High School, Dicle University, Bismil, Diyarbakir, TURKEY*
[2]*Department of Field Crops, Faculty of Agriculture Cukurova University, Adana, TURKEY*
[3]*Department of Field Crops, Faculty of Agriculture, Dicle University, Diyarbakir, TURKEY*
[*]*Corresponding author: msalihsayar@hotmail.com*

ABSTRACT

This study was conducted to determine genotype–environment interactions and the stability status of twelve Hungarian vetch (*Vicia pannonica* Crantz.) genotypes in terms of dry-matter yield and seed yield under the ecological conditions of the Southeastern Anatolia Region of Turkey. The experiments were performed in five locations in the region during the 2008-2009 and 2009-2010 growing seasons. The experiments were performed according to a complete randomized block design with three replications. Genotype–environment interactions were found to be highly significant ($P < 0.01$) for dry-matter yield and seed yield, indicating that the Hungarian vetch genotypes' dry-matter yield and seed yield were significantly affected by the year and condition of the location. The stability of the genotypes was estimated using the mean yield of genotypes (x_i), regression coefficient (b_i), regression deviation mean square ($S^2 d_i$), determination coefficient (R^2), and regression line intercept (a). Stability analysis indicated that although the most stable genotype was the Ege Beyazi-79 cultivar in terms of dry-matter yield, the Oguz-2002 cultivar was the most stable in terms of seed yield.

Keywords: Dry-matter yield, genotype–environment interactions, Hungarian vetch (*Vicia pannonica* Crantz.), seed yield, stability parameters

[*] *Part of the dissertation Thesis submitted by the Corresponding Author to the Institute of the Natural and Applied Sciences of the Cukurova University in partial fulfillment of the requirements of the Ph.D degree in Field Crops.*

INTRODUCTION

Hungarian vetch (*Vicia pannonica* Crantz.) is adapted to the environments of large areas of the world (Magness et al., 1971). The species is one of the most promising annual vetch species, and its cultivation is especially recommended for places with harsh winter conditions (Acikgoz, 1988; Tahtacioglu et al., 1996; Nizam et al., 2011). Winter temperatures in the Southeastern Anatolia region can fall far below 0°C in some years. This situation makes it risky to cultivate forage crops species such as common vetch (*Vicia sativa* L.) and forage pea (*Pisum sativum* var. *arvense* L.), which are vulnerable to harsh winter conditions. For this reason, the cultivation of these species can only occur through spring sowings. In rainfed conditions, especially when there has been a dry spring, the dry-matter yield and seed yield of winter sowings are significantly higher than those of spring sowings for annual legume species. With the anticipated drought caused by global warming, Hungarian vetch is of great importance in this respect.

Yield stability is an interesting feature of today's plant breeding programs, owing to the high annual variation in mean yield, especially in the arid and semi-arid areas (Mohammadi et al., 2012). Producers are most interested in a cultivar that gives consistent yields under different growing conditions; thus, plant breeders usually try a series of genotypes in multi-environments, before a new improved variety is released for production to farmers (Naghavi et al., 2010). Genotype–environment interactions (GEI) can be defined as the response of genotypes to different environments. Genotype–environment interactions are extremely important in the development and evaluation of plant varieties because they can reduce genotypic stability values in diverse environments (Hebert et al., 1995).

An understanding of genotype–environment interactions requires information on the existence and magnitude of the response of individual lines to their environments, but awareness of such interactions provides no quantitative measurements that indicate the stability of individual lines (Abd El- Moneim and Cocks, 1993; Bozoglu and Gulumser, 2000). Recently, interest has focused on regression analysis, an approach originally proposed by Yates and Cochran (1938) and later modified by Finlay and Wilkinson (1963), where stability as a linear relationship between the yield of genotypes over many environments is given by the regression coefficient (b_i), and a genotype with $b_i = 1$ can be considered stable. Eberhart and Russell (1966) further developed the idea by implementing the regression deviation mean square (S^2d_i) as a measure of stability. Genotypes with low (close to 0) deviation from the regression (S^2d_i) value and high (above average) mean efficiency are regarded as stable. Pinthus (1973) presented the coefficient of determination (R_i^2) as the quantity of variation explained by the regression as a portion of the total variation. A high coefficient of determination (R_i^2) (Pinthus, 1973; Teich, 1983) and positive high regression line intercept (a) (Smith, 1982) are also desired criteria in terms of genotypic stability.

A number of stability studies have previously been carried out on different crops in Turkey (Sabanci, 1996; Albayrak et al., 2005; Akcura et al., 2005; Acikgoz et al., 2009; Yucel et al., 2009, Nizam et al., 2011). However, no stability study has been performed for Hungarian vetch in the Southeastern Anatolia region. The objectives of this study were to (1) evaluate the dry-matter yield and seed yield capacity of Hungarian vetch genotypes (G) in different environments (E); (2) identify and assess the G × E interactions; and (3) determine the stability of these interactions using different stability parameters.

MATERIALS AND METHODS
Materials

Twelve Hungarian vetch (*Vicia pannonica* Crantz.) genotypes, including six commercial cultivars and six promising lines, were used as the genetic materials in this study. Five of the lines used, Line-3, Line-10, Line-15, Line-18, Line-55, were supplied from the Eastern Anatolia Agricultural Research Institute, Erzurum, Turkey. The remaining Line 2109 was selected from a breeding program performed in GAP International Agricultural Research and Training Center (GAP IARTC) Diyarbakir' Turkey. In addition to these lines, other cultivars used were Tarm Beyazi-98, Budak, Anadolu Pembesi-2002, Ege Beyazi-79, Oguz-2002 and Beta.

The cultivars were supplied by their breeders' institutions. Accordingly, Tarm Beyazi-98, Anadolu Pembesi-2002, Oguz-2002 were provided by the Central Research Institute for Field Crops, Ankara, Turkey; Ege Beyazi-79 was supplied by the Aegean Agricultural Research Institute, İzmir, Turkey; the Budak cultivar was obtained from the Transitional Zone Agricultural Research Institute, Eskisehir, Turkey; and Beta, which originated in Hungary, was supplied by Serta Agriculture Production Import Export Trade Limited Company, Ankara, Turkey.

The locations where experiments were conducted are given in Table 1. The Southeastern Anatolia region is one of Turkey's seven census-defined geographical regions' and the region is characterized by a continental climate. In this region, summers are dry and hot, whereas winters are cool and rainy. The experiments were conducted under rainfed conditions at five locations having different climate and soil characteristics during two consecutive growing seasons (2008–2009 and 2009–2010) in the Southeastern Anatolia region of Turkey.

Table 1. The environments and some climatic and agronomic information of the locations.

Code	Growing seasons	Locations	Altitude (m)	Soil properties	Sowing date	The average temperature (°C)	Total rainfall (mm)
E1	2008-2009	Diyarbakir	603	pH=7.86 clay-silt	14.11.2008	12.4	455.0
E2	2009-2010	Diyarbakir	607	pH=7.85 clay-silt	20.11.2009	14.3	517.9
E3	2008-2009	Cınar	701	pH=7.84 clay-silt	17.11.2008	12.9	366.3
E4	2009-2010	Cınar	675	pH=7.85 clay-silt	24.11.2009	15.0	417.0
E5	2008-2009	Ergani	995	pH=7.76 clay-silt	07.11.2008	13.8	768.8
E6	2009-2010	Ergani	936	pH=7.77 clay-silt	19.11.2009	14.6	963.6
E7	2008-2009	Cungus	970	pH=7.78 sandy-silt	07.11.2008	9.7	725.0
E8	2009-2010	Cungus	915	pH=7.79 sandy-silt	19.11.2009	11.2	825.2
E9	2008-2009	Hazro	815	pH=7.65 clay-silt	06.11.2008	11.9	927.4
E10	2009-2010	Hazro	808	pH=7.64 clay-silt	17.11.2009	13.8	1055.6

*Data from the Regional Directorate of Meteorology, Diyarbakir, Turkey.

The experiments were conducted according to a randomized complete block design with three replications. Each plot consisted of six rows 5 m in length, and rows were spaced 20 cm apart. The seeding rates were 220 seeds m^{-2} (Munzur et al., 1992). Seeds were sown using an experimental drill. The environment, geographical

coordinates of location, growing season, soil properties, rainfall, temperature, and sowing dates at each location

during the growth periods are summarized in Table 1. In the experiments, half of each plot was harvested in May to calculate dry-matter yield, and the other half was harvested in June to calculate seed yield. Dry-matter yield and seed yield were determined according to the technical instructions for leguminous forage crops published by the Seed Registration and Certification Centre, Ankara, Turkey, in 2001.

Statistical analysis and procedures

We computed the combined analysis of variance on phenotypic data from trials in 10 environments (Comstock, Moll, 1963). The genotypic responses to environmental changes were assessed using a linear regression coefficient (b_i) and the variance of the regression deviations (S^2d_i) using the following formulas proposed by Finlay and Wilkinson (1963) and Eberhart and Russell (1966).

$$b_i = 1 - \frac{\sum_i (Xij - \overline{Xi}. - \overline{X}.j - \overline{X}..)(\overline{X}.j - \overline{X}..)}{\sum_j (\overline{X}.j - \overline{X}..)^2}$$

$$S^2_{di} = \frac{1}{E-2}\left[\sum_i (Xij - \overline{Xi}. - \overline{X}.j - \overline{X}..) - (b_i - 1)^2 \sum_i (\overline{X}.j - \overline{X}..)^2\right],$$

where Xij is the dry-matter or seed yield of genotype i in environment j, $\overline{Xi}$ is the mean yield of genotype i, $\overline{X}.j$ is the mean yield of environment j, $\overline{X}.$ is the grand mean, and E is the number of environments. The coefficient of determination (R_i^2) (Pinthus, 1973) was computed from individual linear regression analyses. Also, the regression line intercept (*a*) was evaluated as a stability parameter (Eberhart, Russell, 1966), and the significance of the regression coefficient (the yield of a single genotype on the mean environment), and the grand means of dry-matter and seed yields, were tested by employing the *t*-test (Steel, Torrie, 1960). The confidence intervals were estimated based on the formula given below:

Confidence interval $= \overline{X} \pm t\text{- value} \times s_x.$

For dry-matter yield and seed yield, regression curves of twelve Hungarian vetch genotypes were developed using the equation y = b_i × x + *a* by making use of an environmental index.

All statistical analyses were performed using the MSTAT–C statistical computer package software program, version 3.00/EM (Freed et al., 1989). The means were compared using a Duncan test at a 0.05 probability level. The grand mean, regression coefficient, and their confidence intervals were taken into account when the stability status of the genotypes was evaluated over nine different environments (Figure 1).

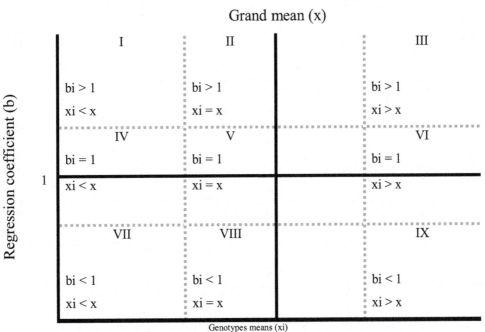

Grand mean (x)

I = Poor adaptability to favorable environmental conditions
II = Average adaptability to favorable environmental conditions
III = Better adaptability to favorable environmental conditions
IV = Poor adaptability to all environmental conditions
V = Average adaptability to all environmental conditions
VI = Better adaptability to all environmental conditions
VII = Poor adaptability to unfavorable environmental conditions
VIII = Average adaptability to unfavorable environmental conditions
IX = Better adaptability to unfavorable environmental conditions

Figure 1. The mathematical explanation of stability environments

RESULTS

In this study, twelve Hungarian vetch genotypes were studied in five different locations for two years. The variation among environments in both dry-matter yield and seed yield was significant ($P < 0.05$) (Table 2). Mean dry-matter yield varied from 3.973 t ha^{-1} in environment 7 to 7.804 t ha^{-1} in environment 1. Seed yield ranged from 0.653 t ha^{-1} in environment 3 to 1.104 t ha^{-1} in environment 9 (Table 2).

Table 2. Mean, min. and max. yields of dry matter and seed yields in the environments.

Environments	Growing seasons	Locations	Dry matter yield (t h^{-1})			Seed yield (t h^{-1})		
			Mean *	Min.	Max.	Mean*	Min.	Max.
E1	2008-2009	Diyarbakir	7.804 a	7.053	8.710	1.016 b	0.780	1.500
E2	2009-2010	Diyarbakir	5.458 cd	4.517	6.130	0.851 de	0.662	1.350
E3	2008-2009	Cınar	6.955 b	6.033	7.710	0.653 h	0.463	1.030
E4	2009-2010	Cınar	5.778 c	4.810	6.910	0.719 g	0.547	0.869
E5	2008-2009	Ergani	5.715 c	4.803	7.653	0.816 ef	0.498	1.080
E6	2009-2010	Ergani	5.061 e	4.200	6.887	0.787 f	0.614	0.962
E7	2008-2009	Cungus	3.973 f	2.423	6.843	0.878 d	0.801	1.107
E8	2009-2010	Cungus	5.716 c	4.633	7.303	0.939 c	0.738	1.085
E9	2008-2009	Hazro	7.226 b	5.823	8.543	1.104 a	0.965	1.280
E10	2009-2010	Hazro	5.211 de	3.477	7.577	0.866 de	0.648	1.260

*Means followed by different letters within a column insignificant differences at the level of $P < 0.05$ for Duncan Range Test

The mean dry-matter yield and seed yield of the twelve Hungarian vetch genotypes ranged respectively from 5.315 t ha^{-1} to 6.999 t ha^{-1}, and 0.748 t ha^{-1} to 1.113 t ha^{-1}. The highest dry-matter yield and seed yield were obtained from Anadolu Pembesi-2002 (3) and Oguz -2002 (9) cultivars, respectively (Table 4).

The results from variance analysis for dry-matter yield and seed yield are shown in Table 3. For dry-matter yield, years, locations, year–location interaction, replications, genotypes, location–genotype interaction, and year–location–genotype interaction were highly significant ($P < 0.01$). Also, the year–genotype interaction was found to be significant ($P < 0.05$). On the other hand, except for the replications, which were not significant ($P > 0.05$), all other interactions were found to be highly significant ($P < 0.01$) for seed yield (Table 3). For both dry-matter yield and seed yield, the second-order interactions (genotype × year × location) was highly significant ($P < 0.01$). This indicates that each location in each year could be treated as a separate environment for the both traits.

Table 3. Analysis of variance for dry matter yield and seed yield in Hungarian vetch genotypes.

Source of variation	df	Dry matter yield			Seed yield		
		Sum of square	Mean square	F value	Sum of square	Mean square	F value
Years (Y)	1	71.262	71.262	128.636**	0.337	0.337	30.7272**
Locations (L)	4	160.502	40.125	72.4308**	4.109	1.027	93.6301**
Y × L	4	188.236	47.059	84.9467**	1.342	0.335	30.5818**
Replications	20	22.283	1.114	2.0111**	0.193	0.010	0.8782ns
Genotypes (G)	11	97.270	8.843	15.9621**	4.499	0.409	37.2859**
Y × G	11	13.688	1.244	2.2463*	0.282	0.026	2.3335**
L × G	44	91.311	2.075	3.746**	2.102	0.048	4.3545**
L × G × Y	44	67.686	1.538	2.7768**	1.343	0.031	2.7831**
Error	220	121.876	0.554		2.413	0.011	
General	359	834.115			16.620		

*: $P < 0.05$ at significance; **: $P < 0.01$ at significance; ns: not significant.

Dry-matter yield

An analysis of variance revealed that genotype–environment interactions were highly statistically significant ($P < 0.01$) for dry-matter yield (Table 3), and regression coefficients ranged from 0.283 to 1.325 for dry-matter yield (Table 4). This large variation in regression coefficients reflects the different responses of different genotypes to environmental changes. With respect to dry-matter yield, the Tarm Beyazi-98 and Budak cultivars showed average adaptability to favorable environmental conditions ($b_i > 1$ and $xi = x$). The varieties that obtained the highest dry-matter yield were Anadolu Pembesi-2002 and Oguz-2002. Due to their small b_i values, they were accepted as having better adaptability to unfavorable environmental conditions ($b_i < 1$ and $xi > x$). These cultivars were relatively better adapted to poor environments and were insensitive to environmental changes. Therefore, the cultivation of such cultivars under unfavorable conditions can be recommended with respect to their dry-matter yield (Table 4, Figure 2 and Figure 3).

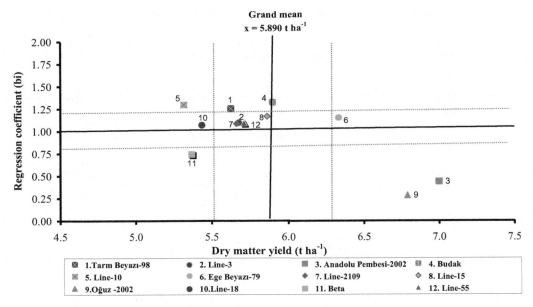

Figure 2. The relationship between the regression coefficients and mean dry matter yield (t ha^{-1}) for twelve Hungarian vetch genotypes.

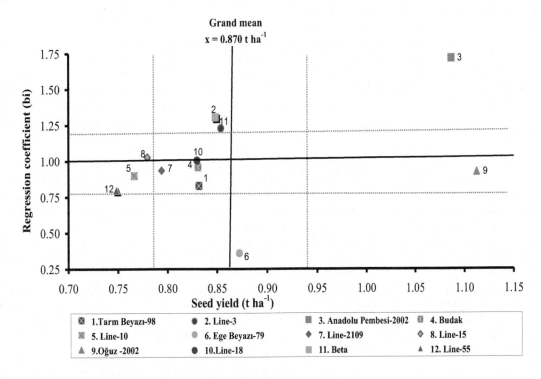

Figure 3. The relationship between the regression coefficients and mean seed yield (t ha^{-1}) for twelve Hungarian vetch genotypes.

Table 4. Stability parameters of Hungarian vetch genotypes for dry matter yield and seed yield

Code	Genotypes	Dry matter yield					Seed yield				
		x_i (t ha^{-1})	b_i	S^2d_i	a	R_i^2	x_i (t ha^{-1})	b_i	S^2d_i	a	R_i^2
1	Tarm Beyazi-98	5.625	1.256*	41.233**	-1.771	0.848	0.832	0.830	0.554*	0.115	0.714
2	Line-3	5.678	1.096	28.550**	-0.778	0.860	0.854	1.23*	1.461**	-0.207	0.675
3	Anadolu Pembesi	6.999**	0.436**	63.469**	4.433	0.304	1.0866**	1.717**	3.397**	-0.395	0.636
4	Budak	5.898	1.325**	100.871**	-1.907	0.717	0.831	0.959	0.803**	0.003	0.697
5	Line-10	5.315**	1.298*	18.396*	-2.327	0.930	0.7663*	0.900	0.518	-0.010	0.758
6	Ege Beyazi-79	6.33*	1.147	20.162*	-0.424	0.905	0.872	0.360**	0.332	0.561	0.440
7	Line-2109	5.663	1.085	51.978*	-0.730	0.768	0.794	0.936	1.721**	-0.014	0.506
8	Line-15	5.863	1.168	25.286**	-1.017	0.887	0.779*	1.029	0.426	-0.109	0.833
9	Oguz -2002	6.790**	0.283**	67.596**	5.122	0.148	1.113**	0.927	1.491**	0.313	0.537
10	Line-18	5.436*	1.070	27.111**	-0.864	0.860	0.830	1.006	0.877**	-0.038	0.699
11	Beta	5.368**	0.743*	61.971**	0.993	0.565	0.848	1.306**	1.063**	-0.278	0.763
12	Line-55	5.712	1.094	32.321**	-0.730	0.844	0.748**	0.801	0.552*	0.058	0.700
Average		5.890	1.000				0.837	1.000			
Confidence limits (0.05)		± 0.349	± 0.213	± 15.784			± 0.074	± 0.206	± 0.544		
Confidence limits (0.01)		± 0.486	± 0.300	± 22.274			± 0.105	± 0.209	± 0.768		

* Significant difference at $P < 0.05$; ** Significant difference at $P < 0.01$.
(x_i): The yield mean, (b_i): Regression coefficient, (S^2d_i): Regression deviation mean square, (a): Regression line intercept, R_i^2: Coefficient of determination

With its high dry-matter yield and a regression coefficient that did not differ significantly from 1.0, the Ege Beyazi-79 cultivar showed better adaptability to all environmental conditions. For dry-matter yield, four lines (Line-3, Line-18, Line-55, and Line-2109) showed average adaptability to all environmental conditions (the regression coefficients did not differ significantly from 1.0), with yields nearly equal to or higher than the grand mean. These varieties along with Ege Beyazi-79 can be considered as the most widely adaptable and stable lines in terms of dry-matter yield for the Southeastern Anatolia region (Table 4, Figure 2).

Seed yield

The genotype–environment interaction was highly significant ($P < 0.01$) for seed yield (Table 3). In this study, the regression coefficient for seed yield ranged from 0.36 to 1.71. With respect to seed yield, despite the fact that Beta and Line-3 showed average adaptability to favorable environmental conditions ($b_i > 1$ and $x_i = x$), Tarm Beyazi-98, Budak, Line-18, and Line-2109 genotypes showed average adaptability to all environmental conditions ($b_i = 1$, $x_i = x$). Although the Anadolu Pembesi-2002 cultivar showed better adaptability to favorable environmental conditions ($b_i > 1$; $x_i > x$), Ege Beyazi-79 showed average adaptability to unfavorable environmental conditions ($b_i < 1$; $x_i = x$). Also, Oguz-2002 showed better adaptability to all environmental conditions ($b_i = 1$; $x_i > x$).

The coefficients of determination (Pinthus, 1973) ranged from 0.148 to 0.930 and from 0. 440 to 0.833 for dry-matter yield and seed yield, respectively. In terms of dry-matter yield, the highest R_i^2 value was found for Line-10; with regard to seed yield, the highest value R_i^2 was identified in Line-15. However, the lowest R_i^2 values were recorded for Oguz-2002 and Ege Beyazi-79 for dry-matter yield and seed yield, respectively (Table 4).

Regression line intercept (a) values ranged from -2.327 to 5.122 and from -0.395 to 0.561 for dry-matter yield and seed yield, respectively. The highest intercept value was recorded in Oguz-2002 for dry-matter yield. Also, Ege Beyazi-79 had the highest intercept value for seed yield. In contrast, Line-10 and Anadolu Pembesi-2002 were found to have the lowest line intercept values for dry-matter yield and seed yield, respectively (Table 4).

DISCUSSION

Genotype–environment interactions were found to be highly significant not only for dry-matter yield but also for seed yield (Table 3). Similarly, Yucel et al. (2009) and Nizam et al. (2011) found significant genotype–environment interactions in some vetch species in terms of dry-matter yield. Also, many researchers have found genotype–environment interactions to be significant for seed yield in different forage crops (Sabanci, 1996; Albayrak et al., 2005; Acikgoz et al., 2009; Nizam et al., 2011). This indicates that these traits differed between locations and planting years. Several researchers stated that genotype–location and genotype–location–year interactions were more important than genotype–year interaction (Akcura et al., 2005 and Ezzat et al., 2010). Becker and Leon (1988) also indicated that the assessment of stability in many locations and years could increase the reliability of both important traits. Here, the mean squares indicated that the effect of location was more important than that of year for all traits (Table 3), and similar results were reported by Ezzat et al. (2010).

In stability analysis, genotypes with high mean yield, a regression coefficient equal to unity ($b_i = 1$), and a small regression deviation mean square ($S^2d_i = 0$) are considered stable (Finlay and Wilkinson 1963; Eberhart and Russell 1966). Additionally, a higher R_i^2 value (Pinthus, 1973) and higher regression line intercept value (Smith, 1982) indicate a reliable stability.

In this study, regression coefficient values for Tarm Beyazi-98, Line-10 and Budak genotypes for dry-matter yield were significantly above unity ($b_i > 1$); also, seed yields in Line-3, Beta, and Anadolu Pembesi-2002 had high regression coefficient values, significantly above unity ($b_i > 1$) (Table 4, Figure 2, Figure 3). Accordingly, these genotypes can be said to be sensitive to environmental change and to have greater specificity of adaptability to high-yield environments (Wachira et al., 2002; Kılıc and Yagbasanlar, 2010). On the other hand, the b_i values of Oguz-2002, Anadolu Pembesi-2002, and Beta cultivars for dry-matter yield and those of Ege Beyazi-79 for seed yield were both significantly below unity ($bi < 1$) (Table 4, Figure 2, Figure 3). Therefore, these genotypes can be considered as having greater resistance to environmental change increased specificity of adaptability to low-yield environments (Wachira et al., 2002; Kılıc and Yagbasanlar, 2010).

The regression deviation mean square (S^2d_i) values of Ege Beyazi-79, Line-15, and Line-10 were smaller than the other genotypes and were not significantly different from zero in terms of seed yield (Table 4). Therefore, these genotypes are able to conserve seed-yield traits in differing environments (Eberhart and Russell, 1966).

However, no genotype had a S^2d_i value significantly different from zero in terms of dry-matter yield (Table 4), although, the S^2d_i values of Line-10 and Ege Beyazi-79 were closer to zero than were those of the other genotypes in terms of dry-matter yield (Table 4).

The coefficient of determination is often considered better for measuring the validity of the linear regression than is S^2d_i because its value ranges between zero and one. A greater R_i^2 value is desired because higher R_i^2 values indicate favorable responses to environmental changes. In the present study, Line-15, Ege Beyazi-79, and Line-10 genotypes had higher R_i^2 values for dry-matter yield. On the other hand, Line-10, Beta, and Line-15 genotypes had higher R_i^2 values for seed yield compared with the other genotypes (Table 4). This indicates that when environmental conditions improve, these genotypes will produce more dry-matter yield and seed yield than will those with lower R_i^2 values.

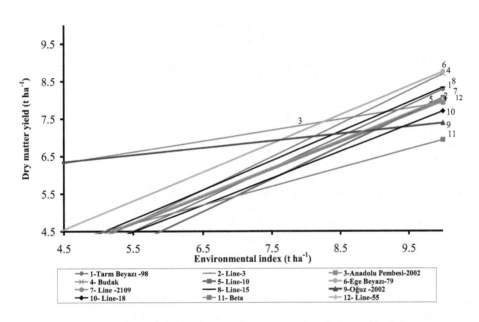

Figure 4. Comparison of Hungarian vetch genotypes by their expected dry matter yield estimated from their regression (stability) equations.

For stability, a positive and higher regression line intercept (a) is desired (Teich, 1983; Ozcan et al., 2005; Kılıc and Yagbasanlar, 2010). In this study, Oguz-2002, Anadolu Pembesi-2002, and Beta cultivars had positive and higher regression line intercepts (a) in terms of dry-matter yield when compared with the other genotypes. Furthermore, Ege Beyazi-79, Oguz-2002, and Tarm Beyazi-98 displayed positive and higher regression line intercepts (a) for seed yield. Genotypes with positive and higher regression line intercepts (a) give above-average dry-matter yield and seed yield under unfavorable environmental conditions, and these genotypes are well adapted to unfavorable environmental conditions. In contrast, Line-10, Budak, and Tarm Beyazi-98 had negative regression line intercepts (a) for dry-matter yield, and Anadolu Pembesi-2002, Beta, and Line-3 genotypes

had negative regression line intercepts (a) values for seed yield. These genotypes give below-average dry-matter yield and seed yield under poor environmental conditions (Table 4, Figs 2, 3, 4 and 5).

Ege Beyazi-79 was the most stable genotype in terms of dry-matter yield. Its regression coefficient was near unity, and it had relatively low S^2d_i and high R_i^2 (90%), thus confirming its stability. However, its low line intercept (a) value indicated that this cultivar has low dry-matter yield potential under unfavorable environmental conditions. Similarly, Nizam et al. (2011) reported that the Ege Beyazi-79 variety could be considered widely adapted for conditions in the Thrace region with a b_i value for dry-matter yield equal to 1 and a low S^2d_i value. On the other hand, b_i values for Anadolu Pembesi-2002 and Oguz-2002 cultivars were below 1, and they had high S^2d_i

values combined with low R_i^2 values; therefore, these genotypes were found to be unstable in terms of dry-matter yield, but their high regression line intercepts indicated that these cultivars have high dry-matter yield potential under unfavorable environmental conditions (Table 4, Figs 2, 3, 4 and 5).

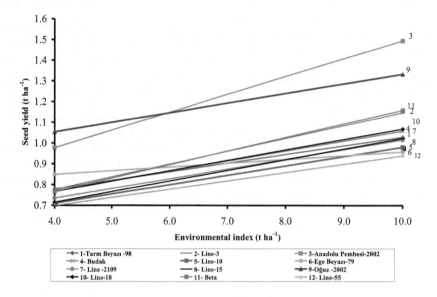

Figure 5. Comparison of Hungarian vetch genotypes by their expected seed yield estimated from their regression (stability) equations.

The highest seed yields were obtained for Anadolu Pembesi-2002 and Oguz-2002 cultivars. However, due to the high regression coefficient, Anadolu Pembesi-2002 is less likely to repeat this feature when compared with Oguz-2002. Therefore, Anadolu Pembesi-2002 showed better adaptability to favorable environmental conditions. On the other hand, with its high mean seed yield, which was significantly higher than the grand mean yield, and with a regression coefficient value not significantly different from 1.0, Oguz-2002 showed better adaptability to all environmental conditions. Thus, Oguz-2002 can be considered the most widely adaptable and stable variety in terms of seed yield for the Southeastern Anatolia region. In fact, one of the most interesting results was that although Line-15 performed well in terms of almost all of the stability parameters (b_i, S^2d_i, R_i^2), is not recommended for seed cultivation due to its lower mean seed yield when compared with the grand mean seed yield (Table 4 and Figure 3).

CONCLUSION

This study was carried out during the 2008–2009 and 2009–2010 growing seasons in five different locations. Genotypes–environment interactions were investigated, and the interactions were found to be highly significant ($P \leq 0.01$) for both dry-matter yield and seed yield. The stability analysis in this study showed that among the twelve genotypes, the Ege Beyazi-79 cultivar was found to be the most stable for dry-matter yield, and the Oguz-2002 cultivar was found to be the most stable genotype for seed yield.

ACKNOWLEDGEMENTS

This research was supported by the Directorate of GAP International Agricultural Research and Training Center (GAP IARTC), Diyarbakir, and Cukurova University Academic Research and Project Units. The authors wish to thank all of them for their support.

LITERATURE CITED

Abd El-Moneim, AM., P.S. Cocks. 1993. Adaptation and yield stability of selected lines of *Lathyrus* spp. under rainfed conditions in West Asia. Euphytica 66: 89-97.

Acikgoz, E. 1988. Annual forage legumes in the arid and semi-arid regions of Turkey. In: D. B. Beck, and L. A. Materori (eds), Nitrogen fixation by legumes in Mediterranean agriculture, ICARDA, Aleppo, pp. 47-54.

Acikgoz, E, A. Ustun, İ. Gul, A.E. Anlarsal, A.S. Tekeli, İ. Nizam, R. Avcioglu, H. Geren S. Cakmakcı, B. Aydinoglu, et al. 2009. Genotype × environment interaction and stability analysis for dry matter and seed yield in field pea (*Pisum sativum* L.). Span J Agric Res. 7(1): 96-106.

Akcura, M., Y. Kaya, S. Taner. 2005. Genotype-environment interaction and phenotypic stability analysis for grain yield of durum wheat in Central Anatolian Region. Turk J Agric For. 29: 369-375.

Albayrak, S., O. Tongel, M. Guler. 2005. stability analysis and determination of seed yield and yield componets of candidate vetch (*Vicia sativa* L.) varieties in Middle Black Sea Region. The Journal of Agricultural Faculty of Ondokuz Mayıs University. 20 (1):50-55.

Becker, H.C., J. Leon. 1988. Stability analysis in plant breeding. Plant Breeding. 101: 1-23.

Bozoglu, H., A. Gulumser. 2000. Determination of genotype x environment interactions of some agronomic characters in dry bean (*Phaseolus vulgaris* L.). Turk J Agric For. 24: 211-220.

Comstock, R.E., R. Moll. 1963. Genotype × environment interactions in Hanson, W.D. and Robinson, H.F. (eds). Statistical Genetics and Plant Breeding. NASNRC Pub., Washington, DC, pp. 164-196.

Eberhart, S.A., W.A. Russell. 1966 Stability parameters for comparing varieties. Crop Sci 6: 36-40.

Ezzat, E.M., M.A. Ali, A.M. Mahmoud. 2010. Agronomic performance, genotype x environment interactions and stability analysis of grain sorghum (*Sorghum bicolor* L. Moench) Asian J Crop Sci. 2 (4): 250-260.

Finlay, K.W., G.N. Wilkinson. 1963 The analysis of adaptation in plant-breeding programme. Aust J Agric Res. 14: 742-754.

Freed, R., S.P. Einensmith, S. Guetz, D Reicosky, V.W. Smail, P. Wolberg. 1989. User's Guide to MSTAT-C analysis of agronomic research experiments, Michigan State Uni. USA.

Hebert, Y., C. Plomion, N. Harzic. 1995. Genotypic × environment interaction for root traits in maize as analysed with factorial regression models. Euphytica. 81: 85-92.

Kılıc, H., T. Yagbasanlar. 2010. Genotype x environment interaction and phenotypic stability analysis for grain yield and several quality traits of durum wheat in the South-Eastern Anatolia Region. Not. Bot. Hort. Agrobot. Cluj. 38 (3) 2010, 253-258.

Magness, J.R., G.M. Markle, C.C. Compton. 1971. Food and feed crops of the United States. International Research Project IR-4, IR Bul. 1 Bul. 828 New Jersey Agric. Exp. Stn. International Research Project, NJ, USA.

Mohammadi, M., R. Karimizadeh, N, Sabaghnia, M.K. Shefazadeh. 2012. genotype × environment interaction and yield stability analysis of new improved bread wheat genotypes. Turk. J. Field Crops. 17(1): 67-73.

Munzur, M., A. Tan, H. Kabakcı. 1992. Effect of different seeding rates on forage and seed yields of some annual forage legumes species. 1991/1992 annual study reports, the Central Research Institute for Field Crops, Ankara, Turkey.

Naghavi, A., O. Sofalian, A. Asghari, M. Sedghi. 2010. Relation between freezing tolerance and seed storageproteins in winter bread wheat (*Triticum aestivum* L.). Turk. J. Field Crops. 15:154–158.

Nizam, İ., M.G. Cubuk, E. Moralar. 2011. Genotype × environment interaction and stability analysis of some Hungarian vetch (*Vicia pannonica* Crantz.) genotypes. African J Agric Res (AJAR). 6(28): 6119-6125.

Ozcan, H., N. Aydin, H.O. Bayramoglu. 2005. Yield stability and correlation among the stability parameters in wheat. Journal of Agric. Sci. 11 (1): 21-25.

Pinthus, M.J. 1973. Estimates of genotypic value a proposed method. Euphytica 22: 345-351.

Sabanci, C.O. 1996. Genotype x environment interactions for seed yield in common vetch (*Vicia sativa* L.). Anadolu J. AARI. 6 (1): 25-31.

Smith, E.L. 1982. Heat and drought tolerant wheats of the future. In: Proc. Natl. Wheat Res Conf Betswille M.D. 26-28 Oct. National Association of Wheat Growers Foundation Washington, DC.

Steel, G.D., J.H. Torrie. 1960. Principles and procedures of statistics with special reference to biological sciences. New York; McGraw-Hill.

Tahtacioglu, L., M. Avci, A. Mermer, H. Seker, C. Aygun. 1996 Adaptation of some winter vetch varieties under ecological conditions of Erzurum. Turkey III. Grassland and Forage Crops Congress, 17-19 June 1996, Erzurum, pp: 661-667.

Teich, A.H. 1983. Genotype × environment interaction variances in yield of winter wheat. Cereal Res Commun. 11: 15-20.

Yates F., W.G. 1938. The analysis of groups of experiments. J Agric Sci Camb. 28: 556-580.

Yucel C., H. Hızlı, H.K. Fırıncıoglu, A. Cil, A.E. Anlarsal. 2009. Forage yield stability of common vetch (*Vicia sativa* L.) genotypes in the Cukurova and GAP Regions of Turkey. Turk J Agric For 33: 119-125.

Wachira F., N.G. Wilson, J. Omolo, G. Mamati. 2002. Genotype × environment interactions for tea yields. Euphytica. 127: 289-296.

12

LINE × TESTER ANALYSIS AND ESTIMATING COMBINING ABILITIES FOR YIELD AND SOME YIELD COMPONENTS IN BREAD WHEAT

Deniz ISTIPLILER*, Emre ILKER, Fatma AYKUT TONK, Gizem CIVI, Muzaffer TOSUN

Ege University, Faculty of Agriculture, Department of Field Crops, İzmir, TURKEY
*Corresponding author: deniz.istipliler@ege.edu.tr

ABSTRACT

Estimating of combining ability is useful to assess genotypes and elucidate the nature and magnitude of gene actions involved. Aim of this study was to determine the combining abilities of some wheat genotypes for yield and some yield related traits by using line × tester mating design. Four advanced wheat lines and one cultivar were used as lines and four wheat genotypes were used as testers. The specific combining ability (SCA) effects were generally found higher than general combining ability effects (GCA) in terms of the agronomic traits studied. As a result, low ratios of $\sigma^2_{GCA}/\sigma^2_{SCA}$, $(\sigma^2_D/\sigma^2_A)^{1/2}$ and low narrow sense heritabilities showed that non-additive effects controlled the traits studied. Hence, the selection process for superior individual plants should be postponed to further generations like F_4 or F_5.

Keywords: Bread Wheat, General combining ability, Line × Tester analysis, Specific combining ability, *Triticum aestivum* L.

INTRODUCTION

Bread wheat (*Triticum aestivum* L.) is the most important cereal crop in the world and plays a crucial role in human and animal nutrition. It is grown more land area worldwide than other field crops and unrivalled in its range of cultivation. In Turkey, it was cultivated on approximately 8 million hectares and 18 and 20 million tons total production is changing every year. Besides, the population in Turkey is increasing day by day, so improving the highly productive wheat genotypes is substantial. This can be achieved by using maximum genetic potential in wheat germplasm.

The knowledge of combining ability is useful to assess differences among the genotypes and also, elucidate the nature and magnitude of gene actions involved. It has an important role to select parents and crosses and it helps to decide breeding methods to be followed to choose desirable individuals (Salgotraet al., 2009).

Plant breeders focus on development of high yielding wheat cultivars by crossing good general combining lines and selecting desirable transgressive segregants from resulting hybrids for grain yield and other traits. Some researchers determined that the general combining ability effects for yield and other characters have played a significant role in selecting parents for grain yield (Kant et al., 2001; Akbar et al., 2009).

Line × tester analysis is one of the most powerful tools for predicting the general combining ability (GCA) of parents and selecting of suitable parents and crosses with high specific combining ability (SCA) (Rashid et al., 2007). Line × tester analysis provides information about combining ability effects of genotypes and also, knowledge regarding genetic mechanism controlling yield components. Information of general and specific combining abilities influencing yield and its components has become increasingly important to plant breeders to select appropriate parents for developing hybrid cultivars especially in cross pollinated crops. Many researchers have studied the combining abilities and gene actions of bread wheat hybrid populations by using line × tester analysis for some traits (Saeed et al., 2001; Krystkowiak et al., 2008; Jain and Sastry, 2012).

The aims of this study were to determine combining abilities of the genotypes, to assess the nature and magnitude of gene actions involved and to predict narrow sense heritability for yield and some yield components in a line × tester crossing design in bread wheat.

MATERIALS AND METHODS

The research was carried out at the experimental area of Department of Field Crops, Faculty of Agriculture, Ege University in Izmir during the 2010-2011 and 2011-2012 growing seasons. The soil of the experimental site has a heavy soil structure with clay-silt soil at 0-20 cm depth and clay-loamy structure at 20-40 cm depth.

Four CIMMYT advanced lines (129, 340, 342, 47) and one registered variety (Sagittario) were used as lines and four wheat genotypes (line 3 and 4 from CIMMYT, Basribey and Ziyabey, registered varieties) were used as testers. The testers are well adapted genotypes to this environment also having early maturity. On the other hand, flag leaves of the lines have ability to longer stay green than the testers.

The nine parents were crossed to produce 20 F_1 hybrids according to the line × tester mating design developed by Kempthorne (1957) during 2010-2011 growing season. The 20 crosses and 9 parents were evaluated in Randomized Complete Block Design with three replications during 2011-2012 growing season. In each replication, parents and F_1 hybrids were sown in two rows of 1 m length with a spacing of 20 cm between rows and 5 cm between plants. Recommended cultural practices were followed to raise a good crop. 120 kg N ha^{-1} and 60 kg P ha^{-1} fertilizer were applied equally at sowing time and during the stem elongation period.

Yield and some yield components were measured in parent and F_1 hybrid plants. Grain yield was measured in grams per row and thousand kernel weight was calculated based on mean value (g) of 4 random samples of 100 kernels from each plot. Other observations were recorded on ten random plants for the characters spike length (cm), plant height (cm) and number of spikes per row then converted to square meter and. Data recorded were subjected to analysis of variance according to Steel and Torrie (1980), to determine significant differences among genotypes. The combining ability and gene effects were studied (Singh and Chaudhary, 1979) using Microsoft Excel. The t-test was used to test whether the combining ability effects were different from 0.

RESULTS

All genotypes that were used in the research showed differences with respect to the traits used in the experiment (Table1). This case showed that there was a significant variation among lines, testers and hybrids, hence it is possible that to calculate the general and specific combining abilities in the population. The lines used in the research exhibited a variation in terms of all studied traits except plant height (Table 1). Also, the interaction between lines and testers were stated as significant for all traits except plant height and spike length. It can be seen that all testers that used in the research generally had similar values in terms of the studied traits (Table2). However, genotype Ziyabey pointed out in terms of high value of spike number for per m^2 and grain yield per row. Furthermore, it was monitored that all lines used in the research have quite different values from each other and there was a big variation between them in terms of the traits measured (Table 2). Otherwise, the hybrids had close values for plant height and spike length traits. However, significant differences were observed among the same hybrids (Table 1).Maximum thousand kernel weight value (49.1 g) was obtained from the combination of 129 × Ziyabey, while minimum value was obtained from the combination of Sagittario × 3 (Table 2). For the thousand kernel weight, there were five combination sexceeded their parental values (129 × 4; 129×Basribey, 129 ×Ziyabey, 340 × 3 and 347 × 3). These data showed that heterotic effects were emerged with high frequency for thousand kernel weight. Spike number for per m^2 trait was closely linked with tillering capacity, in this study, minimum spike number for per m^2 value was found for the combination of 340 ×Ziyabey as 50 pcs/m^2, while maximum number was 128 pcs/m^2 for the combination of 347 ×Basribey. Also,most of the hybrids generally had lower values of spike number for per m^{2}than the parents (Table 2). For some combinations, the values of grain yield per row were determined as close to parent values and for some combinations it was higher than those of the parents. This situation showed that heterotic effects emerged highly in point of grain yield per row in these hybrids. For grain yield per row, the combinations of 129 × 3 and 347 × Basribey had the values of 165.5 g and 209.6 g, respectively and showed higher performance than tester Ziyabey (151 g) which presented the best performance among all testers. It was found that both of these two combinations had superior dominancy effect for grain yield per row.

Table 1. Mean squares for yield and yield components used in the study.

Source of Variation	DF	Plant height (cm)	Spike length (cm)	Thousand kernel Weight (g)	Spikes number per m^2	Yield per row (g)
Replication	2	28.1	1.9*	0.3	61.6	20.2
Genotypes	28	81.1**	2.2**	72.3**	1560.8**	4353.5**
Parents	8	169.5**	3.1**	84.8**	1219.5**	5244.3**
Parents and Hybrids	1	203.5**	36.7**	184.6**	962.3**	29811.6**
Hybrids	19	48.3	0.9	61.1**	1324.5**	3954.0**
Lines	4	34.8	2.2**	124.0**	1043.8**	7087.7**
Testers	3	47.5	0.6	54.8**	2107.7**	5753.9**
Line × Tester	12	53.0	0.6	41.7**	1222.3**	2459.4**
Error	56	18.7	0.4	1.1	41.5	16.7

* :Significant at 1 per cent level, **: Significant at 5 per cent level

Table 2. Mean values of lines, testers and hybrids with respect to studied traits.

Genotypes	Plant height (cm)	Spike length (cm)	Thousand kernel Weight (g)	Spikes number per m^2	Yield per row (g)
Lines					
129	96.6	9.7	44.0	115	146.9
340	95.7	10.7	34.0	89	104.7
342	92.3	10.3	41.1	106	121.6
347	87.7	8.1	30.0	151	39.5
Sagittario	73.0	7.7	29.0	100	32.4
Testers					
3	80.2	9.6	39.2	92	104.2
4	86.2	10.4	42.7	106	113.2
Basribey	83.5	9.1	39.8	102	120.1
Ziyabey	85.8	9.6	38.3	135	151.0
Hybrids					
129×3	88.7	11.1	41.0	95	165.5
129×4	84.9	11.1	45.0	92	149.4
129×Basribey	88.4	10.7	46.3	98	117.6
129×Ziyabey	87.6	10.5	49.1	87	104.4
340×3	88.8	10.9	44.7	84	148.9
340×4	91.5	11.1	42.0	114	129.6
340×Basribey	89.1	11.3	41.0	76	121.2
340×Ziyabey	85.1	10.1	42.7	50	76.9
342×3	86.4	11.6	37.0	73	121.5
342×4	89.9	10.0	36.0	107	151.2
342×Basribey	76.3	10.6	39.1	71	104.1
342×Ziyabey	91.3	10.5	43.0	119	101.5
347×3	89.4	9.7	45.0	76	125.2
347×4	92.5	10.5	41.4	109	147.3
347×Basribey	89.1	10.3	33.3	128	209.6
347×Ziyabey	83.6	10.1	39.4	91	123.7
Sagittario×3	84.3	9.8	31.0	62	86.2
Sagittario ×4	85.4	10.4	39.2	97	140.5
Sagittario ×Basribey	79.7	9.8	35.8	99	49.8
Sagittario ×Ziyabey	89.7	9.8	43.2	54	78.6
LSD$_{0.05}$	*7.08*	*1.04*	*1.70*	*10.61*	*6.69*

General combining ability (GCA) effects of the lines, testers or individual line performances are useful tools for selecting the hybrid parents in cross pollinated plants (Hallauer et al., 1988; Longin et al., 2013). For observed traits, GCA effects of the lines and testers used in the research can be seen at Table 3. Negligible positive and negative effects were observed between lines and testers for plant height. Besides, it can be seen that one of the testers called Basribey had negative and significant GCA effect. Since, all parents used in the research had the plant height reduced genes, there were remarkable similarity in the plant height measurements among hybrids (Table2). Spike length was found as one of important components of grain yield by various researchers (Kahaliq et.al., 2004; Shahid et al., 2002) and the positive GCA effects were generally observed for spike length. Genotype 129 showed a positive and significant GCA among testers, while genotype 347 and Sagittario had negative and significant GCA effects. This situation showed that in the combinations which include these two genotypes had

relatively short spikes (Table 2).All parents except genotype 4 had significant GCA effects for thousand kernel weight. However, it can be realized that negative and significant GCA effects decreasing the thousand kernel weight are more than positive GCA effects. The line 129, which had the highest positive and significant GCA effect, increased the thousand kernel weight in the combinations in which it took a place (Table 2). Spike number for per m^2 is one of the most important traits in relation with productive spike number, and it determines the yield of the plant. Therefore, the tester called number 4 and the line called 347 were the most suitable parents for increasing this trait due to their high GCA effects. Saeed et al (2001), specified some bread wheat genotypes based on GCA effects for enhancing the tillering trait. The line number 347 and the tester number 4 significantly increased the grain yield per row for the combinations in which they took a part. It was found that these two genotypes have positive and significant GCA effects.

Table 3. General combining abilities (GCA) effects of yield and yield components.

Parents	Plant height (cm)	Spike length (cm)	Thousand kernel Weight (g)	Spikes number per m^2	Yield per row (g)
Lines					
129	0.39	0.41[*]	4.59[**]	1.28	17.83[**]
340	1.54	0.34	1.84[**]	-7.56[**]	-5.06[**]
342	-1.14	0.17	-1.97[**]	4.45[*]	-4.61[**]
347	1.53	-0.36[*]	-1.00[**]	12.49[**]	27.26[**]
Sagittario	-2.32	-0.56[**]	-3.46[**]	-10.66[**]	-35.42[**]
Testers					
3	0.41	0.11	-1.01[**]	-10.31[**]	5.27[**]
4	1.72	0.12	-0.05	15.57[**]	19.42[**]
Basribey	-2.49[**]	0.07	-1.63[**]	2.81	2.51[*]
Ziyabey	0.35	-0.30	2.69[**]	-8.07[**]	-27.20[**]

* :Significant at 1 per cent level, **: Significant at 5 per cent level

Although, specific combining ability (SCA) is generally suggested for cross pollinated species such as corn or rye (Hallauer and Miranda 1981; Tomerius 2001; Longin et al., 2007; Longin et al., 2013),SCA effects can be used to select homozygous lines that show transgressive segregation. It was observed that 342 × Ziyabey and Sagittario × Ziyabey combinations had positive and significant SCA effects for plant height (Table 4). Besides

that, both combinations had the plant height values that close to each other about 90 cm which is the appropriate value for the conditions of our region (Table2).

Positive and significant SCA effect for spike length was observed only in the combination of 342 × 3 (Table 4). Also, this combination had the highest spike length among the all combinations (Table 2).

Table 4. Spesific combining abilities (SCA) effects of yield and yield components.

Hybrids	Plant height (cm)	Spike length (cm)	Thousand kernel Weight (g)	Spikes number per m^2	Yield per row (g)
129×3	0.77	0.06	-3.34[**]	15.98[**]	18.20[**]
129×4	-0.53	0.05	-4.30[**]	-9.90[**]	4.05[*]
129×Basribey	3.81	-0.05	2.62[**]	-8.81[**]	4.32[*]
129×Ziyabey	-0.26	-0.10	1.02[*]	5.74	-10.47[**]
340×3	-0.26	-0.01	3.15[**]	13.14[**]	24.45[**]
340×4	1.11	0.08	-0.55	17.60[**]	-8.97[**]
340×Basribey	3.01	0.37	0.03	-7.98[*]	-0.47
340×Ziyabey	-3.86	-0.44	-2.63[**]	-22.77[**]	-15.02[**]
342×3	0.05	0.78[**]	-0.77	-9.19[**]	-3.33
342×4	2.19	-0.76[**]	-2.74[**]	-1.07	12.23[**]
342×Basribey	-7.18[**]	-0.17	1.98[**]	-24.31[**]	-18.01[**]
342×Ziyabey	4.93[*]	0.15	1.52[**]	34.57[**]	9.11[**]
347×3	0.34	-0.53	6.25[**]	-14.57[**]	-31.50[**]
347×4	2.10	0.22	1.64[**]	-7.44[*]	-23.55[**]
347×Basribey	2.97	0.03	-4.80[**]	24.31[**]	55.65[**]
347×Ziyabey	-5.41[**]	0.27	-3.09[**]	-2.31	-0.60
Sagittario×3	-0.91	-0.30	-5.29[**]	-5.36	-7.82[**]
sagittoria×4	-1.08	0.36	1.95[**]	3.80	32.34[**]
Sagittario×Basribey	-2.61	-0.19	0.17	36.79[**]	-41.50[**]
Sagittario×Ziyabey	4.60[*]	0.13	3.17[**]	-15.23[**]	16.98[**]

* :Significant at 1 per cent level, **: Significant at 5 per cent level

DISCUSSION

It can be seen in the Table 4 that there were nine combinations which had positive and significant SCA in terms of thousand kernel weight. The SCA effects were found higher than the GCA effects for this trait (Table 5). Contrary results were reported by Krystkowiak et al., (2008), they studied with 76 wheat hybrids and they found

higher GCA effects for thousand kernel weight. Since, test weight is one of the indicators of flour yield (Pomeranz, 1964) and thousand kernel weight positively correlated with test weight (Warechowska, 2013), the wheat genotypes which has high thousand kernel yield are preferred in the flour industry. Hence these nine combinations are preferable for flour industry with their thousand kernel weight values around 40 g and higher. The hybrid number 347 × 3 had the best SCA to increase thousand kernel weight (Table 4).

In the case of spike number for per m^2 trait, six combinations had positive and significant SCA effects

(Table 4). It was understood that these combinations can generate high amount of productive tillers (Table 2). The number of 347 × Basribey was found as the best hybrid to the number of spike in square meter (Table 4).

Table 5. Genetic component estimations.

	Plant height (cm)	Spike length (cm)	Thousand kernel Weight (g)	Spikes number per m^2	Yield per row (g)
$Б^2_{GCA}$	-0.14	0.01	0.57	2.98	43.62
$Б^2_{SCA}$	11.45	0.06	3.53	393.61	814.25
$Б^2_A$	-0.28	0.02	1.13	5.97	87.24
$Б^2_D$	11.45	0.06	13.53	393.61	814.25
$Б^2_{GCA}/Б^2_{SCA}$	0.01	0.19	0.04	0.01	0.05
$(Б^2_D/Б^2_A)^{1/2}$	6.47	1.63	3.46	8.12	3.05
h^2_{ns}	0.02	0.10	0.08	0.01	0.10

There were nine combinations with respect to grain yield per row which had positive and significant SCA effects (Table 3). It is remarkable that two of them (347 × Basribey and 129 × 3) had quite high grain yield per rows (Table 2).

It was found that the lines 125 and 347 and the tester 4 had the high values in terms of the traits studied in the experiment. Furthermore, these lines had enhancer effects for grain yield per rows in the combinations that they took part in. Also, the combinations 129×3, 129×4, 340×3, 340×4, 342×4, 347×4 and Sagittario×4 came into prominence as promising combinations in terms of the traits evaluated in the research and it can be possible to obtain high yielding lines by carrying these combinations to next generations.

The GCA effects were lower than SCA effects in terms of all traits evaluated in the research in contrast to Titan et al.,(2012). They used 6 wheat lines and seven testers and they tested 42 F_1 combinations for two seasons. They observed that SCA effects were generally lower than GCA effects. Besides, Sharma et. al., (2006) and Borghi and Perenzin (1994) stated that σ^2_{GCA} variance was of greater importance than σ^2_{SCA} for some traits. The difference in the results reported by researchers may be attributed to differences of parental materials used hybridization and to genotype × environments. In our results, it was understood that, non additive gene effects were dominant the genetic control of this traits (Table 5). Likewise, dominant genetic variance was found higher than additive variance for all traits. These results showed that $\sigma^2_{GCA}/\sigma^2_{SCA}$ portion was lower than one and (σ^2_D/σ^2_A) portion, which is an indicator of dominancy degree, higher than one (Table 5). Hence, it can be seen that non-additive genetic effects are controlling the inheritance of studied traits. It was understood that selection for the traits inherited with this manner should be performed in the further generations like F_4 or F_5.Fellahiet al., (2013), reported the importance of non-additive gene action for the plant height, spike length, number of fertile tillers, thousand kernel weight and kernel yield. They

recommended that selection of superior plants should be postponed to later generations due to preponderance of non-additive type of gene actions for all studied traits. Similar results of predominance of non-additive gene action for all studied traits have been reported by Vermaet al., (2007) for barley.

The efficiency of the selection is related with the size of narrow sense heritability in the segregating populations. The heritability degrees were found very low for the traits studied in the research (Table 5). Similar results were found by Erkul et. al., (2010). They estimated the genetic effects by using the generation means in six parameters model and low heritability with low genetic advance was found. This situation showed that the additive variance is very low in this population and the selection must be applied in the further generations.

Line × tester interaction contributed to combinations' variances was found much more than lines and testers, individually. The contributions of the lines to spike length were defined as highest, while to thousand kernel weight and grain yield per row were found closer to line × tester interaction. The contributions of testers were found lower than those of line × tester interactions for all traits (Table 6). Hence, line × tester interactions provide much more variation for the appearing of the traits. It is remarkable that hybrid combinations had higher values than their parents with respect to plant height, spike number for per m^2 and grain yield per row traits.

The $\sigma^2_{GCA}/\sigma^2_{SCA}$ ratio varies depending on the allele frequencies between parental populations (Reifet al., 2007; Longinet al., 2013). The lines selected from different gene pools had favorable $\sigma^2_{GCA}/\sigma^2_{SCA}$ ratio because of their high GCA effects (Labateet al.,1997).In this study, low ratios of $\sigma^2_{GCA}/\sigma^2_{SCA}$, $(\sigma^2_D/\sigma^2_A)^{1/2}$ and low narrow sense heritabilities supported the involvement of non-additive effects with predominance of non-additive type of gene actions (Table 5). Lines and the interaction of line × testers contributed more to variation of the expression of studied traits.

Table 6. The contribution rates of lines, testers and line × tester interaction for hybrid generation (%).

	Plant height (cm)	Spike length (cm)	Thousand kernel Weight (g)	Spikes number per m^2	Yield per row (g)
Lines	15.15	51.05	42.74	16.59	37.74
Testers	15.51	10.12	14.16	25.13	22.98
Line×Tester	69.34	38.83	43.10	58.28	39.28

As the conclusion, the line number 129 was good combiner for the spike length, grain yield per row and thousand kernel weight. The line number 347 and the tester number 4 were good combiners for grain yield per row and number of spike in square meter. The number of 347 × Basribey was found as the best hybrid to the number of spike in square meter and grain yield per row. The hybrid number 347 × 3 had the best SCA to increase thousand kernel weights. The predominance of non-additive type of gene actions clearly showed that selection of superior plants should be delayed to F_4 or F_5 generations.

LITERATURE CITED

Akbar, M., J. Anwar, M. Hussain, M.H. Qureshi and S. Khan. 2009. Line×tester analysis in bread wheat (*Triticum aestivum L.*). J.Agric. Res. 47(1): 411.

Borghi, B.And M. Perenzin.1994.Diallel analysis to predict heterosis and combining ability for grain yield, yield components and bread-making quality in bread wheat (*T. aestivum*). Theoretical and Applied Genetics 89: 975-981.

Erkul, A., A. Unay and C. Konak. 2010. Inheritance of Yield and Yield Components in a Bread Wheat (*Triticum aestivum L.*) Cross. Turk J Field Crops, 15:137-140.

Fellahi, Z.E.A., A.Hannachi, H.Bouzerzour and A.Boutekrabt. 2013.Line×tester mating desing analysis for grain yield and yield related traits in bread wheat (*Triticum aestivum L.*). International Journal of Agronomy. 2013:9.

Hallauer, A.R. andJ.B. Miranda. 1981. Quantitative genetics in maize breeding. Iowa State University Press, Ames, pp 267–298.

Hallauer, A.R.,W.A. Russell andK.R. Lamkey. 1988. Corn breeding. In: Sprague GF, Dudley JW, editors. Corn and corn improvement, 3rd ed. Agron Monogr. 18 ASA, CSSA, SSSA, Madison, WI, pp 469–565.

Jain, S.K. and E.V.D. Sastry. 2012.Heterosis and combining ability for grain yield and its contributing traits in bread wheat (*Triticumaestivum L.*). Journal of Agriculture and Allied Science 1:17-22.

Kant, L.,V.P. Mani andH.S. Gupta. 2001.Winter×springwheat hybridization a promising avenue for yield enhancement. Plant Breeding 120:255-259.

Kempthorne, O. 1957.An Introduction to Genetic Statistics, John Wiley & Sons, New York, NY, USA.

Khaliq, I., N. Parveen, M.A. Chowdhry. Correlation and Path Coefficient Analyses in Bread Wheat International Journal Of Agriculture & Biology 1560–8530/2004/06–4–633–635Krystkowiak,K., T. Adamski, M.Surma and Z. Kaczmarek. 2008. Relationship between phenotypic and genetic diversity of parental genotypes and the specific combining ability and heterosis effects in wheat (Triticum aestivum L.). Euphytica 165: 419-434.

Labate, J.A., K. R. Lamkey, M. Lee and W. L. Woodman. 1997. Molecular genetic diversity after reciprocal recurrent selection in BSSS and BSCB1 maize populations. Crop Sci. 37: 416-423.

Longin, C.F.,M. Gowda, J. Mühleisen, E. Ebmeyer, E. Kazman, R. Schachschneider, J. Schacht M. Kirchhoff, Y. Zhao and J. C. Reif. 2013. Hybrid wheat: quantitative genetic parameters and consequences for the design of breeding programs. Theor. Appl. Genet. 126(11):2791-801.

Longin, C. F. H., H. F. Utz, A. E. Melchinger and J. C. Reif. 2007. Hybrid maize breeding with doubled haploids. II. Optimum type and number of testers in two-stage selection for general combining ability. Theor Appl Genet 114:393–402.

Pomeranz, Y. 1964. Wheat Chemistry and Technology (edited), St. Paul, Minnesota. American Association of Cereal Chemists.

Rashid, M.,A.A. Cheema and M. Ashraf. 2007.Line×Tester analysis in basmati rice. Pakistan Journal of Botany. 39(6): 2035-2042.

Reif, J.C., F. Gumpert, S. Fischer and A. E. Melchinger. 2007. Impact of genetic divergence and dominance variance in hybrid populations. Genetics 176: 1931-1934.

Saeed, A., M. A. Chowdhry, N. Saedd, I. Khalıq and M. Z. Johar. 2001. Line×Tester analysis for some morpho-physiological traits in bread wheat. International journal of Agriculture & Biology 3(4):444-447.

Salgotra, R. K., B. B. Gupta and S. Praveen. 2009. Combining ability studies for yield and yield components in Basmati rice. An International Journal on Rice. 46(1):12-16.

Shahid M., F. Mohammad, M. Tahir. 2002. Path coefficient analysis in wheat. Sarhad J. Agrirc. 18: 385-388.

Sharma, H.C., M.K. Dhillan and B. V. S. Reddy. 2006. Expression of resistance to *Atherigona soccata* in F_1 hybrids involving shoot fly-resistant and susceptible cytoplasmic male-sterile and restorer lines of sorghum. Plant breeding 125: 473-477.

Singh, R. K. and B. D. Chaudhary. 1979. Biometrical Methods in Quantitative Genetic Analysis. Second edition, Ed. Kalyani Publishers 191-201., New Delhi, India.

Steel, R. G. D. and J. H. Torrie. 1980. Principles and procedures of statistics. 2nd edition. McGraw Hill Book Company Inc., New York.

Titan, P., V. Meglic and J. Iskra. 2012. Combining ability and heterosis effect in hexaploid wheat group. Genetika 44: 595-609.

Tomerius, A.M. 2001. Optimizing the development of seed-parent lines in hybrid rye breeding. PhD, University of Hohenheim, Germany.

Verma, A. K., S. R. Vishwakarma and P. K. Singh. 2007. Line x Tester analysis in barley (*Hordeum vulgare* L.) across environments. Barley Genetics Newsletter 37: 29-33.

Warechowska, M., J. Warechowski, A. Markowska. 2013. Interrelations between selected physical and technological properties of wheat grain. Technical Science, 16(4), 281-290

13

SUBSTITUTION POSSIBILITY OF SOME BIOFERTILIZERS FOR MINERAL PHOSPHORUS FERTILIZER IN PEA CULTIVATION

H. Ibrahim ERKOVAN[1]*, M. Kerim GULLAP[1], Kamil HALILOGLU[1], Ali KOC[2]

[1] *AtaturkUniversity, Faculty of Agriculture, Department of Field Crops, Erzurum, TURKEY*
[2] *OsmangaziUniversity, Faculty of Agriculture, Department of Field Crops, Eskisehir TURKEY*
*Corresponding Author: erkovan@atauni.edu.tr

ABSTRACT

The interest in the use of biofertilizer as alternative to mineral fertilizer increase continuously due to increasing mineral fertilizer cost and heavy metal accumulation in the soil such as cadmium. The objective of this study was to assess the effects of four biofertilizer (N_2-fixing, P-solubilizing, N_2 fixing-P solubilizing, commercial biofertilizer) with and without mineral phosphorus fertilizer on yield and quality of forage pea(*Pisumsativum spp. arvense* L.). The application of biofertilizeraffected significantly dry matter yield (DM), crude protein (CP), neutral detergent fiber (NDF) and phosphorus contain. The use of mineral fertilizer increased only dry matter yield. The effect of biofertilizer on pea yield and quality varied significantly depending on year. These results indicated that understanding of factors such as biofertilizer, mineral fertilizer and environment will enable us to use biofertilizer as an alternative to mineral fertilizer to optimize productivity and sustainability of pea production.

Keywords: Biofertilizer, mineral fertilizer, phosphorus, yield and quality, pea

INTRODUCTION

Peas are cultivated widely as rotation or second crops for forage and pulse production in semi arid environments. Both seeds and forages of pea are rich in protein and mineral content (Acikgoz et al., 1985). The productivity of peaslike in the other legume crops are restricted by phosphorus deficiency. Thus, producers rely on mineral phosphorus fertilizer to achieve sustainable production. However, prices of chemical fertilizer increase continuously due to increasing energy cost whichrestricted theirutilization economically. On the other hand, phosphorus fertilizers are not environmental friendly input in agriculture due to cadmium content (Al Fayiz et al., 2007). Recently, there has been interest in more environmentally sustainable agricultural practices (Orson 1996). A considerable numbers of bacteria species that are associated with the rhizosphere are able to exert a beneficial effect on plant growth (Rodriguez and Fraga, 1999). These microorganisms secrete different type organic acid (Illmer and Schinner, 1992), thus lowering the pH in the rhizosphere and consequently dissociate the bound form of phosphate (Rodriguez and Fraga, 1999). Phosphorus biofertilizers also help increase nitrogen fixation and availability of some microelements such as Fe, Zn, etc.Generally, only 0.1% of total P in soil is available to plants (Scheffer and Schachtschabel, 1992). The way of increase to P available to plants is enzymatic decomposition or microbial inoculation (Illmer and Schinner, 1992). Hence, bacteria might be partially substitute chemical fertilizer or they can be used.

Many researchershave isolated nutrient solubilising or fixing microorganism in different soils, plant rhizosphere, root or intercellular spaces of plants (Halder et al. 1990; Illmer and Schinner, 1992; Sahin et al. 2004; Cakmakci et al. 2007) and they are described'plant growth promoting rhizobacteria' (PGPR)promoting plant growth either increasing nutrient intake or changes enzymatic or hormone synthesis, even some strains had pathogen control by having antibiotic effect (Xie et al. 1996; Glick et al. 1998; Zaccaro et al. 1999; Stirk et al. 2002). There are many successful examples of bacteria applicationabout increasing yield in clover, wheatgrass, perennial ryegrass (Holl et al., 1988), sugar beet (Sahin et al., 2004), barley (Salantur et al., 2005), chickpea (Elkoca et al., 2008), pea (Osman et al., 2010),hungarian vetch (Yolcu et al., 2012).

Plant growth promoting rhizobacteria changes chemical compounds of the applied plants. In general, PGPR application encourage an increase in crude protein content (Peix et al. 2001; Osman et al. 2010; Yolcu et al. 2012), decreases in cellulosic content (Mishra et al. 2010; Yolcu et al. 2012) and an increase in some minerals such as P, Ca, K(Peix et al. 2001; Elkoca et al. 2010; Osman et al. 2010) but decrease some others such as S, Cu, Zn (Yolcu et al., 2012) in dry matter.

In general, there are currently no adequate knowledgeabout the effect of PGPR on the yield and chemical components of forage peas. Objectives of this study were to determine the effects of phosphorus (with and without) and bacteria application on (1) dry matter yield of pea, (2)feeding quality of forage,(3) mineral contents of forage (4) and substitution possibility of biofertilizes for mineral phosphorus fertilizerapplication in pea cultivation in semi arid conditions.

MATERIALS AND METHOD

The field experiment was conducted at the experimental station of Faculty of Agriculture, University of Ataturk, Erzurum ($39^0 51^1$N and $41^0 61^1$E, 1850 m above sea level). The soil of experimental area was loamy with organic matter content of 1.92%, with lime 4.65% and pH of 7.24. Corresponding available P_2O_5 and K_2O contents were 27.3 kg ha^{-1}and 120.0 kg ha^{-1}, respectively.In Erzurum, winters are long and extremely cold and summers are cool, short and arid. Long-term annual mean temperature is 5.0^0C, rainfall is 405 mm and relative humidity is 66.5% in the study area.Total annual precipitation and mean annual temperature were 437.8 mm and 5.8 ^{0}C in 2009 and 475.9 mm and 7.9 ^{0}C in 2010, respectively in the experiment years. The monthly distribution of precipitation and monthly average temperature were presented in Table 1.

Table 1. Monthly temperature and precipitation values of experimentalyears (2009 and 2010)and long-term average (1990-2010)

	Mean air temperature ^{0}C			Total precipitation (mm)		
	2009	2010	Long-term	2009	2010	Long-term
January	-12.1	-4.3	-10.6	2.3	52.2	16.7
February	-3.1	-1.8	-9.4	18.8	14.8	20.5
March	-0.7	3.1	-2.8	51.1	82.2	35.2
April	4.3	5.6	5.2	42.3	54.2	60.1
May	10.0	10.4	10.4	43.2	63.6	66.7
June	14.7	15.6	14.8	76.2	50.5	41.9
July	17.2	19.5	19.1	29.2	55.5	24.5
August	17.1	20.3	19.3	22.8	9.0	14.8
September	12.4	17.0	13.9	43.7	8.8	20.2
October	8.7	9.2	7.7	51.0	72.2	44.1
November	1.8	1.8	-0.2	41.4	0.0	28.1
December	-1.1	-1.9	-7.2	15.1	12.9	22.8
Total/Mean	5.8	7.9	5.0	437.1	475.9	395.6

The experiment was arranged a randomized complete block design with three replications. Treatment consist of 0 or 50 kg P_2O_5 ha^{-1}, which suggested doses of phosphorus fertilizer in annual legumes cultivation in the region (Serin and Tan 2011). Triple supper phosphate form of the phosphorus fertilizer were used and five different type biofertilizers were (a) control (C), (b) N_2-fixing (NF) (*Bacillus subtilis* OSU-142), (c) P-solubilizing (PS) (*Bacillus megaterium* M3), (d) N_2 fixing-P solubilizing (NF+PS) (*Burkholderia cepacia* GC sup.B) and (e) commercial biofertilizer (CB) (Bio-one) was developed by Texas University which contain *Azotobacter vinelandi* living aerobic condition and *Clostridium pasteurianum* living anaerobic condition.

The biofertilizer were applied sterilized seeds before sowing and phosphorus fertilizer was broadcasted plots surface before sowing and it was incorporated into soil using hand harrow. Forage pea (*P. sativum ssp. arvense* L. cv Taskent) was sown by hand with 100 seeds per m^2 (Uzun and Acikgoz 1998) in May 20th 2009 and May 15th 2010. The plot size was 1.5 m by 5 m = 7.5 m^2, consisting of 5 rows spaced 30 cm apart. Weed control was done by hand hoeing in the beginning of June. The plots were irrigated 3 times in 2009 and 2010 with flooding system when plant colour turns dark green due to lack of moisture in the soil during the growing season.

Harvesting was performed after taking out one row from each side of the plots and a 0.5 m area from beginning or end of each rowin July 29th2009 and August 03th 2010. Dry matter yield was determined in cutting samples at the pod filling stage and samples were oven-dried at the 68 ^{0}C until reaching a constant weight. After weighting, samples were grounded to pass through a 2 mm sieve and analysed for chemical characteristics.Total N content of the samples was determined by the Kjeldahl method and multiplied by 6.25 to give the crude protein content (Jones 1981). Neutral detergent fiberand ADF content were measured using an ANKOM fiberanalyzer following the procedure described by Anonymous(1995). Mineral content (Ca, P, K andMg) was determined using an Inductively Couple Plasma spectrophotometer (Perkin-Elmer, Optima 2100 DV, ICP/OES, Shelton, CT, USA) (Mertens 2005).

All data were subjected to analysis of variance usingthe Statview package (SAS Institute1998). Means were separated using Duncan's multiple range tests.

RESULTS

Both phosphorus and biofertilizer application affected dry matter yield significantly but the differences in dry matter yield between years were non-significant(Table 2).Optimum phosphorus doses increased 0.50 t ha⁻¹in dry matter yield. In biofertilizer application, the highest dry matter yield was obtained from NF treatment (6.98 t ha⁻¹).Commercial biofertilizer application also gave statistically similar result to NF.The dry matter yield ofthesetwo treatments (NF+PS and PS) was lower than control (Table 2).Phosphorus fertilizer application increased dry matter yield significantly in the first

year,while it was not significant in the second year. This difference was responsible for Y x P interaction (Figure 1a).In the first year, the highest dry matter yields were obtained from both CB and NF treatments but CB had the lowest at the second year.However, NF application had the highest and stable yield increases in dry matter production in both years. The different response of dry matter production to biofertilizer between years conduced Y x BF interactions (Figure 1b).

Table 2. Analysis of variance results with main effects and interactions of biofertilizer and phosphorus fertilizer application on dry matter (DM), crude protein content (CP), neutral detergent fibre (NDF) and acid detergent fibre (ADF)

Treatments	DM (t ha⁻¹)	CP (%)	NDF (%)	ADF (%)
2009	5.71	15.57 B	36.99 B	27.24 B
2010	6.08	18.66 A	46.03 A	34.46 A
Average	**5.89**	**17.11**	**41.51**	**30.85**
P₀	5.60 B	17.40	41.37	31.08
P₅₀	6.18 A	16.83	41.65	30.62
Average	**5.89**	**17.11**	**41.51**	**30.85**
C	5.80 B	17.09 AB	41.58 B	31.88
NF	6.98 A	16.32 B	40.54 B	30.74
PS	5.47 BC	18.56 A	44.49 A	31.34
NF+PS	5.83 C	16.97 AB	40.88 B	31.30
CB	6.38 AB	16.63 B	40.06 B	28.97
Average	**5.89**	**17.11**	**41.51**	**30.85**
Y	ns	**	**	**
P	**	ns	ns	ns
BF	**	**	*	*
Y x P	**	ns	ns	ns
Y x BF	**	**	*	*
P x BF	ns	**	**	**
Yx P x BF	ns	ns	**	ns

ns: non-significant, *: p<0.05, **: p<0.01.
Means in the same column with different letters are significant.

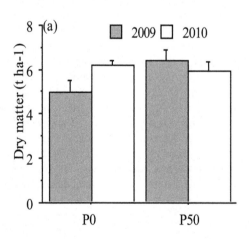 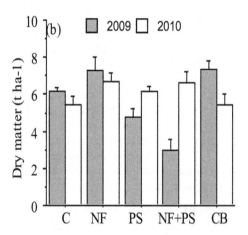

Figure 1. Dry matter yield offorage pea as affected by (a) year x phosphorus, (b) year xbiofertilizer (Bars indicated ± s.e.)

Crude protein content was higher in the second year than in the first year (Table 2). Main effect of phosphorus fertilizer application was insignificant but effect ofbiofertilizer application was significant on the CP (Table 2). Although CP content was higher in all biofertilizer applications in the second year compared to

first year results, no significant differences between years was observed in CB applications (Table 2). Thus, Y x BF interaction was significant (Figure 2a). In addition, NF+PS application with phosphorus had an opposite effect. Hence,P x BF interaction was significant (Figure 2b).

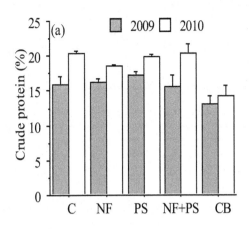
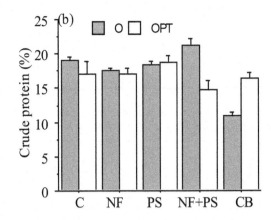

Figure 2. The effect of treatments on crude protein content of forage pea (a) year x biofertilizer,(b) phosphorus x biofertilizer (Bars indicated ± s.e.)

Neutral detergent fiber content was higher in the second year compared to first year results. The main effect of phosphorus fertilizer on NDF content was insignificant. Whereas, the main effect of biofertilizer application was significant(Table 2). Whereas, NDF content harvested PS applied plots was higher than the others. Hence, Y x BF interaction for NDF was significant (Figure 3a).Phosphorus x BF interaction was also significant due to different response of CB to phosphorus fertilizer application (Figure 3b).In addition this interactionsY x P x BF interaction was significant (Figure 3c).

Acid detergent fiber content was higher in the second year than in the first year but main effect of both phosphorus fertilizer and biofertilizer application was insignificant (Table 2).In the first year, biofertilizer applied plots had higher ADF content than alone phosphorus applied plots but in the second year, the highest ADF content was recorded in biofertilizer applied plots. Hence, Y x BF interaction was significant (Figure 4a).Different responses of ADF content to biofertilizer and phosphorus combination caused P x BF interaction (Figure 4b).

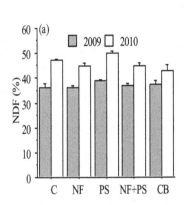
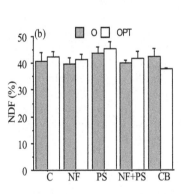
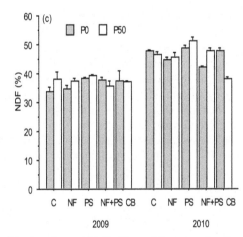

Figure 3.NDF content of forage pea as affected by (a) yearx biofertilizer, (b) phosphorusx biofertilizer, (c) yearx phosphorus xbiofertilizer (Bars indicated ± s.e.)

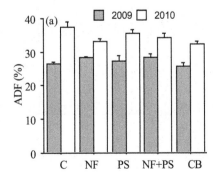
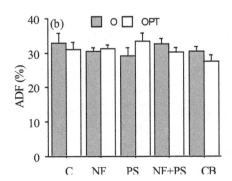

Figure 4. The effect of different fertilizer resources on ADF content of forage pea (a) year x biofertilizer,(b) phosphorus x biofertilizer (Bars indicated ± s.e.)

Year effect was significant on Ca, K and Mg concentrations (Table 3). Calcium content varied from 1.90% to 2.39% and it also varied significantly between the years (p<0.01) (Table 3). Phosphorus content was affected significantly by bacteria application (p<0.05). Commercial biofertilizer and NFapplication causes significant decreases in P content compared to PS application. In the first year, the plant harvested phosphorus fertilizer applied plots had higher P content than untreated plots. But there were no significant differences in the second year. Hence, Y x P and Y x BF interactions were significant for P content (p<0.05)

Table 3. Analysis of variance results with main effects and interactions of biofertilizer and phosphorus fertilizer application on mineral concentration (Ca; Calcium, P; Phosphorus, K; Potassium and Mg;Magnesium)

Treatments	Ca (%)	P (%)	K (%)	Mg (%)
2009	1.80 B	0.29	2.05 B	0.33 B
2010	2.36 A	0.31	3.09 A	0.45 A
Average	**2.13**	**0.30**	**2.57**	**0.39**
P_0	2.19	0.31	2.53	0.38
P_{50}	2.07	0.29	2.61	0.40
Average	**2.13**	**0.30**	**2.57**	**0.39**
C	2.10	0.31 AB	2.25	0.39
NF	1.99	0.28 B	2.45	0.36
PS	2.18	0.32 A	2.70	0.40
NF+PS	2.00	0.31 A	2.67	0.38
CB	2.39	0.27 B	2.79	0.42
Average	**2.13**	**0.30**	**2.57**	**0.39**
Y	**	ns	**	**
P	ns	ns	ns	ns
BF	ns	*	ns	ns
Y x P	ns	*	ns	ns
Y x BF	ns	*	*	ns
P x BF	ns	ns	ns	ns
Yx P x BF	ns	ns	ns	ns

ns: non-significant, *: p<0.05, **: p<0.01.
Means in the same column with different letters are significant.

(Figure 5a and b).Potassium content was affected significantly among the years (p<0.01). K content was higher in the second year than in the first year. There were no significant effects of CB and phosphorus fertilizer applications on K content in the experiment (Table 3). In the second year, there were no significant effects of CB application on K content, whereas, the hay harvested fromPS, NF and control applied plots had higher K content than the other treatments. As a result of this different response of K content to CB treatment, Y x BF interaction was significant (Figure 6). Neitherbiofertilizer nor phosphorus fertilizer application had significant effect on Mg content but the year effect was significant. The hay harvested in the first year had lower Mg content (0.33 %) than the second year (0.45 %)(Table 3).

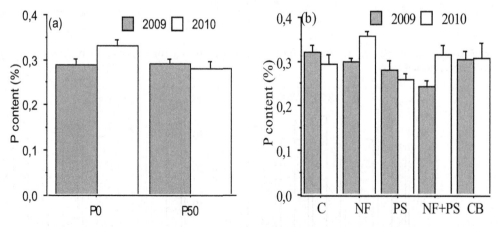

Figure 5.P content of forage pea as affected by (a) yearx phosphorus, (b) year x biofertilizer(Bars indicated ± s.e.)

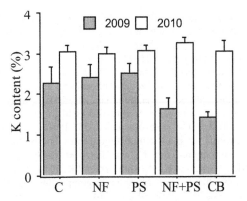

Figure 6. K content of forage pea as affected by (a) year xbiofertilizer(Bars indicated ± s.e.)

DISCUSSION

The data obtained from this study indicated that phosphorus fertilizer and biofertilizer application affected dry matter production of pea crop but there were great differences among biofertilizer treatments. The different response to biofertilizer with respect to dry matter production could be related to microbial content of biofertilizer. Nutrient availability and uptake might be changed as a result of changing rhizosphere microbial activity depending on biofertilizer application. Thus, dry matter production response to treatment must have been changed. Different responses to different microbial inoculant application were reported by the other studies (Elkoca et al. 2008; Osman et al. 2010; Yolcu et al. 2012). Legumes show better response to phosphorus application (Miller and Reetz, 1995; Erkovan et al. 2010).However, while the availability of the nutrient increase in the soil, the response to fertilizer decrease (Read and Ashford, 1968). In the experiment, phosphorus application increased dry matter production significantly in the first year but it was insignificant in the second year. These differences might be originated from microbial content of the experimental soil and the adding biofertilizer cause significantly changes in microbial content ofrhizosphere and as a result of this changes fluctative effect of biofertilizer on dry matter production was recorded due to changing nutrient uptake (Cakmakci et al., 2007). These biofertilizer could attribute to plant growth and yield by providing biologically fixed nitrogen, solubilized immobilized phosphorus and produced phytohormones (Hewedy 1999).

Chemical content of pea crops were affected significantly by years. Higher crude protein content was recorded second year in the experiment.The weather was warmer in the second year than in the first year. Since warmer weather restricted photosynthetic period, carbohydrate accumulation occurs in vegetative tissue as a result of this effect, therefore crude protein content could be higher in the first year. Because initially protein skeleton is constituted in the cell thereafter carbohydrate accumulation occurs (Osman et al., 2010).While sole phosphorus application had no significant effect on CP content, the combine effect biofertilizer and phosphorus fertilizer application varied depending on application combinations. For example, CB plus phosphorus fertilizer

applications resulted in significant increases in CP content compared to sole application of CB.Phosphorus fertilizer or PS application causes significant changes in chemical content and it generally increases crude protein and mineral content (Peix et al., 2001). Because biofertilizer supports phytohormones production, which stimulate nutrients absorption as well as photosynthesis process, as a result of this protein content increases (Cakmakci et al., 2007). But phosphorus fertilizer in the experiment did not affect crude protein content. However, PS bacteria causes significantly increase in crude protein content in hay.Crude protein content is generally increasedby existence of phosphorus which plant can intake easily but environmental factors responsible for very wide changes in crude protein content are not fully understood (Reichert and McKenzie, 1982).The protein content of pea appears to be highly variable. Reichert and MacKenzie (1982) found that protein content varied between 14.5 to 28.5% in genetically identical pea plants grown under same conditions such as year and field.

Since the second year was warmer than the first year, the NDF and ADF contentwere higher in the second year in the experiment. P-solubilizing causes increases in NDF and ADF content. These increases might berelated to increases in cellulosic component synthesis in the plant depending on phosphorus supply (Avila et al., 2011). Since warm weather hastened plant growth in the second year, carbohydrate accumulation decreased in the plant, hence, except phosphorus, investigated minerals content were increased in the second year.Warm weather causes generally an increase in cellulosic content (Osman et al., 2010).Commercial biofertilizer application with phosphorus fertilizer conduced a significant decrease in NDF content compared to alone application, whereas the other biofertilizer applications did not show interactive effect with phosphorus treatment. Acid detergent fiber content was higher in NF and PS applied plots fertilized with phosphorus than unfertilized with phosphorus. Especially, PS application with phosphorus fertilizer caused evidently increases in ADF content compared to alone application of it.As plant growth advanced, mineral acquisition decrease and carbohydrate content increase as long as plant growth.These results indicate that there is an antagonistic effect between biofertilizer and mineral fertilizer depending on microorganism species based on NDF and ADF. Similar results reported also by Mehrvarz and Chaichi (2008).

Earlier ceasing of plant growth always causes higher mineral content compared to normal growing plants (Cakmakci et al., 2007). The decrease in P content of phosphorus fertilizer applied plots might be related to acquisition rate of applied fertilizer phosphorus. Under unfavourable conditions phosphorus fixation increase quickly in the soil. However, PS application causes an increase in P content.These results implied that when phosphorus availability increased in the soil, the plants acquired more P, and consequently, P content inplant tissue increased. Similar supportive results were also reported by other researches(Peix et al. 2001; Cakmakci et al. 2007; Dasci et al. 2010). Our findings implied that efficiency of phosphorus fertilizer is strongly related to soil condition that affects availability of phosphorus fertilizer because PS application have positive effect for phosphorus acquisition for plant due to increase availability of phosphorus.

In conclusions, inorganic phosphorus application had slightly positive effect on hay production in forage pea under Erzurum ecological condition (a high altitude semi arid environment). The effect of mineral phosphorus fertilization closely related to availability rather than the applied amount because there was not any increase in P content of plant under mineral phosphorus application, whereas, it was significant increases under PS application. NF and CB application causes an increase in dry matter. Hence, in order to achieve higher hay production in pea cultivation NF inoculation can be suggested. But in the microbial fertilizer except for NF, there is need of more improvement study with respect to increase stability of their effect underdifferent environment because the responses of microbial fertilizers changed depending on years. In these studies, understanding of interaction between microbial fertilizer and soil microbial content will enable us to use microbial fertilizer as an alternative to mineral fertilizer. Because biofertilizer x environment interaction isvery common phenomenon in the microbial fertilizer studies. These results indicated that biofertilizer could be partially but not completely substitute chemical fertilizer in pea cultivation.

LITERATURE CITED

Anonymous. 1995. Acid detergent and neutral detergent fiber using ANKOM's fiber analyzer F200. Ankom Technology Corporation, Fairport, NY.

Acikgoz, E., V. Katkat, S. Omeroglu and B. Okan. 1985.Mineral elements and amino acid concentrations in field pea and common vetch herbages and seeds. J Agron and Crop Sci. 55: 179-185.

Al-Fayiz, Y.S., M.M. El-Garawany, F.N. Assubaie and M.A. Al-Eed. 2007.Impact of phosphate fertilizer on cadmium accumulation in soil and vegetable crops. Bull Environ Contam Toxicol. 78: 358-362.

Avila, F.W., V. Faquin, J.L. Araujo, D.J. Marques, P.M.R. Junior, A.K.S. Labato, S.J. Ramos and D.P. Baliza. 2011. Phosphate supply affects phosphorus nutrition and biochemical responses in maize plants. Australian J Crop Sci. 5: 646-653.

Cakmakci, R., M. Erat, U. Erdogan and M.F. Donmez. 2007. The influence of plant growth–promoting rhizobacteria on growth and enzyme activities in wheat and spinach plants. J Plant Nutr Soil Sci170:288–295.

Cakmakci, R., M.F. Donmez and U. Erdogan. 2007. The effect of plant growth promoting rhizobacteria on barley seedling growth, nutrient uptake, some soil properties, and bacterial counts. Turk J Agric For. 31: 189-199.

Dasci, M., M.K. Gullap, H.I. Erkovan and A. Koc. 2010.Effects of phosphorus fertilizer and phosphorus solubilizing bacteria applications on clover dominant meadow: II. Chemical composition. Turkish J Field Crops 15: 18-24.

Elkoca, E., F. Kantar and F.Sahin. 2008. Influence of nitrogen and phosphorus solubilizing bacteria on the nodulation, plant growth and yield of chickpea. J Plant Nutrition 31: 157–171.

Elkoca, E., M. Turan and M.F. Donmez. 2010. Effect of single, dual and triple inoculations with bacillus subtilus, bacillus

megaterium and rhizobium leguminosarum by Phasoli on nodulation, nutrient uptake, yield and yield parameters of common bean (Phaseolus vulgaris L. cv. Elkoca-05'). J Plant Nutrition 33: 2104-2119.

Erkovan, H.I., M.K. Gullap, M. Dasci and A. Koc. 2010. Effects of phosphorus fertilizer and phosphorus solubilizing bacteria application on clover dominant meadow: I. Hay yield and botanical composition. Turkish J Field Crops 15: 12-17.

Glick, B.R., D.M. Penrose and J. Li. 1998. A model for the lowering of plant ethylene concentrations by plant growth-promoting bacteria. J Theor Biol. 190: 63-68.

Halder, A.K., A.K. Mishra and P.K. Chakrabartty. 1990. Solubilization of phosphatic compounds by Rhizobium. Indian J Microbiol. 30: 311-314.

Hewedy, A.M. 1999. Influence of single and multi bacterial fertilizer on the growth and fruit yield of tomato. Egypt J Appl Sci 14: 508-523.

Holl, F.B., C.P.Chanway,R.Turkington and R.A.Radley. 1988. Response of crested wheatgrass (AgropyroncristatumL.), perennial ryegrass (LoliumperenneL.) and white clover (Trifolium repensL.) toinoculation with Bacillus polymyxa. Soil Biol Biochem. 20: 19-24.

Illmer, P. and F. Schinner. 1992. Solubilization of inorganic phosphates by microorganisms isolated from forest soils. Soil Biol Biochem. 24: 389-395.

Jones, D.I.H. 1981. Chemical composition and nutritive value.In: Sward Measurement Handbook, ed.Handson, J. Baker, R.D. Davies, A. Laidlows, A.S. Leawer, J.D., 243-265, The British Grassland Society Devon, UK.

Mehrvarz, S. and M.R. Chaichi. 2008. Effect of Phosphate solubilizing microorganisms and phosphorus chemical fertilizer on forage and grain quality of Barley. American-Eurasian J Agric & Environ Sci 3: 855-860.

Mertens, D., 2005.AOAC official method.In: Metal in Plants and Pet Foods. Official Methods of Analysis, ed.Horwitz, W. Latimer, G.W., 3-4, AOAC-International Suite, Gaitherburg, MD, USA.

Miller, D.A. and H.F. Reetz. 1995. Forage fertilization. In: Forages: An Introduction to Grassland Agriculture, ed. Barnes, R.F. Miller, D.A. and Nelson, C.J., 71-87, Iowa State Univ. Press, Ames.

Mishra, A., K. Prasad and G. Rai. 2010. Effect of bio-fertilizer inoculations on growth and yield of dwarf field pea (Pisumsativum L.) in conjunction with different doses of chemical fertilizers. J Agron. 9: 163-168.

Orson, J.A. 1996. The sustainability of intensive arable systems: Implications for rotational policy. Aspects of Applied Biology 47: 11-18.

Osman, M.E.H., M. El-Sheekh, H. El-Naggar and S.F. Gheda. 2010. Effect of two species of cyanobacteria as biofertilizers on some metabolic activities, and yield of pea plant. Biol Fertil Soil 46: 861-875.

Peix, A., A.A. Rivas-Boyero, P.F. Mateos, C. Rodriguez-Barrueco, E. Martinez-Molina and E. Velazquez. 2001. Growth promotion of chickpea and barley by a phosphate solubilising strain of Mesorhizobiummediterraneum under growth chamber conditions. Soil Biol and Biochem. 33: 103-110.

Read, D.W.L. and R. Ashford. 1968. Effect of varying levels of soil and fertilizer phosphorus and soil temperature on the growth and nutrient content of Bromegrass and Reed Canarygrass.Agron J. 60: 680-682.

Reichert, R.D. and S.L. McKenzie. 1982. Composition of peas (Pisumsativum) varying widely in protein content. J Agric and Food Chem.30: 312-317.

Rodriguez, H. and R. Fraga. 1999. Phosphate solubilising bacteria and their role in plant growth promoting. Biotec Advences 17: 319-339.

Sahin, F., R. Cakmakci and F. Kantar. 2004. Sugar beet and barley yields in relation to inoculation with N_2-fixing and phosphate solubilising bacteria. Plant and Soil 265: 123-129.

Salantur, A., A. Ozturk, S.Akten,F.Sahin and M.F.Donmez. 2005.Effect of inoculation with non-indigenous and indigenous rhizobacteria of Erzurum (Turkey) origin on growth and yield of spring barley. Plant and Soil 275: 147-156.

SAS Institute. 1998 Statistical Analysis System Institute: StatView Reference Manual. SAS Institute, Cary, NC.

Scheffer, F. and P. Schachtschabel. 1992.Lehrbuch der Bodenkunde. Ferdinand EnkeVerlag, Stuttgart, Germany.

Serin, Y. and M. Tan. 2011. Forage Legumes. Publication No: 190, University of Ataturk, Faculty of Agriculture Press, Erzurum.

Stirk, W.A., V. Ordog, J. Van Staden and K. Jager. 2002.Cytokinin and auxin like activity in cyanophyta and microalgae. J Appl Phycol. 14: 215-221.

Xie, H., J.J. Pasternak and B.R. Glick. 1996. Isolation and characterization of mutants of the plant growth-promoting rhizobacterium*Pseudomonas putida* GR12-2 that overproduce indolacetic acid. Curr Microbiol. 32: 67-71.

Uzun, A. and E. Acikgoz. 1998. Effect of sowing season and seeding rate on the morphological traits and yields in pea cultivars of different leaf types.J Agron and Crop Sci. 181: 215-222.

Yolcu, H., A. Gunes, M.K. Gullap and R. Cakmakci. 2012. Effects of plant growth-promoting rhizobacteria on some morphologic characteristics, yield and quality contents of Hungarian vetch. Turkish J Field Crops 17: 208-214.

Zaccaro, M.C., G.Z. De Caire, M.S. De Cano, R.M. Palma and K. Colombo. 1999. Effect of cynobacterial inoculation and fertilizers on rice seedlings and postharvest soil structure. Comm Soil Sci Plant Anal. 30: 97-107.

14

INHERITANCE OF GRAIN YIELD AND ITS CORRELATION WITH YIELD COMPONENTS IN BREAD WHEAT (*Triticum aestivum* L.)

Jinbao YAO*, Xueming YANG, Miaoping ZHOU, Dan YANG, Hongxiang MA

Provincial Key Laboratory of Agrobiology, Jiangsu Academy of Agricultural Sciences, Nanjing, CHINA
*Corresponding author: yaojb@jaas.ac.cn

ABSTRACT

The inheritance of grain yield plant[-1] and its correlation with other yield components were investigated in a diallel cross of seven wheat parents during the crop season of 2011-2012. Mean square of GCA effect was 2.90 for grain yield plant-1, which was highly significant (P<0.01), indicating that additive effect played important role in the inheritance of the trait. SCA effect was also highly significant (P<0.01) for grain yield plant-1 (0.68), suggesting that the trait was also controlled by non-additive effect. The estimates of GCA showed that the best combiner for grain yield plant-1 was Ningmai 9. The additive-dominance model was partially adequate for grain yield plant-1 and it was controlled by the over dominance type of gene action. Ningmai 8 possessed maximum dominant genes, whereas Yangmai 9 had maximum recessive genes. Grain yield plant-1 might be controlled by two groups of genes and exhibited moderately high value of narrow sense heritability (h2N=69.51%). The statistical analysis revealed that grain yield plant-1 was positively and significantly correlated with tillers plant-1 (rp=0.584, rg=0.595) and number of grains spike-1 (rp=0.528, rg=0.507) at phenotypic and genotypic levels.

Key words: diallel cross, grain yield, inheritance, Triticum aestivum L.

INTRODUCTION

Wheat (Triticum aestivum L.) is an important cereal crop next to rice in both acreage and production constituting about 22% of the staple food in China（Lu et al., 2010）. Wheat has been cultivated in an area of 24.2 million hectares with the total production of 115 million metric tons in the year of 2010-2011 (Shen, 2012). The average yield of wheat in China is 4.75 t ha-1, which is low as compared to other leading wheat producers in the world like Germany and France where average yields are 7.4 t and 7.2 t /ha, respectively. The yield is generally insufficient to fulfill the domestic requirements due to the increase in population (Xiao, 2006). Therefore, it is necessary to develop the new wheat cultivars, having wider genetic base capable of producing better yield under a wide range of agro-climatic conditions.

The grain yield of wheat is determined by three yield components: productive spikes per unit area, number of grains spike-1 and 1000-grain weight (Tian et al., 2012). The grain yield and its components are controlled by many genes, whose expression is greatly influenced by the varying environments (Groos et al., 2003). In most of the diallel studies of wheat, grain yield plant-1 seemed to be controlled by over dominance type of gene action (Singh and Sharma, 1976; Hussain et al., 2008; Akram et al., 2009; Ojaghi and Akhundova, 2010; Ahmad et al., 2011). However, researchers like Riaz and Chowdhry (2003);

Samiullah et al. (2010) and Farooq et al. (2010) observed partial dominance with additive type of gene action for grain yield plant-1. Zhang and Xu (1997) reported that grain yield plant-1 could result from additive and dominant genes with the possibility of epistatic genetic effects. Heritability estimate is a valuable parameter for determining the magnitude of genetic gain from selection. Low, medium, and high narrow sense heritability estimates were reported for grain yield plant-1 (Mckendry et al., 1998; Novoselovic et al., 2004; Liu and Wei, 2006; Erkul et al., 2010; Ahmad et al., 2011). Grain yield in wheat is a complex trait including number of fertile tillers plant-1, number of grains spike-1 and 1000-grain weight, and it is closely associated with its components (Khaliq et al., 2004; Munir et al., 2007; Ali et al., 2008).

The objective of this study was to investigate combining ability, and gene action for grain yield plant-1, and its correlations with yield components in a population of the 7×7 incomplete diallel cross experiment in wheat. The result of this study can be used in the selection of desirable parents for an effective breeding program to develop the new wheat varieties with high yield potential.

MATERIALS AND METHODS

Plant materials

The experimental material comprised of seven wheat varieties: Ningmai 8, Ningmai 9, Yangmai 158, Yangmai

9, Jimai 17, Zheng 9023, and Yumai 18. The parents were chosen based on their broad genetic background and large variations for grain yield plant-1 and yield components (Table 1). These genotypes were crossed in an incomplete diallel fashion during April, 2011. For each of the cross, 15 spikes were emasculated and bagged to avoid contamination with foreign pollen. Pollination with the pollen collected from the specific male parent was done in the morning when the ovaries became receptive. At maturity the seeds from each cross were harvested and stored separately.

Table 1. Genetic background and yield traits of the seven parents

Parents	Pedigree	Released year	Grain yield plant^{-1}(g)	Tillers plant^{-1}	No. grains spike^{-1}	1000-grain weight(g)
Ningmai 8	Yangmai 5/Yang 86-17	1996	7.43	4.53	51.17	32.51
Ningmai 9	Yang 86-17/Xifeng	1997	9.03	4.97	55.52	32.86
Yangmai 158	Yangmai 4/st1472-506	1993	9.39	4.80	48.96	39.75
Yangmai 9	Jian 3/Yangmai 5	1996	7.80	4.77	47.98	34.12
Jimai 17	Lingfen 5064/Lumai 13	1999	7.12	5.13	39.40	35.70
Zheng 9023	[83(2)3-3/84(14)43//Xiaoyan 6/Xinong 65]F$_3$/3/Shannong 13	2001	8.05	4.93	37.81	43.45
Yumai 18	Zhengzhou 761/Yanshi 4	1990	7.17	4.93	42.43	34.72

Experimental design

The seeds of 7 parents and 21 F1 hybrids were sown in the experimental field, Jiangsu Academy of Agricultural Sciences, Nanjing, China, in the first week of November in 2011 and in a Randomized Complete Block Design with three replications. Plant-to-plant and row-to-row spacing was 6.5 and 25 cm, respectively. Within each block, each genotype occupied a plot of two rows of 2 m-long. Standard cultural operations including weeding, fertilizers and disease control were carried out uniformly. At maturity, in late May 2012, ten plants were randomly selected from each of the parents and F1 progenies to determine fertile tillers plant-1, number of grains spike-1, 1000-grain weight and grain yield plant-1.

Statistical analysis

To assess the differences among parents and F1 progenies, the data were subjected to analysis of variance using the SAS software (SAS Institute Inc. 2002 & 2003). The general and specific combining ability values were estimated using Method II, Model I of Griffing's (1956). Two scaling tests (Mather and Jinks, 1982) were applied to test the validity of the additive-dominance model. Further, the Hayman (1954) method was used for estimation of gene actions. Correlation coefficients between grain yield plant-1 and yield components were determined using the data processing system (Tang and Feng, 2002).

RESULTS AND DISCUSSION

Analysis of variance

Analysis of variance revealed highly significant differences (P≤0.01) among the genotypes for grain yield plant-1 (Table 2). These results permitted further analysis of combining abilities. Both general combining ability (GCA) and specific combining ability (SCA) variances were highly significant for grain yield plant-1 (Table 2), indicating the importance of both additive and non-additive gene effects. These results were in agreement with earlier findings (Wang et al., 2003; Farooq et al.,

2006; Hussain et al., 2012). However, Akram et al. (2011) illustrated that the additive effects were more important for the genetic control of grain yield plant-1.

Table 2. Analysis of variance for combining ability for grain yield plant^{-1}

Source of variation	df	SS	MS	F
Replications	2	7.329	3.664	4.339[*]
Genotypes	27	95.131	3.523	4.172[**]
Error	54	45.609	0.845	
GCA	6	17.424	2.904	10.315[**]
SCA	21	14.287	0.680	2.416[**]
Error	54	15.203	0.282	

*, ** Significance at P≤0.05 and P≤0.01 levels, respectively

Performance of parents and combining ability

Mean grain yield plant-1 and GCA effects of parents were given in Table 3. Significant differences were found for grain yield plant-1 among parents. Yangmai 158 and Ningmai 9 had significantly higher grain yield plant-1 than other five parents and they can be considered as high yielding parents. Zheng 9023 had moderately high yield capacity while Ningmai 8, Yangmai 9, Jimai 17 and Yumai 18 had medium and low yield capacity. Estimates of GCA effects of parents ranged from ☐ 0.625 to 0.915 (Table 3). The highest positive-valued GCA was exhibited for Ningmai 9 followed by Yangmai 158. The GCA value of Ningmai 9 was significantly higher than that of Zheng 9023, Yangmai 9, Jimai 17, Yumai 18 and Ningmai 8 except Yangmai 158. This result indicated that Ningmai 9 was the best combiner for grain yield plant-1 and may serve as genetic sources in breeding programs to increase grain yield. The highest negative-valued GCA was found for Ningmai 8 followed by Yumai 18. The positive and significant correlation (r = 0.900, P≤0.01) between GCA and parental performance (Table 3) suggested that selection of parents for grain yield plant-1 could be made on the basis of their performance per se. The high grain yield plant-1 of certain crosses (Ningmai 8 × Jimai 17, Ningmai 9 × Zheng 9023, and Ningmai 9 × Yangmai9) showed strong positive SCA effects (Table 4). Because in most cases at least one good combining parent was included in these crosses, their progenies had higher grain yield plant-1 than the overall means and yielded desirable transgressive segregations.

Table 3. Mean grain yield plant^{-1} and general combining ability (GCA) effects

Parents	GCA	Mean (g)
Ningmai 9	0.915 aA	9.031 aA
Yangmai 158	0.480 abAB	9.387 aA
Zheng 9023	0.143 bcBC	8.045 bcABC
Yangmai 9	0.006 bcBCD	7.799 cBC
Jimai 17	− 0.325 cdCD	7.123 cC
Yumai 18	− 0.594 dD	7.169 cC
Ningmai 8	− 0.625 dD	7.434 cC
r (GCA, mean)	0.900**	

The values followed by different capital or small letters within the same column are significantly different at the 0.01 and 0.05 probability levels, respectively. ** Significance at P≤0.01 level. r correlation coefficient

Assessment of grain yield plant-1 for Additive-Dominance model

The data were assessed for Additive-Dominance (AD) model by exploiting various adequacy parameters given in Table 5. According to Mather and Jinks (1982) the data will only be valid for genetic interpretation if the value of regression coefficient (b) must deviate significantly from zero but not from the unity. The regression analysis revealed that regression coefficient for grain yield plant-1 departed significantly from zero but not from unity, suggesting the absence of non-allelic interactions in genetic behavior of grain yield plant-1 which in turn attested the data valid for AD model for the trait. The appropriateness of the model data analysis was also verified by the analysis of variance of (Wr+ Vr) and (Wr − Vr). The lack of significant variation in the (Wr− Vr) arrays suggested that any kind of epistasis was not involved in the phenotypic expression of the trait. Although the value of regression coefficient (b) proved the fitness of the data of grain yield plant-1 for AD model, mean square value of (Wr+ Vr) for the trait indicated no significant deviation, thus emphasizing partial validity of the trait. This was also confirmed by Ahmad et al. (2011); Farooq et al. (2011) and Nazeer et al. (2011). The partially adequate model for grain yield plant-1 may be due to the presence of non-allelic interaction, linkage and non independent distribution of the genes in the parents as suggested by Mather and Jinks (1982).

Table 4. Effects of specific combining ability for grain yield plant^{-1} in 21 crosses

Parents	Ningmai 9	Yangmai 158	Yangmai 9	Jimai 17	Zheng 9023	Yumai 18
Ningmai 8	− 0.267	− 0.843	− 0.439	1.377	0.274	0.052
Ningmai 9		0.641	0.999	− 0.152	1.023	0.877
Yangmai 158			0.701	0.664	− 0.693	0.196
Yangmai 9				0.132	0.733	− 0.176
Jimai 17					0.381	− 0.424
Zheng 9023						0.287

$S.E.(\hat{S}_{ij}-\hat{S}_{ik})$: Standard error of differences for SCA effect among crosses with a communal parent; $S.E.(\hat{S}_{ij}-\hat{S}_{kl})$: Standard error of differences for SCA effect among crosses without a communal parent. $S.E.(\hat{S}_{ij}-\hat{S}_{ik}) = 0.708$, $LSD_{0.05} = 1.415$, $LSD_{0.01} = 1.882$; $S.E.(\hat{S}_{ij}-\hat{S}_{kl}) = 0.662$, $LSD_{0.05} = 1.324$, $LSD_{0.01} = 1.760$

Table 5. Adequacy test of additive-dominance model for grain yield plant^{-1}

Parameters	Grain yield plant^{-1}
Joint regression (b)	0.963±0.253
Test for b=0	3.806**
Test for b=1	0.146^{NS}
Mean squares of Wr+ Vr between arrays	1.334^{NS}
Mean squares of Wr− Vr between arrays	0.230^{NS}
Fitness of the data to Additive-Dominance model	Partial

** Significance at P≤0.01 level, NS=non significant

Genetic components of variation for grain yield plant-1

Genetics of grain yield plant-1 was evaluated by calculation of the genetic components of variation D, H1, H2 and F (Table 6). Additive (D) and non-additive (H1 and H2) components were significant, indicating that both additive and dominance effects were important components of genetic variation for grain yield plant-1. However, dominance (H1 and H2) effects were greater than additive (D), suggesting that non-additive gene action played a predominant role in controlling the genetic mechanism of the trait. The (H1/D)0.5 was more than unity which confirmed the greater contribution of non additive genes in the inheritance of grain yield plant-1. These results were in accordance with those of Arshad and Chowdhry (2003); Hussain et al. (2008); Akram et al. (2009); Nazeer et al. (2010); Ojaghi and Akhundova (2010). Preponderance of dominance effects for grain yield plant-1 suggested that the selection for the trait in early generations may not be useful and it had to be delayed till late segregating generations. Asymmetrical distribution of dominant genes was confirmed by unequal estimates of H1 and H2, which was further supported by the value of H2/4H1 (0.219). The F value, which estimates the relative frequency of dominant to recessive alleles in the parents, was negative. This suggests the excess of recessive alleles present in the parents, which was further supported by the small value (< 1.0) of [(4DH1)0.5+F]/ (4DH1)0.5-F]. The mean dominance effect of the heterozygote locus (h2) was significant, suggesting that heterosis breeding could be rewarding for this trait. Significant environmental component (E) indicated that the trait was highly affected by environmental conditions (Ahmad et al., 2011). The number of gene groups differentiating the parents (k) was 1.984, suggesting the inheritance of grain yield plant-1 was controlled approximately by two groups of genes (Dere and Yildirim, 2006; Nazeer et al., 2010). Estimates of narrow sense heritability (h2N) showed moderately high for grain yield plant-1. Such moderately high heritable value for grain yield plant-1 was also reported by other researchers (Novoselovic et al., 2004; Yang and Cao, 2005; Ajmal et al., 2009; Akram et al., 2009; Farooq et al., 2010). However Mckendry et al. (1998); Liu and Wei (2006); Erkul et al. (2010); Ojaghi and Akhundova (2010) found that the narrow sense heritability for grain yield plant-1 was low. Differences in the genetic material and analytical technique used in this study could account for these differences.

Table 6. Estimate of genetic parameters of grain yield plant^{-1} in a 7×7 diallel cross of wheat

Genetic parameters	Grain yield plant^{-1}
D	$0.487^{**}\pm0.118$
F	$-0.742^{**}\pm0.282$
H_1	$1.663^{**}\pm0.283$
H_2	$1.455^{**}\pm0.250$
h^2	$2.886^{**}\pm0.168$
E	$0.315^{**}\pm0.042$
$(H_1/D)^{1/2}$	1.848
$H_2/4H_1$	0.219
$[(4DH_1)^{1/2}+F/((4DH_1)^{1/2}-F)$	0.416
K	1.984
h^2_N (%)	69.51
r[(Wr+Vr), Pr]	-0.171

** Significance at P≤0.01 level. r correlation coefficient

Graphical (Vr/Wr) representation for grain yield plant-1

The Vr/Wr graph (Fig. 1) showed that the regression line intercepted the Wr-axis below the point of origin, suggesting that the trait was controlled by the over dominance type of gene actions and this was superbly maintained by the results provided by the higher values of dominance components H1 and H2 over the additive one D. Similar results have earlier been reported by Asif et al., (1999); Chowdhry et al., (2002); Arshad and Chowdhry, (2003); Kashif and Khaliq, (2003); Saleem et al., (2005). However, these results were not in accordance with Riaz and Chowdhry (2003); Samiullah et al. (2010). The distribution of array points along the regression line (Fig. 1) revealed that parent Ningmai 8 contained maximum dominant genes being closest to the origin while parent

Yangmai 9 carried maximum recessive genes being farthest from the origin. The other two parents Jimai 17 and Yumai 18 also had relatively high frequency of recessive genes. Negative r-value (r = - 0.171) between parental values (Pr) with (Wr + Vr) indicated that the parents with high grain yield plant-1 may carry dominant genes. Similar results were reported by Inamullah et al. (2006); Ojaghi and Akhundova (2010) while Dere and Yildirim (2006) reported that the parents with high grain yield plant-1 may carry recessive genes.

Correlation analysis

Correlation coefficients between grain yield plant-1 and yield components were shown in Table 7. In general, correlation coefficients at genotypic level were higher than those of phenotypic level. It might be due to depressing effect of environment on character association as reported earlier for wheat crop (Proda and Joshi, 1970). Grain yield plant-1 had a highly significant positive genotypic correlation with tillers plant-1 (r = 0.595, P≤0.01) and number of grains spike-1 (r = 0.507, P≤0.01), and it showed positive and non-significant correlation with 1000-grain weight at both genotypic and phenotypic levels, suggesting that increase in tillers plant-1 and number of grains spike-1 would increase grain yield plant-1. These results were in agreement with those of Ali et al. (2008). Khaliq et al. (2004); Munir et al. (2007) reported positive and significant genotypic and phenotypic correlation of grain yield plant-1 with yield components.

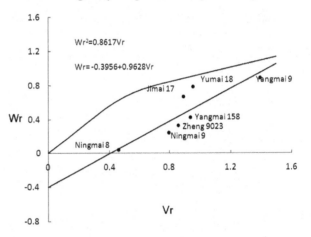

Fig 1. Vr versus Wr graph for grain yield plant^{-1}

Among the yield components, Tillers plant-1 was positive and non-significant correlation with number of grains per spike and 1000-grain weight at genotypic level. Positive and non-significant correlation between tillers plant-1 and number of grains spike-1 at both genotypic and phenotypic levels was also reported by Khan and Dar (2010). Positive and non-significant correlation between tillers plant-1 and 1000-grain weight was found at genotypic level, however the correlation was highly significant (Khokhar et al. 2010), whereas Kashif and Khaliq (2004) reported negative and significant association of tillers plant-1 with 1000-grain weight. There was negative and significant correlation between number of grains spike-1 and 1000-grain weight at both genotypic and phenotypic levels. Similar findings have also been reported by Khan et al. (2010). However, the results were contrary with the findings of Ashfaq et al. (2003).

CONCLUSION

The results revealed that there was significant genetic variation for grain yield plant-1 among the genotypes. Significant GCA and SCA effects for grain yield plant-1 implied the role of both additive and non-additive gene actions in the genetic control of the trait. Ningmai 9 appeared to be a promising parent for improvement of grain yield plant-1. The additive-dominance model was partially adequate for grain yield plant-1. Dominant genes coupled with moderately high narrow sense heritability were involved in the inheritance of grain yield plant-1, suggesting that selection for the trait in early generations (F2 and F3) may not be effective. Grain yield plant-1 had significantly positive genotypic correlation with tillers plant-1 and number of grains spike-1.

Table 7. Phenotypic (r_p) and genotypic (r_g) correlation coefficient between grain yield plant^{-1} and yield components

Characters	r	Tillers plant^{-1}	Number of grains spike^{-1}	1000-grain weight
grain yield plant^{-1}	r_p	0.584**	0.528**	0.251
	r_g	0.595**	0.507**	0.261
Tillers plant^{-1}	r_p		0.107	− 0.062
	r_g		0.132	0.002
Number of grains spike^{-1}	r_p			− 0.552**
	r_g			− 0.614**

** Significance at $P \leq 0.01$ levels.

ACKNOWLEDGMENTS

This work was partially supported by the research projects (BY2012208, CX122026) funded by the Department of Science and Technology, Jiangsu, and Public Finance Office, Jiangsu, respectively.

LITERATURE CITED

Ahmad, F., S.Khan, S.Q.Ahmad, H.Khan, A.Khan and F.Muhammad. 2011. Genetic analysis of some quantitative traits in bread wheat across environments. Afr. J. Agric. Res. 6:686-692.

Ajmal, S.U., N.Zakir and M.Y.Mujahid. 2009. Estimation of genetic parameters and character association in wheat. J. Agric. Biol. Sci. 1:15-18.

Akram, Z., S.U.Ajmal, G.Shabbir, M. Munir and M.Cheema. 2009. Inheritance mechanism of some yield components in bread wheat. Pak. J. Agric. Res. 22:1-8.

Akram, Z., S.U.Ajmal, K.S.Khan, R.Qureshi and M.Zubair. 2011. Combining ability estimates of some yield and quality related traits in spring wheat (Triticum aestivum L.). Pak. J. Bot. 43:221-231.

Ali, Y., B.M.Atta, J.Akhter, P.Monneveux and Z.Lateef. 2008. Genetic variability, association and diversity studies in wheat (Triticum aestivum L.) germplasm. Pak. J. Bot. 40: 2087-2097.

Arshad, M. and M.A.Chowdhry. 2003. Genetic behaviour of wheat under irrigated and drought stress environment. Asian J. Plant Sci. 2:58-64.

Ashfaq, M., A.S.Khan and Z.Ali. 2003. Association of morphological traits with grain yield in wheat (Triticum aestivum L.). Int. J. Agri. Biol. 5:262-264.

Asif, M., I.Khaliq, M.A.Chowdhry and A.Salam. 1999. Genetic mechanism for some spike characteristics and grain yield in bread wheat. Pak. J. Boil. Sci. 2:948-951.

Chowdhry, M.A., A.Ambreen and I.Khaliq. 2002. Genetic control of some polygenic traits in Aestivum species. Asian J. Plant Sci. 1: 235-237.

Dere, S. and M.B.Yildirim. 2006. Inheritance of grain yield per plant, flag leaf width, and length in an 8×8 diallel cross population of bread wheat (T. aestivum L.) Turk. J. Agric. For. 30:339-345.

Erkul, A., A.Unay and C.Konak. 2010. Inheritance of yield and yield components in bread wheat (Triticum aestivum L.) cross. Turk. J. Field Crops 15:137-140.

Farooq, J., I.Habib, A.Saeed, N.N.Nawab, I.Khaliq and G.Abbas. 2006. Combining ability for yield and its components in bread wheat (Triticum aestivum L.). J. Agri. Soc. Sci. 2:207-211.

Farooq, J., I.Khaliq, M.A.Ali, M.Kashif, A.Rehman, M.Naveed, Q.Ali, W.Nazeer and A.Farooq. 2011. Inheritance pattern of yield attributes in spring wheat at grain filling stage under different temperature regimes. Aust. J. Crop Sci. 5:1745-1753.

Farooq, J., I.Khaliq, A.S.Khan and M.A.Pervez. 2010. Studing the genetic mechanism of some yield contributing traits in wheat (Triticum aestivum). Int. J. Agri. Biol. 12:241-246.

Griffing, B. 1956. Concept of general and specific combining ability in relation to diallel crossing system. Aust. J. Biol. Sci. 9: 463-493.

Groos, C., N.Robert, E.Bervas and G.Charmet. 2003. Genetic analysis of grain protein-content, grain yield and thousand-kernel weight in bread wheat. Theor. Appl. Genet. 106:1032-1040.

Hayman, B.I. 1954. The theory and analysis of diallel crosses. Genetics 39:789-809.

Hussain, F., R.A.Sial and M.Ashraf. 2008. Genetic studies for yield and yield related traits in wheat under leaf rust attack. Int. J. Agri. Biol. 10:531-535.

Hussain, T., W.Nazeer, M.Tauseef, J.Farooq, M.Naeem, K.Mahmood, M.Iqbal, A.Hameed, H.M.Nasrullah and S.Freed. 2012. Inheritance of some spike related polygenic traits in spring Wheat (Triticum aestivum L). Afr. J. Agric. Res. 7: 1381-1387.

Inamullah, H.Ahmad, F.Mohammad, Siaj-Ud-Din, G.Hassan and R.Gul. 2006. Diallel analysis of the inheritance pattern of agronomic traits of bread wheat. Pak. J. Bot. 38:1169-1175.

Kashif, M. and I.Khaliq. 2003. Mechanism of genetic control of some quantitative traits in bread wheat. Pak. J. Boil. Sci. 6: 1586-1590.

Kashif, M. and I.Khaliq. 2004. Heritability, correlation and path coefficient analysis for some metric traits in wheat. Int. J. Agri. Biol. 6:138-142.

Khaliq, I., N.Parveen and M.S.Chowdhry. 2004. Correlation and path coefficient analysis in bread wheat. Int. J. Agri. Biol. 6:633-635.

Khan, A.J., F.Azam and A.Ali. 2010. Relationship of morphological traits and grain yield in recombinant inbread wheat lines grown under drought conditions. Pak. J. Bot .42:259-267.

Khan, M.H. and A.N.Dar. 2010. Correlation and path coefficient analysis of some quantitative traits in wheat. Afr. Crop Sci. J. 18:9-14.

Khokhar, M.I., M.Hussain, M.Zulkiffal, N.Ahmad and W.Sabar. 2010. Correlation and path analysis for yield and yield contributing characters in wheat (Triticum aestivum L.). Afr. J. Plant Sci. 4:464-466.

Liu, X.Y. and Y.Z.Wei. 2006. Genetic study on winter wheat yield characters. Chin. J. App. Environ. Biol. 12:1-4.

Lu, B., B.Ding, X.T.Lu, Z.W.Yu, G.C.Zhao and F.S.Wan. 2010. Arrangement planning of Chinese wheat ascendant regions. Chinese J. Agric. Resouces and Regional Planning 31:6-12.

Mahter, K. and J.L.Jinks. 1982. Biometrical Genetics. 3rd Edition. Chapman and Hall, London. UK.

Mckendry, A.L., P.B.E.Mcvetty and L.E.Evans.1998. Inheritance of grain protein concentration, grain yield, and related traits in spring wheat (Tnticum aestivum L.). Genome 30:857-864.

Munir, M., M.A.Chowdhry and T.A.Malik. 2007. Correlation studies among yield and its components in bread wheat under drought conditions. Int .J. Agri. Biol. 9:287-290.

Nazeer, W., Z.Ali, A.Ali and T.Hussain. 2010. Genetic behaviour for some polygenic yield contributing traits in wheat (Tnticum aestivum L.). J. Agric. Res. 48:267-277.

Nazeer, W., J.Farooq, M.Tauseef, S.Ahmed, M.A.Khan, K.Mahmood, A.Hussain, M.Iqbal and H.M.Nasrullah. 2011. Diallel analysis to study the genetic makeup of spike and yield contributing traits in wheat (Triticum aestivum L.). Afr. J. Biotechnol. 10: 13735-13743.

Novoselovic, D., M.Baric, G.Drezner, J.Gunjaca and A.Lalic. 2004. Quantitative inheritance of some wheat plant traits. Genet. Mole. Biol. 27:92-98.

Ojaghi, J. and E.Akhundova. 2010. Genetic analysis for yield and its components in doubled haploid wheat. Afr. J. Agric. Res. 5:306-315.

Paroda, R.S. and A.B.Joshi. 1970. correlations, path-coefficients and the implication of discriminate function for selection in wheat (Triticum aestivum L.). Heredity 25:383-392.

Riaz, R. and M.A.Chowdhry. 2003. Genetic analysis of some economic traits of wheat under drought condition. Asian. J. Plant Sci. 2:790-796.

Saleem, M., M.A.Chowdhry, M.Kashif and M.Khaliq. 2005. Inheritance pattern of plant height, grain yield and some leaf characteristics of spring wheat. Int. J. Agri. Biol. 7:1015-1018.

Samiullah, A.S.Khan, A.Raza and S.Sadique. 2010. Gene action analysis of yield and yield related traits in spring wheat (Triticum aestivum). Int. J. Agri. Biol. 12:125-128.

SAS. 2002-2003. User's guide; Statistics, version 9.1. SAS Institute Inc. Cary, NC, USA.

Shen, H.Y. 2012. Analysis and prospect of China wheat market in 2011. Food and Oil 3:39-42.

Singh, R.B. and G.S.Sharma.1976. Induced polygenic variations in relation to gene action for yield and yield components in spring wheat. Can. J. Genet. Cytol. 18:217-223.

Tang, Q.Y. and M.G.Feng. 2002. Practical statistics and data processing system. Beijing: Science Press.

Tian, J.C., Z.Y.Deng, R.B.Hu and Y.X.Wang. 2012. Yield components of super wheat cultivars with different types and the path coefficient analysis on grain yield. Acta Agron. Sin. 32:1699-1705.

Wang, C.L., H.T.Gao, S.Z.Wang, G.H.Duan, S.H.Wu, X.P.Zhang and S.Z.Lu. 2003. Genetic analysis of quality and yield characters in winter wheat. J. Triticeae Crops 23:26-28.

Xiao, S.H. 2006. Trend of wheat breeding and food security in China. Sci. Tech. Rev. 24:5-7.

Yang, S.Q. and Y.P.Cao. 2005. Genetic studies on three grain weights character in winter wheat. J. Henan Institute Sci. Tech. 33: 4-6.

Zhang, L.H. and M.F.Xu. 1997. An analysis of genetic effects on harvest index and several other agronomic characteristics of wheat. Acta Agri. Nucle. Sin. 11:135-140.

15

QUALITY AND HYGIENIC CONDITIONS OF WHITE LUPIN SILAGE, AFFECTED BY FORAGE STAGE OF GROWTH AND USE OF SILAGE ADDITIVES

Agnieszka FALIGOWSKA[1]*, Marek SELWET[2], Katarzyna PANASIEWICZ[1], Grażyna SZYMANSKA[1]

[1]Poznan University of Life Sciences, Department of Agronomy, POLAND
[2] Poznan University of Life Sciences, Department of General and Environmental Microbiology, POLAND
*Corresponding author: faliga@up.poznan.pl

ABSTRACT

A two-factor field experiment with white lupin cv. Butan was carried out. The first factor was the green forage harvest date (the flat pod stage – Cut 1 and the stage of green ripe seeds – Cut 2), while the second one – application of silage additives: biological (strains of lactic acid bacteria) and chemical (a mixture of organic acids), and the control treatment (without additives). In Cut 2 higher fresh matter (FM) and dry matter (DM) yields were obtained. Silage inoculated with the biological additive contained a significantly greater count of lactic acid bacteria. Both additives reduced counts the *Clostridium* bacteria, yeasts and mould fungi. The silage with the chemical additive had a three-fold higher content of water-soluble carbohydrates (WSC), while the biological additive increased lactic acid (LA) levels. White lupin can be used as a silage raw material, but plants before ensiling should be partially wilted and silage additives should be applied.

Key words: harvest date, lupin silage, silage additives

INTRODUCTION

White lupin (*Lupinus albus* L.) is an annual legume belonging to the *Fabaceae* family, it is used for human consumption, as green manure and forage crop (Huyghe, 1997). Forages are major constituents of dairy and beef cattle diets (Mustafa et al., 2002). The purchase of compound feed represents a substantial part of variable costs in on-farm ruminant production, thus the use of protein-rich alternative forage crops, grown on-farm, needs to be considered (Frasel et al., 2001). Annual legumes and cereals such as common vetch, hairy vetch, grasspea, oat (Dumont at al., 2005), barley and triticale (Rojas at al., 2004), are potentially the most viable fodder sources (Karadag and Buyukburc, 2003), while the use of lupin in animal nutrition may increase profitability of production (McNaughton, 2011). According to (Idziak et al., 2013), based on FAOSTAT data corn is one of the most commonly cultivated plants worldwide. Maize silage is a high-quality forage that is used on many dairy farms and on some beef cattle farms (Budakli Carpici et al., 2010; Iptas and Yavuz, 2008; Kusaksiz, 2010), whereas Doležal et al. (2008) reported that some researchers, e.g. Carruthers et al. (2000) and Egorov et al. (2001), studied the potential for lupin application as a silage raw material also in mixtures with cereals and grasses. Voytekhovich (2000) argued that narrow-leaved lupin silage is of better quality in terms of its nutritive value than white lupin silage. In contrast, Fraser et al. (2005a,b) reported that both lupins can be successfully ensiled as the whole-crop. An appropriate harvest date of a forage crop has a significant effect on silage quality. Delaying of harvest adversely affects the ensiling ability due to an increase in buffer capacity and a decrease of sugar contents. However, the green fodder ensiling capacity may be improved by adding different substances and preparations. They are designed to improve the conditions of lactic fermentation and aerobic stability, to reduce the amount of silage juices and the content of undesirable spores, such as *Clostridium*, and also to improve the collection, palatability and digestibility of fodder. The study conducted by Borreani et al. (2009) showed that field pea, faba bean and lupin may be successfully ensiled after a wilting period under good weather conditions and with the addition of a lactic acid bacteria inoculant.

The experiment was to test the suitability of ensiling white lupin harvested at two different growth stages and to analyze the quality of silage produced using silage additives.

MATERIALS AND METHODS

Experimental site: A field experiment with white lupin cv. Butan was conducted at the Teaching and Experimental Station in Gorzyń (52°33'53 N, 15°53'42 E), belonging to the Poznań University of Life Sciences, Poland. The trial was carried out from 2005 to 2007 on grey-brown podzolic soil under ordinary growing conditions. Four replicate plots of 20 m² were prepared by ploughing and power-harrowing. A fertiliser (P₂O₅ 60 kg ha⁻¹ and K₂O 80 kg ha⁻¹) was applied to the seedbed. No irrigation or fertiliser was applied after sowing. In early April, white lupin seeds were inoculated with *Bradyrhizobium* bacteria (cv. Butan) and drilled at a rate of 225 kg ha⁻¹. Weeds were controlled by post-emergence treatment with 2.0 1 ha⁻¹ of linuron (Agan Chemical Manufacturers Ltd.). Two effects were studied: 1) harvesting dates (stages) and 2) application of silage additives, corresponding to three treatments: wilted crop ensiled either with no additives (0), with a microbial inoculant (B) Polmasil, which contained strains of lactic acid bacteria: *Enterococcus faecium* M74, *Lactobacillus casei*, *Lactobacillus plantarum* and *Pediococcus* spp. at a concentration of 10⁹ CFU (Polmass S.A., Poland), and with a chemical additive (CH) KemiSile 2000, which contained in %: formic acid 55, propionic acid 9, benzoic acid 5, ammonium formate 24, and benzoic acid ester 7 (Kemira OY, Finland). The two harvesting dates and stages were: flat pod (Cut 1) and green ripe seed (Cut 2). At each harvest date the crop at a stubble height of 4-6 cm was cut from the plot area using a plot harvester, and sub-samples of the crop were collected to determine their chemical composition. At each stage of growth the harvested crop was wilted in the field for 24 h.

Ensiling

The wilted crop was chopped with an experimental mechanical chopper to a length of 20–30 mm. Then the inoculant/additive treatment was applied by means of a hand sprayer. About 2.2 kg of the crop was ensiled in sterile 5 dm³ jars (150 mm diameter × 280 mm height). The crop mass was thoroughly mixed before being placed in the mini-silos and then it was compacted in each silo. Four jars - replications of each treatment - were prepared in this way.

The material was stored in the dark at a temperature of 22-25 °C for ten weeks. After that time the jars were opened and representative samples of the ensiled material were collected for analyses of their nutritive value and basic fermentation characteristics. Each year green matter was ensiled in 24 jars (2 harvest dates x 3 (2 silage additives + the control) x 4 replications). The chemical composition of silage was determined by analysing an average sample in each combination (a total of 6 average samples per year); the years were replications. One microbial sample was collected from each jar and its chemical composition was determined: 3 series x 24 (a total of 72 samples per year).

Chemical analysis

The basic composition of forage was determined according to AOAC (1990). The content of water-soluble carbohydrates (WSC) was determined according to the methodology given by McDonald and Henderson (1964), ammonia nitrogen (N-NH₃) (Conway 1962). The pH values were determined, using the pH Meter by Hann Instruments, in a suspension prepared from 10 g of silage and 90 cm³ of deionised water, homogenized for 20 minutes. The concentration of fatty acids was determined using a gas chromatograph equipped with the Supelco FID detector, a 80/100 Chromosorb® WAW glass column of 2 m, I.D. 2 mm with GP filling of 10% SP-1200/1% H₃PO₄ and a Varian 8200 CX autosampler. The carrier gas was hydrogen (flow rate = 30 cm³ min⁻¹), oven temperature was 120°C, injection temperature was 250°C and detector temperature was 300°C. Fluka acid patterns were the reference standards.

Microbiological analysis

The count of *Clostridium* bacteria was determined on MERCK TSC® Agar, the count of lactic acid bacteria – on MERCK ATP Agar, *Enterobacteriaceae* – at a base Fluorocult® LMX Broth, modified according to Manofi,and OSSMER from MERCK solidified with DIFCO agar. The count of mould fungi was determined on a bengal rose agar base, and the counts of yeasts on a wort agar (BTL spółka z o. o., Zakład Enzymów i Peptonów in Łódź). Culture plate was made by successive dilutions.

Statistical analysis

All data were processed using analysis of variance (ANOVA) with the SAS package (SAS Institute, 1999). The means of treatments were compared by means of Tukey's least significant difference test (LSD) at P<0.01 and P<0.05. Field experiments were arranged as a split-plot randomized complete block design with four replications. All data were subjected to analysis of variance based on the general linear model for repeated measurements.

RESULTS AND DISCUSSION

The analysis of variance indicated that there were statistically significant differences between forage crop harvest date and individual years of the study for FM and DM yields and DM content in unwilted and wilted crop (Table 1). The highest yield of green crop (30.1 t ha⁻¹) was recorded in the year 2005, which had the most beneficial weather conditions when lupin vegetation proceeded without major disruptions in water supply. Harvest of lupin in Cut 2 contributed to an increased FM yield and DM yield, and wilting of the crop increased the DM content. The NDF and WSC contents were similar for both harvest dates in the three years, but CP content was on average lower in Cut 2. In studies conducted by Mihailović et al. (2008), FM yield and DM yield of white lupin were also differentiated over the years and amounted respectively to 21.3-50.3 t ha⁻¹ and 3.6-8.6 t ha⁻¹, depending on the cultivar. In the case of pea, FM yield can reach 24.4-30.1 t ha^{-1}, DM yield 4.4-5.5 t ha^{-1} (Turk et al., 2011) and DM content – 318-360 g kg^{-1} (Borreani et al., 2006). The DM yield of soybean intercropped with corn is also higher (Reta Sanchez et al., 2010). The FM yield of narrow-leaved lupin may be 36.6-37.0 t ha^{-1}, and its DM content after wilting is 182-231 g kg^{-1} (Fraser et al., 2005b). In the experiment carried out by Borreani et al. (2009), following a wilting period, the DM content of field pea, faba bean and white lupin increased from 482 to 618 g kg^{-1}, from 237 to 295 g kg^{-1} and from 142 to 173 g kg^{-1}, respectively. The significance of the appropriate choice of harvest date is presented in an experiment conducted by Fraser et al. (2001), who examined, among other things, the effect of harvest date on the suitability of pea and faba bean for ensiling. It turned out that the best term for pea was 12 weeks after sowing, and for faba bean – 14 weeks, when the FM yield, DM content and DM yield were the largest. In the opinion of Turk and Albayrak (2012), harvesting at the late stages caused a reduction in forage quality. Contents of CP decreased with the progress in plant growth, while DM yield, CP yield, and NDF contents increased.

Table 1. The effect of harvest date on fresh matter yield (FM), content of dry matter (DM), dry matter yield and chemical composition of lupin forage in successive years

Parameter	Harvest date (H)	Years (Y)			Significance		S.E.D.	
		2005	2006	2007	H	H × Y	H × Y within H	H × Y between H
FM yield (t ha^{-1})	Cut 1	20.9	15.8	3.8	**	**	2.90	17.25
	Cut 2	39.2	21.0	11.6				
	Mean	30.1	18.6	7.7				
DM yield (t ha^{-1})	Cut 1	3.7	5.3	1.2	**	**	0.95	8.71
	Cut 2	11.3	6.1	3.9				
	Mean	7.5	5.7	2.6				
DM (g kg^{-1}) unwilted forage	Cut 1	158	173	176	**	**	3.15	120.1
	Cut 2	253	182	189				
	Mean	205	177	183				
DM (g kg^{-1}) wilted forage	Cut 1	177	292	321	**	**	8.54	192.0
	Cut 2	288	334	334				
	Mean	233	313	327				
CP (g kg^{-1})	Cut 1	154.0	172.9	108.0	-	-	-	-
	Cut 2	141.0	162.8	115.7				
NDF (g kg^{-1})	Cut 1	261.8	221.3	232.3	-	-	-	-
	Cut 2	266.1	218.3	288.2				
WSC (g kg^{-1})	Cut 1	399.9	394.4	438.7	-	-	-	-
	Cut 2	422.1	386.4	416.5				

Cut 1 flat pod stage; Cut 2 - stage of green ripe seeds; CP - crude protein; NDF - neutral-detergent fibre; WSC - water-soluble carbohydrates; * significant at P<0.05.; ** significant at P<0.01.

According to Gallo et al. (2006), silage quality depends on weather conditions during harvest. In our experiment significant interactions were found between harvest date and the applied silage additive (Table 2). Under the influence of the microbial inoculant the count of lactic acid bacteria significantly increased by about 22.5% in the first dates of harvest. The two additives decreased the number of mould fungi in both the first and second date. On average, in silage from the first harvest date the count of lactic acid bacteria was by 5.8% higher, that of mould fungi was higher by 1.9%, while the count of yeasts was lower by 2%. It was found that the number of lactic acid bacteria under the influence of the microbial inoculant significantly increased by 12.3%, whereas the content of undesirable Clostridium bacteria (16.1-59.2%), yeast (2-6%) and fungi (4.5-14%) significantly decreased as a result of application of both additives. At the same time, it should be mentioned that the chemical additive was more effective, as it significantly decreased also the count of Enterobacteriaceae bacteria. According to Faligowska and Selwet (2012), in yellow lupin silage the microbial inoculant caused a marked increase in the level of lactic acid bacteria. Both additives caused a noticeable decrease in the content of undesirable bacteria from the Enterobacteriaceae family, Clostridium, as well as yeast and mould fungi, still the chemical additive was also more effective than the microbial inoculant.

The chemical composition of white lupin silage was not affected by diverse harvest dates (Table 3). In the case of additives, their addition differentiated only the content of LA and WSC, and decreased the content of N-NH$_3$. When compared to the control, silage with the chemical additive contained three times more WSC. However, silage with the microbial inoculant contained about 50% more LA. CP content was not significantly differentiated in white lupin silage, ranging from 141.7-157.2 g kg^{-1} DM. The crop harvested in Cut 2 contained more WSC, but probably required a higher consumption of WSC in the process of respiration in the early stages of fermentation, because the Cut 2 silage contained about half as much WSC. The results of silage composition, specifically the low level of acetic acid and the very low level of butyric acid, above all suggest a lactic acid homofermentative process both in the control and in silages containing silage additives (McDonald et al.' 1991).

Table 2. The effect of harvest date and application of additives on the microbiological composition of silage (log 10 JTK g^{-1})

Parameter	Harvest date (H)	Inoculation treatment (I)			Significance			S.E.D	
		0	B	CH	H	I	H × I	H × I within H	H × I between H
Lactic acid bacteria	Cut 1	6.88	8.43	6.96					
	Cut 2	6.94	7.09	6.99	**	**	**	0.171	0.175
	Mean	6.91	7.76	6.98					
Enterobacteriacea	Cut 1	3.76	3.60	2.65					
	Cut 2	3.55	3.41	2.50	NS	**	NS	0.423	0.393
	Mean	3.66	3.51	2.58					
Clostridium	Cut 1	3.34	3.02	1.15					
	Cut 2	3.47	2.70	1.62	NS	**	NS	0.621	0.655
	Mean	3.41	2.86	1.39					
Yeasts	Cut 1	4.96	4.83	4.67					
	Cut 2	5.04	4.98	4.73	**	**	NS	0.094	0.096
	Mean	5.00	4.90	4.70					
Mould fungi	Cut 1	4.01	3.85	3.51					
	Cut 2	4.00	3.79	3.37	**	**	**	0.055	0.049
	Mean	4.00	3.82	3.44					

Cut 1 – flat pod stage; Cut 2 - stage of green ripe seeds; 0 - control; B - microbial inoculant; CH - chemical additive;
NS – non-significant; * significant at P<0.05.; ** significant at P<0.01.

Table 3. The effect of harvest date and application of additives on the chemical composition of silage

Parameter g kg^{-1} DM	Harvest date (H)	Inoculation treatment (I)			Significance			S.E.D.	
		0	B	CH	H	I	H × I	H × I within H	H × I between H
DM	Cut 1	220.9	250.6	245.7	NS	NS	NS	57.00	81.10
	Cut 2	218.6	217.9	226.3					
N-NH$_3$	Cut 1	87.5	32.5	75.0	NS	**	NS	18.0	21.2
	Cut 2	85.0	35.0	78.2					
CP	Cut 1	152.5	141.7	157.2	NS	NS	NS	22.16	19.77
	Cut 2	151.4	143.1	148.2					
NDF	Cut 1	319.4	303.1	290.8	NS	NS	NS	45.51	79.20
	Cut 2	351.4	326.2	321.7					
Ash	Cut 1	104.6	102.3	97.0	NS	*	NS	11.20	59.74
	Cut 2	84.1	76.4	67.0					
Fat	Cut 1	24.4	35.6	23.9	NS	NS	*	7.48	17.47
	Cut 2	33.9	32.8	33.9					
WSC	Cut 1	13.7	17.6	42.5	NS	**	NS	14.57	33.13
	Cut 2	7.3	9.4	20.2					
LA	Cut 1	8.5	12.5	5.0	NS	**	NS	4.39	6.02
	Cut 2	8.5	13.1	7.2					
AA	Cut 1	2.3	2.0	5.1	NS	NS	NS	3.98	3.92
	Cut 2	2.7	2.2	2.5					
BA	Cut 1	1.1	0.2	0.5	NS	NS	NS	1.09	1.01
	Cut 2	0.9	0.2	0.1					
pH	Cut 1	4.6	4.1	4.4	NS	**	NS	0.29	0.66
	Cut 2	4.5	3.9	4.2					

Cut 1 – flat pod stage; Cut 2 - stage of green ripe seeds; 0 - control; B - microbial inoculant; CH - chemical additive; DM - dry matter;
N-NH$_3$ - ammonia-N; CP - crude protein; NDF - neutral-detergent fibre; WSC - water-soluble carbohydrates; LA - lactic acid; AA - acetic acid; BC - butyric acid; NS – non-significant; * significant at P<0.05.; ** significant at P<0.01.

In the experiment conducted by Fraser et al. (2005a), harvest date had a significant effect on DM, N-NH$_3$, LA, AA, and WSC concentration in white lupin silage. Fraser et al. (2001) also studied the suitability of pea and field bean as silage materials. It turned out that changes in plant ripeness had little effect on the chemical composition of green forage, but harvest dates differentiated, among other things, the DM content, N-NH$_3$, CP, WSC, LA and pH of silage. Borreani et al. (2006) reported that the stage of growth affected the LA and AA, WSC concentrations in pea silage. In the experiment conducted by Borreani et al. (2009), BA was detected in silages, except for wilted silages made from field pea and white lupin, inoculated with *Lactobacillus plantarum*. As a result, BA was over 25 g kg^{-1} DM in the control silages with the DM content lower than 300 g kg^{-1} DM. The pH and fermentation products were also greatly influenced by the crops and the application of silage additives. Doležal et al. (2008) found that the chemical additive decreased LA, AA, ethanol, N-NH$_3$ and pH, while it raised the CP content of yellow lupin silage. Fraser et al. (2005a) reported that inoculation with *Lactobacillus plantarum* significantly reduced pH and N-NH$_3$, AA, and CP concentrations, while it increased the DM and WSC concentrations in white lupin silages. Similarly, when investigating suitability of pea and faba bean silage Fraser et al. (2001) found that inoculation increased the LA concentration and reduced the pH and N-NH$_3$ and AA concentrations in the silages. Microbial inoculation lowered the pH and N-NH$_3$ values and increased the LA concentrations in all tested pea silages, except for the silages from the earliest harvest date (Borreani et al., 2006).

CONCLUSIONS

White lupin can be used as a silage material, but plants before ensiling should be partially wilted and silage additives should be applied.

LITERATURE CITED

Association of Official Analytical Chemists. 1990. Official Methods of Analysis AOAC. Arlington, USA.

Borreani, G., L. Cavallarin, S. Antoniazzi and E. Tabacco. 2006. Effect of the stage of growth, wilting and inoculation in field pea (Pisum sativum L.) silages. I. Herbage composition and silage fermentation. J. Sci. Food Agric. 86: 1377-1382.

Borreania, G., A. Revello Chiona, S. Colombinib, M. Odoardi, R. Paoletti and E. Tabaccoa. 2009. Fermentative profiles of field pea (*Pisum sativum*), faba bean (*Vicia faba*) and white lupin (*Lupinus albus*) silages as affected by wilting and inoculation. Anim. Feed Sci. Technol. 151: 316-323.

Budakli Carpici, E., N. Celik and G. Bayram. 2010. Yield and quality of forage maize as influenced by plant density and nitrogen rate. Turk. J. Field Crops, 15(2): 128-132.

Carruthers, K., B. Prithiviraj, Q. Fe, D. Cloutier, R.C. Martin and D.L. Smith. 2000. Intercropping of corn with soybean, lupin and forages: silage yield and quality. J. Agron. Crop Sci. 185 (3): 177-185.

Conway, E.J. 1962. Microdiffusion analysis and volumetric error. Crosby Lockwood. London.

Doležal, P., L. Zeman and J. Skládanka. 2008. Effect of supplementation of chemical preservatives on fermentation process of lupin silage. Slovak J. Anim. Sci. 1: 30-38.

Dumont, L.J., G.R. Anrique and C.D. y Alomar. 2005. Effect of two systems of dry matter determination on the chemical composition and quality of direct cut oat silage at different phenological stages. Agric. Tec. (Chile) 65:388-396.

Egorov, I.F. and N.P. Myskov. 2001. Silage from narrow–leaved lupin and its mixtures. In: Kormoproizvodstvo, 5: 27-28.

Faligowska, A. and M. Selwet. 2012. Quality and hygienic condition of yellow lupine silages depending on the harvest date of green forage and additive to ensilaging. Nauka Przyr. Technol., http://www.npt.up-poznan.net/pub/art_6_15.pdf, (Accessed October 17 2011) (in Polish).

Fraser, M.D., R. Fychan and R. Jones. 2001. The effect of harvest date and inoculation on the yield, fermentation characteristics and feeding value of forage pea and field bean silages. Grass Forage Sci. 56: 218-230.

Fraser, M.D., R. Fychan and R. Jones. 2005a. The effect of harvest date and inoculation on the yield and fermentation characteristics of two varieties of white lupin (*Lupinus albus*) when ensiled as a whole-crop. Anim. Feed Sci. Technol. 119: 307-322.

Fraser, M.D., R. Fychan and R. Jones. 2005b. Comparative yield and chemical composition of two varieties of narrow-leafed lupin (*Lupinus angustifolius*) when harvested as whole-crop, moist grain and dry grain. Anim. Feed Sci. Technol. 120: 43-50.

Gallo, M., L. Rajčáková and R. Mlynár. 2006. Effect of different dry matter and biological additives application on fermentation process in red clover silages. Slovak J. Anim. Sci. 39 (1-2): 89-92.

Huyghe, Ch. 1997. White lupin (*Lupinus albus* L.). Field Crops Res. 53: 147-160.

Idziak, R., W. Skrzypczak, H. Waligóra and Z. Woznica. 2013. The effect of mesotrione applied with adjuvants on weed control efficacy and forage sorghum tolerance. Turk J. Agric. For. 37: 265-270.

Iptas S. and M. Yavuz. 2008. Effect of pollination levels on yield and quality of maize grown for silage. Turk J. Agric. For. 32: 41-48.

Karadag, Y. and U. Buyukburc. 2003. Effects of seed rates on forage production, seed yield and hay quality of annual legume-barley mixtures. Turk. J. Agric. For. 27: 169-174.

Kusaksiz, T. 2010. Adaptability of some new maize (*zea mays* l.) cultivars for silage production as main crop in Mediterranean environment. Turk. J. Field Crops, 15(2): 193-197.

McDonald, P., A.R. Henderson and S.J.E. Heron. (1991). The Biochemistry of Silage. 2nd edn. Chalcombe Publications, Marlow.

Mcdonald, P. and A.R. Henderson. 1964. Determination of water-soluble carbohydrates in grass. J. Sci. Food Agric. 15: 395-398.

McNaughton, D. 2011. The use of lupins as an alternative to imported soya in UK livestock systems. In: Proceedings of the 13th International Lupin Conference Poznań (Poland) l. 30

Mihailović, V., G.D. Hill, A. Mikić, B. Ćupina and S. Vasiljević. 2008. White lupin as a forage crop on alkaline soils. In: Lupins for Health and Wealth, Proceedings of the 12th International Lupin Conference, Fremantle, Western Australia. International Lupin Association, Canterbury, New Zealand.

Mustafa, A.F., P. Seguin, D.R. Ouellet and I. Adelye. 2002. Effects of cultivars on ensiling characteristics, chemical composition and ruminal degradability of pea silage. J. Dairy Sci. 85: 3411-3419.

Reta Sánchez, D.G., J.T. Espinosa Silva, A. Palomo Gil, J.S. Serrato Corona, J.A. Cueto Wong and A. Gaytán Mascorro. 2010. Forage yield and quality of intercropped corn and soybean in narrow strips. Span. J. Agric. Res. 8 (3): 713-721.

Rojas G.C., S.A. Catrileo, B.M., Manríquez and F.F. y Calabí. 2004. An evaluation of the cutting stage on triticale (X Triticosecale Wittmack) for silage. Agric. Tec. (Chile) 64:34-40.

Turk M., S. Albayrak and O. Yuksel. 2011. Effect of seeding rate on the forage yields and quality in pea cultivars of differing leaf types. Turk. J. Field Crops, 16(2): 137-141.

Turk, M. and S. Albayrak. 2012. Effect of harvesting stages on forage yield and quality of different leaf types pea cultivar. Turk. J. Field Crops, 17(2): 111-114.

Voytekhovich, I. 2000. Ensilage of narrow–leaved lupin. Vestsi Akademii Agrarnych Nauk Respubliki Belarus 3: 46-49.

16

EFFECT OF SEEDING RATE ON YIELD AND QUALITY OF NON-CHEMICAL FENNEL (*Foeniculum vulgare* Mill.) CULTIVATION

Ayse Betul AVCI

Ege University, Odemis Vocational School, Izmir, TURKEY
Corresponding author: ayse.betul.avci@ege.edu.tr

ABSTRACT

This study was conducted to determine the effects of different seeding rates on yield and quality characteristics of fennel without using any chemical fertilizers and pesticides, in Atabey-Isparta ecological conditions during 2010 and 2011 vegetation periods. The experiment was established as randomized blocks design with three replications and three seeding rates (10, 15 and 20 kg ha^{-1}) were applied. Plant height, fruit yield and biological yield were significantly affected by seeding rate. Fruit yields ranged between 230.35 and 790.96 kg ha^{-1} and the highest yield was obtained from 15 kg ha^{-1} seeding rate. Essential oil contents ranged between 1.60 and 2.46%. The main constituent of the essential oil was identified as *trans*-anethole ranged between 84.48% and 97.79%.

Keywords: *Foeniculum vulgare*, fruit yield, essential oil, *trans*-anethole.

INTRODUCTION

From past to present, plenty of natural resources such as herbal products are used as medicine raw materials. However, natural components are not preferred by the manufacturers because of more expensive obtaining than that of their synthetics. Yet, using of synthetic products causes adverse effects on human health as well as they have therapeutic effects. In this regard, it can be said that increasing population and advancing technology threaten the human health and nature at an unpredictable extent. Due to such reasons, plants that are used in the medicine production are required to cultivate under organic conditions without using chemical inputs like fertilizer and pesticides, instead of collecting from nature or cultivated by using conventional agricultural practices. Therefore, herbal active ingredient can be obtained without undergoing any chemical change. Medicinal and aromatic plants such as sage, anise, thyme, carob, rosemary and fennel are organically grown either for domestic market or for foreign market although it varies on a yearly basis. Organic fennel consumption in Turkey is about 1243.39 tons (Bayram et al., 2010).

Fennel (*Foeniculum vulgare* Mill.) is commonly used as analgesic, anti-depressive, anti-inflammatory, digestive, treatment for disorders and spasmolytic (Basgel and Erdemoglu, 2006). In addition, fennel has carminative, flavoring, antioxidant, antibacterial, antifungal and mosquito repellent properties (Garg et al., 2010).

Fennel is a member of Apiaceae family and grows naturally in Northern (Davis, 1978; Baytop, 1999; Tanker et al., 2007), Western and Eastern Anatolia Regions of Turkey (Ozyilmaz and Yilmaz, 2009; Kizil et al., 2001). It can be annual or perennial (Ceylan, 1997). Although fennel agriculture is done in various provinces in our country such as Bursa, Denizli, Gaziantep, Manisa and Antalya, it is produced around Burdur province the most (Baydar, 2009). There are two important varieties of fennel which are *Foeniculum vulgare* var. *vulgare* (bitter fennel) and *Foeniculum vulgare* var. *dulce* (sweet fennel) (Akgul, 1993). Mentioned researcher reported that sweet fennel fruits contain 2-4% essential oil which the constituents are *trans*- anethole 60-80%, 5-10% fenchone, limonene, methyl chavicol, α- phellandrine, anisaldehyde, *cis*-anethole (due to the toxic effects, presence of *cis*-anethole is not desirable), anisic acid, anicketon, monoterpenes and alcohols.

Various studies were conducted on fennel by many researchers such as sowing methods and times (Ayub et al., 2008), N fertilization and plant density (Nakhaei et al., 2012), sowing date and seeding rate (Arabaci and Bayram, 2005) and sowing date and row spacing (Ahmad et al., 2004). Kandil (2002) also reported that generally fertilization with chemical fertilizers were given higher fennel growth parameters than organic fertilization, nevertheless essential oil content and the main components of the fennel fruit oil were not affected significantly by fertilization with organic and inorganic sources. Thus, the purpose of this study was to determine the effect of different seeding rate on some yield characteristics for fennel without using any chemical fertilizers and pesticides.

MATERIALS AND METHODS

Sweet fennel (*Foeniculum vulgare* Mill.) seeds were obtained from village populations from the producers of Burdur province were used as the study material. The experimental design was randomized blocks with three replications on the experiment. Field research was conducted in Atabey Vocational School of Suleyman Demirel University during 2010 and 2011 vegetation periods. No agricultural activity has been done in the trial area for six years. Different seeding rates (1, 1,5, 2 kg da⁻

¹) were applied in this study which was conducted for a two years period. The parcels were planned in six rows of 3 m length with 40 cm between rows. Sowing date of the first year was 2 April 2010 and the second year was 31 March 2011. Harvesting date of the first year was 16 September 2010 and second year was 7 September 2011. Additional nitrogen fertilization was not made due to the determined total nitrogen values were at a sufficient level (Lindsay and Norwell, 1969; FAO, 1990; Tovep, 1991; Gunes et al., 1996). Weed control and irrigation were made when necessary. In parcels 4 middle rows with 2.5 m length were harvested. Cuscuta (*Cuscuta sp.*) was encountered on some of the parcels in the first growing season, and contaminated plants were taken away from the parcels. In addition, *Graphosoma lineatum [italicum]* (L.) insect was encountered in fruit setting periods of both years, but biological and chemical control were not made.

Table 1. Results of soil analyze of the experiment field

Structure Analysis	for 0-20 cm			for 20-40 cm		
	clay	silt	sand	clay	silt	sand
	22.80%	16.89%	60.31%	20.20%	14.43%	65.37%
N	0.08316%			0.09576%		
P	9.66 ppm			9.83 ppm		
K	390 ppm			360 ppm		
pH	7.54			7.52		
eC	2.05 μs / 24.9°C			0.35 μs / 24.7°C		

Soil samples taken from 0-20 cm and 20-40 cm depth of the experiment field were analyzed in the laboratory of Soil Science Department of Agriculture Faculty of Suleyman Demirel University. According to that, soil structure of the experiment field was determined as sandy clay loam with low alkaline reactivity. Analysis results belonging to the experiment field were provided in the

Table 1. Structure (texture) analysis was determined according to Day Hydrometer Method (Day, 1956), total Kjeldahl Method of Nitrogen Estimation (Kjeldahl, 1883), phosphorus sodium bicarbonate method (Olsen et al., 1954), potassium neutral 1 N Ammonium acetate method (Carson, 1980). pH and ec were determined by 1:1 soil:water mixture (Peech et al., 1947).

Table 2. Some of the climate data of Atabey district for 2010 and 2011 years

Month	Relative Humidity (%)		Temperature (°C)		Total Rainfall (mm)	
	2010	2011	2010	2011	2010	2011
March	60.1	68.7	8.3	5.9	33,2	50,4
April	59.9	67.8	11.5	10.0	47,0	54,7
May	55.4	65.9	16.4	14.0	32,4	43,1
June	61.4	56.2	19.2	19.3	64,5	62,2
July	47.1	40.5	24.6	24.9	40,1	1,8
August	37.8	37.2	26.4	24.4	0,2	0,6
September	54.0	40.2	20.3	19.8	30,0	-
October	72.5	57.0	11.9	10.9	79,1	-
Mean/Total	56.0	54.2	17.3	14.9	326.5	212.8

Climate data regarding the years during which the experiment was conducted were provided in Table 2. According to that, it was observed that the relative humidity, average temperature and total rainfall in 2010, which is the first year of the experiment, were higher compared to the second year. Additional irrigation was done due to lack of rainfall or no rain.

In the study, plant height (cm), number of fruited branches (units(s)/plant), number of umbels

(units(s)/plant), number of umbellets (units(s)/umbel), number of fruits (units(s)/umbel), fruit yield (kg ha⁻¹) and biological yield (kg ha⁻¹) was calculated randomly chosen 10 plants without considering edge effect. The number of umbellets was calculated by counting the umbellets in all umbels of each plant and taking their averages (units(s)/umbel). The number of fruits was calculated by counting fruits in all umbels of each plant and taking their averages (units(s)/umbel). The thousand fruit weight (g) was calculated by counting and weighing 100 units of fruit with 5 replications from each parcel and comparing it with thousand fruit weight. In order to determine the essential oil content in both years, 10 g of air-dried fruits that were obtained from each parcel were weighed, crushed in the grinder and by adding 100 ml distilled water, calculated as milliliter/100 g (%) by water distillation with three replications in Neo-Clevenger apparatus (Wichtl, 1971). Essential oil constituents were analyzed by (%) gas chromatography method, and GC/MS analysis was carried out by utilizing Shimadzu GC/MS-QP 5050 A in Suleyman Demirel University Experimental and Observational Student Practice and Research Center. CP Wax 52 CB (50 m x 0.25 mm *i.d.*, film thickness 1.2 μm) capillary column and Helium as a carrier gas were used. The temperature program reached from 60°C to 220°C with 2°C increases in temperature in a minute, and was applied by maintaining 220°C for 20 minutes. Temperature of the injector was 240°C. Mass spectra were used at 70 eV. After the compounds were ionized in gas chromatography column and separated, mass spectrum of each of them were obtained. Evaluation procedures were conducted by using "Wiley, Nist and Tutor" libraries.

Every characteristic other than the essential oil constituents were subject to analysis of variance in TARIST packaged statistics software according to experimental design of randomized blocks. Differences were determined by F test and mean values were compared according to LSD test.

RESULTS AND DISCUSSION

Plant height

Significant differences were observed between seeding rates, years and their interactions (Table 3). Plant heights varied between 37.30 and 45.80 cm in the first year and 52.83and 56.43 cm in the second year. When the general averages of plant heights were investigated in terms of seeding rates, the highest values were recorded from 20 and 15 kg ha^{-1} seeding rates, respectively. In interaction, the highest plant heights were 45.80 and 44.47 cm from 15 and 20 kg ha^{-1} seeding rates in the first year, respectively. In the second vegetation year, the highest

plant height was found as 56.43 cm in 20 kg ha^{-1} seeding rate. Arabaci and Bayram (2005) were obtained the highest plant heights from 25 and 15 kg ha^{-1} seeding rates, respectively. It was expected that plants grown densely were taller because of not to be found enough space for spreading (Ahmad et al., 2004) and it was clearly seen in the second year values. Ahmad et al. (2004) also reported that the maximum plant height was observed from 40 cm row spacing compared to 70 cm. The plant heights were compared in terms of the year averages, the first year value was 42.52 cm and the second was 54.60 cm. Plant heights of this study were similar with Kizil et al. (2001) and Tunçturk (2008); but lower than Karaca and Kevseroglu (2001), Ahmad et al. (2004) and Arabaci and Bayram (2005). Plant height is a genotypic characteristic and differences occurred between studies may be due to the seed materials from different origins, ecological conditions and agronomic practices.

Number of fruited branches

While different seeding rates were not affecting the number of fruited branches per plant, years and their interactions were affected (Table 3). Number of fruited branches varied between 5.30 and 6.55 branches per plant in the first year, and between 4.97 and 5.87 in the second year. According to two year averages, the highest number of fruited branches was obtained in the first year as 5.67 per plant. In interaction, the highest number of fruited branches was observed in the first year with 6.55 in 10 kg ha^{-1} seeding rate was the application of sparse sowing, in the second year there were no differences between the seeding rates. Arabaci and Bayram (2005) reported that seeding rates were not statistically significant on number of fruited branches per plant, but the interaction between seeding dates and seeding rates were significantly affected on number of fruited branches per plant. Mentioned researchers determined the highest number of fruited branches on 1 April sowing date in 5 and 15 kg seeding rates per hectare and these results were similar to the results of present study in terms of seeding rates. The values of number of fruited branches were in accordance with the findings of Kizil et al. (2001) and Arabaci and Bayram (2005).

Table 3. Mean values of plant height, number of fruited branches and number of umbels for fennel in different seeding rate

Seeding Rate (kg ha^{-1})	Plant height (cm)			Number of fruited branches (unit(s)/plant)			Number of umbels unit(s)/plant		
Years	2010	2011	Mean	2010	2011	Mean	2010	2011	Mean
10	37.30 b	52.83 b	45.07 b	6.55 a	5.03 a	5.79	6.16 a	10.40 a	8.28
15	45.80 a	54.53 ab	50.17 a	5.15 b	4.97 a	5.06	5,90 a	7,95 b	6.93
20	44,47 a	56.43 a	50.45 a	5.30 b	5.87 a	5.59	4,90 a	9,83 a	7.37
Mean	42.52 b	54.60 a	48.56	5.67 a	5.12 b	5.40	5,99 b	9,08 a	7.54
LSD	Y$_{(5\%)}$: 1,88 SR$_{(5\%)}$: 2,31 YxSR$_{(1\%)}$: 3,26			Y$_{(5\%)}$: 0.50 SR: ns YxSR$_{(5\%)}$:0.86			Y$_{(5\%)}$: 0.88 SR: ns YxSR$_{(1\%)}$: 1.52		

Number of umbels

As it can be seen in Table 3, different seeding rates had insignificant effect on number of umbels per plant. On the other hand years and their interactions had significant effect on number of umbels. It was determined that the first year values ranged between 4.90 and 6.16 and in the second year it was between 7.95 and 9.83 per plant. Several researchers revealed that the number of umbels per plant is affected by plant density and when plant density increases, the umbel numbers decrease (Falzari et al., 2005; Ozyilmaz, 2007; Nakhaei et al. 2012). Similar results were obtained especially in the first year of the present study. It can be say that when plants are sowed densely they have competition with each others for unit area, nutrient elements, water and light, therefore the number of umbels decreases. The averages of umbel numbers were examined for years, the second year value was found higher than the first year with 9.08 and it was found statistically different from the first year. The reason of this may be due to the high amount of rainfall in July which probably affected pollination negatively, and therefore the number of umbels reduced. In interaction, there were no differences between the seeding rates for the first year, but in the second year the highest values were found in 10 and 20 kg ha^{-1} seeding rates with 10.40 and 9.83, respectively.

The results of this study were similar with Nakhaei et al. (2012) and lower than Tunçturk (2008) and Arabaci and Bayram (2005). These differences between the studies may be due to the different ecological conditions, agronomic practices and seed origins.

Number of umbellets

Seeding rates, years and their interactions had no significant effect on number of umbellets per umbel (Table 4). Number of umbellets varied between 6.35 and 7.37 in the first year and between 6.68 and 7.34 in the second year. Researchers revealed that when row spacing increases, number of umbellets was decreases (Ozyilmaz, 2007; Nakhaei et al., 2012). Different from these researchers, in present study this situation was no observed. Although there were no statistical differences between the number of umbellets, the highest value was found with an average of 7.27 from 15 kg ha^{-1} seeding rate, and the second year value as 7.06 was found higher than the first year. Higher values were determined than present study by Ozyilmaz (2007) and Nakhaei et al. (2012). The differences between the other studies may be due to for similar reasons as number of umbels per plant.

Number of fruits

The effects of different seeding rates and year x seeding rate interaction were not found statistically significant on number of fruits per umbel, while year was effective. It was seen in Table 4, number of fruits per umbel ranged between 18.82 and 23.64 in the first year and between 37.11 and 42.58 in the second year. When the general averages of two years were compared; the highest number of fruits per umbel was obtained in the second year of the experiment with 39.91. The difference between the vegetation periods may be due to the rainfall in July had negative effects on pollination and consequently fruit set. Falzari et al. (2005) also revealed that fruit set was being reduced by a lack of synchrony between pollen production and stigma receptivity. It may be caused for this reason, number of fruits per umbel was determined lower than other researchers (Arabaci and Bayram, 2005; Ozyilmaz, 2007; Tunçturk, 2008).

Table 4. Mean values of number of umbellets, number of fruits and fruit yield for fennel in different seeding rate

Seeding Rate (kg ha^{-1})	Number of umbellets (unit(s)/umbel)			Number of fruits (unit(s)/umbel)			Fruit yield (kg ha^{-1})		
Years	2010	2011	Mean	2010	2011	Mean	2010	2011	Mean
10	6.90	6.68	6.79	23.64	37.11	30.38	343.9	557.3	450.6 b
15	7.37	7.16	7.27	24.55	42.58	33.57	397.4	799.6	598.5 a
20	6.35	7.34	6.85	18.82	40.03	29.43	263.5	587.5	425.5 b
Mean	6.87	7.06	6,97	22.34 b	39.91 a	31.13	334.9 b	648.2 a	491.53
LSD	Y: ns SR: ns YxSR:ns			Y$_{(5\%)}$: 4.21 SR: ns YxSR:ns			Y$_{(1\%)}$: 82.7 SR$_{(1\%)}$: 101.3 YxSR: ns		

Fruit yield

Significant differences were observed between seeding rates and year for fruit yield, whereas their interactions were insignificant (Table 4). It was seen in the table; values ranged between 263.5 and 397.4 kg ha^{-1} in the first year and between 557.3 and 799.6 kg ha^{-1} in the second year. The average of fruit yields was examined for seeding rates; the highest fruit yield was obtained from 15 kg seeding rate with 598.5 kg per hectare was found significantly different than the other applications. Arabac and Bayram (2005) revealed that 15 kg seeding rate was provided the highest fruit yield with 206.8 kg ha^{-1} for fennel. The average of fruit yields were examined for years, the highest yield was determined in the second year with 648.2 kg ha^{-1} and it was found significantly different from the first year values. As it can be seen from Table 4, the second year values were found two times higher than the first year. This was an expected situation for fruit yield because number of fruits per umbel was lower in the first vegetation year as well. This difference can be explained due to the rainfall in July may have negative effects on pollination and consequently fruit set.

In previous studies, different results were determined for fennel with different seeding density applications. In this study, the fruit yield values were found parallel with Avci and Amir Nia (2007), Tunçturk (2008) and Nakhaei et al. (2012), lower than Kizil et al. (2001), Arabaci and Bayram (2005) and Ozyilmaz (2007), however higher than Ahmad et al. (2004).

Biological yield

Significant differences were observed between seeding rates and year for biological yield, whereas their interactions were insignificant (Table 5). The biological yields were varied between 901.6 and 1322.2 kg ha^{-1} for the first year and between 1340.5 and 2271.6 kg ha^{-1} for the second year. The average of biological yield was examined for different seeding rate; the highest yield was obtained from application of 15 kg per hectare with 1796.9 kg. The average of biological yield was investigated for years; the highest value was determined as 1719.4 kg ha^{-1} for second year.

In previous studies, different results were determined for fennel with different seeding rate and results of this study were found lower than others (Avci and Amir Nia, 2007; Ozyilmaz, 2007). It was may be due to the studies were conducted under different ecological conditions, agronomic practices and different origins of seed materials.

Thousand fruit weight

The different seeding rates and year x seeding rate interaction were insignificant effect on thousand fruit weight, while year was statistically significant (Table 5). Thousand fruit weights ranged between 6.27 and 6.57 g in the first year of the experiment and second year values varied between 7.87 and 8.72 g. The average of thousand fruit weight was obtained as 6.47 g for the first year and 8.10 g for the second year. These differences between the years may be due to the excessive July rainfall in the first year have negative effect on pollination and fruits may have not grown enough thus fruits may have been smaller than the second year. The thousand fruit yield values were found similar with Marotti et al. (1993), Ceylan (1997), Kizil et al. (2001), Arabaci and Bayram (2005) and Tunçturk (2008).

Table 5. The mean values of biological yield, thousand fruit weight and essential oil content for fennel in different seeding rate

Seeding Rate (kg ha^{-1})	Biological yield (kg ha^{-1})			Thousand fruit weight (g)			Essential oil content (%)		
Years	2010	2011	Mean	2010	2011	Mean	2010	2011	Mean
10	1205.1	1340.5	1273.5 b	6.57	8.57	7.57	1.72	2.46	2.09
15	1322.2	2271.6	1796.9 a	6.27	7.87	7.07	1.62	2.09	1.86
20	901.6	1544.6	1223.1 b	6.54	7.87	7.21	1.60	2.04	1.82
Mean	1143 b	1719.4 a	1431. 2	6.47 b	8.10 a	7.28	1.65 b	2.20 a	1.92
LSD	Y$_{(1\%)}$: 299.1 SR$_{(5\%)}$: 366.3 YxSR:ns			Y$_{(5\%)}$: 0.39 SR: ns YxSR:ns			Y$_{(1\%)}$: 0.36 SR: ns YxSR:ns		

Essential oil content

Different seeding rates and year x seeding rate interaction were not made significant effect on essential oil content, while year was effective (Table 5). Arabaci and Bayram (2005) also revealed that seeding rates were not made significant effect on essential oil content of fennel. Ozyilmaz (2007) stated that row spacing, sowing rates and their interactions were affected on essential oil content and emphasized that the highest content was obtained from 30 cm row spacing and 250 seeds/m^2. In this study, essential oil contents ranged between 1.60 and 1.72% in the first year and between 2.04 and 2.46% in the second year. The average of the essential oil contents was examined for vegetation periods; the first year content was found as 1.65% and the second was 2.20%. Essential oil of fennel fruits is related to fruit length (Ceylan, 1997) and essential oil is excreted from exudate canals present between costa (Zeybek, 1985). The difference between the years may be due to the fruit lengths. In the second year,

fruit lengths may be taller and therefore thinner than the first year, because of the number of umbels and umbellets per plant and the number of fruits per umbel were higher.

The essential oil contents of the present study were lower than the other researchers values were found by Cosge et al. (2008), Arabaci and Bayram (2005) and Telci et al. (2009) but similar with Kizil et al. (2001), Ozyilmaz and Yilmaz (2009), Avci (2010), Tunçturk et al. (2011). The differences between the results of other researchers may be due to the seed materials of different origins, different ecological conditions under which the plants are cultivated, different sowing times, fertilizer applications, harvesting at different maturity stages and difference of agronomic practices conducted at growing stage.

Essential oil constituents

In the first year five constituents and in the second year nine constituents were identified in fennel essential oil (Table 6). The replications of essential oil were combined for determination of percentage of constituents, thus no statistical evaluation was made to determine the effects of different seeding rates and year on essential oil constituents. It was seen in Table 6; the main constituent of the fennel essential oil was *trans*-anethole and the average of the first year content (96.38%) was determined higher than the second year (86.44%). In the first year, the subsequent constituent was p-allyl anisole (2.46%) and it was limonene (6.35%) in the second year. α-pinene, sabinene, β-myrcene and β-cis-ocimene were not exist between the constituents of the first year while they were in the second year.

In previous *trans*-anethole was identified as the main constituent of the essential oil of fennel by the researchers (Avci, 2010; Telci et al., 2009; Cosge et al.,2008; Avci and Amir Nia, 2007; Kapoor et al., 2004; Kan, 2006; Anwar et. al, 2009). When *trans*-anethole values of previous studies were compared with present study Avci (2010) and Avci and Amir Nia (2007) were found similar values, but Telci et al. (2009), Cosge et al. (2008), Kapoor et al. (2004), Kan (2006), Anwar et. al (2009) were determined lower values. These differences between the results may be due to the seed materials of different origins, different ecological conditions, different sowing times, fertilizer applications, harvesting at different maturity stages and agronomic practices.

Table 6. Chemical composition of fennel essential oil in different seeding rates (%)

| Constituents | Rt | 2010 | | | | 2011 | | | |
| | | Seeding Rate (kg ha⁻¹) | | | | Seeding Rate (kg ha⁻¹) | | | |
		10	15	20	Mean	10	15	20	Mean
α-Pinene	6.3	-	-	-	-	0.39	0.29	0.21	0.30
Sabinene	9.4	-	-	-	-	0.15	0.11	-	0.13
β-Myrcene	11.0	-	-	-	-	0.15	0.13	-	0.14
Limonene	13.0	0.31	0.10	0.11	0.17	7.49	6.45	5.11	6.35
β-cis-Ocimene	14.3	-	-	-	-	0.73	0.68	0.5	0.64
Fenchone	24.2	0.59	0.59	0.69	0.62	1.30	1.60	1.36	1.42
p-Allyl anisole	40.8	2.86	1.50	3.01	2.46	5.10	5.39	3.27	4.59
trans-Anethole	50.5	95.50	97.79	95.85	96.38	84.48	85.31	89.53	86.44
p-Anisaldehyde	62.7	0.71	-	0.31	0.51	-	-	-	-

CONCLUSION

Fennel is one of the important medicinal and aromatic plants because it is used in drug production and as herbal tea. Therefore, cultivation of this plant without using any chemical fertilizers and pesticides is very essential. Otherwise, active ingredients and compositions may change and some chemical residues may occur. For this reason, this study was intended to determine the appropriate seeding rate for fennel cultivation without using any chemicals. In present study, the fruited branch, umbel, umbellet and fruit numbers per plant, thousand fruit weight and essential oil content values were not adversely affected on seeding rates; whereas plant height, fruit yield and biological yields were influenced. In conclusion, the highest fruit yield was obtained on 15 kg ha⁻¹ seeding rate and organic fertilizers could be recommended for increasing of fruit yield.

ACKNOWLEDGEMENT

The author would like to thank Ebru FITIL KAVUT and Onder OZAL for seeding and soil analyses.

LITERATURE CITED

Ahmad, M., S.A. Hussain, M. Zubair and A. Rab, 2004. Effect of different sowing seasons and row spacing on seed production of fennel (Foeniculum vulgare). Pakistan Journal of Biological Sciences. 7(7): 1144-1147.

Akgul, A., 1993. Spice Science and Technology. Association of Food Technology Publ. 15, Ankara.

Arabaci, O. and E. Bayram, 2005. The effect of different sowing date and seeding rate on yield and some important characteristics of fennel (Foeniculum vulgare Mill.). 6th Turkish Field Crops Congress. 5-9 September 2005, Antalya, 529-534.

Anwar, F., M. Ali, A.I. Hussain and M. Shahid, 2009. Antioxidant and antimicrobial activities of essential oil and extracts of fennel (Foeniculum vulgare Mill.) seeds from Pakistan. Flavour Fragr. J. 24: 170-176.

Avci, A.B. and R. Amir Nia, 2007. The effect of different line interval and seeding rate on some yield and quality characteristics of fennel (Foeniculum vulgare mill.). 7th Turkish Field Crops Congress, 25-27 June 2007, Erzurum, 793-797.

Avci, A.B., 2010. Organic farming opportunities of anise, coriander and fennel in Burdur region. Book of Proceedings, International Conference on Organic Agriculture in Scope of Environmental Problems. 03-07 February 2010, Famagusta, Cyprus Island, 290-291.

Ayub, M., M.A. Nadeem, A. Tanveer, M. Tahir, M.T.Y. Saqib and R. Nawaz, 2008. Effect of different sowing methods and times on the growth and yield of fennel (Foeniculum vulgare Mill.). Pak. J. Bot., 40(1): 259-264.

Basgel, S. and S.B. Erdemoglu, 2006. Determination of mineral and trace elements in some medicinal herbs and their infusions consumed in Turkey. Science of The Total Environment, 359: 82– 89.

Baydar, H., 2009. Medicinal and Aromatic Plants Science and Technology. 3. ed. Suleyman Demirel University Publ. 51, Isparta.

Baytop, T., 1999. Phytotherapy in Turkey. From past to present. 2. ed. Nobel Medical Publ İstanbul.

Bayram, E., S. Kirici, S. Tansi, G. Yilmaz, O. Arabaci, S. Kizil and I. Telci, 2010. Enhancement possibilities of production of medicinal and aromatic plants. 5th Turkish Technical Congress of Agricultural Engineering, 11-15 January 2010, Ankara.

Carson, P.L., 1980. Recommended potassium test. in: Recommended chemical soil test procedures for the North Central Region. Rev.ed. North Central Regional Publication 221. North Dakota Agric. Exp. Stn.North Dakota State University, Fargo. USA.

Ceylan, A., 1997. Medicinal Plants-II (Essential Oil Plants). Agriculture Faculty of Ege University Pub 481. İzmir.

Cosge, B., B. Gurbuz, H. Kendir and A. Ipek, 2008. Composition of essential oil in sweet fennel (Foeniculum vulgare mill. var. dulce) lines originated from Turkey. Asian J. Chem. 20 (2): 1137-1142.

Davis, P.H., 1978. Flora of Turkey and The East Aegean Islands, vol.4. Edinburgh University Press, Edinburgh, pp 352-377.

Day, P.R., 1956. Report of the Committee on Physical Analysis, 1954-1955. Soil Science Society of America. Soil Sci. Soc. Am. Proc. 20: 167-169.

Falzari, L.M., R.C. Menary and V.A. Dragar, 2005. Reducing fennel stand density increases pollen production, improving potential for pollination and subsequent oil yield. HortScience. 40 (3): 629-634.

FAO., 1990. Micronutrient, assessment at the country level: an international study. FAO Soils Bulletin 63. Rome.

Garg, C., S.A. Khan, S.H. Ansari and M. Garg, 2010. Efficacy and safety studies of Foeniculum vulgare through evaluation of toxicological and stadardisation parameters. Int J Pharm Pharm Sci. 2 (2): 43-45.

Gunes, A., M. Aktas, A. Inal and M. Alpaslan, 1996. Physical and chemical characteristics of Konya Closed Basin territories. Agriculture Faculty of Ankara University Pub 1453. Ankara.

Kandil, M.A.M.H, 2002. The effect of fertilizers for conventional and organic farming on yield and oil quality of fennel (Foeniculum vulgare Mill.) in Egypt. Braunschweig: FAL, 94 p, Landbauforsch Volkenrode SH 237.

Kan, Y., M. Kartal, S. Aslan and N. Yildirim, 2006. Composition of essential oil of fennel fruits cultivated at different conditions. J. Fac. Pharm, Ankara, 35 (2): 95–101.

Kapoor, R., B. Giri and G. Mukerji, K. 2004. Improved growth and essential oil yield and quality in Foeniculum vulgare mill on mycorrhizal inoculation supplemented with P-fertilizer. Bioresource Technology, 93: 307–311.

Karaca A. and K. Kevseroglu, 2001. The research of some important phenology and morphology characters of coriander (Coriandrum sativum L.) and fennel (Foeniculum vulgare Mill.). 4th Turkish Field Crops Congress, 17-21 September 2001. Tekirdag, 243-248.

Kizil, S., N. Arslan and A. Ipek, 2001. The effect of different sowing dates on yield and yield components of fennel (Foeniculum vulgare Mill. var. Dulce). 4th Turkish Field Crops Congress, 17-21 Eylul 2001, Tekirdag, 331-334.

Kjeldahl, J. 1883. Neue Methode zur Bestimmung des Stickstoffs in organischen Korpem. Z. Anal. Chem. 22: 366-382.

Lindsay, W.L. and W.A. Norwell., 1969. Development of a DTPA micronutrient Soil Test. Soil Sci. Am. Proc. 35: 600-602.

Marotti, M., V. Dellacecca, R. Piccaglia and E. Giovanelli, 1993. Agronomic and chemical evaluation of three "Varieties" of Foeniculum vulgare Mill.. ISHS Acta Horticulturae 331, pp. 63-69.

Nakhaei, A., S. G., Moosavi, R., Baradaran and A.A., Nasrabad, 2012. Effect of nitrogen and plant density levels on yield and yield components of fennel (Foeniculum vulgare L.). IJACS. 4 (12), 803-810.

Olsen, S.R, C.V. Cole, F.S. Watanabe and L.A. Dean, 1954. Estimation of available phosphorus in soils by extraction with sodium bicarbonate. US.Dept. of Agric.Cric. 939 p.

Ozyilmaz, B., 2007. Effect of different row spacing and seeding rate on yield, yield component and some quality properties of fennel (Foeniculum vulgare Mill. var. dulce). Gaziosmanpasa University, Graduate School of Natural and Applied Sciences Master Thesis, 84 p.

Ozyilmaz, B. and G. Yilmaz, 2009. The effects of row spacing and seeding rate on yield, yield component and some quality properties of fennel (Foeniculum vulgare Mill. var. dulce). 8th Turkish Field Crops Congress, 19-22 October 2009, Hatay, 69-72.

Peech, M., L.T. Alexander, L.A. Dean and J.F. Reed, 1947. Methods of soil analysis for soil fertility investigations. U.S. Dept. Agr. Circ. 757 p.

Tanker, N., M. Koyuncu and M. Coskun, 2007. Pharmaceutical Botany. 3rd Ed. Pharmacy Faculty of Ankara University Pub 93, Ankara.

Telci, I., E. Bayram and B. Avci, 2006. Changes in yields, essential oil and linalool contents of Coriandrum sativum varieties (var. vulgare Alef. and var. microcarpum DC.) harvested at different development stages. Europ. J. Hort. Sci. 71(6): 267–271.

Telci, I., I. Demirtas and A. Sahin, 2009. Variation in plant properties and essential oil composition of sweet fennel (Foeniculum vulgare Mill.) fruits during stages of maturity. Industrial Crops and Products. 30: 126–130.

Tovep., 1991. Productivity Inventory of Turkey Soils. T.C. The Ministry of Agriculture and Rural Affairs.

Tuncturk, M., 2008. Effects of different nitrogen doses on the agricultural and chemical properties of fennel (Foeniculum vulgare Mill.). Asian J. Chem. 20 (4): 3209-3217.

Tuncturk, R., M. Tunçturk and D. Turkozu, 2011. The effects of different phosphorus and nitrogen doses on the yield and quality of fennel (Foeniculum vulgare Mill.) in Van ecological conditions. YYU J Agr Sci. 21(1):19-27.

Wichtl, M., 1971. Die Pharmakognostisch-Chemische Analyse Akad. Verlagsgesellschaft, Frankfurt, 479 p.

Zeybek, N., 1985. Pharmaceutical Botany, Seeds Indoor Plants (Angiospermae) Systematics and Important Substance. Pharmacy Faculty of Ege University 1. Izmir.

17

THE EFFECTS OF LOCATION AND THE APPLICATION OF DIFFERENT MINERAL FERTILIZERS ON SEED YIELD AND QUALITY OF POT MARIGOLD (*Calendula officinalis* L.)

Radosav JEVDOVIC[1], Goran TODOROVIC[2], Miroslav KOSTIC[1], Rade PROTIC[3],*
Slavoljub LEKIC[4], Tomislav ZIVANOVIC[4], Mile SECANSKI[2]

[1]*Institute for Medicinal Plant Research "Dr Josif Pancic", Belgrade, REPUBLIC OF SERBIA.*
[2]*Maize research Institute, Zemun Polje, Belgrade-Zemun, REPUBLIC OF SERBIA,*
[3]*Institute for Science Application in Agriculture, Belgrade, REPUBLIC OF SERBIA.*
[4]*Faculty of Agriculture, University Belgrade, REPUBLIC OF SERBIA.*
*Corresponding author: gtodorovic@mrizp.rs

ABSTRACT

The four-replicate trial with a local variety Domaći oranž (*Local orange*) was set up according to the randomised complete-block design in four locations (Gorobilje, Arilje, Pančevo and Kačarevo) and two variants of fertilizing (200 kg ha^{-1} KAN with 27 % of nitrogen and 400 kg ha^{-1} NPK 15:15:15) and the control without fertilizing. According to the three factorial analysis of variance for all observed traits it was determined that there were very significant differences within growing locations and fertilizing variants and their interaction (L x F). The highest seed yield (672.84 kg ha^{-1}) was detected in the variant with 400 kg NPK ha^{-1}. The significantly lower seed yield (579.84 kg ha^{-1}) was obtained in the variant with KAN at the rate of 200 kg ha^{-1} and the control variant (344.88 kg ha^{-1}). The highest total seed germination of 91.84% and the 1000-seed weight (6.83 g) were obtained in the variant with 400 kg NPK ha^{-1}. Total seed germination (85.87) and the 1000-seed weight (5.82 g) obtained in the control were significantly higher than total seed germination (83.31%) and the 1000-seed weight (5.42 g) obtained in the variant with 200 kg ha^{-1} KAN.

Key words: pot marigold, location, mineral fertilizers, seed, germination, 1000-seed weight.

INTRODUCTION

Marigold (*Calendula officinalis* L.) is a decorative and medicinal plant. It belongs to the family of Asteracea, sub-family Asteroidea (*Tubuliflorae*) (Kojić and Pekić, 1995). It is used more in folk medicine than scientific medicine (Tucakov, 1996). This species contains a sufficient amount of active substances, such as: saponosoides, essential oils, flavonides, sterols, polysaccharides, sesquiterpene lactones, etc. (Kovačević, 1995). Flower heads (*Calendula flos*) and seeds (*Calendula semen*) are most often used in the pharmaceutical industry. It has anti-viral, anti HTV and anti-oxidant effects (Muley et al., 2009)

In order to provide a high-quality raw material of high and stable yields, pot marigold is increasingly plantation grown. The fertilization with N, P and K fertilizers is of the greatest significance for the balanced nutrition in the seed crop production. The application of different fertilization treatments had a considerable effect on the different vegetative growth properties of *Calendula officinalis* L. plants compared to the unfertilized control (Hussein et al., 2011). Numerous studies (Golcz et al., 2006; Dzida and Jarosz, 2006; Biesiada and Kuś, 2010) have shown that nitrogen fertilization results in a significant increase in quantity and quality of herbal plant yields. According to Rahmani et al., 2009, nitrogen had a significant effect on all plant properties of calendula (1000-seed weight, seed yield, head diameter and the number of seeds per head) achieved after application of 90 kg N ha-1. Also, the results showed that applications of N fertilizer increased seed yield of calendula, because nitrogen, which is a primary constituent of proteins, is extremely susceptible to loss when considering that average recovery rates fall in the range of 20 to 50% for dry matter production systems in plants. Nitrogen is a major nutrient that influences plant yields and protein concentrations. When the amount of available soil N limits the yield potential, additions of N fertilizers can substantially increase plant yields (Olson and Swallow, 1984; Grant et al., 1985).

The aim of the present study was to determine to which extent different types of mineral fertilizers and growing locations affect the pot marigold seed yield and quality.

MATERIALS AND METHODS

The local variety of pot marigold Domaći oranž was observed. The trial was set up on different types of soils in the Zlatibor region (Gorobilje and Arilje) and the southern Banat region (Pančevo and Kačarevo) during 2006 and 2007. The soil in Gorobilje, i.e. Arilje, is brown earth, i.e. alluvium, respectively. Marshy black soil prevails in Pančevo, while chernozem is present in Kačarevo. Chemical properties of the studied types of soil were presented over locations (Table 1). Mean monthly air temperatures and precipitation sums for 2006 and 2007 in locations of Pančevo and Kačarevo obtained from the state meteorological station Banatski Karlovac (latitude - 45° 03', longitude - 21° 02' and altitude - 89 m AMSL) are presented in Figures 1 and 3. On the other hand, Figures 2 and 4 present meteorological data for the locations of Arilje and Gorobilje registered by the state meteorological station Požega (latitude - 43° 50', longitude - 20° 02', and altitude - 310 m AMSL).

Table 1. Chemical properties of the soil over studied locations

Location Soil type	Kačarevo Chernozem	Pančevo Marshy black	Gorobilje Alluvium	Arilje Brown earth
pH(KCl)	7.30	6.56	6.30	6.89
CaCO$_3$ (mg kg^{-1})	12.30	1.41	3.81	4.78
Humus (%)	3.14	2.97	3.00	2.11
P$_2$O$_5$ (mg kg^{-1})	20.50	4.10	9.70	8.47
K$_2$O (mg kg^{-1})	17.00	37.30	35.50	29.36

Two variants of fertilizing and the control (C) without fertilizing were applied in all locations during both years. The N fertilizer KAN (27% N) at the rate of 200 kg ha^{-1} was applied in the first variant (F$_1$) prior to sowing, while mixed NPK (15:15:15) fertilizer at the rate of 400 kg ha^{-1} was applied in the second variant (F$_2$) during autumn primary tillage.

The four-replicate trials were set up according to the randomised complete-block design. The elementary plot size amounted to 20 m^2. Sowing was done on optimum dates during the third decade of April, while sowing rate was 8 kg seeds ha^{-1}.

Harvest was done in the full seed maturity stage. The harvested seed was dried down to 10 % moisture and cleaned from admixtures. Seed germination was tested in Petri dishes on filter paper in four replicates, each consisting of 100 seeds, constantly at 20 °C. The first, i.e., second count was done after seven, i.e. 14 days, respectively, according to the ISTA Rules (Handbook 2006). The seed yield was estimated in kg per hectare, and the 1000-seed weight was determined.

The test of significance of differences among estimated average values of observed factors (year, location and variants of fertilizing) was done by the application of the model of the analysis of variance for factorial trials set up according to the randomised complete-block design. All evaluations of significance were performed on the basis of the F- and LSD-test at the probability levels of 5% and 1%.

RESULTS AND DISCUSSION

The mean monthly air temperatures, obtained from both meteorological stations, were insignificantly higher during the 2007 growing season than during the 2006 growing season (Figures 1 and 2). The mean monthly precipitation sums in April, June and August in 2006 were significantly higher than the mean for the same months in 2007, while the mean monthly precipitation sums in March, May, September and November in 2007 were higher than the mean for the same months in 2006 (Figures 3 and 4). Climate conditions during the growing season were favourable for pot marigold cultivation in both years of investigation.

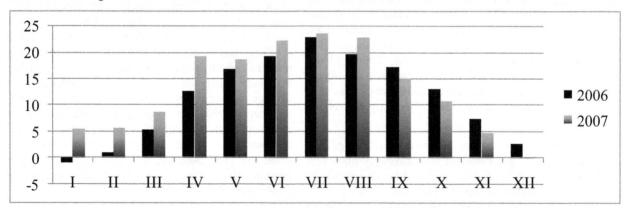

Figure 1. Mean monthly air temperatures (°C) recorded by the meteorological station Banatski Karlovac (Pančevo and Kačarevo)

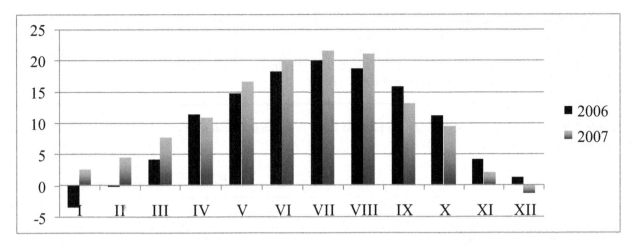

Figure 2. Mean monthly air temperatures (^{o}C) recorded by the meteorological station Požega (Arilje and Gorobilje)

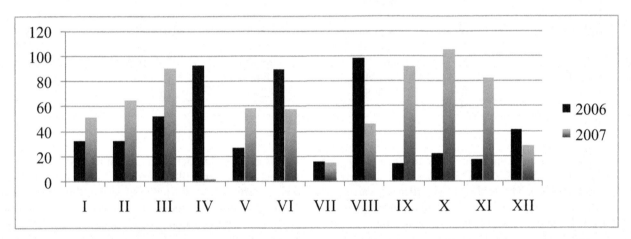

Figure 3. Mean monthly precipitation sums (mm) recorded by the meteorological station Banatski Karlovac (Pančevo and Kačarevo)

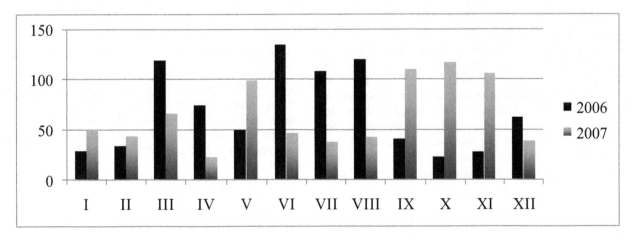

Figure 4. Mean monthly precipitation sums (mm) recorded by the meteorological station Požega (Pančevo and Kačarevo)

According to the three factorial analysis of variance for all observed traits, very significant differences within growing locations and variants of fertilizing and their interaction (L x F) were determined. Very significant differences in total germination between studied years were established, while there were no significant differences for other traits over years. The Y x L, Y x F and Y x L x F interactions were very significant only for the trait 1000-seed weight (Table 2).

Table 2. Three factorial analysis of variance for seed yield, first account, total seed germination and 1000-seed yield of pot marigold

		Seed yield	Seed germination		1000-seed weight
			First count	Total germination	
Source of variation	d.f.	F value			
Replication	3	0.0217	1.4547	2.3278	1.2469
Year (Y)	1	0.3183	3.1243	16.4685**	0.8653
Location (L)	3	66.0981**	230.2583**	343.4380**	205.2433**
Y x L	3	0.1407	0.3349	0.4398	8.6497**
Fertilizing (F)	2	514.3688**	69.5336**	107.4187**	161.6098**
Y x F	2	0.0987	2.8581	2.6514	11.2552**
L x F	6	4.0173**	6.9606**	5.0710**	19.9920**
Y x L x F	6	0.2242	1.6484	2.3886	6.0701**
Error	69	1776.882	8.335	5.708	0.001
Total	95				

** $P \leq 1\%$

There was no significant difference in average seed yields obtained in 2006 (535.00 kg ha^{-1}) and in 2007 (530.10 kg ha^{-1}). The seed yield significantly varied over locations and types of soil. The higher seed yield was obtained under conditions without the fertilizer application on chernozem in Kačarevo and alluvium in Arilje than on marshy black soil in Pančevo and brown earth in Gorobilje. Chernozem and alluvium are lighter, looser soils richer in humus and therefore they are more favourable for the production of pot marigold and its seed. The highest, i.e. lowest seed yield, on the average, was recorded in Kačarevo (581.91 kg ha^{-1}), i.e. in Gorobilje (429.21 kg ha^{-1}), respectively. The seed yield in Gorobilje was significantly lower than seed yields in other locations (Table 3). Fertilizing significantly affected the seed yield increase, hence the highest yield (672.84 kg ha^{-1}) was detected in the variant with 400 kg NPK ha^{-1}. The significantly lower seed yield (579.84 kg ha^{-1}) was obtained in the variant with KAN at the rate of 200kg ha^{-1}. The seed yield (344.88 kg ha^{-1}) recorded in the control was significantly lower than seed yields obtained in both variants with the fertilizer application. The results are in accordance with results obtained by Doddagoudar et al., (2002) in China aster. The higher seed yield obtained with fertiliser NPK 15:15:15 may be attributed to a greater number of flowers, filled seeds and a greater 1000-seed weight. This was mainly due to availability of adequate quantity of nutrients for better filling up of seeds, which resulted in the increased seed weight. These findings are in agreement with the findings of Mantur (1988) and Doddagoudar et al., (2002) in China aster, Hugar (1997) in gaillardia, Sigedar et al., (1991) and Shivakumar (2000) in marigold and Akkannavar (2001) in ageratum. In contrast to the application of nitrogen fertilizer KAN, the distribution of NPK 15:15:15 contributes to the incorporation of not only nitrogen into the soil, but also of phosphorus and potassium, which are most important for plant growth and flowering. They also play a key role in the production of higher flower and seed yield of ornamentals (Kashif, 2001).

The average seed yield of the L x F interaction varied from 290.00kg ha^{-1} (control in Gorobilje) to 733.37kg ha (in the variant with 400 kg NPK ha^{-1} in Kačarevo). Seed yields recorded in Kačarevo, Arilje and Pančevo in the variant with 400 kg NPK ha^{-1} were significantly higher than remaining yields obtained in the L x F interaction (Table 3).

Table 3. Average seed yield and 1000-seed weight of pot marigold over years, locations and fertilizing variants

Location	Fertilizing	Seed yield (kg ha⁻¹)			1000-seed weight (g)		
		2006	2007	Average	2006	2007	Average
Gorobilje	C	280.50	299.50	**290.00**	4.37	4.53	**4.45**
	F1	471.00	452.25	**461.62**	3.93	4.26	**4.09**
	F2	539.75	532.25	**536.00**	6.11	6.21	**6.16**
Average		**430.42**	**428.00**	429.21	**4.80**	**5.00**	**4.90**
Arilje	C	372.00	365.75	**368.87**	6.77	5.00	**5.88**
	F1	599.25	618.25	**608.75**	4.56	4.77	**4.66**
	F2	724.25	720.00	**722.12**	6.89	6.98	**6.93**
Average		**565.17**	**568.00**	566.58	**6.07**	**5.58**	**5.82**
Pančevo	C	342.50	337.25	**339.87**	6.01	6.06	**6.03**
	F1	623.75	611.75	**617.75**	5.64	5.81	**5.72**
	F2	703.00	696.75	**699.87**	6.76	6.89	**6.82**
Average		**556.42**	**548.58**	552.50	**6.14**	**6.25**	**6.19**
Kačarevo	C	384.25	378.00	**381.12**	6.83	7.05	**6.94**
	F1	643.00	619.50	**631.25**	6.71	7.73	**7.22**
	F2	736.75	730.00	**733.37**	7.42	7.42	**7.42**
Average		**588.00**	**575.83**	581.91	**6.99**	**7.40**	**7.19**
Average	C	344.65	345.12	**344.88**	5.99	5.66	**5.82**
	F1	584.25	575.44	**579.84**	5.21	5.64	**5.42**
	F2	675.94	669.75	**672.84**	6.79	6.87	**6.86**
Average (Y)		**535.00**	**530.10**		**6.00**	**6.06**	

	L	Lsd $_{0.05}$ = 24.28	Lsd $_{0.01}$ = 32.23	L	Lsd $_{0.05}$ = 0.18	Lsd $_{0.01}$ = 0.24
	F	Lsd $_{0.05}$ = 21.02	Lsd $_{0.01}$ = 27.92	F	Lsd $_{0.05}$ = 0.16	Lsd $_{0.01}$ = 0.21
	Y x L	Lsd $_{0.05}$ = 42.05	Lsd $_{0.01}$ = 55.83	Y x L	Lsd $_{0.05}$ = 0.26	Lsd $_{0.01}$ = 0.34
				L x F	Lsd $_{0.05}$ = 0.32	Lsd $_{0.01}$ = 0.42
				Y x Đ	Lsd $_{0.05}$ = 0.22	Lsd $_{0.01}$ = 0.30
				Y x L x F	Lsd $_{0.05}$ = 0.45	Lsd $_{0.01}$ = 0.59

The average 1000-seed weight (6.06 g) recorded in 2007 was significantly higher than the average 1000-seed weight (6.00 g) obtained in 2006. Furthermore, the average 1000-seed weight varied from 4.90 g in Gorobilje to 7.19 g in Kačarevo. As in the case of other studied traits, the highest 1000-seed weight (6.83 g) was obtained in the variant with 400 kg NPK ha⁻¹. The 1000-seed weight (5.82 g) recorded in the variant without fertilizing was significantly higher than the 1000-seed weight (5.42 g) in the variant with 200 kg KAN ha⁻¹. The value of the Y x L interaction was the highest for the 1000-seed weight (7.40 g) obtained in the location of Kačarevo in 2007 and was significantly higher than remaining values of this trait of the Y x L interaction. The Y x F interaction showed that the 1000-seed weight obtained in the variant with 400 kg fertilizers ha⁻¹ in 2007 (6.87 g) did not significantly differ from the 1000-seed weight (6.79 g) obtained in 2006. All other values of Y x F interactions for the 1000-seed weight were significantly lower than the stated values. The value of the 1000-seed weight of the L x F interaction varied from 4.09 g in Gorobilje in the variant with 200 kg KAN ha⁻¹ to 7.42 g in Kačarevo in the variant with 400 kg NPK ha⁻¹. The Y x L x F interaction showed that the highest 1000-seed weight of 7.73 g was obtained in the second variant with 200 kg KAN ha⁻¹ in Kačarevo in 2007. This value did not significantly differ from the value of 7.42 g recorded in the variant with 400 kg NPK ha⁻¹ in Kačarevo in 2006 and 2007. The value of the Y x L

x F interaction was significant only for the 1000-seed weight in comparison to values of remaining traits (Table 2).

The location, fertilizing variants and their interactions affected the first count (Table 2). The average value of this trait varied from 70.21% (Gorobilje) to 91.00% (Kačarevo). The first count ranged from 79.15 % (200 kg ha⁻¹) to 87.43% (400 kg NPK ha⁻¹). Similarly to the seed yield, the first count (93.50 %) was the highest in the variant with 400 kg NPK ha⁻¹ in the location of Kačarevo. The first count in the same fertilizing variant in the locations of Arilje and Pančevo did not significantly differ from the first count obtained in Kačarevo, while all other values of the L x F interaction were significantly lower. The average first count (90.25%) obtained in the variant with 200 kg KAN ha⁻¹ in Kačarevo did not significantly vary from the variant with 400 kg NPK ha⁻¹ (93.50%, Table 4).

Total seed germination (95.58%) determined in Kačarevo was significantly higher than total seed germination in other studied locations. The lowest total germination (74.42%) obtained in Gorobilje was significantly lower than total seed germination established in Arilje (90.58%) and Pančevo (87.45%). The highest total seed germination of 91.84% was obtained in the variant with 400 kg NPK ha⁻¹. Total seed germination obtained in control (85.87%) was significantly higher than

Tablo 4. Average values of the first count and total seed germination of pot marigold over years, locations and fertilizing variants

First count (%)					Total germination (%)		
Location	Fertilizing	2006	2007	Average	2006	2007	Average
Gorobilje	C	68.50	69.75	**69.12**	71.50	74.00	**72.75**
	F1	67.50	68.75	**68.12**	70.50	72.00	**71.25**
	F2	73.00	73.75	**73.37**	78.25	80.25	**79.25**
Average		**69.67**	**70.75**	70.21	**73.42**	**75.42**	74.42
Arilje	C	89.75	85.00	**87.37**	92.75	89.75	**91.25**
	F1	79.00	81.75	**80.37**	83.75	86.50	**85.12**
	F2	90.75	93.00	**91.87**	93.50	97.25	**95.37**
Average		**86.50**	**86.58**	86.54	**90.00**	**91.17**	90.58
Pančevo	C	80.00	81.25	**80.62**	83.25	86.75	**85.00**
	F1	76.75	79.00	**77.87**	81.50	84.50	**83.00**
	F2	90.25	91.75	**91.00**	93.50	95.25	**94.37**
Average		**82.33**	**84.00**	83.16	**86.08**	**88.83**	87.45
Kačarevo	C	89.00	89.50	**89.25**	94.75	94.25	**94.50**
	F1	87.50	93.00	**90.25**	90.75	97.00	**93.87**
	F2	94.50	92.50	**93.50**	98.25	98.50	**98.37**
Average		**90.33**	**91.67**	91.00	**94.58**	**96.58**	95.58
Average (F)	C	**81.81**	**81.37**	**81.59**	**85.56**	**86.19**	85.87
	F1	**77.68**	**80.62**	**79.15**	**81.62**	**85.00**	83.31
	F2	**87.12**	**87.75**	**87.43**	**90.87**	**92.81**	91.84
Average (Y)		**82.21**	**83.25**		**86.02**	**88.00**	
	L	Lsd $_{0.05}$= 1.66	Lsd $_{0.01}$ = 2.21	L	Lsd $_{0.05}$ = 1.38	Lsd $_{0.01}$ = 1.83	
	F	Lsd $_{0.05}$= 1.44	Lsd $_{0.01}$ = 1.91	F	Lsd $_{0.05}$ = 1.19	Lsd $_{0.01}$ = 1.58	
	Y x L	Lsd $_{0.05}$= 2.88	Lsd $_{0.01}$ = 3.82	Y x L	Lsd $_{0.05}$ = 2.38	Lsd $_{0.01}$ = 3.16	

total seed germination (83.31%) recorded in the variant with 200 kg KAN ha^{-1}. In Kačarevo, total seed germination of 98.37% in the variant with 400 kg NPK ha^{-1} was significantly higher that other obtained values of the L x F interaction (Tab 4). The higher seed quality (1000-seed weight and germination) observed in the variant with fertilizer NPK 15:15:15 might be due to a proper development of seed and also a higher number of filled seed weight per flower, which in turn might have supplied adequate food reserves during germination. This kind of improvement in seed quality attributes was also reported by Mantur (1988), Doddagoudar et al., (2004) and Gnyandev (2006) in China aster, Hugar (1997) in gaillardia, Shivakumar (2000) in marigold and Akkannavar (2001) in agaratum.

CONCLUSION

The significant effect of the location, year, fertilizing and the L x F interaction was determined for all observed traits on the basis of the three factorial analysis of variance. Very significant differences over years of investigation were determined for total seed germination, while there was no significant difference over years for remaining traits. The Y x L, Y x F and Y x L x F interactions were very significant only for the trait 1000-seed weight.

The highest seed yield, first count, total count and the 1000-seed weight were recorded in Kačarevo, while the lowest values of all studied traits were obtained in Gorobilje.

Higher average values of all observed traits were obtained in the plots treated with mineral fertilizers than in control plots. The application of NPK fertilizer at the rate of 400 kg ha^{-1} resulted in the highest average seed yield, 1000-seed weight, first count and total seed germination in the location of Kačarevo.

According to the Y x L x F interaction, it was determined that the highest 1000-seed weight was obtained in the first fertilizing variant in Kačarevo in 2007. This value did not significantly differ from the 1000-seed weight obtained in the second fertilizing variant in Kačarevo in 2006 and 2007.

ACKNOWLEDGEMENT

The research was supported through the Project III N° 46008 by the Ministry of Science and Technological Development of the Republic of Serbia.

LITERATURE CITED

Akkannavar, B. R., 2001. Influence of nitrogen, phosphorus, spacing and growth retardants on seed yield and quality of ageratum. M.Sc. (Agri.) Thesis, University of Agricultural Sciences, Dharwad.

Biesiada, A., A. Kuś, 2010. The effect of nitrogen fertilization and irrigation on yielding and nutritional status of sweet basil (*Ocimum basilicum* L.). Acta Sci. Pol., Hortorum Cultus, 9 (2): 3-12.

Doddagoudar, S. R., B. S. Vyakaranahal, M. Shekhargouda, A. S. Naliniprabhakar, V. S. Patil, 2002. Effect of mother plant nutrition and chemical spray on growth and seed yield of china aster cv. Kamini. Seed Research, 30(2) : 269-274.

Doddagoudar, S. R., B. S. Vyakaranahal, M. Shekhargouda, 2004. Effect of mother plant nutrition and chemical spray on seed germination and seedling vigour index of China aster cv. Kamini. Karnataka Journal of Agricultural Sciences, 17(4) : 701-704.

Dzida, K., Z. Jarosz, 2006. Yielding and chemical composition of *Origanum majorana* L. depending on different nitrogen-potassium fertilization. Acta Agroph. 7(3): 561-566.

Gnyandev, B., 2006. Effect of pinching, plant nutrition and growth retardants on seed yield, quality and storage studies in china aster *(Callistephus chinensis (L.) Nees.)*, M.Sc. (Agri.) Thesis, University of Agricultural Sciences, Dharwad.

Golcz, B., K. Politycka, K. Seidler- Łożykowska, 2006. The effect of nitrogen fertilization and stage of plant development on the mass and quality of sweet basil leaves (*Ocimum basilicum* L.). Herba Pol. 52 (1/2):22-30.

Grant, C.A., E.H. Stobbe, G.J. Racz , 1985. The effect of fall-applied N and P fertilizers and timing of N application on yield and protein content of winter wheat grown on zero-tilled land in Manitoba. Can. J. Soil Sci. 65:621–628.

ISTA, International Seed Testing Association (2006): Handbook.

Hugar, A. H., 1997. Influence of spacing, nitrogen and growth regulators on growth, flower and seed yield in Gaillardia (*Gaillardia pulchella* var. Picta Fouger). Ph.D. Thesis, University of Agricultural Sciences, Dharwad.

Hussein, M.M., RA. Sakr, L.A. Badr, K.M.A.L. Mashat, 2011. Effect of some fertilizers on botanical and chemical characteristics of pot marigold plant (*Calendula officinalis* L.), Journal of Horticultural Science and Ornamental Plants, 3 (3), 220-231.

Kashif, N., 2001. Effect of NPK on growth and chemical effect on Vase-life of Zinnia. MSc Thesis. PMAS Arid Agr. Univ. Rawalpindi, Pakistan, p.23.

Kojić, M., S. Pekić , 1995. Botanika, "Nauka", Beograd.

Kovačević, N., 1995. Fiziološki aktivni sastojci biljaka familije *Asteraceae*, Arhiv za farmaciju, god. 45. br. 5, str.183-198, Beograd.

Mantur, S.M., 1988. Studies on nutrition, growth regulators and soil salinity on flower and seed production in China aster. Ph.D. Thesis, University of Agricultural Sciences, Dharwad.

Muley, B.P., S.S. Khadabadi, N.B. Banarase, 2009. Phytochemical constituents and pharmacological activities of (Calendula officinalis) Linn (Asteraceae): a review. Tropical J. Pharmaceutical Res., 8: 455-465.

Rahmani, N,, J. Daneshian, H.A. Farahani, 2009. Effects of nitrogen fertilizer and irrigation regimes on seed yield of calendula (*Calendula officinalis* L.), Journal of Agricultural Biotechnology and Sustainable Development Vol. 1(1) pp. 024-028, October.

Olson, R.V., C.W. Swallow, 1984. Fate of labeled nitrogen fertilizer applied to winter wheat for five years. Soil Sci. Soc. Am. J.,48:583–586.

Shivakumar, C. M., 2000. Effect of mother plant nutrition, plant density and maturity on seed yield and quality in marigold. M.Sc. (Agri). Thesis, University of Agricultural Sciences, Dharwad.

Sigedar, P.D., K.W. Anserwadekar, M.S. Rodge, 1991. Effect of different level of nitrogen, phosphorous and potassium on growth and yield of *Calendule officinallis* L., South Indian Horticulture, 39(4): 308-311.

Tucakov, J., 1996. Lečenje biljem, "Rad ", Beograd.

18

USING REMOTE SENSING AND SOIL PHYSICAL PROPERTIES FOR PREDICTING THE SPATIAL DISTRIBUTION OF COTTON LINT YIELD

Javed Iqbal[*1], *John J. Read*[2], *and Frank D. Whisler*[3]

[1] *Institute of Geographical Information Systems, School of Civil & Environmental Engineering, National University of Sciences and Technology, Islamabad, Pakistan*
[2] *U.S. Department of Agriculture, Agricultural Research Services*[2], *Genetics and Precision Agriculture Research Unit, Mississippi, USA*
[3] *Department of Plant and Soil Sciences, Mississippi State University, Mississippi, USA*
[*]*Corresponding author: javed@igis.nust.edu.pk*

ABSTRACT

This field crop research study addresses the potential of image based remote sensing to provide spatially and temporally distributed information on timely basis for site-specific cotton crop management. Universal applicability of site specific crop management is hampered by lack of timely distributed and economically feasible information on soils and crop conditions in the field and their interaction. The objectives of this study were to demonstrate (1) how site-specific lint yield and associated soil physical properties in a cotton (*Gossypium hirsutum L.*) production field are related to changes in NDVI across the growing season, and (2) when multispectral images should be collected to optimize the cost and efficiency of remote sensing as a tool for site-specific management of the cotton crop. Temporal multispectral images data acquired comprised 10 dates (1998) and 17 dates (1999) during growing seasons, respectively with analysis focused on 24 areas of interest (AOI) (each 2 x 8 m) located in two transects on a 162-ha farm field. Along each transect, soil textural classification ranged from sandy loam to silt loam. At an early growth stage [~300-600 degree days (DDs) after emergence], low NDVI and plant density were associated with soils having low saturated hydraulic conductivity (k_s) and characterized as drainage ways. Among the AOI's, maximal NDVI was reached at approximately 1565 DD in 1998 and 1350 DD in 1999. A strong range of Pearson correlation (r^2=0.65 – 0.83) between lint yield and NDVI during flowering stage (~800-1500 DDs) supports the utility of NDVI maps for site-specific application. However, values for NDVI did not correlate well with lint yields beyond 1500 DDs [fruit (boll) opening stage] and decreased sharply on sites with sandy soil texture. Visual separation of seasonal trends in the NDVI vs. DD relationship was also related to sandy soil vs. silt loam soil texture and seasonal rainfall difference between years. Based on the statistical relationship between NDVI vs. DD it was concluded that acquisition of a single imagery during peak bloom period would be sufficient for predicting the spatial distribution of lint yield and will also be economically feasible. Results of this study indicate that spatial variability in soil physical properties induced variability in crop growth and yield. Similar methodology could be adopted for site-specific management of other crops.

Keywords Alluvial soils; Degree days; Precision agriculture, Soil texture; Volumetric water content

INTRODUCTION

Farm profitability depends on proper management on timely basis of inputs, such as fertilizers, crop varieties, pesticides, and irrigation. The site specific crop management system is designed to target crop and soil inputs according to within field spatial variability/requirements to optimize profitability and reduce the environmental impact (Shaver et al., 2011; Hochman et al., 2013). Geoinformatic techniques (remote sensing & GIS) can meet many of the data requirements in precision agriculture, because it appears to provide a means to timely assess field-level spatial variability in crop and soil conditions (Hatfield et al., 2008). To encourage the adoption of remote sensing techniques to be utilized by growers/consultants, agricultural scientists should define a timeline for acquiring imagery specific to major field crops for specific crop inputs/management. Due to the indeterminate growth habit of cotton, timely information regarding crop vigor and yield assessment is particularly important for the management of production inputs.

The reflectance spectra of a vegetation canopy is affected by plant chemical composition and canopy level architecture, that is, the arrangement and/or density of leaves, flowers, stems, and their shadows against a soil background (Wiegand, et al. 1991; Wall et al., 2008). Two reflectance indices, the near infrared (NIR) to red ratio (NIR/R) and the normalized difference vegetation index

(NDVI), have been widely used for assessment of crop biophysical conditions, such as, water and nutrient status, leaf nitrogen concentration, leaf area index (LAI) and above ground biomass (Basso et al., 2012; Lofton et al., 2012). The principle is that chlorophyll and carotenoid pigments have major effects on canopy reflectance in the visible wavelengths; whereas, changes in reflectance in the near-infrared (NIR) region of the spectra are related to differences in leaf structure and canopy development (Shaver et al., 2011). In this sense, the ability of NDVI to capture changes in crop development over time has made it a useful parameter for predicting tiller development in wheat (*Triticum aestivum* L.) (Flowers et al., 2001; Duchemin et al., 2006), improving models that simulate crop phenology (Boken et al., 2005), and large-scale forecasting of crop yields (Kastens et al., 2005).

Reflectance measurements in cotton have been related to changes in N status (Winterhalter et al., 2011; Zhang et al., 2012), lint yield (Yang et al., 2004; Zhao et al., 2007) and yield-limiting insect populations (Zhang et al., 2012). In a Mississippi study with Upland cotton, Zhao et al. (2007) reported final lint yield correlated well with NDVI measurements based on a hand-held spectroradiometer. In a California study, Plant et al. (2000) reported significant correlation between lint yield and NDVI values integrated over the growing season. They found the spatial-temporal pattern of NDVI tends to indicate the presence of N stress and were closely coincident with the onset of plant water stress. Additionally, the relative nitrogen vegetation index was not superior to NDVI as an indicator of N stress. Similarly, Li et al. (2001) established significant cross correlation of cotton reflectance in red, NIR and NDVI with soil water and the percentage of sand and clay at different locations within a production field. They further suggested variable-rate N applications could be based upon spatial variability of crop and soil reflectance characteristics. Ritchie and Bednarz, (2005) proposed a spectrometric method for quantifying cotton defoliation (a chemical treatment to remove leaves prior to mechanical harvest), based on the relationship between narrow-waveband NDVI values and leaf area index. They suggested reflectance indices based on red edge (705 nm to 720 nm) measurements and argued that it can offer accurate and consistent estimates of when to use chemical defoliants in cotton production, which could potentially increase defoliation efficiency and decrease costs.

The practice of acquiring weekly or biweekly remote sensing imagery data for research purposes may be economically feasible but may not be economically practical for growers. Since the bottom line for the farmer is profitability. Therefore, the objectives of this study were to (1) resolute when multispectral imagery should be collected to minimize the cost and optimize the efficiency of remote sensing data as a tool for site-specific management of the cotton crop and (2) demonstrate how site-specific lint yield and associated soil physical properties in a cotton production field are related to changes in NDVI across the growing the seasons.

MATERIALS AND METHODS

Study Site Description

The field study was conducted at a private farm (34° 00' 41" N, 90° 55' 39" W) located near the Mississippi River on the northern edge of Bolivar County, MS, USA (Fig. 1). The study site was a 162 ha, cotton production field, having a 2-m elevation range and irrigated with central-pivot system. Climatic data for the site were recorded using an automatic weather station (Fig. 2). A soil survey was conducted to identify the major soil types in the field. The soils in Mississippi Delta are alluvial in nature. Three major soil types were found in the study area i.e. Commerce silt loam (fine-silty, mixed, superactive, nonacid, thermic Aeric Fluvaquents); Robinsonville sandy loam (coarse-loamy, mixed, superactive, nonacid, thermic Typic Udifluvents); and Souva silt loam (very fine, smectitic, thermic choromic Epiaquents). Commerce silt loams are characterized as somewhat poorly drained, moderately slow in water permeability, medium in runoff with 1% slope, and typically water-saturated in layers below 0.5 to 1.2 m during the months of December-April. Robinsonville sandy loam soils are characterized as well drained, slow to medium in runoff with water table fluctuates between 1.2 - 1.8 m during the months of January-April. Souva silt loam soils are somewhat poorly drained. These soils are usually found in the depressions & drainage ways of the fields and water table usually fluctuate within 1-2 m (USDA-NRCS, 1951).

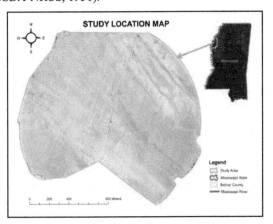

Fig. 1. Study area location map.

Soil Data Analysis

In previous studies (Iqbal et al., 2005) reported on spatial variability of soil physical properties by sampling 209 soil profiles (Fig. 3) in 18 parallel transects to a depth of 1.0 m based on a 91.4 m interval. The present study reports results for two of these field transects located on the west (T$_3$) and east side (T$_{14}$) of the field (Fig.3). These two transects were selected because it approximately represent 97 % of the soil-mapping units of the field. The geographic coordinates of each site were recorded using an eight-channel, differential Global Positioning System (GPS) receiver (March-II-E, CMT Inc., Corvallis, OR).

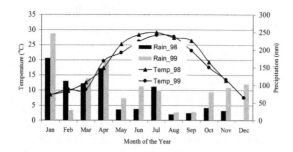

Figure 2. Monthly precipitation and average daily temperature during the experimental period.

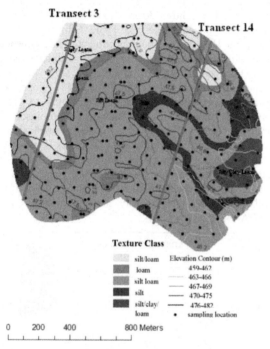

Fig. 3. Field boundary map and the location of 12 sampling sites, at a spacing of approximately 91 m, within each of two transect.

At each location, a tractor-mounted hydraulic soil sampling machine (Giddings no. 10-T Model GST, Giddings Machine Co. Inc., Fort Collins, CO) was used to collect undisturbed, soil profiles to a depth of 1 m as described by Iqbal et al. (2005). The profile was removed from the sampling tube and the depth of each horizon was recorded based on soil morphological characteristics. The different soil profile horizons were excised, mixed separately, and a subsample placed in a waxed, cardboard box. Soil horizon samples were air dried, crushed, sieved (2 mm), and analyzed for particle size distribution using the hydrometer method (Gee and Bauder, 1986) and organic C content using dry combustion method (Rabenhorst, 1988). Additionally, two cores were collected to 1-m depth near the first profile core using the Giddings probe equipped with either a 7.62-cm by 7.62-cm (diameter by depth) or a 7.62- by 1-cm sampling rings. Undisturbed soil cores were obtained from depths representative of a surface, subsurface, and deep soil

horizon depths. Soil cores were transported to the laboratory for determination of bulk density, saturated soil hydraulic conductivity, and volumetric soil water content. The 7.62- x 1-cm soil cores were used to determine soil volumetric water content at seven pressure heads of 1, 10, 33, 67, 100, 500, and 1500 kPa using a pressure plate apparatus (Klute, 1986). The data was used to construct a soil moisture release curve, which describes the relationship between soil water tension and moisture content and is influenced largely by soil texture. The saturated soil hydraulic conductivity for each undisturbed soil profile core/depth was determined using the falling head method (Klute and Dirksen, 1986). Soil available water content of coarse-textured soils was expressed as the difference between volumetric soil water content at field capacity (-10 kPa) and water available at permanent wilting point (-1500 kPa); whereas, for fine-textured soils it was expressed as the difference between soil water content at -33 kPa and -1500 kPa (Jury et al., 1991).

Conventional crop management practices were applied by the grower during the two growing seasons (1998 and 1999). Cotton (cv NuCotn 33B) was planted in 1-m row spacing on April 1, 1998 and April 30, 1999. The seeding rate was 11-13 kg ha⁻¹, which would produce a final population of approximately 98,500 plants ha (40,000 acre⁻¹). Split applications of fertilizer N were side-dressed as urea ammonium nitrate solution [Ensol: 32% N in NH_4NO_3 $CO(NH_2)$] using an 8-row applicator in late May (72 kg N ha⁻¹) and early July (62 kg N ha⁻¹). Cotton was harvested on September 8, 1998 and September 22, 1999 using a four-row picker equipped with a GPS and field-calibrated yield monitor (± 2%). To ensure accuracy at the 12 sampling sites within each field transect, the picker was maneuvered to precise locations in order to harvest a 16-m section of the rows centered on each site (4-m width x 16-m length). Once the seed cotton was harvested, it was transferred to a weigh wagon and the biomass determined. A subsample (~1000 kg) was ginned using a 10-saw gin to determine lint percentage (lint weight/seed cotton weight), which was used to estimate lint yield on a site-by-site basis. Lint percentage ranged from 38 to 41% across the 24 field locations.

Analysis of Cotton Crop Remote Sensing Data

The NASA commercial remote sensing program acquired imagery at 7-10 d intervals during the growing season, April to September. A three-band digital camera system mounted in the belly of a fixed wing aircraft operated at an average speed of 204 km h⁻¹ (110 knots) obtained the multispectral imagery from an altitude of 1824 m above ground level, which rendered a 1 m² spatial resolution per pixel. The three bands were centered at 540 nm (green, chlorophyll reflectance), 695 nm (red, chlorophyll absorption), and 840 nm (near-infrared – NIR tissue reflectance) with a 10 nm spectral resolution.

Imagery data comprised 10 acquisition dates in 1998 and 17 acquisition dates in 1999. Image pixel values were extracted from a 2 by 8 m rectangular area of interest (AOI) centered on each of the 24 study sites using

ERDAS IMAGINE (ERDAS, 1997). Each AOI was oriented at an angle of 164° to align with the crop rows. Sample sites and AOI's were at the same locations each year. The average DN value of each AOI (n=16) was used to derive the site-specific normalized difference vegetation index (NDVI) using the formula:

$$NDVI = \frac{(NIR_{840} - Red_{695})}{(NIR_{840} + Red_{695})}$$

The concept of degree days (DD, daily heat units) was used to relate potential plant growth to the temporal pattern of NDVI for each AOI. The DD concept can be used to better predict the physiological development, growth stages and maturity of a crop than calendar date, and is calculated as follows:

$$DD = \frac{T_{max} - T_{min}}{2} - T_{base}$$

where, T_{max} is the maximum daily temperature, T_{min} is the minimum daily temperature, and T_{base} is the base temperature of 15.6 °C (Vories et al., 2011). The principle is that starting at 15.6 °C (60 °F), cotton doubles its growth rate for every 10 °C (18 °F) increase in ambient temperature. Degree days were accumulated daily from the date of plant emergence to determine plant growth stages, which is crucial for making management decisions in cotton production.

To compare the spatial distribution of plant growth between two growing seasons, 1998 and 1999, imagery data of NDVI were overlaid on a three-dimensional map of elevation for the 162-ha field. Additionally, the elevation data were analyzed using ArcView GIS software (ESRI, 1998) in order to derive natural pattern of water flow across the field landscape. Once this 'hydrology network' was derived, the results were overlaid on the three-dimensional NDVI map in order to visualize its relationship to main drainage areas and channels in the field.

RESULTS AND DISCUSSION

During the growing season, April – September, approximately 36% less rainfall was recorded in 1998 (376 mm) than 1999 (594 mm) (Fig. 2). In 1998, relatively low rainfall during May and June resulted in a 150 mm rainfall deficit for the period. Water plays a crucial role in controlling cotton crop development to achieve a favorable balance between vegetative growth and reproductive development. During this period the plant is flowering and a number of young bolls are establishing. These young bolls, if successful, will have a high contribution to final yield. However, it is during this stage that they are highly sensitive to moisture stress. Turner et al (1986) showed that water deficits influence photosynthesis, leaf expansion, the retention of flowers and bolls, and ultimately the yield of cotton.

Additionally, ambient air temperatures during the May-August period were somewhat warmer in 1998 than

1999. Because cotton is typically at pre-bloom (flowering) stage of development in early July (~450 DDs), decreased soil water potential during this critical period may cause transpiration to fall below the potential evaporation rate, could likely cause stress and reduce crop biomass. Based on site-specific NDVI measurements in July, relatively low rainfall in 1998 apparently led to increased variability in NDVI, as compared to 1999 when rainfall was similar to the long-term average (Figs. 2 and 4).

Remote Sensing of Cotton Crop

Variability in site-specific NDVI was evident in both years which were in agreement with Plant et al. (2000). NDVI profile during July readily captured differences in cotton growth and development both across the two years and among the 12 sampling sites (Fig. 4). Beginning in July, the cotton crop usually grows vigorously and develops a relatively high leaf area index. Site-specific NDVI values in July ranged from 0.33 to 0.52 in 1998 (mean = 0.44 ± 0.05), and from 0.53 to 0.62 in 1999 (mean = 0.60 ± 0.02), with higher NDVI typically observed in Transect 3, as compared to Transect 14. Five of the twelve sites in Transect 3 have soils with sandy loam texture class with relatively high saturated hydraulic conductivity (Fig.4 and Table 1).

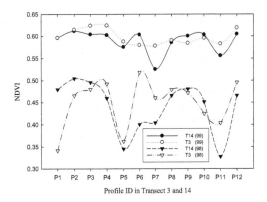

Fig. 4. Normalized difference vegetation index (NDVI, based on digital number values) of cotton on July 18, 1998 and July 29, 1999 at 12 sampling sites (P1 – P12) for soil physical properties in each of two transects, 'T3' and 'T14'.

Unlike Transect 3, nine of the twelve sites in Transect 14 have soils with silt loam soil texture class, which had relatively low values for soil saturated hydraulic conductivity, suggesting relatively low soil available water content at lower depths. Consistent with these observations, remote sensing imagery in July indicated relatively low NDVI for sites 7 and 11 in Transect 14 in 1999, and for sites 5, 7, and 11 in both of the transects in 1998 (Fig. 4). The sites in Transect 3 having low NDVI were associated with sandy-loam and silt-loam soil textures; whereas, the sites in Transect 14 were associated with silt-loam textures of the surface soil (Table 1). The Robinsonville soil series (sandy loam soils) was found in 7 of the 12 sites in Transect 3 and were characterized by high sand content (40 to 70 %), high hydraulic conductivity (15 to 80 cm day[-1]) and low volumetric water content at plant wilting point (generally < 17%) (Table1). By comparison, surface soils along Transect 14 were dominated by Commerce and Souva soil series (silt loam to silt clay loam) that were characterized by high clay

content (12to 20 %), low hydraulic conductivity (0.14 to 7.0 cm day[-1]) and high volumetric water content at the wilting point (generally 20 to 26%).

Table 1. Surface soil (0-30 cm depth) bulk density (BD), saturated hydraulic conductivity (K_{sat}), clay content, sand content, textural classification, organic matter (OM), volumetric water content at field capacity (FC), volumetric water content at wilting point (WP) and soil available water content (SAWC) at 24 sampling sites in two field transects in Livingston Field, Perthshire Farm, Mississippi.

Soil type[a]	Site	BD g cm[-3]	Ksat cm day[-1]	Clay %	Sand %	Texture	OM %	Volumetric water (%)[b]		
								FC	WP	SAWC
						Transect 3				
So	P01	1.16	37.37	24.4	18.5	silt loam	1.42	19.69	8.97	10.72
So	P02	1.16	42.55	21.5	16.5	silt loam	1.66	15.70	9.89	5.81
So	P03	1.01	26.49	14.0	11.8	silt loam	1.42	14.04	6.75	7.28
Cc	P04	1.23	62.54	12.7	36.5	silt loam	1.10	32.57	22.19	10.37
Ra	P05	1.19	79.10	7.5	70.4	sandy loam	0.51	42.82	15.90	26.92
Rl	P06	1.28	8.41	10.1	49.7	loam	0.90	34.50	23.25	11.25
Cl	P07	1.24	11.05	10.1	45.1	loam	0.90	24.24	13.69	10.55
Ra	P08	1.35	8.10	17.8	58.4	sandy loam	0.77	30.72	18.82	11.91
Ra	P09	1.19	16.48	15.5	58.9	sandy loam	0.86	41.91	26.62	15.29
Ra	P10	1.21	72.95	6.3	58.8	sandy loam	0.66	18.14	7.06	11.07
Rl	P11	1.41	77.30	10.1	42.7	loam	0.86	28.44	13.22	15.22
Ra	P12	1.28	18.66	8.8	57.0	sandy loam	0.82	28.79	17.08	11.71
						Transect 14				
Cs	P01	1.28	0.06	1.9	10.2	silt	1.47	14.69	6.56	8.14
Cc	P02	1.14	32.03	3.2	26.9	silt loam	1.00	17.04	7.32	9.71
Cc	P03	1.09	7.58	1.3	33.6	silt loam	1.18	25.26	15.38	9.88
Cc	P04	1.21	104.17	2.2	27.9	silt loam	1.12	16.46	9.30	7.16
So	P05	1.11	3.42	19.4	4.5	silt loam	1.29	42.00	32.99	9.01
Cc	P06	1.19	0.14	21.1	10.1	silt loam	1.36	32.19	24.00	8.19
Cc	P07	1.07	0.33	25.1	2.5	silt loam	1.92	36.36	20.68	15.68
Cc	P08	1.17	0.46	8.3	14.8	silt loam	1.12	26.98	11.78	15.21
Cc	P09	1.17	165.65	2.2	32.0	silt loam	1.20	37.96	26.14	11.82
Cc	P10	1.08	7.00	12.8	19.4	silt loam	1.41	36.16	21.32	14.85
Ra	P11	1.16	22.28	4.4	74.6	loamy sand	0.47	30.59	21.26	9.34
Cl	P12	1.11	2.76	8.8	43.9	loam	0.87	30.71	24.53	6.18

[a] Soils comprise three types (or series), a Commerce silt loam (Cc), a Robinsonville sandy loam (Ra), and a Souva silt loam (So). Sub types were identified due to difference in the surface-soil texture and included Robinsonville loam (Rl), Commerce silt (Cs), Commerce loam (Cl).
[b] Field capacity is the amount of soil water content held by the soil after excess water has drained (which is equivalent to the water content at -33 kPa pressure head); the wilting point can be defined as the soil water content at which plants wilt, when transpiration exceeds the rate of water uptake by roots (which is equivalent to the water content at -1500 pressure head); and soil available water content is the portion of water content that can be readily absorbed by plant roots (which is equivalent to the difference in water content between pressure heads of -33 and -1500 kPa).

Seasonal curves of the relationship between site-specific NDVI values and DDs were used as surrogates of crop canopy development (Fig. 5). Difference between years in the NDVI response curves illustrate difference in crop vigor due to annual rainfall, as well as difference in leaf area development, as the field was planted approximately 30 days earlier in 1998 than 1999. Negative, but progressively increasing NDVI values were observed until approximately 875 DDs in 1998 (~ June 17) and approximately 750 DDs in 1999 (~ July 2). In 1998, distinctive NDVI curves were evident between 300 and 600 DDs, when cotton is typically in pre-bloom (or

squaring) stage of plant development (Reddy et al., 1993). Low NDVI at pre-bloom stage was probably due to low plant density, plants of lower vigor or growth between years, or both of these effects. Relatively low rainfall in May-June 1998 probably also contributed to site-specific difference in NDVI during the pre-bloom growth stage. Among the 12 AOI's, visually separation of the seasonal trends in the NDVI vs. DD relationship was more obvious across Transect 3, where the soils were sandy in texture, as compared to Transect 14, where the soils were silt loam in texture (Table 1). With regard to the potential effects of soil attributes on NDVI, most of the sites located on sandy soils in Transect 3 had low NDVI and a more abrupt decline in NDVI values was observed as the plants matured (beyond ~1500 DDs), as compared to sites on silt-loam soils in Transect 14. Low water holding capacity of sandy soils in Transect 3 apparently limited plant growth and development at some sites, leading to earlier termination of flowering (cut-out), which led to early crop maturity. In a Texas study on the relationship between cotton lint yield and narrow wavebands (3.6 nm bandwidth) of red and NIR reflectance (centered at 630 and 830 nm, respectively), images for two fields in mid-June after peak vegetative development revealed plants in areas with high sandy soil texture had low NDVI, low plant density and low canopy cover (Yang et al., 2004). These results and those from related field studies (Li et al., 2001; Bronson et al., 2005) suggest that low water- and nutrient-holding capacity of the surface soils (0-30 cm depth) with sandy loam soil texture may cause the observed spatial distribution of growth and yield development in this cotton production field. Apparently, yield was reduced because irrigation provided by the center pivot was less than plant transpiration.

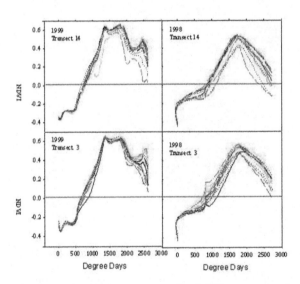

Fig. 5. Relationship between normalized difference vegetation index (NDVI) and growing degree days (DD, 15°C base temp) at 12 sampling sites in 1998 and 1999, where multispectral imagery data was used to calculate NDVI with 1-m resolution for each 2 x 8 m area of interest. Note the distinctive pattern for NDVI curves between 300 and 600 DD.

NDVI for Crop Management on Different Soils

Strong Pearson correlation coefficients between NDVI and final lint yield were obtained across 24 AOI's at peak-bloom, corresponding to image acquisition dates of 18 July 1998 and 29 July 1999 (Fig. 6). In 1998, NDVI values reached a maximum at ~1565 DDs (July 18) and explained approximately 83% of the variability in final lint yield ($r^2 = 0.83$; $p < 0.001$). A decreasing trend in NDVI beyond approximately 1565 DDs (Fig. 5) was associated with a decreasing trend in the correlation

between NDVI and yield (Fig. 6). In sugarcane (*Saccharum officinarum* L.), Guérif and Duke (2000) used reflectance data to calibrate the 'SUCROS' model and improve its relationship with final biomass yield. They concluded that yield estimation was dependent on stage of plant development (i.e., the timing of remote sensing data) and the best relationship was obtained when the data covered the whole period of leaf area development, including the maximal leaf area index.

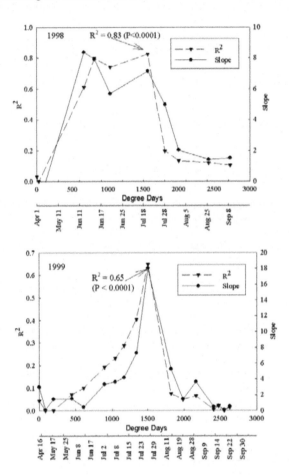

Fig. 6. Changes in R^2 and slope of the linear relationship between normalized difference vegetation index (NDVI) and lint yield across 24 sampling sites as a function of degree days (DD, 15 °C base temp) in 1998 (a) and 1999 (b).

In 1999, final yield and NDVI were related linearly on 29 July ($r^2 = 0.65$; $p < 0.001$), when the slope of the relationship was maximal at approximately 18 NDVI units per degree day.

Similarly, Li et al. (2001) reported proximal measurements of broad-band NDVI (26-32 nm bandwidth) in August, when leaf area and NDVI were maximal and explained approximately 64% of the variation in final lint yield. The relationship between NDVI and lint yield varied in different studies. In a remote sensing study of two production fields (16-22 ha), Yang et al. (2004) used hyperspectral reflectance observations in June after the crop achieved its maximum canopy cover. They rectified NDVI and final yield on a 8-m grid and also aggregated 102 hyperspectral bands (3.63 nm bandwidth) to mimic four broad bands of the Landsat ETM+ sensor, (blue, 450–515 nm; green, 525–605 nm; red, 630–690 nm;, and NIR, 775–900 nm). In that study, broad-band NDVI explained 30% and 38% of the variability in yield for the two fields, respectively, and similar Pearson correlation was obtained between yield and narrow-band NDVI. Additionally, Yang et al. (2004) reported stepwise regression models based on several significant narrow-waveband NDVIs explained 61 to 69% of lint yield variability. While these results agree with other studies that accuracy of yield estimation in cotton increases as plants mature (Plant et al., 2000; Zhao et al., 2007) and NDVI can have broader uses other than just yield prediction since different management decisions occur at different times of the season.

Figure 7 illustrates areas of the 162-ha field with low NDVI highlighted with 'grey' to 'yellow' pixels and encircled in red. These areas were also low in plant density and associated with a main drainage way with inclined topography (a channel produced by the flow of surface runoff). Additionally, these and other low NDVI areas of the field were observed in both the main and tertiary drainage ways. Low yields were recorded in 1999 at sampling sites located in the main drainage ways and areas adjacent to drainage ways. Apparently, decreased crop performance in or around the drainage ways that were transected by Transect 14 profiles 5 and 11 were related to a thinning of the fertile surface soil and exposure of the subsurface soil, which is relatively hard and impervious to water. Because percolation is impeded in this area's soil, water is not retained and generates more as surface runoff, particularly during heavy rainfall or irrigation events which also removes applied N and ultimately create nutrients deficiency in those impermeable drainage ways. Lower NDVI values of N-deficient cotton are associated with lower plant biomass relative to cotton with sufficient N (Li et al., 2001). To mitigate excessive runoff losses, this area of the field would need to be leveled and sub-soiled in order to improve hydraulic conductivity of the soil.

By combining degree days with knowledge of the NDVI vs. lint yield relationship it is possible to characterize cotton growth stage(s) when the NDVI utility might be the most profitable in agronomic management decisions (Reddy et al., 1993; Hatfield et al., 2008). Results for this relationship at multiple crop growth stages suggest remote sensing to encompass the 'first-square' growth stage (typically around 550 DDs), may aid in decisions concerning fertilizer or irrigation management, particularly under low rainfall conditions i.e. from emergence to flowering stage of growth, leaf area and vegetative structures that will then support future reproductive growth. Multispectral imagery to encompass the 'mid-bloom' (flowering) growth stage (between approximately 1000 and 1500 DDs), would be useful for estimation of cotton lint yield. Such a remote sensing

temporal studies may help to determine the need and/or timing of harvest aides (e.g., chemical defoliants) (Ritchie and Bednarz, 2005), as well as areas of field that should be harvested first. Flowers et al. (2001) reported a significant simple correlation ($r = 0.88$) between tiller density and NIR digital counts in wheat during tillering phase (Zadoks growth stage 25) and suggested that the relationship could be used to direct N fertilizer applications on a site-by-site basis. In either case, site-specific NDVI appeared to provide timely information regarding the spatial distribution of soil properties in the field. These results agree with studies in the Canadian Prairies of Saskatchewan that compared yield estimates for wheat based on either NDVI from 16 years of NOAA AVHRR satellite data (between 1987 and 2002) or the Cumulative Moisture Index, a land-based weather-related measurement (Wall et al. (2008). Those authors found NDVI possessed explanatory power four weeks earlier in the growing season, as compared to the Cumulative Moisture Index.

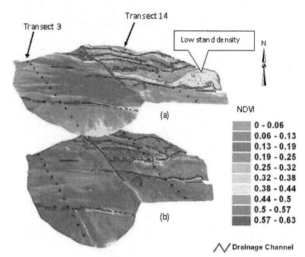

Fig. 7 Three-dimensional map of normalized difference vegetation index (NDVI) for the field on July 28, 1998 (a) and July 29, 1999 (b). Values for NDVI were draped over an elevation map that was produced using a GPS-equipped cotton yield monitor and used to delineate major drainage ways in the field. The area encircled in 'red', which had low NDVI values in 1998 due to low plant density, corresponds to one of the largest drainage ways.

CONCLUSIONS

This 2-yr field study demonstrated an association between the spatial distribution of crop NDVI and point estimates of several soil physical properties, some of which had influenced cotton growth and yield. Additionally, differences between years in NDVI patterns across field could be attributed to differences in seasonal rainfall coupled with inherent spatial variability in soil physical properties. Optimal timing of airborne remote sensing data acquisition for agronomic management was identified by the occurrence of significant correlation during the growth season between lint yield and NDVI across 24 sampling sites. The observation of strong yield vs. NDVI correlation in imagery during July, with low values confined chiefly to the main drainage ways, suggest two to three multispectral images acquired at flowering stage of plant development would be optimal for assessing spatial variability in lint yield on alluvial flood plain soils common in the Mississippi Delta region. Acquisition of multispectral remote sensing data earlier in the growth season may be useful for directing site-specific applications of N fertilizer or plant growth regulators.

ACKNOWLEDGEMENTS

This study was in part supported by The National Aeronautical and Space Administration-funded Geosystems Research Institute (formerly Remote Sensing Technology Center) at Mississippi State University (NASA grant number NCC13-99001). The excellent technical assistance of Kim Gourley and Sam Turner (USDA, Agricultural Research Service) is greatly appreciated.

LITERATURE CITED

Basso, B., Fiorentino, C., Cammarano, D., Cafiero, G., Dardanelli, J., 2012. Analysis of rainfall distribution on spatial and temporal patterns of wheat yield in Mediterranean environment. Eur J Agron 41: 52-65.

Boken, V. K., Shaykewich, C. F., 2005. Improving an operational wheat yield model using phenological phase-based normalized difference vegetation index. Int. J. Remote Sens. 26: 3877-3897.

Bronson, K. F., Booker, J. D., Keeling, J.W., Boman, R. K., Wheeler, T. A., Lascano, R. J., Nichols, R.L., 2005. Cotton canopy reflectance at landscape scale as affected by nitrogen fertilization. Agron. J. 97: 654-660.

Duchemin, B., Hadria, R., Erraki, S., Boulet, G., Maisongrande, P., Chehbouni, A., Escadafal, R., Ezzahar, J., Hoedjes, J. C. B., Kharrou, M. H., 2006. Monitoring wheat phenology and irrigation in Central Morocco: On the use of relationships between evapotranspiration, crops coefficients, leaf area index and remotely-sensed vegetation indices. Agric. Water Manag. 79: 1-27.

ERDAS, 1997. ERDAS Imagine, Ver. 8.5. Atlanta, GA, USA.

ESRI, 1998. ArcViewGIS, Ver. 3.2. Environmental Systems Research Institute, Redland, CA, USA.

Flowers, M., Weisz, R., Heiniger, R., 2001. Remote sensing of winter wheat tiller density for early nitrogen application decisions. Agron. J. 93: 783-789.

Gee, G. W., Bauder, J. W., 1986. Particle size analysis. In: A. Klute (Ed.), Methods of Soil Analysis, Part 1, 2nd ed. Agron. Monogr. 9, ASA, Madison, WI, pp.383-409.

Guérif, M., and C.L. Duke, 2000. Adjustment procedures of a crop model to the site specific characteristics of soil and crop using remote sensing data assimilation Agriculture, Ecosystems and Environment 81: 57–69.

Hatfield, J.L., Gitelson, A.A., Schepers, J.S., Walthall, C.L., 2008. Application of spectral remote sensing for agronomic decisions. Agron. J. 100: 117-131.

Hochman, Z., Carberry, P.S., Robertson, M.J., Gaydon, D.S., Bell, L.W., McIntosh, P.C., 2013. Prospects for ecological intensification of Australian agriculture. Eur J Agron 44: 109-123.

Iqbal, J., Thomasson, J. A., Jenkins, J. N., Owens, P. R., Whisler, F. D., 2005. Spatial variability analysis of soil physical properties of alluvial soils. Soil Sci. Soc. Am. J. 69: 1338-1350.

Jury, W. A., Gardner, W. R., Gardner, W. H., 1991. Soil physics. 5th ed. John Wiley & Sons, New York.

Kastens, J. H., Kastens T. L., Kastens, D. L. A., Price, K. P., Martinko E. A., Lee, R., 2005. Image masking for crop yield forecasting using AVHRR NDVI time series imagery. Remote Sens. Environ. 99: 341-356.

Klute, A. 1986. Water retention: Laboratory methods. In: A. Klute (Ed.), Methods of Soil Analysis, Part 1, 2nd ed. Agron. Monogr. 9, ASA, Madison, WI, pp. 635-660.

Klute, A., Dirksen, C., 1986. Hydraulic conductivity and diffusivity: Laboratory methods. In: A. Klute (Editor), Methods of Soil Analysis, Part 1, 2nd ed. Agron. Monogr. 9, ASA, Madison, WI, pp. 687-732.

Li, H., Lascano, R. J., Barnes, E. M., Booker, J., Wilson, L. T., Bronson, K. F., Segarra, E., 2001. Multispectral reflectance of cotton related to plant growth, soil water and texture, and site elevation. Agron. J. 93: 1327-1337.

Lofton, J., Tubana, B.S., Kanke, Y., Teboh, J., Viator, H., Dalen, M., 2012. Estimating Sugarcane Yield Potential Using an In-Season Determination of Normalized Difference Vegetative Index. Sensors-Basel 12: 7529-7547.

Plant, R. E., Munk, D. S., Roberts, B. R., Vargas, R. L., Rains, D. W., Travis, R. L., and Hutmacher, R. B., 2000. Relationships between remotely sensed reflectance data and cotton growth and yield. Trans. ASAE, 43: 535-546.

Reddy, K. R., Hodges, H. F. McKinion, J. M., 1993. A temperature model for cotton phenology. Biotronics, 2:47-59.

Ritchie, G. L., Bednarz, C. W., 2005. Estimating defoliation of two distinct cotton types using reflectance. J. Cotton Sci. 9: 182-189.

Rabenhorst, M.C., 1988. Determination of organic carbon and carbonate carbon in calcareous soils using dry combustion. Soil Sci. Soc. Am. J. 52: 965-969.

Shaver, T.M., Khosla, R., Westfall, D.G., 2011. Evaluation of two crop canopy sensors for nitrogen variability determination in irrigated maize. Precis Agric 12: 892-904.

Turner, N.C., Hearn, A.B., Begg, J.E., Constable, G.A., 1986. Cotton (Gossypium-Hirsutum-L) - Physiological and Morphological Responses to Water Deficits and Their Relationship to Yield. Field Crop Res 14: 153-170.

USDA-NRCS, 1951. Bolivar County Mississippi. USDA-Natural Resources Conservation Service Publication Number 5, U.S. Government Printing Office, Washington, DC, USA. Online http://websoilsurvey.nrcs.usda.gov/app/HomePage.htm. Accessed 27 September 2011.

Wall, L., Larocque, D., Leger, P. M., 2008. The early explanatory power of NDVI in crop yield modeling. Int. J. Remote Sens. 29: 2211-2225.

Wiegand, C. L., Richardson, A. J., Escobar, D. E., Gerbermann, A. H., 1991. Vegetation indices in crop assessments. Remote Sens. Environ. 35: 105-119.

Winterhalter, L., Mistele, B., Jampatong, S., Schmidhalter, U., 2011. High throughput phenotyping of canopy water mass and canopy temperature in well-watered and drought stressed tropical maize hybrids in the vegetative stage. Eur J Agron 35: 22-32.

Yang, C., Everitt, J. H., Bradford, J. M., Murden, D., 2004. Airborne hyperspectral imagery and yield monitor data for mapping cotton yield variability. Precis. Agric. 5: 445-461.

Zhang, H., Hinze, L.L., Lan, Y., Westbrook, J.K., Hoffmann, W.C., 2012. Discriminating among Cotton Cultivars with Varying Leaf Characteristics Using Hyperspectral Radiometry. T Asabe 55: 275-280.

Zhao, D., Reddy, K. R., Kakani, V. G., Read, J. J., Koti, S., 2007. Canopy reflectance in cotton for growth assessment and lint yield prediction. Europ. J. Agronomy 26: 335-344.

19

YIELD AND QUALITY TRAITS OF SOME PERENNIAL FORAGES AS BOTH SOLE CROPS AND INTERCROPPING MIXTURES UNDER IRRIGATED CONDITIONS

Mehmet Salih SAYAR[1]*, Yavuz HAN[2], Halil YOLCU[3], Hatice Yücel[4]

[1]*Crop and Animal Production Department, Bismil Vocational Training School, Dicle University, Bismil, Diyarbakir, TURKEY*
[2]*GAP International Agricultural Research and Training Center, Diyarbakir, TURKEY*
[3]*Kelkit Aydın Doğan Vocational Training School, Gumushane University, Kelkit, Gumushane, TURKEY*
[4]*Eastern Mediterranean Agricultural Research Institute, Dogankent, Adana, TURKEY*
* Corresponding author: msalihsayar@hotmail.com

ABSTRACT

A field study was conducted to evaluate the yield and quality traits of sole lucerne (L), sole bromegrass (B), sole tall fescue (T), sole orchardgrass (O), sole ryegrass (R), and lucerne + bromegrass + tall fescue (L+B+T) and lucerne + bromegrass + tall fescue + orchardgrass + ryegrass (L+B+T+O+R) intercropping mixtures at the GAP International Agricultural Research and Training Centre under the irrigated conditions during 2009, 2010, and 2011 in the Southeastern Turkey. Dry matter yield (DMY), crude protein content (CPC), crude protein yield (CPY), acid detergent fiber (ADF), neutral detergent fiber (NDF), dry digestible matter (DDM), dry matter intake (DMI), total digestible nutrients (TDN), and relative feed value (RFV) were determined in this study. The L+B+T intercropping mixture and sole lucerne provided higher yields than the other crops tested. Sole lucerne had higher protein and quality contents than the other sole perennial forages and intercropping mixtures. The L+B+T intercropping mixture had a higher yield and quality than the other sole perennial forages and intercropping mixtures, with the exception of sole lucerne.

Keywords: ADF, crude protein, dry matter yield, intercropping mixture, relative feed value

INTRODUCTION

The Southeastern Anatolian Region comprises nine provinces and is one of the driest region of Turkey. Dry farming systems have been practiced in most of the agricultural areas of this region until last decades (Bengisu, 2011). The Southeastern Anatolian Project (GAP) was initiated in 1989 in this region to increase the area of irrigated agricultural land and electricity production (Sahin and Tasligil, 2013). A large number of hydroelectric power plants have been completed, and irrigation studies are progressing in this region (Sahin and Tasligil, 2013). Large areas of agricultural land recently have started to be irrigated in this area as a result of project work (Bengisu, 2011).

The region has significant potential in terms of the presence of livestock, but the yield of these animals is not at the desired level (Sakarya et al., 2008). There are several reasons for this situation, but one of the most important is the insufficient production of quality roughage (Sakarya et al., 2008). Total roughage that is produced in forage crops cultivated areas and pasture-rangelands can meet only 33% of the available roughage requirements in this region (Sayar et al., 2010). Farmers

have a significant opportunity to meet their roughage requirements due to the increased productivity resulting from the improved irrigation. They have started to produce forage under irrigated conditions in areas where irrigation works have been completed through the project. However, the new irrigation conditions have necessitated changes to established crop patterns for successful forage production. There is limited knowledge regarding which forage crop species and mixtures of crops can be grown under the new irrigated conditions. The determination of which new perennial forage species and mixtures of crops are most convenient is important for improving forage production.

It is hypothesized that lucerne, bromegrass, tall fescue, orchardgrass, ryegrass and mixtures of these crops can be successfully cultivated for yield and quality under the new irrigated conditions. Therefore, the aim of this study was to evaluate the yield and some quality traits of sole lucerne, sole bromegrass, sole tall fescue, sole orchardgrass, sole ryegrass, and lucerne + bromegrass + tall fescue (L+B+T) and lucerne + bromegrass + tall fescue + orchardgrass + ryegrass (L+B+T+O+R) intercropping mixtures under the irrigated conditions.

MATERIALS AND METHODS

Field experiment and growth conditions

This study was conducted for 3 consecutive years (2009, 2010, and 2011) at an experimental site in the GAP International Agricultural Research and Training Centre (GAP IARTC), Diyarbakır, Turkey (37°56'41.0"N, 40°15'16.8"E and altitude 607 m). Sole lucerne, sole bromegrass, sole tall fescue, sole orchardgrass, sole ryegrass, and lucerne + bromegrass + tall fescue (L+B+T) and lucerne + bromegrass + tall fescue, + orchardgrass + ryegrass (L+B+T+O+R) intercropping mixtures were sown on 20 March 2009 under irrigated conditions. The experimental design was a Randomized Complete Block with three replications. The common and scientific, cultivar names and the seeding rates of the sole forage crops and their mixtures are given in Table 1. Some characteristic traits of the experimental field soil are presented in Table 2. The research plots consisted of six rows of 6-m length, and the rows were spaced 30 cm apart. Nitrogen and phosphorus fertilizers were applied to the soil at a rate of 60 kg ha^{-1} before seeding.

Table 1. Common, scientific and cultivar names and seeding rates of the sole forage crops and their mixtures

Common name	Scientific name	Cultivar name	(kg ha^{-1})
Lucerne (L)	Medicago sativa L.	Elçi	30 kg
Bromegrass (B)	Bromus inermis Leys	Population	40 kg
Tall Fescue (T)	Festuca arundinacea Schreb.	Shelby	40 kg
Orchardgrass (O)	Dactylis glomerata L	Amba	30 kg
Ryegrass (R)	Lolium perenne L	XLT	40 kg
20% L +40% B +40% T			6 kg + 16 kg + 16 kg
25% L + %20B + 25%T + 15%O + 15%R			7.5 kg + 10 kg + 8 kg + 4.5 kg + 6 kg

Table 2. Some chemical and physical properties of the research soil.

Depth	Color	pH	Saturation (%)	Organic Matter (%)	CaCo$_3$ (%)	P$_2$O$_5$ (kg ha^{-1})	K$_2$0 (kg ha^{-1})	Structure
0-30 cm	red brown	7.8	62	1.4	13.7	28	480	Clay-loamy

All plots were irrigated once a week with sprinkler irrigation. Climatic data for the experimental location are shown in Table 3. Lucerne and mixtures including lucerne were harvested at 10% of the flowering stage. Other forage crops were harvested at the beginning of their flowerings. All plots were mowed four times in 2009 and six times in 2010 and 2011.

Table 3. Climatic data of the location in 2009, 2010, 2011 and long-term average (1975-2012) at Diyarbakır location, Turkey

	J	F	M	A	M	J	J	A	S	O	N	D	
Years					**Total Precipitation (mm) (Monthly)**								**Mean**
2009	12.4	70.0	63.9	43.9	9.1	25.8	1.6	0.0	25.2	62.4	55.6	87.2	457.1
2010	113.4	40.2	68.7	22.4	31.6	11.2	0.0	0.0	0.4	63.0	0.0	48.0	398.9
2011	40.0	49.9	46.6	209.0	80.1	13.6	0.6	0.0	9.2	11.8	73.0	40.2	574.0
1975-2012	62.8	67.8	67.3	67.7	39.6	9.0	0.4	0.4	4.3	32.1	51.1	67.4	469.9
					Mean air temperature (°C) (Monthly)								**Total**
2009	1.4	5.6	7.9	11.8	18.2	25.9	29.5	28.6	22.9	18.5	9.8	7.1	15.6
2010	5.4	6.6	11.1	14.2	20.4	27.2	32.3	32.0	27.0	18.1	11.1	6.5	17.7
2011	3.5	4.7	9.0	13.0	17.7	25.5	31.4	30.7	25.0	16.4	6.4	2.3	15.5
1975-2012	1.6	3.6	8.6	13.8	19.2	26.3	31.2	30.3	24.7	17.1	9.0	3.7	15.8
					Mean relative humidity (%) (Monthly)								**Mean**
2009	73.3	82.5	73.8	71.3	51.8	32.2	26.1	19.8	33.0	42.0	83.5	80.9	55.9
2010	80.9	79.9	66.6	60.4	49.3	29.1	19.6	17.5	27.4	56.0	41.1	68.9	49.7
2011	73.4	69.5	56.4	75.7	67.6	38.0	22.5	21.7	30.2	41.6	58.8	73.9	52.4
1975-2012	75.1	70.8	65.5	64.4	56.4	37.4	28.1	27.7	33.5	51.2	66.4	74.7	54.3

Plant and soil analysis

Dry matter yields were determined after green samples (0.5-kg biomass from each plot) were oven-dried at 70□C for 48 h and then ground to pass through a 1-mm sieve for the analysis of crude protein. The Kjeldahl method was used to determine the total N and the crude protein content of both sole forages and mixtures were determined as 6.25 □ N (AOAC, 1995). Acid detergent fiber (ADF) and neutral detergent fiber (NDF) analyses were undertaken according to ANKOM (1997). Dry digestible matter (DDM), dry matter intake (DMI), total digestible nutrients (TDN), and relative feed value (RFV) were calculated according to the equations of Schroeder (1994), as follows:

DDM% = 88.9 - (0.779 □ ADF%)

DMI% = 120 / NDF

TDN% = 96.35 - (ADF% □ 1.15)

RFV = (DDM% □ DMI%) / 1.29

Also quality classes of the forages were determined according to Lacefield (1988).

Statistical analysis

All statistical analyses of data were performed using the JMP 5.0.1 statistical software package (SAS Institute, 2002), and the differences between means were compared using a least significant difference (LSD) test at the 0.05 probability level (Steel and Torrie 1980).

RESULTS AND DISCUSSION

The annual effects on dry matter yield (DMY), crude protein content (CPC), crude protein yield (CPY), ADF, NDF, DDM, DMI, TDN, and RFV were expected to be important and therefore, the results were expressed by year and averaged over the 3 years of the study.

Dry matter yield

The highest dry matter yield in 2009 was found in sole lucerne (17.19 t ha^{-1}) (Table 4). Sole lucerne (30.21 t ha^{-1}), L+B+T (28.88 t ha^{-1}) and the L+B+T+O+R intercropping mixture (26.71 t ha^{-1}) had higher dry matter yields in 2010. In 2011, the highest dry matter yield was achieved in the L+B+T intercropping mixture (32.21 t ha^{-1}) and the L+B+T+O+R intercropping mixture (29.09 t ha^{-1}), this followed by sole lucerne (27.52 t ha^{-1}). The L+B+T intercropping mixture and sole lucerne had higher dry matter yields (24.99 t ha^{-1} and 24.97 t ha^{-1}, respectively) than those of the other crops, based on the mean for the 3 years 2009, 2010, and 2011 (Table 4). Similarly, Kir and et al. (2010) reported that dry matter yields of sole lucerne changed from 8.75 t ha^{-1} to 30.05 t ha^{-1} among the means of seven years. Also, Koc et al. (2004) reported that there were differences between sole tall fescue and a tall fescue + lucerne intercropping mixture in terms of the dry matter yield. Balabanli et al. (2010) reported differences in the dry matter yield amongst various intercropping mixtures. In addition, Karadag and Buyukburc (2004) and Yolcu et al. (2009a) determined differences in the dry matter yield between sole and intercropping mixtures of forages.

Table 4. Dry matter yield, crude protein content and crude protein yield of some perennial forages as sole crops and intercropping mixtures

Treatments	Dry matter yield (t ha^{-1})				Crude protein content (%)				Crude protein yield (t ha^{-1})			
	2009	2010	2011	Mean	2009	2010	2011	Mean	2009	2010	2011	Mean
Lucerne	17.19 a	30.21 a	27.52 b	24.97 a	21.5 a	19.3 a	20.0 a	20.3 a	3.70 a	5.84 a	5.50 a-b	5.01 a
Bromegrass	4.15 e	8.82 b	6.29 c-d	6.42 c	17.4 b	13.8 d	14.7 c-d	15.3 c	0.72 e	1.21 d-e	0.92 c	0.95 d-e
Tall Fescue	3.90 e	6.27 b	7.88 c-d	6.02 c	14.0 c	11.6 e	13.3 d	13.0 d	0.55 e	0.73 e	1.04 c	0.77 e
Orchardgrass	3.60 e	9.57 b	4.56 d	5.91 c	16.0 b-c	14.4 c-d	15.8 c	15.4 c	0.58 e	1.38 d	0.72 c	0.89 e
Ryegrass	7.02 d	7.88 b	8.51 c	7.81 c	16.5 b	15.1 b-d	15.7 c	15.7 c	1.16 d	1.16 d-e	1.34 c	1.22 d
L + B + T	13.90 b	28.88 a	32.21 a	24.99 a	20.0 a	16.7 b	18.2 b	18.3 b	2.78 b	4.80 b	5.86 a	4.48 b
L + B + T + O + R	11.29 c	26.71 a	29.09 a-b	22.36 b	19.9 a	15.8 b-c	17.4 b	17.7 b	2.25 c	4.18 c	5.06 b	3.83 c
Mean	8.72 b	16.90 a	16.58 a	14.07	17.9 a	15.2 c	16.4 b	16.5	1.68 b	2.76 a	2.92 a	2.45
CV (%)	9.88	12.4	13.1	8.0	7.3	6.0	5.4	3.7	11.7	10.5	15.2	6.7
LSD (0.05)	1.53**	3.73**	3.87**	2.00**	2.3**	1.7**	1.6**	1.1**	0.35**	0.52**	0.79**	0.29**

Crude protein content and crude protein yield

Crude protein content of sole lucerne (21.5%), L+B+T (20.0%) and L+B+T+O+R (19.9%) intercropping mixtures were higher than the four sole grasses species in 2009 (Table 4). However, In both 2010 and 2011, the highest crude protein content was found only in sole lucerne (19.3% and 20.0%, respectively). In both years the L+B+T intercropping mixture had the next highest crude protein content (16.7% in 2010 and 18.2% in 2011). Accordingly; the greatest crude protein content was found in sole lucerne (20.3%) followed by the L+B+T (18.3%) and L+B+T+O+R (17.7%) intercropping mixtures, based on the mean for the 3 years 2009, 2010,

and 2011 (Table 4). Findings determined in this study related to crude protein content of sole perennial forage species complied with Serin et al. (1998) (15.7-19.7%) and Albayrak et al. (2011) (18.9%) in sole lucerne, Meyer et al. (1977) in bromegrass (8-19%), Weller and Cooper, (2008) in ryegrass (12.2-17.3%), Sahin et al. (2010) in orchardgrass (7.82%-15.1%). Furthermore; many researchers reported that crude protein content of sole tall fescue forage ranged between 7.7% and 16.8% (Sheaffer and Marten, 1986; Evers et al., 1993; Macadam et al., 1997, Kusvuran and Tansı, 2003; Cınar, 2012). In both 2009 and 2010, sole lucerne had the greatest crude protein yield (3.70 t ha^{-1} and 5.84 t ha^{-1}, respectively) (Table 4), followed by the L+B+T intercropping mixture (2.78 in 2009 and 4.80 t ha^{-1} and 2010). The highest crude protein yield in 2011 was found in the L+B+T intercropping mixture (5.86 t ha^{-1}) and sole lucerne (5.50 t ha^{-1}) followed by the L+B+T+O+R intercropping mixture (5.06 t ha^{-1}). Accordingly; sole lucerne had the highest crude protein yield (5.01 t ha^{-1}) based on the mean for the 3 years of 2009, 2010, and 2011 (Table 4), followed by the L+B+T intercropping mixture (4.48 t ha^{-1}). Similar differences in crude protein content and yield were reported by Yolcu et al. (2009a) between various sole forage crops and intercropping mixtures. In addition, Yucel and Avcı (2009) reported differences in the crude protein content and yield between sole forages and intercropping mixtures. Additionally; in consistent with our research findings many researchers reported that not only sole lucerne sowing had more crude protein content than grasses sowings, but also legume + grass sowings had more crude protein content than sole grasses sowings (Barnett and Posler, 1983; Spandl and Hesterman, 1997; Serin et al., 1998; Albayrak and Ekiz, 2005, Albayrak et al., 2011).

Acid detergent fiber (ADF) and neutral detergent fiber (NDF)

Acid detergent fiber (ADF) and neutral detergent fiber (NDF) concentrations are important quality parameters of forages (Schroeder, 1994; Caballero et al., 1995; Henning et al., 1996; Assefa and Ledin, 2001; Albayrak et al., 2011). Although ADF refers to the cell wall portions of a forage that are made up of cellulose and lignin, NDF refers to the total cell wall, which is comprised of the

ADF fraction plus hemicellulose. As the ADF and NDF percentages increase, quality and digestibility of a forage usually decrease (Lacefield, 1988; Schroeder, 1994; Henning et al., 1996; Joachim and Jung, 1997; Albayrak et al., 2011).

In the study; sole lucerne had the lowest ADF content in 2009 and 2011 (28.3 and 28.1%, respectively). However, there weren't found significant differences among lucerne (29.8%), ryegrass (32.7%), L + B + T (32.9%) and L + B + T + O + R (32.6%) in terms of ADF content in 2010. And ADF contents of the treatments were found lower than the other species (Table 5). Accordingly; the lowest ADF content was found in sole lucerne (28.7%) followed by the L+B+T (32.1%) and L+B+T+O+R (32.4%) intercropping mixtures, based on the mean for the 3 years 2009, 2010, and 2011 (Table 5). On the other hand; sole lucerne had the lowest NDF content in 2009 and 2010 (35.0 and 44.3%, respectively), but there weren't found statistically significant differences among the treatments in terms of NDF contents in 2011. And, the lowest NDF content was found in sole lucerne (41.5%) followed by the L+B+T+O+R (45.1 %) and L+B+T (45.2 %) intercropping mixtures, based on the mean for the 3 years 2009, 2010, and 2011 (Table 5). Similarly; many researchers reported that differences in the ADF and NDF among various sole sowings and intercropping mixtures were found as highly significant (Yucel and Avcı 2009; Yolcu et al., 2009a; Yolcu et al., 2009b, Balabanlı et al., 2010; Albayrak et al., 2011).

Table 5. Acid detergent fiber and neutral detergent fiber content of some perennial forages as sole crops and intercropping mixtures

Treatments	Acid detergent fiber (ADF) (%)				Neutral detergent fiber (NDF) (%)			
	2009	2010	2011	Mean	2009	2010	2011	Mean
Lucerne	28.3 e	29.8 c	28.1 d	28.7 e	35.0 f	44.3 d	45.3	41.5 d
Bromegrass	34.0 b-c	38.0 a-b	34.7 a-c	35.6 b-c	45.5 c	52.6 b-c	49.2	49.1 b
Tall Fescue	37.3 a	40.0 a	37.1 a-b	38.1 a	51.9 a	58.3 a	55.8	55.3 a
Orchardgrass	34.8 b	36.4 b	37.8 a	36.3 a-b	47.8 b	53.6 b	46.9	49.4 b
Ryegrass	32.0 c-d	32.7 c	37.1 a-b	33.9 c-d	46.4 b-c	48.6 c-d	50.0	48.3 b
L + B + T	31.7 d	32.9 c	31.6 c-d	32.1 d	39.1 e	49.7 b-c	46.9	45.2 c
L + B + T + O + R	32.6 c-d	31.7 c	32.9 b-c	32.4 d	41.8 d	49.3 b-c	44.3	45.1 c
Mean	33.0 b	34.5 a	34.2 ab	33.9	43.9 b	50.9 a	48.3 c	47.7
CV (%)	3.6	5.6	7.6	4.0	2.5	4.9	8.7	3.2
LSD (0.05)	2.1**	3.5**	4.6**	2.4**	1.9**	4.5**	ns	2.8**

Dry digestible matter (DDM) and dry matter intake (DMI)

There is inverse relation between ADF percent of a forage and its DDM value, likewise; similar relation has between NDF percent of a forage and its DMI value. Namely, as ADF percent of a forage increase, its dry matter digestibility by livestock decreases, similarly, as NDF percent of a forage increases, intake amount of the forage by livestock decreases (Lacefield, 1988; Schroder, 1994; Henning et al 1996; Jeranyama and Garcia, 2004). In the study; sole lucerne had the highest DDM content in 2009 (66.9%), and DDM contents of sole lucerne, sole ryegrass, L + B + T and L + B + T + O + R intercropping mixtures were higher than DDM contents of sole

bromegrass, sole tall fescue and sole orchardgrass in 2010. Additionally; DDM contents of lucerne and L + B + T were higher than the other treatments in 2011. According to averages of the three years, the highest DDM content was found in sole lucerne (66.5 %), and it was followed by the L+B+T (63.9%) and L+T+B+O+R (63.7%) intercropping mixtures (Table 6). Similarly; the greatest DMI contents were found in sole lucerne (3.4 and 2.7%, respectively) in 2009 and 2010 (Table 6). Sole lucerne had the highest DMI content (2.9%), followed by the L+B+T (2.7%) and L+B+T+O+R (2.7%) intercropping mixtures, based on the mean for the 3 years 2009, 2010, and 2011 (Table 6). Yucel and Avcı (2009) also reported differences in DDM and DMI among sole vetch, triticale, and mixtures containing these crops. Similar differences in DDM and DMI were reported among

different forage crops and intercropping mixtures by Yolcu et al., (2009a).

Table 6. Digestible dry matter and dry matter intake of some perennial forages as sole crops and intercropping mixtures

Treatments	Digestible dry matter (DDM)				Dry matter intake (DMI)			
	2009	2010	2011	Mean	2009	2010	2011	Mean
Lucerne	66.9 a	65.7 a	67.0 a	66.5 a	3.43 a	2.72 a	2.66	2.94 a
Bromegrass	62.4 c-d	59.3 b-c	61.9 b-d	61.2 c-d	2.64 d	2.28 b-c	2.44	2.45 c
Tall Fescue	59.8 e	57.8 c	60.0 c-d	59.2 e	2.31 f	2.06 d	2.16	2.18 d
Orchardgrass	61.8 d	60.6 b	59.4 d	60.6 d-e	2.51 e	2.25 c-d	2.56	2.44 c
Ryegrass	64.0 b-c	63.4 a	60.0 c-d	62.5 b-c	2.59 d-e	2.47 b	2.41	2.49 c
L + B + T	64.2 b	63.3 a	64.3 a-b	63.9 b	3.07 b	2.42 b-c	2.58	2.69 b
L + B + T + O + R	63.5 b-c	64.2 a	63.3 b-c	63.7 b	2.87 c	2.43 b-c	2.72	2.68 b
Mean	63.2 a	62.0 b	62.3 ab	62.5	2.78 a	2.38 c	2.51 b	2.55
CV (%)	1.4	2.4	3.2	1.7	2.16	4.64	8.40	2.74
LSD (0.05)	1.6**	2.7**	3.6**	1.9**	0.11**	0.20**	ns	0.13**

Total digestible nutrients (TDN)

The TDN refers to available the nutrients for livestock in forages, and variation among TDN values depend on the ADF concentration of the forages and, as percent of ADF increases, TDN declines (Albayrak et al. 2011). According to Henning et al. (1996) 61.2% TDN value is enough for most of the production stages of livestock. In the study; the highest TDN content was found in sole

lucerne in 2009, 2010, and 2011 (63.8, 62.1 and 64.1%, respectively) (Table 7). Sole lucerne had the greatest TDN content (63.3%) followed by the L+B+T (59.5%) and L+B+T+O+R (59.1%) intercropping mixtures, based on the mean for the 3 years 2009, 2010, and 2011 (Table 7). Similarly; Albayrak et al. (2011) reported differences in the TDN values among sole perennial and intercropping perennial forages. Also, they cited that the highest the TDN value was obtained from sole lucerne sowing.

Table 7. Total digestible nutrients and relative feed value of some perennial forages as sole crops and intercropping mixtures

Treatments	Total digestible nutrients (TDN)				Relative feed value (RFV)			
	2009	2010	2011	Mean	2009	2010	2011	Mean
Lucerne	63.8 a	62.1 a	64.1 a	63.3 a	177.6 a	138.6 a	138.3 a	151.4 a
Bromegrass	57.2 c-d	52.6 b-c	56.5 b-d	55.4 c-d	127.7 d	104.8 c	117.2 b-d	116.4 c
Tall Fescue	53.4 e	50.4 c	53.7 c-d	52.5 e	107.4 f	92.5 d	100.3 d	100.0 d
Orchardgrass	56.3 d	54.5 b	52.8 d	54.6 d-e	120.3 e	105.5 c	118.2 a-d	114.7 c
Ryegrass	59.6 b-c	58.7 a	53.7 c-d	57.3 b-c	128.4 d	121.5 b	112.2 c-d	120.6 c
L + B + T	59.8 b	58.6 a	60.0 a-b	59.5 b	152.7 b	118.5 b	128.8 a-c	133.2 b
L + B + T + O + R	58.9 b-c	59.9 a	58.6 b-c	59.1 b	141.5 c	121.1 b	133.5 a-b	132.1 b
Mean	58.4 a	56.7 b	57.1 ab	57.4	136.5 a	114.6 c	121.2 b	124.1
CV (%)	2.3	3.9	5.2	2.7	2.2	5.2	9.5	3.2
LSD (0.05)	2.4**	4.0**	5.3**	2.7**	5.4**	10.6**	20.4*	7.1**

Relative feed value (RFV) and the forages quality classes

RFV is an index combining the important nutritional components of intake and digestibility of forages. Although the index has no units, comparisons forage quality of grasses, legumes, and intercropping mixtures can be made by using the index. A forage with 41% ADF and %43 NDF has 100 RFV value. The other forages can be evaluated comparison with this value. As ADF and NDF percents decrease, the RFV value increases (Schroder, 1994). In the study; the highest RFV was found in sole lucerne (151.4) followed by the L+B+T (133.2 %) and L+B+T+O+R (132.1 %) intercropping mixtures, based on the mean for the 3 years 2009, 2010, and 2011

(Table 7). Similarly; Albayrak et al. (2011) and Yolcu et al. (2009a) reported that there were differences in the RFV among different sole sowings and intercropping mixtures.

By taking into consideration the RFV value of forages, Lacefield (1988) devoted the forages to quality classes. In this classification; if a forage RFV value is bigger than 151, it is accepted as the best quality forage. If forage the RFV value is between 151-125, 124-103 and 102-87, it is accepted as first, second and third quality classes respectively (Lacefield, 1988). In our study; based on the mean for the 3 years RFV values; although sole lucerne forage took place in the best (prime) quality class, the forages obtained from L + B + T and L + B + T + O + R intercropping mixtures took place in the first quality class. On the other hand; except for sole tall fescue forage, third quality class, the other sole grasses species forages took place in the second quality class.

CONCLUSION

Substantial changes in the established crop patterns would be required for successful forage production when switching from dry farming to the irrigated agriculture. The determination of the most convenient new perennial forage species and mixtures of the crops is important for improving forage production.

The results of this research revealed that the L+B+T intercropping mixture and sole lucerne had the highest yields. We suggest that sole lucerne should be cultivated primarily under the irrigated conditions, due to its greater protein and quality components than the L+B+T intercropping mixture. Legume and grass intercropping mixtures produce balanced feeds in terms of protein, carbohydrate, and other quality components. Therefore, if intercropping mixtures are planted, an L+B+T intercropping mixture will be preferable due to its higher yield and quality, and easier establishment and management as compared to the L+B+T+O+R intercropping mixture.

ACKNOWLEDGEMENT

The study was supported financially by the General Directorate of Agricultural Research and Policy (TAGEM/TA/07/11/02/001), We are also grateful to the GAP International Agricultural Research and Training Center, Diyarbakır, for their assistance with field trials, and the Eastern Mediterranean Agricultural Research Institute, Adana, Turkey, for their assistance in terms of the dry matter quality analyses.

LITERATURE CITED

Albayrak, S., M. Turk, O. Yuksel, M. Yılmaz. 2011. Forage yield and the quality of perennial legume- grass mixtures under rainfed conditions. Notulae Botanicae Horti Agrobotanica Cluj-Napoca, 39:114-118

Albayrak, S., H. Ekiz. 2005. An investigation on the establishment of artificial pasture under Ankara's ecological conditions. Turk. J. Agric. For. 29:69-74.

Ankom Technology Corporation, 1997. Operator's manual. Ankom 200/220 Fiber Analyzer. Ankom Thec. Corp.

AOAC. 1995. Offical Methods of Analysis. Assoc. of Official Analytical Chemists, Arlington, VA. Ankom Technology Corporation, 1997. Operator's manual. Ankom 200/220 Fiber Analyzer. Ankom Thec. Corp.

Assefa, G. and I. Ledin. 2001. Effect of variety, soil type and fertilizer on the establishment, growth, forage yield, quality and voluntary intake by cattle of oats and vetches cultivated in pure stand and mixtures. Animal Feed Science Technology, 92: 95-111.

Balabanli, C., S. Albayrak, M. Turk, O. Yuksel. 2010. A research on determination of hay yields and silage qualities of some vetch + cereal mixtures. Turk. J. Field Crops, 15: 204-209.

Barnett, F. and G.L. Posler. 1983. Performance of cool-season perennial grasses in pure stands and in mixtures with legumes. Agronomy Journal, 75(4):582-586.

Bengisu, G. 2011. Crop rotation systems for sustainable agriculture in the GAP Region. Alinteri, 20: 33-39.

Caballero AR, E.L. Goicoechea-Oicoechea, P.J. Hernaiz-Ernaiz. 1995. Forage yields and quality of common vetch and oat sown at varying seeding ratios and seeding rates of vetch. Field Crops Res. 41:135-140.

Cınar, S. 2012. determination of yield and quality characteristics of some cultivars and populations of tall fescue (*Festuca arundinaceae* Schreb.) in Cukurova Region. Journal of Agric. Fac. of Gaziosmanpasa Univ. 29 (1): 29–33

Evers, G.W., M. Gabrysch, and C.R. Tackett. 1993. Performance of Cool-Season Perennial Grasses on Poorly Drained Clay Soils. Forage Research in Texas, PR-5080, p. 6-9.

Henning, J.C., G.D. Lacefield, and D. Amaral-Phillips. 1996. Interpreting forage quality reports. Agron. Publ. ID-101., University of Kentucky, Lexington, KY.

Jeranyama, P. and A.D. Garcia. 2004. Understanding Relative Feed Value (RFV) and Relative Forage Qality (RFQ). http://agbiopubs. sdstate.edu/articles/ExEx8149.

Joachim H and G. Jung. 1997. Analysis of forage fiber and cell walls in ruminant nutrition. Journal of Nurtition, 127:810-813.

Karadag, Y., U. Buyukburc. 2004. Forage qualities, forage yields and seed yields of some legume–triticale mixtures under rainfed conditions, Acta Agriculturae Scandinavica, Section B - Soil & Plant Sci. 54: 140-148.

Kir, B., G. Demiroglu, R. Avcioglu, H. Soya. 2010. Effects of grazing on some yield and quality traits of a rotation pasture mixture under mediterranean. Turk. J. Field Crops, 2010, 15(2): 133-136.

Koc, A., A. Gokkus, M. Tan, B. Comaklı and Y. Serin 2004. Performance of tall fescue and lucerne-tall fescue mixtures in highlands of Turkey. New Zealand Journal of Agricultural Research, 47: 61-65.

Kusvuran, A. and V. Tansı. 2003. Determining effect of cutting frequency on some vegetative and generative characteristics of different grasses species in Cukurova conditions. Journal of Agric. Fac. of Cukurova University. 18 (1): 45-54.

Lacefield, G.D., 1988. Alfalfa Hay Quality Makes the Difference. University of Kentucky Department of Agronomy AGR-137, Lexington, KY.

Macadam, J.W., R.E. Whitesides, M.B. Winger and S. Buffer. 1997. Pasture Species for Grazing-Based Dairy Production Under Irrigation in the Intermountain West. Proceedings of the XVIII. International Grassland Congress, Canada, p. 99-100.

Meyer, D.W. J.F. Carter, F.R. and Vigil 1977. Bromegrass fertilization at six nitrogen rates: long and short term effects. North Dakota Farm. Res. 34: 13-17.

Sahin, E., M. Tosun, K. Haliloglu, M. Aydın. 2010. Agricultural and quality properties of oltu ecotypelines of wild orchardgrass (Dactylis glomerata L.). Journal of Agric. Fac. of Süleyman Demirel University, 5 (1):24-35, 2010

Sahin, G., N. Tasligil. 2013. The past the present and the future of Southeastern Anatolia Project (SAP). International Refereed Online Journal of Social Sciences. 36:1-26.

Sakarya, E., Y. Aral, E. Aydın. 2008. The significance of the Southeastern Anatolian Project and livestock in the development of Southeastern Anatolia Region. Vet Hekim Der Derg. 79 : 35-42.

SAS, Institute. 2002. JMP Statistics. Cary, NC, USA: SAS Institute, Inc. 707 p.

Sayar, M.S., A.E. Anlarsal, M. Basbag. 2010. Current situation, problems and solutions for cultivation of forage crops in the Southeastern Anatolian Region. J. Agric. Fac. HR.U., 14: 59-67.

Schroeder, J.W. 1994. Interpreting Forage Analysis. Extension Dairy Specialist (NDSU), AS-1080, North Dakota State University.

Serin, Y, A. Gokkuş, M. Tan, B. Çomaklı and A. Koç. 1998. Determination of suitable forage crop species and their mixturres of meadow establishment. Turk. J. Agric. For. 22:13-20.

Sheaffer, C.C. and G.C. Marten, 1986. effect of mefluidide on cool-season perennial grass forage yield and quality. Agronomy Journal, 78:75-79.

Spandl, E. and O.B. Hersterman, 1997. Forage quality and alfalfa characteristics in binary mixtures of alfalfa and bromegrass or timothy. Crop Sci. 37:1581-1585.

Steel, R.G.D. and J.H. Torrie. 1980. Principles and Procedures of Statistics: A Biometrical Approach. 2. ed. New York: McGraw-Hill Publ. Company.

Yucel, C. and M. Avci. 2009. Effect of different ratios of common vetch (Vicia sativa L.) – triticale (Triticosecale Whatt) mixtures on forage yields and quality in Cukurova plain in Turkey. Bulg. J. Agric. Sci., 15: 323-332.

Yolcu, H., M. Dasci and M. Tan, 2009a. Evaluation of annual legumes and barley as sole crops and intercrop in spring frost conditions for animal feeding I. Yield and Quality. J. Anim. Vet. Adv., 8: 1337-1342.

Yolcu, H., M. Polat and V. Aksakal. 2009b. Morphologic, yield and quality parameters of some annual forages as sole crops and intercropping mixtures in dry conditions for livestock. J. of Food, Agriculture & Environ. 7: 594-599.

Weller, R.F., A. Cooper. 2008 Seasonal changes in the crude protein concentration of mixed swards of white clover/perennial ryegrass grown without fertilizer N in an organic farming system in the United Kingdom. Grass and Forage Sci. 56: 92-95

20

EFFECT OF WEED CONTROL METHODS ON HAY YIELD, BOTANICAL COMPOSITION AND FORAGE QUALITY OF A MOUNTAIN PASTURE

Selahattin CINAR*[1], Mustafa AVCI[2], Serap KIZIL AYDEMIR[3]

[1]Department of Field Crops, Faculty of Agriculture, Gaziosmanpasa University, Tokat, TURKEY
[2]East Mediterranean AgricultureResearch Institue, Adana, TURKEY
[3]Variety Registration and Seed Certification, Ankara, TURKEY
*Corresponding author: scinar01@hotmail.com

ABSTRACT

This research was conducted to determine the effect of control methods on the weed composition of a pasture located at high altitude zone of the Cukurova Region, Turkey. The experiment was designed as randomized complete block with three replicates for three years. The mowing, fertilization, applications of 2.4-D, Picloram+2.4-D, Paraquat and Glyphosate were studied as weed control methods. The highest dry matter yield was obtained from the Picloram+2.4-D treatments. Dry matter yields in all treatments were greater as compared to the control. Grass contribution to the hay yield in the Picloram+2.4-D application was statistically significant (P<0.01) higher than the other treatments. Paraquat and glyphosate decreased the crude protein yield, while glyphosate increased higher crude protein and relative feed value contents compared with the other treatments.

Key words: botanical composition, pasture, weed control, yield and quality

INTRODUCTION

Rangelands are the most important feed sources of animal husbandry in Turkey. Arable land area had sharply increased between 1950 and 1960 in the county, this situation negatively affected the rangelands due to decline in the rangeland areas. The increased number of livestocks along with the decreased rangeland area led to overgrazing and deterioration of rangeland botanical composition. Mismanagement of rangelands caused 90% loss of the original vegetation on rangelands in Turkey (Genckan et al., 1990). Decline in pastures due to the heavy grazing and mismanagements has to be controlled by proper rehabilitation and management techniques in order to meet the needs of increased population in Turkey.

The consequence of mismanagement of pastures in Turkey invaded weeds. Weeds reduce feed quality, animal production, and in some cases lead to the poisoning. Thus weed populations in rangelands should be controlled and reduced. Several weed control methods are widely practiced on pasture such as mowing (Vallentine, 1980; Tanner et al., 1988; Mc Daniel and Taylor 2003), chemical applications (Passera et al., 1992; Gokkus and Koc, 1996) and fertilization (Jacobs and Sheley, 1999; Altin, 1992; Altin et al., 2005; Vallentine, 1980). However, very few research studies have been done to improve such lands in our region. Researches for weed control with herbicides, mowing and fertilization were very limited in Turkey. Therefore, this study was conducted to determine the effects of different weed control treatments on a mountain pasture.

MATERIAL AND METHODS

Experimental Area

A field experiment was conducted for 3 consecutive years during 2007-2010 on a natural pasture at Karakilic village of Karaisalı town in Adana province of Turkey. The altitude of experimental area was 1530 m (37°19 N, 34°56 E) and topography was flat. Soil texture was silty clay with slightly alkaline. The experiment was established in a clay soil with pH 6.87, organic matter content 4.7%, available P content 7.6 ppm and Zn 0.7 ppm (Anonymous, 2007).

The climate is Mediterranean climate with hot and dry summer and heavy precipitation during winter. The coolest month is January with a monthly mean temperature of 8.9 °C and the hottest month is August with 27.6 °C. The lowest total precipitation during experiment was in 2008 (393.0 mm) and highest was in 2009 (954.0 mm). The long term average annual precipitation of the study area is 871.1 mm. The mean values of temperature and relative humidity during the experimental period were close to the long-term averages (Anonymous, 2012).

Forage Yield, botanical composition and quality

The experiment design was completely randomized block with three replications. Seven treatment plots were inserted to each block. Treatments included control, mowing, fertilization, 2.4-D, Picloram+2.4-D, Paraquat and Glyphosate. Phosphorus (50 kg ha[-1]) and nitrogen (100 kg ha[-1]) were applied to all plots except control plots (Altin et al., 2005).

Mowing was applied at budding or blooming stage of weeds (Altin et al., 2005). Herbicides, 2.4-D amine (3200 ml ha[-1]), Picloram +2.4-D amine (1000 ml ha[-1]), Paraquat (5000 ml ha[-1]) and Glyphosate (15000 ml ha[-1]) were applied at the 3-5 leafs stage of the weeds (Vallentine, 1980). The 2.4-D and Picloram +2.4-D were applied to all experimental plots whereas Paraquat and Glyphosate were only to the target plants (Darrell and Leon, 2005). Herbicides were applied in the first and second years of the experiment as one application per year.

The plot sizes were 20 m^2 (4 x 5 m) and 1.5 m space were given between the plots. Four permanent quadrates (70 cm x70 cm=0.5 m^2 in size) were randomly placed in each plot, and the data were obtained from these quadrates. The samples were hand-separated, dried at 70 °C for 48 h and weighed. Samples were analyzed for crude protein contents (CP), neutral detergent fiber (NDF) and acid detergent fiber (ADF) (Van Soest et al., 1985). The ADF values were used to predict the digestible dry matter with the following formula;

Digestible Dry Matter (DDM) = ((88.9-0.779) x % ADF))

Neutral detergent fiber (NDF) was used to predict dry matter intake with the formula described below.

Dry Matter Intake (DMI) = (120 / % NDF).

Relative feed value (RFV) is calculated by multiplying digestible dry matter by dry matter intake and then dividing by 1.29 (Schroeder, 1994).

Statistical analyses

Data were analyzed using MSTATC (V.1.2, Michigan State University, USA). The differences between means were separated by Duncan multiple range test (P ≤0.05), however means of years were compared with the least significant difference (LSD) test (P ≤0.05)

RESULTS AND DISCUSSION

Dry matter yield

The variation in the total precipitation during the experiment resulted significant differences in dry matter yields among the years. Since the total precipitation (954 mm) in the third year was significantly higher than the other two years (558 and 393 mm), the average dry matter yield in the third year was significantly higher compared to the first two years. The average dry matter yield in the second year was significantly lower than that in the first and third years (Table 1).

The weed control methods had significant effects on the dry matter yield. Mowing, 2,4-D and fertilization applications in the first year yield significantly higher dry matter compared with the other treatments. Application of herbicides Picloram+2,4-D along with Paraquat did not significantly change the dry matter yield. The Glyphosate application gave significantly lower dry matter yield compared with the control plots (Table 1).

Table 1. Dry matter yields (kg ha[-1]) obtained from different treatments

Treatment	Dry matter yield (kg ha[-1])			
	1st year	2st year	3st year	Means
Control	1014.0 c*	669.0 bc	1627.0 cd	1103.3 c
Mowing	1521.0 ab	1407.0 ab	2485.0 bc	1804.3 ab
Fertilization	1811.0 a	1600.0 a	2333.0 c	1914.7 a
2.4-D	1830.0 a	914.0 abc	3195.0 ab	1979.7 a
Picloram+2.4-D	1213.0 bc	1612.0 a	3382.0 a	2069.0 a
Paraquat	812.0 cd	511.0 c	1945.0 c	1089.3 c
Glyphosate	490.0 d	158.0 c	1023.0 d	557.0 c
Mean	1241.6 B[+]	981.6 C	2284.3 A	

*,+Values within rows and columns with different letters differ significantly (P≤0.05)

Increase in dry matter yield by the fertilization was also reported by different researchers (Gokkus, 1990; Buyukburc, 1991; Koc et al., 1994; Hatipoglu et al., 2001; Cinar et al., 2005; Hatipoglu et al., 2005). Application of herbicides such as Picloram+2.4-D, Paraquat and Glyphosate affected not only weeds but also valuable pasture plants. Therefore, dry matter yields of with herbicide application exception of 2.4-D were lower compared to the plots with only fertilization or 2.4-D applications. Due to high effectiveness of Picloram+2.4-D on the plants comparing the 2.4-D (Hickman et al., 1990), the plots with 2,4-D gave higher dry matter yield than the plots with Picloram+2,4-D. In the second year of the experiment, the treatments of Picloram+2.4-D, fertilization, mowing and 2,4-D significantly increased the dry matter yield of the pasture. Other treatments did not change the dry matter yield of the pasture compared with the control. In the third year, application of 2.4-D and Picloram+2.4-D significantly increased the dry matter yield.

The treatments significantly increased the dry matter yield of the pasture except Paraquat and Glyphosate. Similar results were also reported by Bovey et al., (1972), Nichols and Mc Murphy, (1969) Gokkus and Koc, (1996) and Roger et al., (2000).

Botanical composition

Contributions of plants with the exception of legumes to the dry matter yield of the pasture significantly changed depending on the years (Table 2).

The average rate of the grasses was significantly higher in the third year compared to the other two years while the other family plants were significantly lower in the third year. The highest grass rates were determined with the Picloram+2.4-D treatments. According the mean values, the highest grass rates were obtained from the Picloram+2.4-D application during the experiment (88.3%, 100.0%, 99.0%, respectively).

Table 2. Ratios of legumes, grasses, and others family plants botanical composition with different treatments (%)

Treatment	Grasses				Legumes				Others Family Plants			
	1st year	2st year	3st year	Means	1st year	2st year	3st year	Means	1st year	2st year	3st year	Means
Control	30.0 bc*	26.7 de	56.3 d	37.7 de	7.7	12.0	11.0	10.2	62.3 bc	61.3 bc	32.7 c	52.1 bc
Mowing	54.7 b	58.0 bc	78.0 bc	63.6 bc	4.3	3.0	1.3	2.9	41.0 c	39.0 cd	20.7 cd	33.6 cd
Fertilization	34.3 bc	33.3 cd	63.7 cd	43.8 cd	5.3	4.3	5.0	4.9	60.4 bc	62.4 bc	31.3 c	51.4 bc
2,4-D	49.7 bc	79.7 ab	89.7 ab	73.0 b	1.3	0.0	0.3	0.5	49.0 bc	20.3 de	10.0 cd	26.4 de
Picloram+2,4-D	88.3 a	100.0 a	99.0 a	95.8 a	0.0	0.0	0.0	0.0	11.7 d	0.0 e	1.0 d	4.2 e
Paraquat	26.3 cd	13.7 de	24.3 e	21.4 ef	3.3	10.3	5.0	6.2	70.4 b	76.0 ab	70.7 b	72.4 b
Glyphosate	0.7 d	3.0 e	3.3 f	2.3 f	2.3	10.0	2.0	1.4	97.0 a	87.0 a	94.7 a	96.2 a
Mean	40.6 B+	44.9 B	59.2 A		3.5	5.7	3.5		55.9 A	49.4 A	37.3 B	

*,+Values within rows and columns with different letters differ significantly (P≤0.05)

The rates of grasses were steadily increased. The average rates of other family plants were steadily decreased on the years. Grasses rates in the botanical composition increased with the decrease in the others family plants (Table 2).

Picloram +2.4-D and 2.4-D applications resulted in increase of grasses rates with the decreased rate of others family plants (Bovey et al., 1972; Gokkus and Koc, 1996; Masters et al., 2002; Ferrell et al., 2004; Grekul et al., 2005).

Crude protein, Crude protein yield

The crude protein ratio of pastures significantly changed depending on the years (Table 3). The averaged crude protein ratio in second year was significantly higher than those in the first and third years.

Table 3. Crude protein ratio (%) and crude protein yield (kg ha^{-1}) obtained from different treatments

Treatment	Crude Protein Ratio (%)				Crude Protein Yield (kg ha^{-1})			
	1st year	2st year	3st year	Means	1st year	2st year	3st year	Means
Control	10.3 d*	13.2 e	10.1 d	11.2 d	104.4 bc*	88.3 b	164.3 d	119.0 bc
Mowing	12.7 c	16.2 c	11.4 c	13.4 c	193.2 a	227.9 a	283.3 bc	234.8 a
Fertilization	13.0 c	15.4 cd	11.6 c	13.3 c	235.4 a	246.4 a	270.6 bc	250.8 a
2.4-D	13.0 c	14.2 de	11.2 c	12.8 c	237.9 a	139.8 ab	357.8 ab	241.8 a
Pic+2.4-D	13.7 bc	14.3 de	13.1 b	13.7 bc	166.2 ab	230.5 a	443.0 a	279.9 a
Paraquat	15.0 ab	19.3 b	12.7 b	15.7 b	121.8 bc	98.6 b	247.0 cd	155.8 b
Glyphosate	16.7 a	21.2 a	14.5 a	17.5 a	81.8 c	33.5 b	148.3 d	87.9 c
Mean	13.5 B+	16.3 A	12.1 B		167.5 B+	160.1 B	276.4 A	

*,+Values within rows and columns with different letters differ significantly (P≤0.05)

The crude protein contents of the pastures were significantly affected by various weed control treatments tested. However the effects of treatments on crude protein ratio significantly changed depending on the years. Glyphosate and Paraquat application of the first year caused significantly higher crude protein ratio than all other treatments. The treatments of Glyphosate in all years significantly increased the crude protein rate of the pasture compared all the applications. The variation in the crude protein ratio by year was due to the variation in the botanical composition of the legume (Table 2). Broad-leaved species have higher crude protein content compared to the others family plants (Vallentine, 1980).

The averaged values of three years indicated that the crude protein ratio was significantly increased with all of the treatments compared to control. Since fertilization is reported the main reason for an increase in crude protein ratio (Gokkus and Koc, 1995, Cinar et al., 2005, Hatipoglu et al., 2005, Mut et al., 2010), the results obtained in the current study can also be attributed to the fertilization.

The crude protein (CP) yield of the pasture significantly changed depending on the years. The averaged crude protein yield in third year was significantly higher than those in the first and second years. The variation in the crude protein yield of the pasture depending on the years was due to the variation in dry matter yield (Table 1) and crude protein ratio.

Nutrient values of hay relatively depend on the botanical composition and harvesting time. The results indicated that CP contents of hay in pastures were lower than 16-18% which requires supplementary feeding to obtain high performance from milk cows (Conrad and Martz, 1985).

Crude protein yield ranged from 81.8 to 237.9 kg ha^{-1} in the first year, from 33.5 to 246.4 kg ha^{-1} in the second year and from 148.3 to 443.0 kg ha^{-1} in the third year. The highest average crude protein yield was obtained from the Picloram+2.4-D, fertilization, 2.4-D and mowing respectively. Crude protein yield depends on dry matter yield and crude protein ratio. Therefore, applications that have a high dry matter yield and crude protein content have higher crude protein yield. Similar results were also found by Ozaslan (1996), Gokkus and Koc (1995), Roger et al., (2000).

ADF, NDF and RFV

The analysis of variance suggested that applications of herbicides had no significant impact on ADF contents, however there were significant differences on ADF contents among means of treatments (Table 4). ADF ranged from 32.0 to 34.2% in the first year, from 31.6 to 33.8% in the second year, from 31.8 to 35.0% in the third year. The highest mean ADF ration was obtained from the control treatment with 33.9%. According to the averaged values of three years, the ADF contents were decreased in all treatments compared with control. Because fertilizer is given to all applications except the control. Fertilization decreases the rate of the ADF (Cinar et al., 2005, Hatipoglu et al., 2005, Mut et al., 2010).

Table 4. % ADF, % NDF and RFV obtained different treatments

Treatment	ADF (%)				NDF (%)				RFV			
	1st year	2st year	3st year	Means	1st year	2st year	3st year	Means	1st year	2st year	3st year	Means
Control	34.2	33.8	33.7	33.9 a	48.4	50.2	52.1	50.2 bc	119.7 c	115.9 c	111.9 b	115.8 c
Mowing	33.3	33.4	33.7	33.5 b	51.0	52.2	54.6	52.6 b	114.8 c	112.1 c	106.7 c	111.2 c
Fertilization	33.4	32.8	33.8	33.3 b	48.0	51.6	52.4	50.7 bc	121.9 bc	114.2 c	111.1 b	115.7 c
2.4-D	33.2	33.0	33.4	33.2 b	51.2	52.7	55.4	53.1 b	114.5 c	111.5 c	105.6 c	110.5 c
Pic+2.4-D	33.0	31.6	31.8	32.1 b	57.0	58.4	62.0	59.1 a	103.1 d	102.4 d	96.2 d	100.6 d
Paraquat	32.1	33.0	35.0	33.4 b	43.4	42.3	54.3	46.7 c	137.0 b	139.0 b	105.6 c	127.0 b
Glyphosate	32.0	33.8	33.7	33.2 b	40.8	39.4	39.4	39.9 d	145.9 a	147.7 a	147.9 a	147.2 a
Mean	33.0	33.1	33.6		48.5 B[+]	49.5 B	52.9 A		122.4 A[+]	120.4 A	112.1 B	

*,+Values within rows and columns with different letters differ significantly (P≤0.05)

Linn and Martin (1999), reported that legumes have higher CP contents and lower ADF and NDF contents compared with grasses therefore digestibility is closely related to cellulose and lignin content of ADF. Caddel and Allen (2012), stated that the most important factor of hay quality is the development stage of hay at harvest. ADF contents varied with species and families in mixtures and the development time and additionally the ratio of ADF affects the digestibility.

The applications of herbicides didn't statistically affect the NDF contents, but the effects of years were found statistically different on NDF contents (P<0.05). Average NDF contents of harvested hay samples for each year for different applications were briefly illustrated in Table 3. The average NDF contents in the third year was significantly higher than the first two years. The variation in the NDF contents of the pasture depending on the years might be due to the variation in ratio of the grasses The ratio of grasses (59.2%) in the third year was significantly higher than those (40.6% and 44.9%) in the other two years. Grasses have higher NDF content than other plant families (Pearson and Ison, 1987).

NDF ranged from 40.8 to 57.0% in the first year, from 39.4 to 58.4% in the second year, from 39.4 to 62.0 % in the third year and from 39.9 to 59.1 % in the three year means. The highest NDF ratio was obtained from the Picloram+2.4-D with 59.1, the lowest NDF ration was obtained from the Glyphosate with 39.9 %. Picloram+2.4-D applications increased the rate of the NDF. NDF ratio in grasses is higher than legumes and other plant families (Pearson and Ison, 1987). Glyphosate applications decreased the rate of NDF. NDF values is lower in broad leaved species is lower than grasses NDF (Pearson and Ison, 1987).

The analyses of variance indicated that treatments generated statistically significant RFV values in the first and second year and as well as for average of all three years. Besides, RFV exhibited statistically significant result depending on the years (Table 3). RFV ranged from 103.1 to 145.9 in the first year, from 102.4 to 147.7 in the second year, from 96.2 to 147.9 in the third year and from 100.6 to 147.2 in the three year means. The results of three years mean values showed that the highest RFV (147.2) was obtained from the Glyphosate treatment while the lowest RFV (100.6) was obtained from the Picloram+2.4-D with.

RFV is an important quality character and measures the overall feed value of forages. RFV is used to compare quality of forage based on the maturity of the plant when harvested. The higher the RFV in all forages is the more digestible and palatable (Schroeder, 1994; Mut et al., 2010). The RFV values in third y compared to the first and second year were lower.

The RFV is an index that is used to predict the intake and energy value of the forages and it is derived from the digestible dry matter (DDM) and dry matter intake (DMI). Forages with an RFV value over 151, between 150-125, 124-103, 102-87, 86-75, and fewer than 75 are considered as prime, premium, good, fair, poor and reject, respectively. Experiment, the RFV value was higher Glyphosate treatment than in the other treatments. The lowest RFV was obtained from the Picloram+2.4-D. Since RFV value was calculated from ADF and NDF, the observed differences were reflective of previously described ADF and NDF differences (Mut et al., 2010).

RFV of Glyphosate and Paraquat treatments are premium. RFV of Picloram+2.4-D treatment is poor. RFV of other treatments are fair.

CONCLUSION

The result demonstrated that applications of Picloram+2.4-D, 2.4-D, fertilization and mowing increased the dry matter yields of the pasture according to averages of three years. Picloram+2.4-D applications increased the grasses and decreased the legumes. Glyphosate applications increased the crude protein ratio. Applications increased the crude protein yield with the exception Paraquat and glyphosate and control. Applications of Glyphosate decreased the NDF content and increased the RFV.

ACKNOWLEDGEMENTS

Thanks are due to TUBITAK (Project of 106O585) for their financial support to this research work.

LITERATURE CITED

Altin, M. 1992. Meadow-Range İmprovement. Trakya University Tekirdag Agriculture Faculty Press No: 152, Tekirdag

Altin, M. A. Gokkus, A. Koc, 2005. Range Development, TUGEM, 468 p. Ankara.

Anonymous, 2007. Chemical Analysis Results, University of Cukurova Laboratory of Soil Science, Faculty of Agriculture, Adana, Turkey.

Anonymous, 2012. Meteorological data for Adana, http://www.wunderground.com/q/zmw:00000.1.17350

Bovey, R.W. R.E. Meyer, L. Morton, 1972. Herbage Production Following Brush Control With Herbicides in Texas. Journal of Range Management, 22(5): 315-317

Buyukburc, U. 1991. A research on improvement of natural pasture in Karayav_an village of Polatlı town with different fertilizer types and levels. Cumhuriyet University, Agricultural Faculty of Tokat 8: 38.

Caddel, J. E. Allen, 2012. Forage quality interpretations, http://virtual.chapingo.mx/dona/paginaCBasicos/f-2117.pdf

Cinar, S. M. Avci, R. Hatipoglu, K. Kokten, I. Atis, T. Tukel, S.K. Aydemir, H. Yucel, 2005. Hanyeri village (Tufanbeyli-Adana) slope ceremonial part of the botanical composition of manure nitrogen and phosphorus, a study on the effects of forage yield and quality of grass, Turkey 6 Field Crops Congress, 5-9 September 2005, Antalya, Turkey, Volume II, p: 873-877.

Conrad, H.R. and F.A. Martz, 1985. Forages for dairy cattle, p: 550-559. In M.E. Heath et al., (ed) Forages: the science of grassland agriculture, 4 th ed. Iowa State University.

Darrell, L. W. Leon, 2005. Weed Control in Pasture&Range. College of Agriculture and Biological Sciences.Cooperative Extension Service. South Dakota State University.

Duzgunes, O. T. Kesici, O. Kavuncu, F. Gurbuz, 1987. Research and experimental methods, Ankara Univ. Agr. Fac. Publications, No: 295, Ankara, Turkey.

Ferrell, J.A. G.E. Macdonald, B.J. Brecke, J.J. Mullahey, T. Ducar, 2004. Weed Management in Pastures and Rangeland.SS-AGR-08 Agronomy Department, Florida Cooperative Extension Service, Institute of Food and Agricultural Sciences, University of Florida.Website at http://edis.ifas.ufl.edu/wg006

Genckan, M.S. R. Avcioglu, H. Soya, O. Dogan, 1990. Problems and solutions of use, maintain and improvement of rangelands in Turkey. Turkey Agriculture Engineer III.Technical Congress 8-12 January 1990, pp. 53-61. Ankara

Grekul, C. W. Cole, D.E. Bork, 2005. Canada thistle (Cirsium arvense) and Pasture Forage Responses to Wiping with Various Herbicides. Weed Technology 19(2) Lawrence. Weed Science Society of America, p: 298-306.

Gokkus, A. 1990. Effect of fertilizing, irrigation and grazing on chemical and botanical composition of meadows at Erzurum plain. Journal Agriculture Faculty, Ataturk University p: 21: 7-24.

Gokkus, A. A. Koc, 1995. Hay yield, botanical composition and useful hay content of meadows in relation to fertilizer and herbicide application. Tr. J. of Agriculture and Forestry, 19: 23-29

Gokkus, A. A. Koc, 1996. Yield and botanical composition of grasslands at different times, the effects of different herbicides, Tr. J. Agric and Forestry, 20, 375-382.

Hatipoglu, R. M. Avci, N. Kilicalp, K. Kokten, S. Cinar, 2001. Research on the effects of phosphorus and nitrogen fertilization on the yield and quality of hay as well as the botanical composition of a pasture in the Cukurova region. Proceedings of 4th. Field Crops Congress, 17-21 September, Tekirdag University Press, Turkey pp.1-6.

Hatipoglu, R. M. Avci, S. Cinar, K. Kokten, I. Atis, T. Tukel, N. Kilicalp, C.Yucel, 2005. A research on effects of nitrogen and phosphorus fertilization on botanical composition, hay yield, and quality in humid part of Hanyeri Village rangeland (Tufanbeyli – Adana). Turkey VI. Field Crops Congress, September 5-9, pp. 867-872.

Hickman, M.V. C.G. Messersmith and R.G. Lym, 1990. Picloram release from leafy spurge roots. Journal of Range Management, 43(5):442-445.

Jacobs, J.S. and R.L.Sheley, 1999. Spotted knapweed, forbs and grass response to 2.4-D and N-fertilizer. Journal of Range Management, 52:482-488

Koc, A. 2000. Turkish Rangelands and Shrub Culture. Rangelands 22: 25-26.

Koc, A. B. Comakli, A. Gokkus, L. Tahtacioglu, 1994. The effects of nitrogen, phosphorus and ungrazed on plant density of Guzelyurt village in Erzurum. Proc. Turkey Field Crops Congress, Vol: Forage and Grassland, Izmir, pp: 78–82

Linn, J. G.N.P. Martin, 1999. Forage quality tests and interpretations, http://extension.umn.edu/distribution/livestocksystems/ID2637.html

Masters, B. J.Hanson, T.Gee, 2002. Pasture Weed Management. Dow AgroSciences, LLC 36309 State Hwy H,Graham, Missouri 64455.

Mc Daniel, K.C. and J.P. Taylor, 2003. Saltcedar recovery after Herbicide burn and mechanical clearing practices Journal of Range Management 56(5): 439-445.

Mut, H. İ. Ayan, Z. Acar, U. Basaran, O. Onal-Asci, 2010. The effect of different improvement methods on pasture yield and quality of hay obtained from the abandoned rangeland, Turkish Journal of Field Crops, 15(2):198-203

Nichols, J.T. and W.E. Mc Murphy, 1969. Range Recovery and Production as İnfluenced by Nitrogen and 2,4-D Treatments. Journal of Range Management, p:116-119

Ozaslan, A. 1996. The effects of ripping, nitrogen and herbicide applications on improvement of vegetation in Erzurum pastures. Ataturk University, Department of Field Crops, (Master Thesis), Erzurum p. 54.

Passera, C.B. O. Borsetto, R.J. Candia and C.R. Stasi, 1992. Shrub control and seeding influences on grazing capacity in Argentina. Journal of Range Management 45(5):480-482.

Pearson, C.J. and P.L. Ison, 1987. Agronomy of Grassland Systems. Cambridge University Press, Cambridge

Roger, L.S. A.D. Celestine, B.H. Mary, S.J. James, 2000. Spotted knapweed and grass response to herbicide treatments. Journal of Range Management. Society for Range Management. Volume 53:176-182 North Dakota University

Schroeder, J.W. 1994. Interpreting Forage Analysis. North Dakota State University NDSU Extension Service AS-1080.

Tanner, G.W. J.M. Wood, R.S. Kalmbacher and F.G. Martin, 1988. Mechanical shrub control on flatwoods range in South Florida. Journal of Range Management 41(3):245-248.

Vallentine, J.F. 1980 Range Development and Improvement, Brigham Young University Press, Provo, Utah. p: 357-358.

Van Soest, P.J. 1985. Composition, fiber quality, and nutritive value of forages. (E Heath, F Barnes, S Metcalfe eds.). Forages, Iowa State University Press, Iowa, s. 412-421.

21

EFFECTS OF WATER STRESS ON LEAVES AND SEEDS OF BEAN (*Phaseolus vulgaris* L.)

Ali Akbar GHANBARI[1] * Seyyed Hassan MOUSAVI[1] Ahmad MOUSAPOUR GORJI[1] Idupulapati RAO[2]

[1]*Seed and Plant Improvement Institute (SPII), Karaj, IRAN*
[2]*Centro Internacional de Agricultura Tropical (CIAT), A. A. 6713, Cali, COLOMBIA*
*Corresponding author: aghanbari2004@yahoo.com

ABSTRACT

To determine the changes of nitrogen contents in the leaves and seeds of common bean (*Phaseolus vulgaris* L.) genotypes under contrasting moisture regimes, two field experiments were conducted as split-plot in a randomized complete block design with four replications. Two levels of irrigation (irrigation after 55-60 and 100-110 mm evaporation from class A pan, respectively) and eight genotypes including white beans (WA4502-1 and WA4531-17), red beans (Akhtar, D81083 and AND1007) and Chitti beans (KS21486, MCD4011 and COS16) were studied in the main- and sub-plots, respectively. Leaf nitrogen (N) and proline contents were measured at two growth stages (pre-flowering and pod filling period). Grain yield, seed N and seed protein contents were measured at harvest. The results indicated that white beans had lower leaf N and seed protein contents than red and Chitti beans under both irrigation regimes. Under drought conditions, AND1007 and COS16 showed markedly higher levels of accumulation of leaf N and proline. Seed protein was higher in Chitti beans. Water deficit reduced the leaf N by 19% and 28% at two growth stages and grain yield by 39.8%. By contrast, proline content of all genotypes was increased by 105%. Seed N and protein contents had the lowest reductions under drought, while increasing N and proline contents in the leaves increased grain yield under this condition. Besides, lower values of seed N and protein is associated with higher yields of genotypes. Totally, based on the grain yield, red beans were more drought-susceptible than white and Chitti groups.

Keywords: Bean, drought, minerals, nitrogen, yield

INTRODUCTION

Beans are food legumes that are consumed by many people worldwide (Broughton et al., 2003). About 60% of the bean growing area in the tropics is affected by terminal or intermittent drought stress (Beebe et al., 2008). New common bean cultivars have been developed through selection and incorporation of various physiological, phenological and morphological characteristics that improve yield under drought conditions (Beaver et al., 2003). De Souza et al. (1997) studied the effect of water deficit on leaf characteristics and concluded that severe drought saccelerated leaf senescence by reducing leaf nitrogen (N) and chlorophyll contents.

Nitrogen is an essential nutrient in the plant production. For many plant species, a strong correlation was observed between leaf N and CO_2 assimilation (Baker and Rosenqvist, 2004). A large part of N in the plant is allocated to leaves and a large amount of leaf N is allocated to photosynthetic system. Photosynthetic activity is related to leaf N and the net photosynthetic rate increases with higher levels of leaf N. Generally, drought decreases leaf N content leading to a decrease in photosynthesis (Nakayama et al., 2007). Excessive production of different types of compatible solutes is a response of plants to drought and other stresses. Proline, as a solute, is widely distributed in plants which accumulate greater than the other amino acids in the stressed plants (Cardenas-Avila et al., 2006). Beebe et al. (2008) believed that proline accumulation may associate with osmotic adjustment resulting inhibition of protein synthesis. Of the several biochemical indices of water deficit injury, proline accumulation and decline in protein synthesis have been reported in many plants (Ashraf and Iram, 2005). According to Sanchez et al. (2007), there is a positive relationship between N availability and proline accumulation.

For importance of N to increase protein in the seeds of common bean, the main objective of the present study was to determine the relationships between grain yield and nitrogenous compounds in the leaves and the seeds of eight common bean genotypes, belonging to three groups: red, white and Chitti, under two contrasting moisture regimes.

MATERIALS AND METHODS

Three groups of common bean (*Phaseolus vulgaris* L.) genotypes consisted of red (Akhtar, AND1007, and D81083), white (WA4502-1 and WA4531-17), and Chitti (COS16, KS21486, and MCD4011) were evaluated under control (irrigation after 55-60 mm evaporation from class A pan) and drought (irrigation after 100-110 mm evaporation from class A pan) conditions at the research farm of Seed and Plant Improvement Institute (SPII), Karaj, Iran. Drought conditions were induced after seedling establishment (from emergence of 3rd trifoliate leaf) to maturity. Split-plot experiments were performed in a randomized complete block design (RCBD). Each year, four replications were used for each treatment and subsequently for each trait. Replications were combined doubles (1 with 3 and 2 with 4), so the number of replications was reduced to two occurrences per year. Replications of the first year and second year were analyzed together. The seeds were sown on 28 June 2009 and 13 June 2010. Irrigation treatments and genotypes were placed into the main- and sub-plots, respectively. Seeds of each genotype were sown at 6 rows of 5 m length with plant spacing 5 cm, separated by 50 cm rows. At two growth stages during crop development, pre-flowering (between V4 and R5 stages) and pod filling duration (R8), five plants of each treatment and in each plant three central leaflets were randomly selected for sampling for leaf N and protein. These leaves were dried at a temperature of 75°C for 48 h and then total N content of samples was determined using the Kjeldahl method. At the flowering (R6) stage, central leaflets were collected from top and middle parts of the plants in each treatment. These samples were transported to the laboratory in the liquid nitrogen and maintained at -80°C. The free proline content in leaf tissue was determined using spectrophotometer according to the method described by Bates et al. (1973). At the harvesting time, seeds were rinsed in distilled water and then N content of samples was determined using the Kjeldahl method. Seed protein content was determined using seed N value multiplied by 6.25. Finally, grain yield from each treatment was determined based on g per plant. Data were analyzed based on experimental design model. Means comparison was performed based on Duncan's multiple range test (P≤0.05). All statistical analyses were performed using SAS (version 9.1) and SPSS (version 16) software.

RESULTS

Leaf N in vegetative (pre-flowering) and reproductive (pod filling) growth stages was significantly influenced by water regimes. At the both growth stages, white beans had lower leaf N contents under both water regimes (Table 1).

Table 1. Leaf nitrogen and proline contents under control (N) and drought (S) conditions.

Genotypes	Leaf N at pre-flowering (%)		Leaf N at pod filling (%)		Leaf proline (μmol g^{-1} FW)	
	N	S	N	S	N	S
Akhtar	2.15 ab	1.73 b	0.87 ab	0.66 b	0.80 ab	1.60 ab
AND1007	2.23 a	1.83 a	0.88 ab	0.73 a	0.73 bc	1.73 a
D81083	2.10 abc	1.65 cd	0.83 bcd	0.54 c	0.63 c	1.55 ab
WA4502-1	1.85 d	1.60 de	0.78 d	0.51 c	0.88 a	1.25 c
WA4531-17	1.95 cd	1.55 e	0.80 cd	0.51 c	0.88 a	1.40 bc
COS16	2.25 a	1.88 a	0.90 a	0.72 a	0.65 c	1.78 a
KS21486	2.00 bcd	1.60 de	0.83 bcd	0.54 c	0.73 bc	1.45 bc
MCD4011	2.15 ab	1.70 bc	0.85 abc	0.63 b	0.73 bc	1.60 ab
Mean	2.08 a	1.69 b	0.84 a	0.60 b	0.75 b	1.54 a
Mean squares (MS)	Irrigation: 1.24 Error $_a$: 0.003 Genotype: 0.06 Irrigation×Genotype: 0.004 Error $_b$: 0.002		Irrigation: 0.44 Error $_a$: 0.003 Genotype: 0.01 Irrigation×Genotype: 0.003 Error $_b$: 0.001		Irrigation: 0.08 Error $_a$: 0.02 Genotype: 0.01 Irrigation×Genotype: 0.06 Error $_b$: 0.005	

Different letters within each column indicate significant difference at p≤0.05.
FW: fresh weight of leaves.

Reduction of leaf N content due to water deficit was greater in pod filling duration. Both water regimes significantly influenced the leaf free proline content. This amino acid increased greater than two fold in the stressed plants. In this study, under control conditions white beans had higher levels of leaf proline while these genotypes showed lower proline contents than red and Chitti groups under drought conditions (Table 1).

Seed N and protein contents were significantly affected by both water regimes. White beans had lower seed N and protein contents than red and Chitti beans under both water regimes. Seed N and protein contents were not that much affected by drought compared with the other traits. Significant genotypic differences were observed for grain yield under both control and drought conditions. One of the Chitti bean genotypes, KS21486 showed the lowest grain yield under both growing conditions. In this study, grain yield reduction due to water deficit was 39.8% (Table 2).

The greatest effect of drought on yield reductions per plant was observed with one of the red beans (Akhtar). One of the Chitti beans, MCD4011 showed greater level of drought resistance with low value of % reduction in grain yield (Table 3).

Table 2. Seed nitrogen and protein contents and grain yield under control (N) and drought (S) conditions.

Genotype	Seed N (%)		Seed protein (%)		Grain yield (g plant^{-1})	
	N	S	N	S	N	S
Akhtar	3.05 bcd	2.95 abc	19.1 bcd	18.4 abc	12.34 d	5.35 e
AND1007	3.10 bcd	2.98 abc	19.4 bcd	18.6 abc	23.15 a	10.71 ab
D81083	3.18 ab	2.93 bcd	19.9 ab	18.3 bcd	8.01 e	5.76 e
WA4502-1	3.03 cd	2.83 d	18.9 cd	17.7 d	17.14 b	11.69 a
WA4531-17	2.98 d	2.88 cd	18.6 d	18.0 cd	15.00 c	8.81 c
COS16	3.30 a	3.05 a	20.6 a	19.1 a	13.11 cd	9.77 bc
KS21486	3.18 ab	2.98 abc	19.9 ab	18.6 abc	5.80 f	3.25 f
MCD4011	3.15 bc	3.00 ab	19.7 bc	18.8 ab	9.12 e	7.09 d
Mean	3.12 a	2.95 b	19.49 a	18.14 b	12.96 a	7.80 b
Mean squares (MS)	Irrigation: 0.23 Error $_a$: 0.03 Genotype: 0.02 Irrigation×Genotype: 0.003 Error $_b$: 0.002		Irrigation: 0.23 Error $_a$: 0.14 Genotype: 1.08 Irrigation×Genotype: 0.14 Error $_b$: 0.10		Irrigation: 229.05 Error $_a$: 1.12 Genotype: 9.36 Irrigation×Genotype: 1.61 Error $_b$: 0.22	

Different letters within each column indicate significant difference at p≤0.05.

Table 3. Reduction (%) of leaf and seed N, grain yield and increase (%) of leaf proline of genotypes induced by drought when the means of control treatment for the same traits were took into consideration 100%.

Genotype	Leaf N (pre-flowering)	Leaf N (pod filling)	Leaf proline	Seed protein	Yield per plant
Akhtar	19.5	24.1	100.0	3.3	56.6
AND1007	17.9	17.0	137.0	3.8	53.7
D81083	21.4	34.9	146.0	7.2	28.1
WA4502-1	13.5	34.6	42.0	6.6	31.8
WA4531-17	20.5	36.3	59.1	3.4	41.3
COS16	16.4	20.0	173.8	7.6	25.5
KS21486	20.0	34.9	98.6	6.3	43.9
MCD4011	20.9	25.9	119.2	4.8	22.3
Mean	18.8	28.5	105.3	6.9	39.8

DISCUSSION

Crops respond differently to environmental stresses such as drought. Improving genetic resistance of crops to drought has been a major challenge for plant breeders. Crop resistance to drought has been attributed to different mechanisms leading to different response types (Chaves et al., 2003). According to our results, WA4502-1 showed the lowest amount of leaf nitrogen content in both vegetative and reproductive stages while the highest values of leaf N in these two growth stages was observed with COS16, indicating better ability of this genotype in acquiring N either from soil or from biological nitrogen fixation (BNF) and in remobilizing N under favorable water regime (control) conditions. In general, white beans had lower contents of leaf N than the other two groups indicating their poor potential for BNF and N metabolism. A large amount of N in the plant is allocated to leaves and a large part of leaf N is invested in the photosynthetic system. Photosynthetic activity is related to leaf N and the photosynthetic rate increases with higher levels of leaf N (Nakayama et al., 2007). In the present study, drought decreased the N accumulation in all genotypes so that this reduction was larger in drought sensitive genotypes. Our results indicated that white beans had lower leaf N contents at pod filling period (R8) and low leaf proline contents than the other two bean groups under water deficit conditions. Sanchez et al. (2007) reported that the relationship between N availability and proline accumulation is usually positive. Given that proline accumulation is one of the mechanisms of crop resistance to stress conditions such as drought (Cardenas-Avila et al., 2006; Chaves et al., 2003), white bean genotypes are considered as drought-susceptible. Our results indicated that there was a general decreasing trend in total seed protein in all genotypes due to water deficit which is in agreement with findings of Ashraf and Iram (2005). According to Fresneau et al. (2007), drought induces changes in a number of physiological and biochemical processes including inhibition of protein synthesis. It has been observed that increased amounts of free proline in wheat cultivars could be associated with more effective mechanisms of dehydration tolerance and drought avoidance. It was reported that in chickpea (*Cicer arietinum*) amino acid content increased under drought conditions apparently due to hydrolysis of proteins (Ashraf and Iram, 2005). Aranjuelo et al. (2011) found that water stressed plants could invest a large quantity of carbon and N resources into the synthesis of osmoregulants in the leaves such as proline for maintaining cell turgor. According to our results, one of the least susceptible genotypes to drought in refer to N content of leaves is AND1007 which showed greater leaf N content at the pod filling period than the other genotypes. This genotype had the lowest reductions in leaf N content at reproductive stage (17%). COS16 with its highest values of leaf N in both growth stages (V4-R5 and R8) and leaf proline content could be considered as drought resistant. This genotype has also high capacity to acquire and remobilize N under both water regimes. Evaluation of leaf N changes between vegetative stage (pre-flowering) and pod filling duration revealed that the greatest reduction in the leaf N in control conditions was observed with MCD4011 while under drought conditions, the greatest and the lowest reduction in leaf N were observed with COS16 and WA4531-17, respectively. These results indicate that N remobilization from leaves was greater in MCD4011 and COS16 under control conditions while it was greater with COS16 than the other genotypes under drought conditions. Also, these results suggest the high capability of COS16 for N remobilization to other sinks such as pods. Ramirez-Vallejo and Kelly (1998) found that under moderate water stress N partitioning was not impaired, but under severe stress N remobilization was reduced in common bean. Drought resistant cultivars may be more efficient in assimilate production and translocation to the seeds (Rosales-Serna et al., 2004). Nakayama et al. (2007) found decreased N accumulation in the leaves of studied cultivars under drought. It is well known that drought impairs the uptake of N in the plants. Also, drought sensitive genotypes accumulate less N than drought resistant genotypes. Previous studies indicated that high performance of common bean genotypes under drought was associated with their ability to mobilize photosynthates toward developing grain and to utilize the acquired N more efficiently for seed production (Beebe et al., 2008; Polania et al., 2008). According to Araujo and Teixeira (2008), remobilization of nutrients such as N from vegetative to reproductive organs plays a fundamental role in the legume grain yield. As shown by Schiltz et al. (2005), the contribution of N remobilization to seeds varies from 70% in peas, 43 to 94% in lentil, 80% in faba bean, and 84% in common bean. Common bean pods and seeds are major sinks for N and its allocation to seeds dominates the reproductive N budget (Araujo and Teixeira, 2008). Under drought conditions, WA4502-1 showed the lowest reductions in leaf N content at vegetative stage (13.5%). This genotype had also the lowest increase in leaf proline accumulation (42%). The greatest reductions in leaf N content at vegetative stage (21.4%), and the lowest reduction in leaf N content at R8 stage (17%) were observed with D81083 and AND1007, respectively. Similar to the observations made by Singh (2007), we also

found that drought reduced N partitioning and fixation. Our results showed that seed N and protein contents had the lowest reductions under drought conditions. COS16 and D81083 showed the highest reductions in seed N under water deficit, indicating high sensitivity of N accumulation in the seeds of these genotypes to drought. Grain yield is the most important trait in many studies. Genotypic differences based on grain yield have been reported for drought resistance in common bean (Teran and Singh, 2002). In our research, AND1007 had the highest grain yield in control treatments, but in the stressed plots WA4502-1 showed higher yield than the others. Water deficit reduced mean grain yield of all genotypes by 39.8% which varied between 56.6% (in Akhtar) and 22.3% (in MCD4011). Singh (2007) and Teran and Singh (2002) found average yield reductions of 52% to 62% in the dry bean varieties under drought conditions. According to results, increasing N and proline contents in the leaves resulted in grain yield increases under drought conditions. Besides, lower values of seed N and protein is associated with higher yields of genotypes. Our results indicated also that based on leaf proline, seed protein and grain yield, AND1007 and COS16 were identified as superior genotypes under drought (Fig. 1).

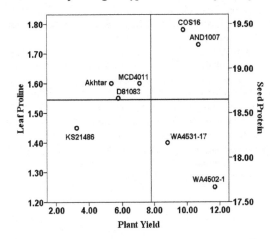

Figure 1. Trilateral relations (plant yield-leaf proline-seed protein) for bean genotypes under drought conditions.

In conclusion, comparisons among the genotypes revealed that white beans were more drought-susceptible than red and Chitti groups in the studied traits except to grain yield. According to our results, the highest mean grain yields under both conditions were observed with AND1007 and WA4502-1. Intra-grouping evaluations showed that WA4502-1 has a relatively better performance under drought when compared with the other white bean genotype. In red beans, AND1007 was superior to others and due to its other desirable attributes it could be a good candidate to introduce to drought-prone areas. In Chitti group, KS21486 is less preferable due to its small seed size and poor market potential while MCD4011 is considered as a promising genotype due to its good market potential for grain and its greater level of drought resistance based on small changes in grain yield under water deficit.

LITERATURE CITED

Aranjuelo, I., G. Molero, G. Erice, J. Christophe Avice, S. Nogues, 2011. Plant physiology and proteomics reveals the leaf response to drought in alfalfa (*Medicago sativa* L.). J. Exp. Bot. 62:111-123.

Araujo, A.P., M.G. Teixeira, 2008. Relationship between grain yield and accumulation of biomass, nitrogen and phosphorus in common bean cultivars. Rev. Bras. Ci Sol. 32:1977-1986.

Ashraf, M., A. Iram, 2005. Drought stress induced changes in some organic substances in nodules and other plant parts of two potential legumes differing in salt tolerance. Flora 200:535-546.

Baker, N.R., E. Rosenqvist, 2004. Applications of chlorophyll fluorescence can improve crop production strategies: an examination of future possibilities. J. Exp. Bot. 55:1607-1621.

Bates, L., R. Waldren, J. Teare, 1973. Rapid determination of proline for water stress studies. Plant Soil 39:205-207.

Beaver, J.S., J.S. Rosas, J. Myers, J.A. Acosta-Gallegos, J.D. Kelly, 2003. Contributions of the bean/cowpea CRSP to cultivar and germplasm development in common bean. Field Crops Res. 82:87-102.

Beebe, S.E., I.M. Rao, C. Cajiao, M. Grajales, 2008. Selection for drought resistance in common bean also improves yield in phosphorus limited and favorable environments. Crop Sci. 48:582-592.

Broughton, W.J., G. Hernandez, M. Blair, S. Beebe, P. Gepts, J. Vanderleyden, 2003. Beans (*Phaseolus* spp.)- model food legumes. Plant Soil 252:55-128.

Cardenas-Avila, M.L., J. Verde-Star, R.K. Maiti, R. Foroughbakhch-P, H. Gamez-Gonzalez, S. Martinez-Lozano, M.A. Nunez-Gonzalez, G. Garcia Diaz, J.L. Hernandez-Pinero, M.R. Morales-Vallarta, 2006. Variability in accumulation of free proline on in vitro calli of four bean (*Phaseolus vulgaris* L.) varieties exposed to salinity and induced moisture stress. ΦYTON 75:103-108.

Chaves, M.M., J.P. Maroco, J.S. Pereira, 2003. Understanding plant responses to drought- from genes to the whole plant. Funct. Plant Biol. 30:239-264.

De Souza, P.I., D.B. Egli, W.P. Bruening, 1997. Water stress during seed filling and leaf senescence in soybean. Agron. J. 89:807-812.

Fresneau, C., J. Ghashghaie, G. Cornic, 2007. Drought effect on nitrate reductase and sucrose-phosphate synthase activities in wheat (*Triticum durum* L.): role of leaf internal CO_2. J. Exp. Bot. Advance Access:1-10.

Nakayama, N., H. Saneoka, R.E.A. Moghaieb, G.S. Premachandra, K. Fujita, 2007. Response of growth, photosynthetic gas exchange, translocation of ^{13}C-labelled photosynthate and N accumulation in two soybean (*Glycine max* L. Merrill) cultivars to drought stress. Int. J. Agric. Biol. 9:669-674.

Polania, J.A., M. Grajales, C. Cajiao, G. Garcia, J. Ricaurte, S. Beebe, I. Rao, 2008. Physiological evaluation of drought resistance in elite lines of common bean (*Phaseolus vulgaris* L.) under field conditions. CIAT, Cali, Colombia.

Ramirez-Vallejo, P., J.D. Kelly, 1998. Traits related to drought resistance in common bean. Euphytica 99:127-136.

Rosales-Serna, R., J. Kohashi-Shibata, J.A. Acosta-Gallegos, C. Trejo-Lopez, J. Ortiz-Cereceres, J.D. Kelly, 2004. Biomass distribution, maturity acceleration and yield in drought stressed common bean cultivars. Field Crops Res. 85:203-211.

Sanchez, E., G. Avila-Quezada, A.A. Gardea, J.M. Ruiz, L. Romero, 2007. Biosynthesis of proline in fruits of green bean plants: deficiency versus toxicity of nitrogen. ΦYTON 76:143-152.

Schiltz, S., N. Munter-Jolain, C. Jeudy, J. Burstin, C. Salon, 2005. Dynamics of exogenous nitrogen partitioning and nitrogen remobilization from vegetative organs in pea revealed by ^{15}N in vivo labeling throughout seed filling. Plant Physiol. 137:1463-1473.

Singh, S.P. 2007. Drought resistance in the race Durango dry bean landraces and cultivars. Agron. J. 99:1219-1225.

Teran, H., S.P. Singh, 2002. Selection for drought resistance in early generations of common bean populations. Can. J. Plant Sci. 82:491-497.

22

YIELD AND QUALITY PERFORMANCES OF VARIOUS ALFALFA (*Medicago sativa* L.) CULTIVARS IN DIFFERENT SOIL TEXTURES IN A MEDITERRANEAN ENVIRONMENT

Yasar Tuncer KAVUT*, Riza AVCIOGLU

Ege University, Faculty of Agriculture, Department of Field Crops, Izmir, TURKEY
*Corresponding author: tuncer.kavut@ege.edu.tr

ABSTRACT

A field trial was conducted at the experimental fields of Ege University in Bornova and Odemis in 2006 and 2007. The effects of two soil textures (heavy and light) and four cultivars (TT-2008, TT-2009, P-5683 and Elci) on the yield and chemical compositions of some alfalfa cultivars were evaluated under the Mediterranean climatic conditions. The experimental design was a Randomized Complete Blocks with four replications. Results indicated that, the effects of soil texture on yield and quality traits of crop material were significant. TT-2008 and P-5683 alfalfa cultivars had better performances than the other cultivars with regard to yield (fresh herbage, dry matter and crude protein) and quality (contents of dry matter, crude protein, crude ash, ADF and NDF values) characteristics. The overall mean of alfalfa yield at light textured soil condition was significantly higher as compared to the heavy textured soil.

Key Words: Alfalfa, Soil texture, Cultivars, Yield, Quality characteristics.

INTRODUCTION

Alfalfa is one of the most important forage crops worldwide due to its excellent forage quality, high forage yield in a wide range of environments, and high adaptability to different climatic conditions (Moreira and Fageria, 2010). From all forage crops, which together with meadows have a major contribution in ensuring the forage base, alfalfa crop (*Medicago sativa* L.) occupies an important position. This plant is distinguished by its forage value, vast cultivation area and high digestibility, and from the point of view of farmers and world's agricultural scientist is considered to be the "Queen of Fodder Herbs". From agrobiological point of view, alfalfa gathers also a number of other peculiarities: good revaluation of irrigation water, high capacity for regeneration after mowing, high rate of competitiveness.

According to archaeological information or ancient philosophers, *Medicago sativa* L. crop has been taken in culture for 4000 years B.C. in regions of southwest Asia (Dale et al., 2012). It has other superior forage qualities and high yields that can be consumed by livestock readily and has high protein content and it is also rich in minerals and vitamins (Soya et al., 2004; Geren et al., 2009). Yield characteristics of alfalfa have been reported as follows; green herbage yield 3520-11660 kg ha^{-1}, dry matter yield 1780-3230 kg ha^{-1} and crude protein yield 246.4-321.3 kg ha^{-1} (Sengul et al., 2003; Abusuwar and Bakri, 2009; Saruhan and Kusvuran, 2011; Albayrak and Turk, 2013; Mala and Fadlalla, 2013).

In several studies, quality properties of alfalfa have been reported as 23.4-24.0%, 18.9-22.7%, 7.5-9.7%, 27.5-42.9% and 39.3-50.3% for dry matter content, crude protein, crude ash, acid detergent fiber (ADF) and neutral detergent fiber (NDF) respectively (Tomic et al., 2006; Geren et al., 2009; Tongel and Ayan, 2010; Saruhan and Kusvuran, 2011; Albayrak and Turk, 2013). Alfalfa leaves serve as a factory for raw, biodegradable plastic beads, other industrial products or better livestock feed, while the stem goes to ethanol production. Alfalfa has an area of 628.641 ha and 12.6 million ton fresh herbage production quantity in Turkey (Anonymous, 2013). Despite all these advantages, the alfalfa planting area in Turkey is limited and far from to meet the need for quality roughage (Avcioglu et al., 2001). Therefore, to increase alfalfa planting areas in the country, it is important to find the alfalfa varieties suitable to different ecological regions mainly with proper soil textures (Cinar and Hatipoglu, 2014).

Soil texture, which affects the soil's ability to hold onto nutrients and water (Anonymous, 2005). Coarse-textured soils generally have high infiltration rates with good penetration by roots. Many studies on coarse textured soil displayed the dry matter yield of alfalfa superiority of to fine textured soils (Rechel et al., 1991;

Geren et al., 2009). In the coastal regions, non-dormant alfalfa cultivars suitable to warm, humid climates are common (Sheaffer et al., 1998). However, in the cool and high plateaus of eastern Turkey dormant and cold tolerant cultivars can be grown successfully (Altinok and Karakaya, 2002). There are many ongoing researches on alfalfa at different ecologies of the world and multi location trials including diversity of soil textures are important for releasing new developed varieties (Meyer, 2005; Avci et al., 2013).

The objective of this study was to determine the suitable alfalfa cultivar for the Ege region, a Mediterranean environment with high forage yield and quality by testing at two locations with different soil texture for 2 years.

MATERIALS AND METHODS

Location of the Experiment

The trial was carried out during the 2006 and 2007 growing seasons at Bornova experimental fields (38°27.236 N, 27°13.576 E, 28 m a.s.l) of the Faculty of Agriculture Department of Field Crops and at Odemis experimental fields (38°13.234 N, 27°57.880 E, 115 m a.s.l) of Odemis Vocational Training College of the Ege University. Meteorological data and the soil properties of locations are presented in Table 1 and 2, respectively. As can be seen in Table 2, heavy soil texture is represented by Bornova whereas light soil by the Odemis location.

Table 1. Some physical and chemical characteristics of the soils of the experimental sites

Characteristics	Bornova	Odemis	Characteristics	Bornova	Odemis
Sand (%)	24.72	68.72	Organic Matter (%)	1.13	1.58
Silt (%)	42.72	24.00	$CaCO_3$ (%)	21.52	1.44
Clay (%)	32.56	7.28	N (%)	0.11	0.16
pH	7.80	7.28	P (ppm)	40.52	20.50

Table 2. Some meteorological data for the experimental sites

Temperature (°C)						
	2006		2007		Long year average	
Monhts	Bornova	Odemis	Bornova	Odemis	Bornova	Odemis
April	16.8	16.2	16.2	14.7	15.1	14.6
May	21.0	20.5	22.4	21.7	20.3	19.9
June	26.2	25.4	27.5	27.1	25.5	25.0
July	28.5	27.9	30.1	29.5	28.0	27.5
Aug	29.2	28.4	29.2	28.8	27.3	26.6
Sept.	23.5	22.8	24.4	23.0	22.9	22.1
Oct.	18.3	17.9	19.7	18.3	18.0	16.9
Nov.	11.4	10.2	12.0	17.9	12.8	11.4
Mean	21.9	21.2	22.7	22.6	21.2	20.5
Precipitation (mm)						
	2006		2007		Long year average	
Aylar	Bornova	Odemis	Bornova	Odemis	Bornova	Odemis
April	27.0	14.1	19.3	35.4	49.5	54.8
May	0.0	21.4	44.1	31.5	30.6	26.6
June	19.2	6.2	0.3	6.8	9.0	10.4
July	0.0	22.0	0.0	0.0	3.2	5.1
Aug	0.0	0.0	0.0	0.0	1.6	1.6
Sept.	133.5	66.9	0.0	7.0	16.1	14.8
Oct.	88.6	77.0	107.7	74.5	37.3	34.1
Nov.	46.7	65.1	138.5	138.5	95.8	85.5
Total	315.0	272.7	309.9	293.7	243.1	232.9

Field applications and the experimental design

The experiment was carried out in a Randomized Complete Block Design with four replications. Four alfalfa cultivars (TT-2008, TT-2009, P-5683 and Elci) were used as plant material. Seeding rate was 20 kg ha^{-1} for all cultivars (Soya et al., 2004). Each plot size was 2 m x 5 m = 10 m^2. Sowing was made by hand, on September

17th, 2005 at Bornova and September 19th, 2005 at Odemis.

The soil was harrowed 10 days before planting, then 30 kg ha^{-1} N and 100 kg ha^{-1} P$_2$O$_5$ were broadcasted and disked to produce a smooth seed bed. Since there were no significant problems of pests, diseases or weeds in the study, no chemical was applied. None of the alfalfa seeds were inoculated with *Rhizobium* bacteria, which existed naturally in the soil. All plants were irrigated throughout the growing season according to morphological appearance of plants in both years.

Measurements and chemical analysis

The plots were cut at the 10-25 % flowering stage, cutting mid 6 rows of plots in order to avoid border effects (net 4.8 m^2), by cutting the plants leaving a 5 cm stubble height (Avcioglu et al., 2001). After 6 cuts in the first and 8 cuts in the second experimental year, harvested fresh forage were weighed and dried to a constant weight at 70°C during 48 h. Crude ash was determined at 550 °C (Bulgurlu and Ergul, 1978). The dried samples were ground in a mill passed through a 1 mm screen. The crude protein content (CPC) was calculated by multiplying the Kjeldahl N concentration by 6.25 (Kacar and Inal, 2008). The neutral detergent fibre (NDF) and acid detergent fibre (ADF) concentrations were measured according to Ankom

Technology (Ankom 2000 Fiber Analyzer, Ankom Technology Corp., Fairport, NY, USA).

Statistical analysis

Statistical analyses were done by using the TOTEMSTAT Statistical program (Acikgoz et al., 2004). The treatment means were compared by the LSD test described by Steel and Torrie (1980).

RESULTS AND DISCUSSION

The results of the analysis of variance of the data combined over 2 years and 2 locations (soil texture) are shown in Table 3. It can be seen in Table 3 that year, cultivar, soil texture had significant F values for the fresh herbage yield (FHY), dry matter yield (DMY), crude protein yield (CPY) and dry matter content (DMC). Year x cultivar interaction was significant for the FHY, DMY and CPY. The F value of cultivar x soil texture interaction was significant for the DMY and the DMC.

Table 3. Results of analysis of variance and the F values of the traits measured

Source of Variation	DF	FHY	DMY	CPY	DMC
Year (Y)	1	340.745 **	368.200 **	284.255 **	89.194 **
Cultivar (C)	3	77.211 **	87.789 **	61.388 **	26.905 **
Soil Texture (ST)	1	10.993 **	27.705 **	20.049 **	32.887 **
YxC	3	18.001 **	13.611 **	10.374 **	2.218 ns
YxST	1	1.004 ns	3.146 ns	3.780 ns	1.871 ns
CxST	3	1.928 ns	3.209 *	2.098 ns	7.694 **
YxCxST	3	0.277 ns	0.042 ns	0.190 ns	0.617 ns
CV (%)		4.99	5.98	7.08	2.35
Source of Variation	**DF**	**CPC**	**CAC**	**ADF**	**NDF**
Year (Y)	1	1.762 ns	0.012 ns	0.003 ns	23.708 **
Cultivar (C)	3	0.578 ns	1.053 ns	16.399 **	14.504 **
Soil Texture (ST)	1	0.001 ns	1.010 ns	0.000 ns	0.970 ns
YxC	3	1.163 ns	0.195 ns	0.673 ns	2.210 ns
YxST	1	0.696 ns	0.100 ns	2.580 ns	4.073 *
CxST	3	0.050 ns	1.152 ns	0.537 ns	1.197 ns
YxCxST	3	0.392 ns	0.139 ns	0.505 ns	1.485 ns
CV (%)		3.22	10.85	4. 10	3.48

Abbreviations: DF: Degrees of freedom; CV: Coefficient of variation, NS: Not significant, *: P<0.05, **: P<0.01, FHY: Fresh herbage yield, DMY: Dry matter yield, CPY: Crude protein yield, DMC: Dry matter yield, CPC: Crude protein content, CAC: Crude ash content, ADF: Acid detergent fiber, NDF: Neutral detergent fiber

The significant variation of year, soil texture, cultivar and year x cultivar interaction indicated the treatment of results separately at heavy textured and light textured soil and to compare the means by using the LSD based on significant F values shown in Table 3.

Yield parameters

The means of the alfalfa cultivars for FHY, DMY and CPY are shown in Table 4. Average fresh herbage yield of P-5683 (139688 kg ha^{-1}) was highest in 2007 while yields of Elci (86781 kg ha^{-1}) and TT-2009 (89688 kg ha^{-1}) were lower in 2006. In two years average, both cultivars of P-5683 (118844 kg ha^{-1}) and TT-2008 (117594 kg ha^{-1}) had the highest fresh herbage yield. These could be due to the different genotypic characteristics of the cultivars tested in the experiment and it appears that adaptable cultivars for

specific conditions should be recommended for proper regions (Altinok and Karakaya, 2002; Abusuwar and Bakri, 2009; Saruhan and Kusvuran, 2011; Cinar and Hatipoglu, 2014). Average fresh herbage yield of alfalfa cultivars grown on light soil (110408 kg ha^{-1}) were also higher compared to heavy soil (105930 kg ha^{-1}) in two years average, evidencing the superiority of light soil conditions over the heavy soil. These results were in agreement with Forbes and Watson (1992) who stated that soil factors, which influence root growth and development and thereby affect the yield of a crop, can be classified as nutritional, biological or physical including soil temperature, aeration and resistance to penetration by roots. As it is expected the fresh herbage yields of all tested alfalfa cultivars, being perennial crops, increased significantly in the second year compared to the first year.

Table 4. Yield performances of four alfalfa cultivars in two different soil textures

| Cultivars | 2006 | | | 2007 | | | 2006-2007 | | |
| | Soil Texture (ST) | | | Soil Texture (ST) | | | Soil Texture (ST) | | |
	Light	Heavy	Mean	Light	Heavy	Mean	Light	Heavy	Mean
Fresh Herbage Yield (kg ha^{-1})									
TT-2008	113000	103688	108344 b	131688	122000	126844 a	122344	112844	117594
TT-2009	90625	88750	89688 b	113825	111188	112506 a	102225	99969	101097
P-5683	97750	98250	98000 b	141000	138375	139688 a	119375	118313	118844
Elci	87688	85875	86781 b	107688	99313	103500 a	97688	92594	95141
Mean	97266	94140	95703	123550	117719	120634	110408	105930	108169
Dry Matter Yield (kg ha^{-1})									
TT-2008	25130	22150	23640 b	32193	27753	29973 a	28661 a	24951 a	26806
TT-2009	19405	18895	19150 b	25913	24558	25235 a	22659 b	21726 b	22193
P-5683	22345	21090	21718 b	33710	31130	32420 a	28028 a	26110 a	27069
Elci	18198	17918	18058 b	23445	21688	22566 a	20821 c	19803 c	20312
Mean	21269	20013	20641	28815	26282	27548	25042	23148	24095
Crude Protein Yield (kg ha^{-1})									
TT-2008	4940	4428	4684 b	6588	5650	6119 a	5764	5039	5401
TT-2009	3865	3745	3805 b	5208	4968	5088 a	4536	4356	4446
P-5683	4400	4170	4285 b	6748	6180	6464 a	5574	5175	5374
Elci	3645	3643	3644 b	4728	4280	4504 a	4186	3961	4074
Mean	4213	3996	4104	5818	5269	5543	5015	4633	4824

There were significant differences with regard to the dry matter yields among the cultivars and average dry matter yields of TT-2008 and P-5683 (28661 kg ha^{-1} and 28028 kg ha^{-1}, respectively) were higher than other tested crop material in light soil whereas Elci had the lowest dry matter yield (19803 kg ha^{-1}) in heavy soil in two years average. Some researchers also pointed out that dry matter yield characteristic is closely related to the overall growth performances of alfalfa crops (such as fresh herbage yield and dry matter content) and there is great variation among various cultivars (Andueza et al., 2001; Abusuwar and Bakri, 2009; Basbag et al., 2009). Average dry matter yield of alfalfa cultivars grown on light soil (25042 kg ha^{-1}) were significantly higher compared to heavy soil (23148 kg ha^{-1}), displaying the favourable textures of light soils for alfalfa root development and growth. Some researchers, confirming our present results, also compared the yield of alfalfa on light and heavy soil under Mediterranean conditions over 3 years and emphasized that alfalfa produced significantly higher dry matter yield in light soil (Avcioglu et al., 2001; Geren et al., 2009).

Average crude protein yield of P-5683 (6464 kg ha^{-1}) was highest in 2007 whereas Elci and TT-2009 (3644 kg ha^{-1} and 3805 kg ha^{-1}, respectively) performed worse than the others in 2006. These crude protein yield performances were the natural results of the fresh herbage yield variations displayed by different alfalfa cultivars tested and soil types. The statements of Sengul et al. (2003); Geren et al. (2009); Avci et al. (2013) confirmed our results. Average crude protein yield of alfalfa cultivars grown on light soil (5015 kg ha^{-1}) were higher than heavy soil (4633 kg ha^{-1}), displaying again the favourable textures of light soils for alfalfa development. In two years average, TT-2008 (5401 kg ha^{-1}) and P-5683 (5374 kg ha^{-1}) cultivars displaying higher crude protein yield (Table 4), namely a better feed quality which is accepted as a

vital peculiarity in alfalfa hay (Al-Suhaibani, 2010; Mala and Fadlalla, 2013) had also higher crude protein yields than other crop material tested, while all cultivars showed a better performance in the light textured soil.

Quality parameters

The results of statistical analysis indicated that only the effects of experimental factors and CxST interaction effect were significant for the dry matter content of alfalfa cultivars, whereas there were not any significant factor or interaction effect on the crude protein and the crude ash content characteristics (Table 5).

Dry matter content results also showed that P-5683 and TT-2008 (23.38 % and 23.34 %, respectively) had higher values than other cultivars in light soils. Average dry matter contents of cultivars were higher in light than heavy soil and also significantly higher in the second experimental year. As stated by many research workers, dry matter content of forage crops is one of the dependable criteria of biomass production and high rate of dry matter content is mostly indicate a better adaptability and yield performance. Our findings confirmed these approaches and other researcher's statements (Avcioglu et al., 2001; Abusuwar and Bakri, 2009; Basbag et al., 2009).

The trends of crude protein and crude ash contents of alfalfa cultivars were similar to each other. Although those nutritional trait variations were not significant, the crude protein and crude ash contents of the cultivars ranged between limited values in both years and 2 years average (19.81-20.16 and 8.40-9.52%, respectively). The reason for having no striking variation in terms of either crude protein or crude ash content could be that all tested alfalfa crops were cut in similar flowering stages (10-25% flowering) during harvest procedures. Present results indicated that the crude protein and crude ash contents of tested crop materials represented the high quality legume hay and the values were in agreement with those of many researchers (Sheaffer et al., 1998; Avcioglu et al., 2001; Avci et al., 2013).

Table 5. Quality performances of four alfalfa cultivars in two different soil textures

Cultivars	2006			2007			2006-2007		
	Soil Texture (ST)			Soil Texture (ST)			Soil Texture (ST)		
	Light	Heavy	Mean	Light	Heavy	Mean	Light	Heavy	Mean
Dry Matter Content (%)									
TT-2008	22.22	21.36	21.79	24.45	22.75	23.60	23.34 a	22.06 a	22.70
TT-2009	21.41	21.29	21.35	22.76	22.09	22.42	22.09 b	21.69 ab	21.89
P-5683	22.85	21.45	22.15	23.91	22.50	23.20	23.38 a	21.97 a	22.68
Elci	20.75	20.87	20.81	21.74	21.84	21.79	21.25 c	21.35 b	21.30
Mean	21.81	21.24	21.52	23.22	22.29	22.75	22.51	21.77	22.14
Crude Protein Content (%)									
TT-2008	19.68	19.94	19.81	20.47	20.37	20.42	20.07	20.16	20.11
TT-2009	19.93	19.82	19.87	20.08	20.23	20.16	20.00	20.03	20.01
P-5683	19.70	19.77	19.73	20.01	19.84	19.93	19.85	19.81	19.83
Elci	20.04	20.32	20.18	20.16	19.72	19.94	20.10	20.02	20.06
Mean	19.83	19.96	19.90	20.18	20.04	20.11	20.01	20.00	20.01
Crude Ash Content (%)									
TT-2008	9.36	8.79	9.07	9.03	9.14	9.08	9.19	8.96	9.08
TT-2009	9.40	9.16	9.28	9.63	9.53	9.58	9.52	9.35	9.43
P-5683	8.72	9.33	9.03	8.80	9.30	9.05	8.76	9.31	9.04
Elci	8.49	9.37	8.93	8.31	9.10	8.71	8.40	9.24	8.82
Mean	8.99	9.16	9.08	8.94	9.27	9.11	8.97	9.22	9.10
ADF (%)									
TT-2008	35.58	34.24	34.91	35.09	36.52	35.80	35.33	35.38	35.36 b
TT-2009	38.20	37.95	38.07	37.51	38.29	37.90	37.85	38.12	37.99 a
P-5683	36.93	35.76	36.34	36.24	35.81	36.02	36.59	35.78	36.18 b
Elci	38.76	39.05	38.91	38.11	38.73	38.42	38.43	38.89	38.67 a
Mean	37.37	36.75	37.06	36.74	37.34	37.04	37.05	37.04	37.05
NDF (%)									
TT-2008	45.14	45.18	45.16	47.20	46.16	46.68	46.17	45.67	45.92
TT-2009	45.49	48.24	46.86	51.07	49.78	50.42	48.28	49.01	48.64
P-5683	44.82	44.47	44.64	48.18	45.59	46.88	46.50	45.03	45.76
Elci	48.48	47.73	48.11	48.79	48.78	48.79	48.64	48.26	48.45
Mean	45.98 b	46.40 b	46.19	48.81 a	47.58 a	48.19	47.39	46.99	47.19

The effects of YxST interaction and Y and C on NDF value were significant, while the other two or three way interactions were not (Table 5). The average NDF value of light soil in 2007 was the highest (48.81 %), whereas average NDF values of both soil types were lower than others in 2006 which was the indication of higher quality (Table 5). In terms of NDF variation among the cultivars, TT-2009 and Elci (48.64% and 48.45%, respectively) had higher values, while P-5683 and TT-2008 (45.76% and 45.92%, respectively) had lower NDF values. Only the effects of different cultivars were statistically significant on ADF content (Table 5). ADF contents of alfalfa cultivars ranged between 35.36-38.67 % and higher ADF values were recorded in Elci and TT-2009 (38.67% and 37.99%, respectively), whereas TT-2008 and P-5683 had lower ADF rates (35.36% and 36.18% respectively) indicating higher feed quality. The NDF and ADF values of alfalfa cultivars tested in the experiment were consistent with the findings of various researchers (Markovic et al., 2007; Dale et al., 2012).

In general, stem cell wall constituents (NDF, ADF, ADL, cellulose, and hemicellulose) are highly positively correlated with each other, but negatively associated with crude protein (Jarrige et al., 1988; Erkovan et al., 2009). NDF is the amount of protective substances obtained from residue after boiling a fodder sample in neutral detergent solution. NDF residue, actually contains very little pectic substances, but may contain negligible amounts of products like starch, nitrogenous substances and tannins (Kiraz, 2011; Selmi et al., 2013). The amount of protective substances residue obtained after boiling the sample feed with detergent solution is called ADF. ADF content is regularly higher than the crude fiber from forage, these features being closely related, since both are an estimate of the amount of cellulose + lignin (Jarrige et al., 1988). Overall results of ADF and NDF contents indicated that TT-2008 and P-5683 cultivars, having lower ADF and NDF values throughout the experimental years, had the higher feed quality. Those cultivars also ranked first according to the Feed Quality Ranking List of American Forage and Grassland Council (Rohweder et al.,

1978). Many researchers working on alfalfa forage quality indicated that protein and ash content are favourable properties to increase the nutritional value of material whereas cellulose and related derivatives are unfavourable contents (Riday and Brummer, 2005; Dale et al., 2012). Considering this approach, it can be concluded that tested cultivars were significantly different in term of cellulose, hemicellulose and lignin content depending on their genetic properties. Riday and Brummer (2005) also stated that alfalfa nutritive value traits rarely exhibit genotype by environment interaction. These nutritional characteristics can be considered as a selection criteria to choose proper cultivars for this type of ecologies and soil types.

CONCLUSION

In conclusion, it could be suggested that alfalfa cultivation on light soil had many advantages including higher biomass production and forage quality as compared to practices on heavy soil under the Mediterranean climatic conditions. TT-2008 and P-5683 cultivars of alfalfa are the most adaptable and promising forage crop material for this type of Mediterranean ecologies. Moreover, these cultivars should be considered as parents the for future breeding programmes in the area.

LITERATURE CITED

Abusuwar, A.O. and E. Bakri. 2009. Effect of water quality and weeding on yield and quality of three alfalfa (*Medicago sativa* l.) cultivars. Austrialian Journal of Crop Science, 3(6), p: 315-321.

Acikgoz, N., E. Ilker and A. Gokcol. 2004. Assessment of biological research on the computer. EU TOTEM, İzmir (in Turkısh).

Al-Suhaibani, N.A. 2010. Estimation yield and quality of alfalfa and clover for mixture cropping pattern at different seeding rates. American-Eurasian Journal Agriculture & Environment Sci., 8 (2), p: 189-196.

Albayrak, S. and M. Turk. 2013. Changes in the forage yield and quality of legume-grass mixtures throughout a vegetation period. Turkish Journal of Agriculture and Forestry, 37, p: 139-147.

Altinok, S. and A. Karakaya. 2002. Forage yield of different alfalfa cultivars under Ankara conditions. Turkish Journal of Agriculture and Forestry, 26, p: 11-16.

Andueza, D., F. Munoz, A. Maisterra and I. Delgado. 2001. Forage yield and crude protein content of lucerne cultivars established in the ebro middle valley. Preliminar results. In: Options Méditerranéennes. Series A: Séminaires Méditerranéens, 45, p: 73-76.

Anonymous. 2013. Turkish statistical institute. http://www.turkstat.gov.tr (Accessed November 24, 2014).

Anonymous. 2005. Fertilising dairy pastures: A manual for use in the Target 10 Soils and Fertilisers Program. Department of Primary Industries, Melbourne, Victoria, Australia, ISBN 1741462193, 2nd ed., 318p.

Avci, M.A., A. Ozkose and A. Tamkoc. 2013. Determination of yield and quality characteristics of alfalfa (*Medicago sativa* l.) varieties grown in different locations. Journal of Animal and Veterinary Advances, 12 (4), p: 487-490.

Avcioglu, R., H. Geren and H. Ozkul. 2001. Seasonal changes of quantitative and qualitative performances of some alfalfa cultivars in the Mediterranean coastal part of Aegean Region. Options méditerranéennes, Numero a-45: Quality in lucerne and medics for animal production, 14-16 September 2001, Zaragoza (Spain), CHIEAM, FAO, p: 205-209.

Basbag, M., R. Demirel and M. Avci. 2009. Determination of some agronomical and quality properties of wild alfalfa (*Medicago sativa* l.) clones in turkey. Journal of Food, Agriculture & Environment Vol. 7 (2), p: 357-359.

Cinar, S. and R. Hatipoglu. 2014. Forage yield and botanical composition of mixtures of some perennial warm season grasses with alfalfa (*Medicago sativa* L.) under Mediterranean conditions. Turk J Field Crops, 19(1), p: 13-18.

Bulgurlu, S. and M. Ergul. 1978. Physical, chemical and biological analyses methods of feeds. Izmir, Ege Univ. Agric. Fac. Publish No: 127 (in Turkish).

Dale, L.M., I. Rotar, A. Thewis, R. Vidican, V. Florian and A. Ciure. 2012. Determination of alfalfa crude fiber, ndf, adf and lignin content by nir spectrometry. Lucrări Ştiintifice, vol. 55, seria Agronomie, p: 245-249.

Erkovan, H.I., M.K. Gullap, M. Dasci and A. Koc. 2009. Changes in leaf area index forage quality and above-ground biomass in grazed and ungrazed rangelands of eastern anatolia region. Journal of Agricultural Sciences, 15 (3), p: 217-223.

Forbes, J.C. and R.D. Watson. 1992. Plants in agriculture. Cambridge University Pres., Cambridge, 355p.

Geren, H., B. Kir, G. Demiroglu and Y.T. Kavut. 2009. Effects of different soil textures on the yield and chemical composition of alfalfa (*Medicago sativa* L.) cultivars under mediterranean climate conditions. Asian Journal of Chemistry, 21 (7), p: 5517-5522.

Jarrige, R. 1988. Alimentation des bovins, ovins et caprins. INRA Paris, 471p.

Kacar, B. and A. Inal. 2008. Plant analysis. Ankara, Nobel Publish No:1241, 892p (in Turkish).

Kiraz, A.B. 2011. Determination of relative feed value of some legume hays harvested at flowering stage. Asian Journal of Animal and Veterinary Advances, 6 (5), p: 525-530.

Mala, A.S.E.E. and B. Fadlala. 2013. Effect of stage of cutting alfalfa (berseem) on crude protein content and dry matter yield. ARPN Journal of Science and Technology, 3 (10), p: 982-985.

Markovic, J., J. Radovic, Z. Lugic and D. Sokolovic. 2007. The effect of development stage on chemical composition of alfalfa leaf and stem. Biotechnology in Animal Husbandry 23 (5-6), p: 383-388.

Meyer, D.W. 2005. Plant height as determinant for harvesting alfalfa. In: Forage Focus, Meyer, D.W. (Ed.). Midwest Forage Association, Minneapolis, USA., p: 12-13.

Moreira, A. and N.K. Fageria. 2010. Liming influence on soil chemical properties, nutritional status and yield of alfalfa grown in acid soil. R. Bas. Ci. Solo, 34, p:1231-1239.

Rechel, E.A., B.D. Meek, W.R. DeTar and L.M. Carter. 1991. Alfalfa yield as affected by harvest traffic and soil compaction in a sandy loam soil. Journal of Production Agriculture (4) 2, p: 241-246.

Riday, H. and E.C. Brummer. 2005. Relationships among biomass yield components within and between subspecies of Alfalfa. http://naldc.nal.usda.gov/naldc/download.xhtml?id=16387&content=PDF (Accessed November 24, 2014).

Rohweder, D.A., R.F. Barnes and N. Jorgensen. 1978. Proposed hay grading standards based on laboratory analyses for evaluating quality. Journal Animal Sci., 47, p: 747-759.

Saruhan, V. and A. Kusvuran. 2011. Determination of yield performances of some lucerne (*medicago sativa* l) cultivars and genotypes under the Southeastern Anatolian region conditions. Journal of the Faculty of Agriculture, Izmir, 48 (2), p: 133-140 (in Turkish).

Selmi, H., Z. Abdelwahed, A. Rouissi, B. Jemmali, L. Tayachi, M. Amraoui and H. Rouissi. 2013. Preliminary nutritional characterization of some shrubs from the north of Tunisia. International Journal of Research in Agriculture and Food Sciences. 1 (1), p: 36-39.

Sengul, S., L. Tahtacioglu and A. Mermer. 2003. Determination of suitable alfalfa (Medicago sativa L) cultivars and lines for Eastern Anatolia Region. Journal of the Faculty of Agriculture, Erzurum, 34 (4), p: 321- 325 (in Turkish).

Sheaffer, C.C., D. Cash, N.J. Ehlke, J.L. Hansen, J.C. Henning, J. Grimsbo Jewett, K.D. Johnson, M.A. Peterson, M. Smith and D.R. Viands. 1998. Entry x environment interactions for alfalfa forage quality. Agronomy Journal, 90, p: 774-780.

Soya, H., R. Avcioglu and H. Geren. 2004. Forage crops, Hasad Press, Turkey, 223p (in Turkish).

Steel, R.G.D. and J.H. Torrie 1980. Principles and procedures of statistics, second edition, New York: McGraw-Hill.

Tomic, Z., Z. Nesic, V. Krnjaja, M. Zujovic and M.M. Petrovic. 2006. Forage production and quality of some new legume cultivars in Serbia. Sustainable grassland productivity: Proceedings of the 21st General Meeting of the European Grassland Federation, Badajoz, Spain, p: 282-284.

Tongel, M.O. and I. Ayan. 2010. Nutriotional contents and yield performances of lucerne (medicago sativa l.) cultivars in southern black sea shores. Journal of Animal and Veterinary Advances, 9 (15), p: 2067-2073.

23

YIELD AND EARLY MATURITY RESPONSE TO FOUR CYCLES OF MODIFIED MASS SELECTION IN PURPLE WAXY CORN

Satang HUSSANUN[1], Bhalang SURIHARN[1, 2], Kamol LERTRAT[1, 2]*

[1]*Department of Plant Science and Agricultural Resources, Faculty of Agriculture,*
Khon Kaen University, Khon Kaen, THAILAND
[2]*Plant Breeding Research Center for Sustainable Agriculture,*
Khon Kaen University, Khon Kaen, THAILAND
Corresponding author: bsuriharn@gmail.com.

ABSTRACT

High yield and early maturity are important characters in corn breeding. The objectives of the study were to evaluate the responses to four cycles created by modified mass selection method to increase yield and early maturity of a purple waxy and to investigate the correlations between yield and among other important traits in a purple waxy corn (*Zea mays* L. var. *ceratina*) population. Four cycles was evaluated for two seasons in 2012/13. A randomized complete block design with four replications was used. The results indicated that the selection method has led to improvement in many characters of this population. C_4 cycle gave the highest whole ear yield of 16.0 t ha^{-1} followed by 15.4 t ha^{-1} of C_3 cycle. Although C_4 also had the highest marketable husked yield (9.9 t ha^{-1}), it had the lowest days to tasseling and days silking. Genetic gains per cycle were 0.68 (P≤0.01) and 0.37 (P≤0.01) for whole ear yield and marketable husked yield, respectively, whereas genetic gain for days to tasseling and silking was -1.8 for both traits. As correlations between marketable husked yield with days to silking and days to tasseling were negative and low, direct selection for marketable husked yield would result in early maturity in this corn population, and several generations of recurrent selection are required.

Keywords: *Zea mays* L. var. *ceratina*, corn breeding, population improvement, correlation, indirect selection

INTRODUCTION

Corn is an important food crop in the world. Waxy corn that is a special corn type can be used as food in several countries and as replacement for dent corn for animal feed (Johnston and Anderson 1992; Schroeder et al., 1998). Moreover, waxy corn starch is unique and has high amylopectin content, which gives high possibilities in industrial use (Klimek-Kopyra et al., 2012). Corn has a wide range of kernel colors such as white, yellow, black and purple. There are numerous special cultivars that contain colored pigments and give rise to numerous varieties of black and purple corn. The dark purple color of corn is caused by high content of anthocyanins located in the pericarp layers and cob. Anthocyanin pigment was found in all parts of purple corn, but it was found at particularly high concentration in the husk and cob (Li et al., 2008). Studies on antioxidant and anticarcinogenic properties of anthocyanin has been conducted mainly in normal purple corn, making it more attractive for nutraceutical and functional food market (Cevallos-Casals and Cisneros-Zevallos, 2003; 2004). Kernels and cobs of purple corn possessed excellent antioxidant activity, and the application of these natural food colorants by the food industry would be increased considerably (Yang and Zhai

2010). Therefore, it is important to breed waxy corn for high anthocyanin.

High yield is still a primary goal of most plant breeding programs (Ferh 1987). Pest resistance, stalk strength, uniformity, kernel quality and early maturity are also important in corn and waxy corn breeding programs. Currently, most corn varieties grown commercially are hybrids. However, some waxy corn varieties in Asian counties are open-pollinated varieties (OPVs). Therefore, improved OPVs are important in the countries where seed production is not well developed, and they also have potential as germplasm sources for hybrid development. OPVs with high yield or good adaptation are excellent germplasm sources for corn breeding programs.

Mass selection is the simplest and inexpensive method for population improvement in cross pollinated crops, and this method could be effective for selection of the traits that can be identified before or at the time of flowering (Vasal et al., 2004). However, the method is not effective for improvement of crop yield or quantitative characters because only female plant is selected. Recently, modified mass selection for prolificacy and ear length has been used effectively for yield improvement in waxy corn (Kesornkeaw et al., 2009; Senamontry et al., 2013).

However, the information on population improvement for high yields and early maturity are still lacking in corn and purple waxy corn. Previously study in field maize indicated that simultaneous improvement of these traits is difficult due to positive correlation between yields and maturity in maize (Beck et al., 1990; Konak et al., 1997).

Vegetable waxy corn with early maturity and high yield is required for specific cropping systems that have a limited time to grow waxy corn with late maturity. Is it possible to select purple waxy corn for early maturity for some extent without significant yield reduction? To respond to this question, mass selection experiment was carried out for four cycles. The objectives of this research were to evaluate the response to modified mass selection for yields and maturity of a purple waxy corn population, KND composite #1, and to investigate correlations between yields and among other important traits.

MATERIALS AND METHODS

Plant materials

Three corn populations (PFC, HJ and SLE) with different kernel and cob colors were used to generate new population for selection experiment. PFC is a field corn population with purple kernels and purple cobs, high yield, late maturity, HJ is a waxy corn with white kernels, white cobs, good eating quality, stay green, good stand ability, high yield and medium maturity, and SLE is an OPV waxy corn population with white kernels, white cobs, good eating quality, medium yield and early maturity.

Population improvement

The base population of waxy corn, Khon Kaen KND composite #1, was developed from three populations (PFC, HJ and SLE) in 2008 to 2009 at the Experimental Farm, the Faculty of Agriculture, Khon Kaen University, Thailand. Two hundred plants in each population were used to create new population. At two or three days after silking, bulked pollens from all plants of HJ were used to pollinate all ears in PFC population in the rainy season 2008, and all pollinated ears in PFC population were harvested. The F_1 generation was further backcrossed to HJ population in the dry season 2008/2009.

The seeds of the backcross generation were mixed with the seeds of SLE population in the same proportion in the rainy season 2009. The bulked seeds were sown for two seasons to create a random mating base population. The seeds showing waxy endosperm were selected from the F_2 generation using potassium iodide (KI). A blue-black color indicates that the seeds are field corn, and red color shows that the seeds are waxy corn. The population was then developed to improve yield and earliness through modified mass selection method.

Modified mass selection was carried out consecutively for four cycles. Selection from C_0 generation to C_4 generation was performed at spacing of 0.80 x 0.25 m which provided population density of 8,000 plants per 0.16 ha^{-1}. Plants of about 10 percent in each cycle were

used for generation advance, and remnant seeds were stored in a refrigerator for further evaluation. Before silk emergence, bulked pollens from selected plants with purple kernels, purple cobs, early maturity, good stand ability, big ears and disease free were used to pollinate all ears of selected plants. C_4 population was completed in the rainy season 2012.

Field Experiment

The experiment was conducted in the late rainy season 2012 and dry season 2012/2013 at the Experimental Farm, Khon Kaen University, Thailand (16°47′ N, 102°81′ E, 200 msl). Five populations of purple waxy corn (C_0 to C_4) were evaluated in a randomized complete block design with four replications for two seasons.

The plot size was four-row plot with five meters in length and spacing of 0.80 x 0.25 m. Conventional tillage was practiced for soil preparation, and 15-15-15 fertilizer of N-P-K as basal dose at the rate of 171 kg ha^{-1} was incorporated into the soil during soil preparation. Two splits of 15-15-15 fertilizer at the rate of 93.75 kg ha^{-1} plus urea (46-0-0) at the rate of 93.75 kg ha^{-1} for first split and 15-15-15 fertilizer at the rate of 125 kg ha^{-1} plus urea at the rate of 62.5 kg ha^{-1} for second split were applied to the crop at 14 days after planting (DAP) and 30 DAP, respectively. At flowering stage, 13-13-21 fertilizer was applied at the rate of 156.25 kg ha^{-1}. Therefore, total dose of fertilizers consisted of 150.65 kg ha^{-1} nitrogen, 78.78 kg ha^{-1} phosphorus and 91.27 kg ha^{-1} potassium, respectively. Irrigation was supplied regularly to avoid drought stress, and insect pests, diseases and weed were appropriately managed to obtain optimum growth and yield of the crop for both seasons.

Data collection

Data were recorded for whole ear yield, marketable husked yield, ear diameter, ear length, plant height, ear height, days to tasseling and days to silking. Days to 50% tasseling and silking were recorded from total number of plants in each plot. After pollination, plant height and ear height were recorded from 10 randomly chosen plants in each plot. Harvest time was determined at 20 days after pollination (R4 growth stage). All ears from the center two rows or 40 plants were harvested and weighed. Ear diameter, ear length and marketable husked yield were recorded from good 10 representative ears in each plot.

Data analysis

Analysis of variance was performed for each character in each season. A combined analysis of variance of two seasons was conducted when error variances of two seasons were homogeneous, and the ratio of error variances between two seasons was smaller than 3 (Gomez and Gomez 1984). Significant differences among cycles were assessed by least significant difference (LSD) at 0.05 probability level, and all analyses were carried out using MSTAT-C software (Russel 1994). Regression analysis was used to evaluate the correlated responses to selection for studied traits. Response to selection was determined by simple linear regression. Difference from zero of b values was determined by T-test.

RESULTS AND DISCUSSION

Analysis of variance

Seasons were significantly different for whole ear yield, days to tasseling and days to silking, and cycles

were significantly different for most (Table 1). Differences in seasons indicated that environment was important for the expression of these traits in waxy corn, and differences in cycles indicated the significant effects of modified mass selection on these traits. However, interactions between season and cycle were not significant for all traits. Low interactions indicated that the cycles responded similarly for all traits in both seasons.

Table 1. Mean square for yield, yield component and agronomic traits of five cycles purple waxy corn population across two seasons 2012/13.

Df	1	6	4	4	24	
Trait	**Mean square**					**C.V. (%)**
	Season (S)	**Rep./S**	**Cycle (C)**	**S x C**	**Pooled error**	
Yield trait						
Whole ear yield (t ha^{-1})	16,650**	1,200	9,708**	640ns	791	6.1
Marketable husked yield (t ha^{-1})	87ns	478	3741**	317ns	664	9.2
Yield components						
Ear diameter (cm)	0.03ns	0.02	0.25**	0.01ns	0.02	3.3
Ear length (cm)	9.4ns	2.8	6.3**	0.4ns	0.5	4.3
Agronomic traits						
Plant height (cm)	1ns	171	678**	15ns	17	1.9
Ear height (cm)	244ns	536	1,625**	458ns	175	11.3
Day to tasseling (day)	518.4**	1.0	63.2**	1.5ns	0.8	1.8
Day to silking (day)	403.2**	2.0	66.9**	1.2ns	1.1	2.1

ns = non significant

** = significant at P < 0.01, respectively

Response to selection

C$_4$ cycle had the highest whole ear yield of 16.0 t ha^{-1} followed by C$_3$ cycle with whole ear yield of 15.4 t ha^{-1} (Figure 1). C$_4$ Cycle also had the highest marketable

husked yield of 9.9 t ha^{-1} followed by 8.9 t ha^{-1} of C$_3$ cycle. Genetic gains per cycle were 0.68 (P≤0.01) and 0.37 (P≤0.01) for whole ear yield and marketable husked yield, respectively (Figure 1).

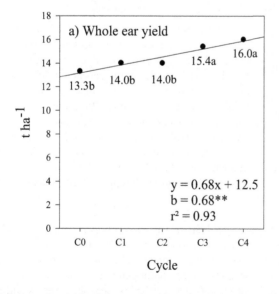
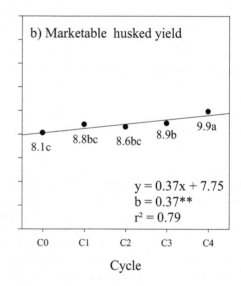

Figure 1. Means for yields of five cycles of purple waxy corn populations. Letters at least one same letter indicate non-significant difference by LSD. ** b-values are significantly different from zero at P < 0.01.

In previous study, modified mass selection could improved prolificacy, yield and ear length in waxy corn (Kesornkeaw et al., 2009; Senamontry et al., 2013). Seven selection methods (mass, modified ear-to row, half-sib with inbred tester, full-sib, S1-progeny, S2-progeny, and

reciprocal full-sib selection) were successful in significantly improving the population per se performance for grain yield in the BS11 maize population (Weyhrich et al., 1998). The results are supported by previous findings and ensure the usefulness of modified mass selection for yield improvement in waxy corn.

Cycle C_4 and cycle C_3 had the biggest ears with diameters of 4.5 and 4.4 cm, respectively (Figure 2). Ear lengths for C_4, C_3, C_2 and C_1 were 18.5, 17.4, 17.0 and 16.8 cm, respectively. Genetic gains of 0.11 and 0.54 cm per cycle were observed for ear diameter and ear length, respectively. The results indicated that yield of waxy corn could be improved by selection for larger and longer ears.

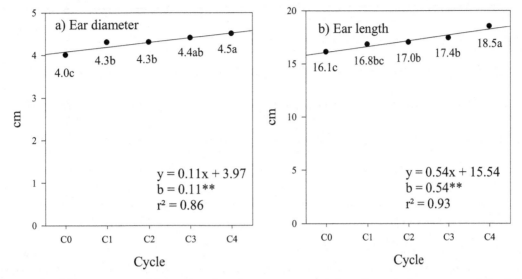

Figure 2. Means for yield components of five cycles of purple waxy corn populations. Letters at least one same letter indicate non-significant difference by LSD. ** b-values are significant different from zero at P < 0.01.

Plant heights for C_4, C_3, C_2, C_1 and C_0 were 201, 207, 215, 220 and 223 cm, respectively, and ear heights were 105, 106, 116, 118 and 104 cm in respective order (Figure 3). Gains per cycle were estimated at -0.57 cm (P≤0.01) for plant height and -8.2 cm (P≤0.01) for ear height. Carena and Cross (2003) found in maize that mass selection at high plant population densities resulted in taller plants with high ears, and selection at high planting densities did not produce population with improved ability to withstand high densities. Plant population density affected crop yield and response to selection through the increase in competition (Fasoula and Tollenaar, 2005). In this study, modified mass selection reduced plant height and ear height, and these effects are preferable because shorter plants are more resistant to lodging. Plant population density used in this study was optimum to produce shorter plants with lower ear placement.

It should be noted that modified mass selection for yield resulted in the reduction in days to tasseling and days to silking in M_4 population (45 and 46 days, respectively), and gain per cycle of selection was estimated at -1.8 for both traits. It had a tendency to obtain early maturity plants in response to selection for yield. The effect of modified mass selection for yield on maturity is preferable as early maturity is more suitable to many cropping systems.

The results in this study did not support previous findings. Selection for yield in maize resulted in the increase in maturity (Beck et al., 1990; Konak et al.,

1997). The differences in the results could be due to the fact that some crops obtain high yield though high biomass, but some crops obtain high yield thought effective partitioning of biomass into harvestable yield.

Correlated response

The correlation coefficients between whole ear yield with marketable husked yield (r=0.71, P≤0.01), ear diameter (r=0.58, P≤0.01) and ear length (r=0.33, P≤0.05) were positive and significant, but whole ear yield was negatively correlated with plant height (r=-0.54, P≤0.01) and ear height (r=-35, P≤0.05) (Table 2). Whole ear yield was not significantly correlated with days to tasseling (r=0.02) and days to silking (r=-0.07). Similar correlation coefficients were observed for marketable husked yield with ear diameter, ear length, plant height and ear height. However, the correlation coefficients between marketable husked yield with days to tasseling (r=-0.22) and days to silking (r=-0.28), although they were not significant, was positive and much higher than those with whole ear yield.

In previous investigations, yield was negatively and significantly correlated with days to silk (Konak et al., 1997), and maize genotypes with early maturity had lower yield of about 20-30% than did the genotypes with late maturity (Beck et al., 1990). in wheat, grain yield was positively correlated with maturity, rate and duration of grain fill, and harvest index (Iqbal et al., 2007). The improved C_4 population had high yield and early maturity through modified mass selection. Harvest index should be a factor determining yield increase in this population.

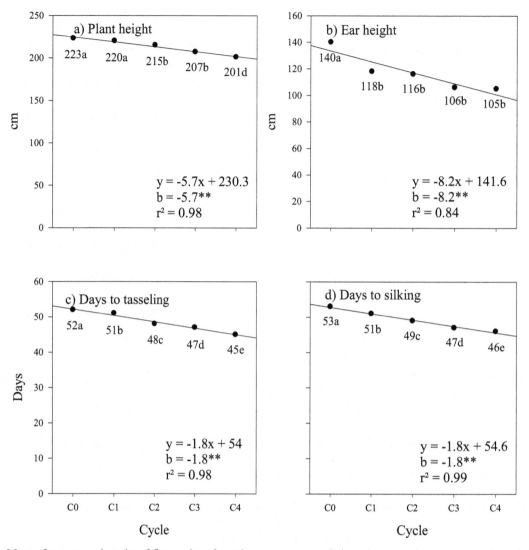

Figure 3. Means for agronomic traits of five cycles of purple waxy corn populations. Letters at least one same letter indicate non-significant difference by LSD. ** b-values are significant different from zero at P < 0.01.

Table 2. Correlations among yields, yield components and agronomic traits (n=40).

	Whole ear yield	Marketable husked yield	Ear diameter	Ear length	Plant height	Ear height	Days to tasseling
Marketable husked yield	0.71**						
Ear diameter	0.58**	0.53**					
Ear length	0.33*	0.43**	0.46**				
Plant height	-0.54**	-0.40*	-0.55**	-0.38*			
Ear height	-0.35*	-0.35*	-0.49**	-0.18	0.65**		
Days to tasseling	0.02	-0.22	-0.28	-0.62**	0.46**	0.26	
Days to silking	-0.07	-0.28	-0.31*	-0.62**	0.53**	0.32*	0.98**

*, ** = significant at P < 0.05 and P < 0.01, respectively

Ear diameter was closely associated with ear length (r=0.46, P≤0.01). However, these traits were negatively associated with plant height, ear height, days to tasseling and days to silking although the correlation coefficients between days to tasseling with ear diameter (r=-0.28) and ear height with ear length (r=-0.18) were not significant. Plant height, ear height, days to tasseling and days to silking were interrelated as most correlation coefficients were positive and significant except for the correlation coefficient between days to tasseling with ear height (r=0.26).

In conclusion, modified mass selection was successful in developing high yield and early maturity population of waxy corn from the base population Khon Kaen KND composite #1. Whole ear yield, marketable husked yield, ear diameter and ear length were significantly increased, whereas modified mass selection resulted in reductions in days to tasseling, days to silking, plant height and ear height. The responses to selection are preferable because the new population have better agronomic traits. Early maturity is more suitable for many cropping system, and shorter plants with lower ear placement are more resistant to lodging. As correlations between whole ear yield with days to tasseling and days to silking were not significant, selection for early maturity for a limited extent would not detrimental to whole ear yield.

Modified mass selection is a rapid and convenient method for population improvement for yield and purple density of kernels and cops in purple waxy corn as a cycle of selection can be completed in one season and both male and female parents can be selected. However, purple color is a complex character, and kernel color segregates in the improved population. More cycles of recurrent selection for purple color is still required to eliminate segregating progenies in the population. This improved population (C_4) can be used as OPVs or germplasm source for high yield and early maturity in purple corn breeding programs.

ACKNOWLEDGMENTS

We are grateful for the financial support provided by the National Center for Genetic Engineering and Biotechnology, Bangkok, Thailand and the Plant Breeding Research Center for Sustainable Agriculture, Khon Kaen University, Thailand. Acknowledgement is extended to Khon Kaen University and the Faculty of Agriculture for providing financial support for manuscript preparation activities.

LITERATURE CITED

Beck, D.L., S.K. Vasal and J. Crossa. 1990. Heterosis and combining ability of CIMMYT's tropical early and intermediate maturity maize (*Zea mays* L.) germplasm. Maydica 35: 279-85.

Cerena, M.J. and H.Z. Cross. 2003. Plant density and maize germplasm improvement in the Northern corn belt. Maydica 48: 105-111.

Cevallos-Casals, B.A. and L. Cisneros-Zevallos. 2003. Stoichiometric and kinetic studies of phenolic antioxidants from Andean purple corn and red-fleshed sweet potato. J. Agric. Food Chem. 51: 3313-3319.

Cevallos-Casals, B.A. and L. Cisneros-Zevallos. 2004. Stability of anthocyanin-based aqueous extracts of Andean purple corn and redfleshed sweet potato compared to synthetic and natural colorants. Food Chem. 86: 69-77.

Fasoula, V.V. and M. Tollenaar. 2005. The impact of plant population density on crop yield and response to selection in maize. Maydica 50: 39-48.

Fehr, W.R. 1987. Principle of Cultivars Development. New York: McMillan publ. company.

Gomez, K.A. and A.A. Gomez. 1984. Statistical Procedure for Agricultural Research. 2. ed. New York: John Wiley & Sons.

Iqbal, M., A. Navabi, D.F. Salmon, R.C. Yang and D. Spaner. 2007. Simultaneous selection for early maturity, increased grain yield and elevated grain protein content in spring wheat. Plant Breeding126: 244-250.

Johnston, L.J. and P.T. Anderson. 1992. Effect of waxy corn on feedlot performance of crossbred yearling steers. Minnesota Beef Cattle Research Report. pp. 1–5.

Kesornkeaw, P., K. Lertrat and B. Suriharn. 2009. Response to four cycles of mass selection for prolificacy at low and high population densities in small ear waxy corn. Asian J. Plant Sci. 8: 425-432.

Klimek-Kopyra, A., A. Szmigiel, T. Zając and A. Kidacka. 2012. Some aspects of cultivation and utilization of waxy maize. Acta Agrobot. 65: 3-12.

Konak, C., A.Ünay, A. Zeybek and E. Acartürk. 1997. Correlation and path analyses in maize (*Zea mays* L.). Turk. J. Field Crops 2: 47-52.

Li, C.Y., H.W. Kim, S.R. Won, H.K. Min, K.J. Park, J.Y. Park, M.S. Ahn and H.I. Rhee. 2008. Corn husk as a potential source of anthocyanins. J. Agric. Food Chem. 56: 11413-11416.

Weyhrich, R.A., K.R. Lamkey and A.R. Hallauer. 1998. Responses to Seven Methods of Recurrent Selection in the BS11 Maize Population. Crop Sci. 38: 308-321.

Russel O.F. 1994. MSTAT—C v.2.1 (a computer based data analysis software). Michigan: Crop and Soil Science Department, Michigan State University.

Schroeder J.W., G.D. Marx and C.S. Park. 1998. Waxy corn as a replacement for dent corn for lactating dairy cows. Ani. Feed Sci. Tec. 72: 111-120.

Senamontry, K., K. Lertrat and B. Suriharn. 2013. Response to five cycles of modified mass selection for ear length in waxy corn. Sabrao J. 45: 332-340.

Vasal, S.K., N.N. Singh, B.S. Dhillon and S.J. Patil. 2004. Population improvement strategies for crop improvement. In: Plant Breeding-Mendelian to Molecular Approaches, ed. Jain H.K., and Kharkwan M.C., 391-406, Narosa Publishing House, New Delhi.

Yang, Z. and W. Zhai. 2010. Identification and antioxidant activity of anthocyanins extracted from the seed and cob of purple corn (*Zea mays* L.). Innovative Food Sci. Emerging Technol. 11: 169-176.

24

FORAGE YIELD AND QUALITY OF COMMON VETCH MIXTURES WITH TRITICALE AND ANNUAL RYEGRASS

Emine BUDAKLI CARPICI*, Necmettin CELIK

Uludag University, Faculty of Agriculture, Department of Field Crops, Bursa, TURKEY
*Corresponding author:ebudakli@uludag.edu.tr

ABSTRACT

The objective of this research was to determine the forage yield, quality and physiological properties of mixtures of common vetch (*Vicia sativa* L.) with triticale (×*Triticosecale* Wittmack) and annual ryegrass (*Lolium moltiflorum*) grown as intermediate winter crop system under the conditions of Southern Marmara Region. Forage yield, light interception, leaf area index, crude protein content, crude protein yield, ADF and NDF contents were measured. The highest forage yield (15.21 t ha-1) was obtained from 50% common vetch + 50% triticale mixture which was followed by 75% common vetch + 25% triticale, 25% common vetch + 75% triticale, 25% common vetch + 75% annual ryegrass mixtures.The results of two-year averages indicated that the 75% common vetch + 25 % triticale mixturewas the best one for good quality and higher forage yield, and its suitability wasrecommended for the experimental ecology.

Keywords: annual ryegrass, common vetch, triticale, quality, intermediate winter crop

INTRODUCTION

Legumes and cereals do not provide satisfactory yields when they are pure seeded. There are some rational reasons of this situation. First of all, legume crops are low-yielding, especially in areas where rainfall is insufficient, and the plant lodging causes some problems during harvest. On the other hand, cereals produce high forage yields but with low protein content which is far from the requirements of many livestock (Rakeih et al., 2010). However, when legumes and cereals are combined in a mixture for forage production they exhibit good results.

Common vetch (*Vicia sativa* L.), an annual legume with climbing habit and high levels of protein, is usually grown in mixture with small grain annual cereals for forage production (Lithourgidis et al., 2006). Different cereals had been tested in mixtures with common vetch and the same or different results had been reported. The most proper cereals for mixture with common vetch were reported to be oat (*Avena sativa* L.) (Cabaollero and Goicoechea, 1986 and Thomson et al., 1990), barley (*Hordeum vulgare* L.) (Thompson et al., 1992 and Roberts et al., 1989), triticale (x *Triticosecale* Wittmack) (Anil et al., 1998). Sowing ratios in a mixture composed of legumes and cereals have yielded different results in different studies. Findings of Hasar and Tukel (1994) showed that 25% common vetch + 75% triticale mixture yielded more dry matter than other ratios of vetch-triticale mixtures. Turemen et al. (1990) reported that the highest

dry matter yield was obtained from pure stand of annual ryegrass. Albayrak et al. (2004) found the highest dry matter and crude protein yields in the mixture including 70% hairy vetch + 30% triticale. Lithourgidis et al.(2006) evaluated pure sowings of common vetch, triticale and oat monocultures as well as mixtures of common vetch with the cereals in two seeding ratios (55:45 and 65:35) for forage yield and quality. They found that in all mixtures, the crude protein content increased as common vetch seeding proportion increased. Pure common vetch had the highest crude protein content. In contrast, triticale and oat pure sowings produced the lowest crude protein content. Kokten et al. (2009) reported that the mixture comprising 20% vetch and 80% triticale gave the highest dry matter yield with23.5% vetch ratio under Adana conditions, while the mixture of 40% vetch + 60% triticale gave the highest dry matter yield with 10% vetch ratio under Kozan conditions in Turkey.

Research findings of above studies showed that more crude protein yield was obtained from the 80-90% vetch ratio. On account of protein content and yield, the vetch-triticale combination containing 90-80% vetch should be preferred to obtain high quality forage. In case of requirement of the forage containing high energy, the mixture containing higher triticale ratio could be preferred.

The objectives of the present study were to evaluate the pure stands of common vetch, triticale and annual ryegrass as well as the mixtures of common vetch with the mentioned cereals in different ratios (75:25, 50:50 and 25:75) for higher forage yield and quality, and to determine the proper mixture rates of vetch-cereal under rainfed conditions of Southern Marmara Region.

MATERIALS AND METHODS

Field studies were conducted during the 2009-2010 and 2010-2011 growing seasons on clay loam soil at the Agricultural Research and Experiment Center of Uludag University, Bursa (40° 11′ N, 29° 04′ E). Soil test values indicated a pH of 7, none saline, low values in lime and

organic matter and rich in potassium. Precipitation distribution and amount differed markedly between the 2009-2010 and 2010-2011 growing seasons (Table 1). Total precipitation in 2009-2010 was 236.8 mm over the long years and it resulted from higher precipitations fell in December, January, February and March of this year. Total precipitation in 2010-2011 growing season was 60.7 mm lower than that of the long years (Table 1). There were almost no differences between temperatures of experimental years and the long years.

Table 1. Monthly precipitation and temperature in 2009-2010, 2010-2011 and long years (1975-2008) in Bursa.

Months	Monthly Precipitation (mm)			Monthly Temperature (°C)		
	Long Years	2009-2010	2010-2011	Long Years	2009-2010	2010-2011
November	85.4	80.6	24.0	10.3	10.1	15.5
December	96.4	119.1	152.6	7.1	9.5	9.8
January	80.3	149.7	72.4	5.4	7.0	5.6
February	66.2	178.9	18.4	5.9	9.3	6.0
March	62.7	115.3	67.4	8.5	9.0	8.3
April	65.2	63.4	76.8	13.0	13.4	10.5
May	43.4	29.4	27.3	17.7	19.2	16.7
Total	499.6	736.4	438.9	-	-	-
Mean	-	-	-	9.7	11.1	10.3

In the experiment, common vetch (*Vicia sativa* L.) cv. Gulhan, triticale(×*Triticosecale* Wittmack) cv. Karma 2000 and annual ryegrass (*Lolium multiflorum*) cv. Caramba were used as plant materials. These species were sown as pure stands and mixtures. For mixtures, common vetch was combined with cereals in double combinations, and thus nine treatments were formed. Treatments were tested in a randomized complete block design with three replications. Plot size was 1.6 x 5 m = 8 m^2 and seeded with eight rows. Seed amounts of species per hectare were based on the seed amounts of their pure sowings. The pure sowing rates of common vetch, triticale and annual ryegrass were 120, 200 and 30 kg ha^{-1}, respectively. Seeding rate of a mixture was calculated by multiplying the seed amount of pure seeding of each species with its ratio in the mixture and the multiplication results of two species were summed. Mixtures were formed by combining vetch seed with those of cereals (triticale and annual ryegrass) in 75, 50 and 25 % ratios.

Sowings were made by hand on 15 November 2009 and 23 November 2010. Prior to sowing, 50 kg ha^{-1} P$_2$O$_5$ of 42-44% triple superphosphate and 30 kg ha^{-1} N of 34% ammonium nitrate were applied as standard fertilizers.

Light interception values were determined using a LI SA191-A Quantum Sensor and were calculated according to the formula of Zaffaroni and Schneiter (1989) at the flowering stage of common vetch. At the same stage, leaf area indices were measured by LAI-2000 (LI-COR, Lincoln). Harvest was made at ground level when the lower pods of common vetches fully formed. After harvest, green forage samples were randomly taken from harvested green forages of each plot and put into cotton

bags. Samples were oven-dried at 78 °C for 48 hours and weighed, then dry weight percentages were calculated. Forage yield of each plot was calculated by multiplying the fresh forage weight of each plot with its dry forage weight percentage. Oven-dried samples of plots were ground and 1 g samplewas used for the total nitrogen determination and 0.5 g for ADF and NDF. ADF and NDF were analysed by sequential detergent analysis method (Goering and Van Soest, 1970) and total nitrogen by Kjedahl method (AOAC, 1984). Crude protein content was calculated by multiplying total nitrogen with 6.25 constant.

The whole data obtained were exposed to variance analysis at 1% and 5% significance level according torandomized complete block experimental design. For the differences among means of each parameter affected from treatments, the LSD test was used at 5% significant level by using MINITAB and MSTAT-C programs.

RESULTS AND DISCUSSION

The forage yields averaged over-two year data were presented in Table 2. The highest forage yield (15.21 t ha^{-1}) was obtained from 50% common vetch + 50% triticale mixture which was followed by 75% common vetch + 25% triticale, 25% common vetch + 75% triticale, 25% common vetch + 75% annual ryegrass mixtures. Yucel and Avci (2009) reported that mixture of common vetch-triticale (50:50) produced about 63% more dry matter yield than pure sowing of common vetch, but about 12% less than that of pure triticale sowing. Numerous workers have determined different ratios for mixtures for maximum dry matter yield (Turemen et al., 1990; Hasar and Tukel, 1994; Albayrak et al., 2004; Kokten et al.,

2009; Yucel and Avci, 2009). These differences may be resulted from different ecological conditions, cultivars and cropping systems.

Table 2. Forage yield (tons ha⁻¹), light interception (%) and leaf area index of pure and mixture seedings of common vetch, triticale and annual ryegrass (Two-year averages).

Combinations of species	Mixture rate	Forage yield (tons ha⁻¹)	Light interception (%)	Leaf area index
Common vetch	100	13.62 a-d	94.7 a	7.5 ab
Triticale	100	12.53 d	83.7 c	4.7 c
Common vecth:Triticale	75:25	14.48 ab	94.8 a	6.5 b
Common vecth:Triticale	50:50	15.21 a	94.2 a	6.7 ab
Common vecth:Triticale	25:75	14.28 a-c	89.3 b	5.3 c
Annual ryegrass	100	9.56 e	82.1 c	5.4 c
Common vecth:Annual ryegrass	75:25	12.75 cd	95.2 a	7.7 a
Common vecth:Annual ryegrass	50:50	13.56 b-d	95.3 a	7.5 ab
Common vecth:Annual ryegrass	25:75	13.95 a-d	95.2 a	6.7 ab

Means bearing by the same letter in a column were not significantly different at the $p < 0.05$ level using the LSD test.

Light interception of species combinations indicated significant differences. All mixtures of common vetch with annual ryegrass, two mixtures of common vetch with triticale (75% + 25% and 50% + 50% common vetch: triticale) and pure common vetch sowing intercepted light at the highest rate. The lowest light interceptions were realized at pure stands of cereals. Another important point was that the light interception in pure vetch stands was higher than those of cereals (Table 2).This result may be arisen from the higher leaf area index of common vetch than cereals. Stands of species combinations formed leaf area indices which were higher in pure stands of common vetch than in those of cereals. The combinations of common vetch with annual ryegrass yielded stands with greater leaf area indices when compared with those of common vetch-triticale combinations (Table 2).

Crude protein content is one of the most important criteria for fodder quality evaluation. The forage crude protein contents of the treatments varied greatly. The crude protein content of herbage of pure seeding vetch was the highest (21.0%) among all herbages obtained from other stands. Pure triticale seeding produced herbage with the lowest crude protein content. When the ratio of common vetch in any mixture increased, so did the crude protein content of that mixture also (Table 3). Similar results have been reported by some researchers studied similar mixtures of cereals and legumes under intermediate winter crop system (Lithourgidis et al.,2006; Kokten et al., 2009; Yucel and Avci, 2009).

Table3. Crude protein content (%),ADF (%) and NDF (%) contents of pure and mixture of seedings common vetch, triticale and annual ryegrass (Two-year averages).

Combinations of species	Mixture rate	Crude protein content (%)	ADF (%)	NDF (%)
Common vetch	100	21.0 a	32.0 d	55.0 bc
Triticale	100	6.9 f	41.7 a	65.5 a
Common vecth:Triticale	75:25	14.8 c	34.0 cd	46.7 d
Common vecth:Triticale	50:50	12.5 d	38.8 b	58.2 b
Common vecth:Triticale	25:75	9.7 e	42.2 a	65.1 a
Annual ryegrass	100	8.3 ef	35.5 c	53.3 bc
Common vecth:Annual ryegrass	75:25	16.8 b	33.5 cd	54.7 bc
Common vecth:Annual ryegrass	50:50	15.4 bc	33.5 cd	51.5 cd
Common vecth:Annual ryegrass	25:75	12.0 d	35.8 c	52.0 c

Means bearing by the same letter in a column were not significantly different at the $p < 0.05$ level using the LSD test.

ADF is an important criterion for quality forage, and it is expected to be present in low levels. In this study, the forages with low ADF content was produced by the pure stand of common vetch, designating high quality forage and followed by 75% common vetch + 25% triticale, 75% common vetch + 25% annual ryegrass and 50% common vetch + 50% annual ryegrass mixtures. Among all the sowings, the pure triticale stand and 25% common vetch + 75% triticale produced forages higher in ADF content (Table 3). Castro et al. (2000) and Yucel and Avci (2009)

reported that increasing the ratio of triticale in a mixture increases% ADF content of forage.

The results of the NDF values determined in the forages indicated differences among different sowing kinds. The lowest levels of NDF values were determined in the forages produced at 75% common vetch + 25% triticale mixture, while the pure triticale and 25% common vetch + 75% triticale mixture produced forages rich in NDF. Lithourgidis et al. (2006) reported that increasing the ratio of vetch in a mixture increases % NDF ratio.

These discrepancies can be attributable to the different cultivars, harvest stages, ecologies and agricultural systems in which the different studies were conducted.

CONCLUSIONS

This study was performed to determine the properintermediate winter crop system of common vetch combined with triticale and annual ryegrass in Southern Marmara Region and to attribute to the fodder problems and budget of livestock producers. For these purposes, common vetch and two annual cereals (triticale and annual ryegrass) were grown either in pure sowings or mixtures in different combinations such as 75:25, 50:50 and 25:75 (common vetch: cereals). The findings of two-year research are as follows: (a) the highest forage yield (15.21 t ha^{-1}) was produced from 50% common vetch + 50% triticale mixture, (b) the highest light interception values were determined at pure stand of common vetch, 75% common vetch + 25% triticale, 50% common vetch + 50% triticale, 75% common vetch + 25%annual ryegrass, 50% common vetch + 50% annual ryegrass and 25% common vetch + 75% annual ryegrass mixtures, (c) the highest leaf area index was obtained from 75% common vetch + 25% annual ryegrass mixture, followed by pure stand of common vetch and 50% common vetch + 50% triticale, 50% common vetch + 50%annual ryegrass and 25% common vetch + 75% annual ryegrass mixtures, (d) forages of pure common vetch contained crude protein higher than those of all the other sowings, (e) crude protein contents of forages determined in pure triticale and annual ryegrass sowings were the lowest among all sowings and (f) triticale grown alone produced higher contents of ADF and NDF in its forages than the other sowings, indicating poor quality of forage.

In brief, as an overall evaluation, 75% common vetch + 25% triticale mixture can be recommended for quality and higher forage yield to be grown economically and safely under ecological condition of Southern Marmara Regionfor intermediate winter crop system

ACKNOWLEDGEMENTS

This research was supported by The Scientific Research Projects Unit of Uludag University (Project No: 2009/16, Project Leader: Dr. Emine Budakli Carpici). The author is grateful to The Scientific Research Projects Unit of Uludag University for financial support.

LITERATURE CITED

A O A C 1984 Official methods of analysis association of official agricultural chemists. 15th Ed:Washington D.C.

Albayrak, S.,M. Guler and M. O. Tongel, 2004. Effects of seed rates on forage production and hay quality of vetch-triticale mixtures. Asian Journal of Plant Sciences, 3 (6): 752-756.

Anil, L.,J. Park, R. H. Phipps and F. Miller, 1998. Temperate intercropping of cereal for forage: a review of potential for growth and utilization with particular reference to the UK. Grass and Forage Science. 53:301-317.

Caballero, R. and E. L. Goicoechea, 1986.Utilization of winter cereals as companion crops for common vetch and hairy vetch. Proceedings of the 11th General Meeting of the European Grass Fed., 379–384.

Castro, M. P., F. Sau and J. Pineiro, 2000. Effect of seeding rates of oats (*Avena sativaL*), wheat (*Triticum aestivum* L.) and common vetch (*Vicia sativa* L.) on the yield, botanic composition and nutritive value of the mixture. CIHEAM-Options Mediterraneennes, pp. 207-211.

Goering, M.K. and P. J. Van Soest,1970. Forage fibre analysis. USDA Agricultural Handbook, USA,379,1-20.

Hasar, E. And T. Tukel, 1994. Effects of mixture rate and cutting times on forage yield, quality and seed yield of mixture components in mixtures ofvetch (*Vicia sativa* L.) andtriticale (*Triticum x Secale*) grownthe lowland conditions of Cukurova. Field Crops Congress, 25-29 April 1994, vol. III:104-106, Izmir. In Turkish with English Abstract.

Kokten, K., F. Toklu,I. Atis and R. Hatipoglu, 2009. Effects of seeding rate on forage yield and quality of vetch (*Vicia sativa* L.) - triticale (*Triticosecale* Wittm.) mixtures under east mediterranean rainfed conditions. African Journal of Biotechnology, 8 (20):5367-5372.

Lithourgidis, A.S.,I. B. Vasilakoglou, K. V. Dhima, C. A. Dordas and M. D. Yiakoulaki,2006. Forage yield and quality of common vetch mixtures with oat and triticale in two seeding ratios. Field Crops Research 99:106–113.

Rakeih N., H. Kayya, A. Larb andN. Habib, 2010. Forage yield and competition indices of triticale and barley mixed intercropping with common vetch and grasspea in the Mediterranean Region. Jordan Journal of Agricultural Sciences, 6(2):194-207.

Roberts, C.A.,K. J. Mooreand K. D. Johnson, 1989. Forage quality and yield of wheat-common vetch at different stages of maturity and common vetch seeding rate. Agronomy Journal, 81: 57–60.

Thompson, D. J.,D. G. Stoutand T. Moore, 1992. Forage production by 4 annual cropping sequences emphasizing barley under irrigation in southern interior British Columbia. Canadian Journal of Plant Science,72:181-185.

Thompson, E. F.,S. Rihawi and N. Nersoyan, 1990. Nutritive value and yields of some forage legumes and barley harvested as immature herbage, hay and straw in North-West Syria. Experimental Agriculture. 26:49-56.

Turemen, S.,T. Saglamtimur, V. Tansi and H. Baytekin, 1990. Performance of annual ryegrass and common vetch in association under different ratios. Cukurova University Journal of the Faculty Agric., 5 (1): 69-78.In Turkish with English Abstract.

Yucel, C. and M. Avci, 2009. Effect of different ratios of common vetch (*Vicia sativa* L.)-triticale (*Triticosecale* Whatt*)* mixtures on forage yields and quality in Cukurova plain in Turkey. Bulgarian Journal of Agricultural Science, 15(4): 323-332.

Zaffaroni, E. and P. J. Schneiter, 1989. Water-use efficiency and light interception of semi-dwarf and standard-height sunflower hybrids grown in different row arrangements. Agronomy Journal, 81:831-836.

25

YIELD AND HERB QUALITY OF THYME (*Thymus vulgaris* L.) DEPENDING ON HARVEST TIME

Beata KRÓL [*1], *Anna KIEŁTYKA-DADASIEWICZ* [2]

[1]*University of Life Sciences in Lublin, Department of Industrial and Medicinal Plants, POLAND,*
[2]*State Higher Vocational School in Krosno, POLAND*
[*]*Corresponding author: beata.krol@up.lublin.pl*

ABSTRACT

Thyme (*Thymus vulgaris* L) is a medical and aromatic plant intensively used in pharmaceutical, food, and cosmetic industries. Our investigations carried out in 2010-2011 were focused on effect of harvest time on yield and herb quality of two thyme cultivars ('Słoneczko' and 'Deutscher Winter'). Plants were harvested in the first year of cultivation in three periods: 140, 160, and 180 days after sowing (i.e. in the third decade of August, second decade of September and first decade of October). The study showed that harvesting time had a significant effect on the yield and quality of thyme. The delayed harvesting resulted in increased plant height and their mass but decreased quality of herb (lesser quantity of essential oil and thymol). The optimal time of harvest appeared to be 160 days after sowing (i.e. in the second decade of September. Harvest in this time ensured fairly good yield of herb and high its quality. The weather conditions prevailing during the vegetation period had a substantial effect on the yield and quality of herb. The 'Deutscher Winter' cultivar produced higher yields in favourable weather conditions, whereas the 'Słoneczko' cultivar exhibited a more stable yield in adverse atmospheric conditions.

Key words: cultivar, essential oil content, thymol, *Thymus vulgaris*

INTRODUCTION

Thyme (*Thymus vulgaris* L.) is a perennial plant belonging to the Lamiaceae family. It is native to Europe and the Mediterranean basin and adaptable to a wide range of environmental conditions (Stahl-Biskup and Sáez, 2002). Thyme is an aromatic medicinal plant of increasing economic importance in Europe, Asia, North Africa and North America. Essential oil of thyme has been reported to be one of top 10 of essential oils (Maghdi and Maki, 2003)

The pharmacological properties of the plant and of its different extracts, in particular the essential oils, has been thoroughly studied and afforded the many industrial mainly as food and cosmetic additive (Sacchetti et al., 2005) and medical applications (Maghdi and Maki, 2003). The oil was reported to have antimicrobial (bacteria and fungi) (Cetin et al., 2011), expectorant (Büechi et al., 2005) activities, most of which are mediated by thymol and carvacrol. Antispasmodic (Begrow et al., 2010) as well as antioxidant (Haraguchi et al., 1996) activities were also reported for the alcoholic extract of the plant.

Thyme is cultivated in many regions of world and is one of the most important medicinal plant in Europe. Thyme plantations usually survive 2 or 3 year, but in some regions, the plants may freeze during cold winters. Therefore in countries, where the winters are severe (Scandinavian countries, Eastern Europe and Canada) thyme is treated as a one year culture (Dambrauskiene et al., 1999; Letchamo et al., 1999; Galambosi et al., 2002).

The yield of herb of thyme can be influenced by environmental factors, as well as by agricultural practice (Stahl-Biskup and Sáez, 2002). Harvesting time is one of the important factors determining the quantitative and qualitative characteristics of thyme (Badi et al., 2004)

The essential oil content and its chemical composition are the most important characteristics of aromatic herbs. Content of essential oil in dry herb of thyme ranges from 0.3 % (Ozguven and Tansi, 1998) to 4.0 % (Carlen et al., 2010). The content of the essential oil depends from several factors, the most important being genetic characteristics, stage of development (Christensen and Grevsen, 2006; Mewes at al., 2008) environmental conditions (Raal et al., 2005; Alizadeh et al., 2011). Agronomic factors (Kołodziej, 2009; Król, 2009) as well as drying and storage conditions (Calin-Sanchez et al., 2013) exerts an influence on the essential oils content too. There are however scarce information on the effect time of harvest on the chemical composition of essential oil.

The aim of this study was to analyze the impact of time of the harvest on the yield and quality of herb two cultivars of thyme utilized as a one year plantation

MATERIALS AND METHODS

The study was carried out at the Experimental Farm of the University of Life Sciences in Lublin (51° 14′ 53″ N, 22° 34′ 13″ E), Poland, during the 2010 and 2011 growing seasons. The experiment was conducted on brown podzolic soil of loess origin, neutral in reaction (pH$_{KCl}$ - 7.1), characterized by high content of phosphorus (85 mg·P kg^{-1}) and average potassium (166 mg K·kg^{-1}) and magnesium (65 mg Mg·kg^{-1}) contents.

The study comprised two cultivars of thyme ('Słoneczko' and 'Deutscher Winter') and consisted of three harvest times: H1 - 140 days after sowing (DAS) - i.e. in the third decade of August; H2 - 160 DAS - i.e. in the second decade of September; and H3 - 180 DAS - i.e. in the first decade of October. The 'Słoneczko' cultivar seeds were supplied by the Institute of Natural Fibres and Medicinal Plants, Poland and the 'Deutscher Winter' cultivar by Bingenheimer Saatgut AG, Germany. The study was conducted in a randomized block design with four replications. Each experimental plot was 4 m long and 2.5 m wide (10 m^2). Seeds were sown directly into the ground in the first decade of April in rows spaced 30 cm (sowing rate 5 kg seeds ha^{-1}). Mineral fertilizers were applied in the amount of N - 60; P - 22; K - 100 kg per ha^{-1} (phosphorus and potassium were applied as triple superphosphate and potassium chloride before sowing the seeds). Nitrogen fertilization was applied (in two doses - half before sowing and half after plant emergence) as ammonium nitrate. Manual weed control and soil loosening in interrow spaces was performed during the growth of plants.

Before harvest, plant height was determined (20 plants measured from each object). After harvest, the herb was weighed and dried (in a drying house at 35°C) and air-dry mass was determined. Next, the herb was rubbed on sieves (mesh diameter 5mm) and thus dry leaves yield was obtained.

The essential oil was extracted from air-dried leaves (30 g) in a glass Clevenger-type distillation apparatus following European Pharmacopoeia (2004) and subjecting the material to hydrodistillation for 3 h. The assays were conducted in triplicate. The extracted essential oil was stored in a dark glass container at a temperature of – 10°C until the time of chromatographic separation. The quantitative and qualitative determination of the essential oil components was made using a gas chromatograph (Varian 4000 GC/MS/MS) equipped with a FID detector and fused silica capillary column (25 m x 0.2 mm). The carrier gas was helium with the splitting ratio of 1:1000 and capillary flow rate of 0.5 ml min^{-1}. A temperature of 50°C was applied for 1 min, and then the temperature was incremented to 250°C at a rate of 4°C min^{-1}; 250°C was applied for 10 min. The qualitative analysis was carried out on the basis of MS spectra, which were compared with the spectra in the NIST Mass Spectral Library (NIST 2002) and with data available in the literature (Joulain and König, 1998; Adams 2001). The identity of the compounds was confirmed by their retention indices, taken from the literature (Joulain and König, 1998; Adams 2001).

The predicted production yields of thyme oils, in litres per hectare, were calculated from the dry leaves yield and the oil content. The numerical results of the experiment were statistically elaborated by the analysis of variance (ANOVA) for three factors - harvest time H (3), cultivars C (2) and years Y (2). The least significant difference test was used to compare differences in means among treatments at the 0.05 probability level.

The basic climatic factors during thyme vegetation are presented in Table 1. In order to determine the variability of meteorological factors and assess their effect on the yield and quality of the plant, the Sielianinov hydrothermal coefficient (k) was calculated based on the equation: $k = P/0.1\sum t$

P – total rainfall in the given period (mm)

$\sum$ t – total of average daily air temperatures from that period (°C) (Skowera and Puła, 2004).

According to the classification presented by Skowera and Puła (2004), there are distinguished the following conditions: extremely dry (k ≤ 0.4), very dry (k ≤ 0.7), dry (k ≤ 1.0), rather dry (k ≤ 1.3), optimal (k ≤ 1.6), rather wet (k ≤ 2.0), wet (k ≤ 2.5), very wet (k ≤ 3.0), and extremely wet (k >3.0).

Thermal conditions in both years of the field experiment were favourable for thyme; the average air temperature in the vegetation season was markedly higher if compared with that in the long-term period. Moisture conditions were also more favourable than in the long-term period. In terms of precipitation, 2010 was more advantageous, not only because of greater amounts but also more uniform distribution of rainfall. In 2011, symptoms of drought (in May) and severe drought (in August and September) occurred as a result of irregular rainfall.

Table 1. Rainfall, air temperature and hydrothermal coefficient during vegetation of thyme in 2010 and 2011 in comparison with multi-year period (1980-2009)

Year	Month							
	April	May	June	July	August	September	October	Mean or Total
Average temperature of air (^{o}C)								
2010	9.4	14.0	17.8	21.2	19.5	12.3	5.6	14.2
2011	10.8	14.3	18.5	18.1	19.0	15.5	8.0	14.8
mean for long term	8.4	14.1	16.8	18.5	18.1	13.5	8.4	13.9
Monthly of rainfall (mm)								
2010	34.1	108.4	44.8	125.7	106.1	88.9	9.2	517.2
2011	44.9	30.7	55.5	282.9	17.8	5.9	23.8	461.5
mean for long term	41.3	63.2	70.6	83.1	68.4	53.5	24.1	404.2
Sielianinov's hydrothermal coefficient								
2010	1.21	2.58	0.83	1.91	1.75	2.41	0.54	1.60
2011	1.38	0.69	1.00	5.04	0.30	0.13	0.95	1.35
mean for long term	1.64	1.45	1.40	1.44	1.22	1.32	0.92	1.34

Source: Laboratory of Agrometeorology, University of Life Sciences in Lublin, Poland

RESULTS AND DISCUSSION

The height of plants oscillated between 23.4 and 32.8 cm and markedly depended on both: harvesting time and weather conditions during vegetation (Table 2). The plants were higher in 2010, when the rainfall was greater and more uniformly distributed than in 2011. (Table 1, 2). At the consecutive harvests the height of plants increased (the significant differences however were only between the first and third harvest). The height of thyme plants varies to a great extent and depends on the genotype (Mewes at al., 2008), environmental conditions (Galambosi et al., 2002) and cultivation practice (Badi et al., 2004) and oscillates between 10 and 50 cm.

Table 2. The influence of harvest time on height of plants of thyme (cm)

Cultivar	Year	Harvest time			Mean
		H1*	H2	H3	
'Słoneczko'	2010	27.8	30.9	31.4	30.0
	2011	23.4	25.1	26.2	24.9
	Mean	25.6	28.0	28.8	27.5
'Deutscher Winter'	2010	28.2	31.4	32.8	30.8
	2011	25.6	26.4	27.2	26.4
	Mean	26.9	28.9	30.0	28.6
Mean for harvest time		26.3	28.5	29.4	28.1

LSD$_{0.05}$ harvest time (H) - 2.53; cultivar (C) - ns.; years (Y) - 4.25; H x C – ns.; C x Y - 3.71; H x Y - 3.34

ns. - no significant differences
*H1 – 140 days after sowing
H2 – 160 days after sowing
H3 – 180 days after sowing

The herb yield depended significantly on the factors studied (Table 3). In both cultivars, the highest yields of fresh and air-dry herb were achieved from plants harvested in the first decade of October (i.e. 180 DAS), while the lowest – during the first harvest (140 DAS). No significant differences were found in the fresh and air dry herb yields between the cultivars, however, the 'Słoneczko' yields were more stable in adverse weather conditions in 2011 year when there were symptoms of drought (Table 3). Chauhan et al. (2011) in diverse climatical conditions the highest yields of fresh and dry mass of herb obtained when thyme was harvested 115 days after sowing. The latest harvesting (180 DAS) caused decreased fresh and dry herb weight, which resulted from stem drying and defoliation. Rey (1991) considers harvest time of thyme as an important factor affecting yields and their quality. He asserts, that optimal time may differ markedly between particular regions.

Table 3. The influence of harvest time of thyme on yield of fresh and air dry herb and ratio fresh to air dry herb

Cultivar	Year	Yield of fresh herb (t ha⁻¹) Harvest time			Mean	Yield of air dry herb (t ha⁻¹) Harvest time			Mean	Ratio fresh/ air dry herb Harvest time			Mean
		H1*	H2	H3		H1*	H2	H3		H1*	H2	H3	
'Słoneczko'	2010	20.9	24.7	25.8	23.8	5.95	7.81	8.74	7.50	3.51	3.16	2.95	3.21
	2011	17.9	18.9	19.5	18.8	5.44	7.71	8.16	7.10	3.29	2.45	2.39	2.71
	Mean	19.4	21.8	22.7	21.3	5.69	7.76	8.45	7.30	3.40	2.81	2.67	2.96
'Deutscher Winter'	2010	21.7	25.0	25.9	24.2	7.35	8.33	9.25	8.31	3.40	3.00	2.80	3.07
	2011	17.5	18.0	18.9	18.1	5.50	7.11	7.59	6.73	3.18	2.53	2.49	2.73
	Mean	19.6	21.5	22.4	21.2	6.43	7.72	8.42	7.52	3,29	2.77	2.65	2.90
Mean for harvest time		19.5	21.7	22.6	21.3	6.06	7.74	8.44	7.41	3.35	2.79	2.66	2.93

LSD$_{0.05}$

harvest time (H)		2.04	1.15
cultivar (C)		ns.	ns.
year (Y)		4.15	0.92
H x C		2.62	1.10
C x Y		4.26	1.20
H x Y		2.13	1.53

ns. - no significant differences
*See table 2

The ratio of the fresh to dry herb weight in our experiment ranged from 2.39 to 3.51 (Table 3). A high value of this trait is disadvantageous, as this involves higher costs of drying. The delay in harvesting thyme decreased water content in the herb, which resulted in a decline in the proportion between the fresh and air-dry weight of the herb (Table 3). According to literature data, the ratio of fresh to dry herb weight in thyme grown in Poland is 2.4 - 3.5 (Kołodziej 2009; Król, 2009). Similar value of this feature (2.6 – 3.7) was reported in experiments conducted in diverse climatic conditions (Badi et al., 2004).

In our experiment, the dry leaves yields ranged between 2.23 and 4.08 t·ha⁻¹, and corresponded with results obtained by other authors in Poland (Table 4) (Kołodziej, 2009; Król, 2009). Similar results were obtained in Finland whereas in Canada thyme yields were lower and did not exceed 1.5 t·ha⁻¹ (Letchamo et al.' 1999). Stahl-Biskup and Sáez (2002) report that yields of thyme in the first year of vegetation amount from 2 to 2.5 t·ha⁻¹, of dry plant material. Investigations conducted by Dudaš and Böhme (2004) showed that the 'Deutscher Winter' gave higher yield than the 'Słoneczko', which in turn was characterised by a better proportion of leaves. In our experiment, there were no differences in the mean yields of dry leaves between these cultivars, but an interaction between the cultivars and years was reported. 'Deutscher Winter' produced higher yields in favourable weather conditions (2010), whereas in unfavourable weather (2011), yields were significantly reduced (by 32%). Also Galambosi et al. (2002) reported strong susceptibility of this cultivar to adverse atmospheric conditions (decrease of yields by 15-35 %).

Table 4. Yield of dry leaves and contribution of leaves in herb of thyme

Cultivar	Year	Yield of dry leaves (t ha⁻¹) Harvest time			Mean	Contribution of leaves (%) Harvest time			Mean
		H1*	H2	H3		H1*	H2	H3	
'Słoneczko'	2010	2.71	3.57	3.83	3.37	47.0	46.5	44.6	46.0
	2011	2.54	3.02	3.27	2.94	41.5	40.7	38.9	40.4
	Mean	2.63	3.29	3.45	3.12	44.3	43.6	41.8	43.2
'Deutscher Winter'	2010	3.42	3.83	4.08	3.78	45.2	44.7	42.8	44.2
	2011	2.23	2.73	2.79	2.58	39.1	37.2	35.5	37.3
	Mean	2.83	3.28	3.43	3.18	42.2	40.9	39.2	40.8
Mean for harvest time		2.73	3.29	3.44	3.15	43.2	42.3	40.5	41.9

LSD$_{0.05}$
harvest time (H) - 0.41; cultivar (C) - n.s; year (Y) - 0.73
H x C - 0.70; C x Y - 0.84; H x Y - 0.65

* See table 2

Both cultivars produced the lowest yield of dry leaves in the objects with the earliest harvest (140 DAS), delaying harvest resulted in significant increase of yield (Table 4).

An important indication of the quality herb of thyme is the percentage of leaves, as the essential oil is accumulated in oil glandules present in these organs but scarce in the stems (they are discarded in the final stage of raw material preparation) (Sharafzadeh et al., 2010). Słoneczko' cultivar is characterised by the highest amounts of leaves among the European varieties of thyme (Dudaš and Böhme, 2004). Similarly, in our research the 'Słoneczko' cultivar contained greater amounts of leaves (43.2% versus 40.8% in the 'Deutscher Winter' cultivar) (Table 4). The share of leaves in air-dry weight was also related to weather conditions and time of harvest. In both cultivars, the highest share of leaves was noted in herb from the first harvest (140 DAS) and the lowest - at the last harvest (180 DAS- Table 4). This was probably caused by defoliation and lignification of stems. However in experiment of Rey (1991) in different climatical conditions the delay in harvesting thyme increased share of leaves.

In our study, the content of essential oil in the dry leaves ranged from 2.23 to 3.61% (Table 5) and was comparable with results of Galambosi et al. (2002), Zawiślak (2007), Marzec et al. (2010) but higher than stated by Badi et al. (2004), Alizadeh et al. (2011) and Christensen and Grevsen (2006). The content of essential oil in thyme plants can be markedly affected by environmental conditions, time of harvest and other agronomical factors (Stahl-Biskup and Sáez, 2002; Raal et al., 2005). In our study, the content of essential oil in the dry leaves of both cultivars was largely determined by the harvest date and weather conditions during the vegetation season (both cultivars responded similarly - Table 5). The herb was harvested at the first date contained the greatest quantity of essential oil (mean 3.44%) whereas the smallest - at the third date (2.67%). The differences between the first and second harvest were negligible while between second and third – considerable. Decreasing content of essential oil accompanying delayed thyme harvest was reported by Badi et al. (2004) and Rey (1991). Also Christensen and Grevsen (2006), obtained more essential oil in thyme herb harvested in September than in October. Galambosi et al. (2002) however report that shorter growing season affects first at all biomass production, and has only slight influence on essential oil content and composition.

Table 5. The effect of harvest time of thyme on the content and yield of essential oil; percent of thymol in oil

Cultivar	Year	Essential oil content (%)			Mean	Yield of essential oil (kg·ha⁻¹)			Mean	Thymol (%)			Mean
		Harvest time				Harvest time				Harvest time			
		H1*	H2	H3		H1*	H2	H3		H1*	H2	H3	
'Słoneczko'	2010	3.48	3.35	2.87	3.23	94	120	110	108	59.2	57.5	55.1	57.3
	2011	3.32	3.05	2.59	2.99	74	92	80	82	55.8	54.3	52.7	54.3
	Mean	3.40	3.20	2.73	3.11	84	105	95	95	57.5	55.9	53.9	55.8
'Deutscher Winter'	2010	3.61	3.43	2.98	3.34	123	131	122	125	54.1	52.7	50.3	52.4
	2011	3.35	2.64	2.23	2.74	75	72	62	70	53.2	52.1	49.9	51.7
	Mean	3.48	3.04	2.61	3.06	99	101	92	98	53.7	52.4	50.1	51.9
Mean for harvest time		3.44	3.12	2.67	3.08	92	103	94	96	55.6	54.2	52.0	53.9
LSD$_{0.05}$													
harvest time (H)			4.42				8.2				3.1		
cultivar (C)			ns.				n.s				3.7		
year (Y)			4.12				25.7				ns.		
H x C			3.92				13.1				3.5		
C x Y			4.65				28.3				ns.		
H x Y			3.73				15.5				3.9		

* See table 2
ns - no significant differences

In our experiment the mean content of essential oil in both cultivars did not differ markedly (Table 5). In study carried out by Dudaš and Böhme (2004) where there were compared several European cultivars, herb of 'Deutscher Winter' contained more essential oil in comparison with 'Słoneczko'. Seidler-Łożykowska (2007) reported that atmospheric conditions prevailing during the growing season were an important factor determining essential oil content in thyme. In our study, both the herb yield and oil content were higher in 2010, which was characterised by

higher precipitation rates and lower temperatures during the growing season, than those in 2011 (Table 4, 5). Berbeć and Kołodziej (2007) found that appropriate water supply ensured high yields but also contributed to reduction of the quantity of active compounds in herb, in our experiment however this opinion was not confirmed.

In our experiment, the oil yield (a resultant of dry leaf yield and essential oil content) was the highest (irrespective of the cultivar) in objects where the plants were harvested 160 days after seed sowing (Table 5). The mean essential oil yields (regardless of the years and harvesting periods) in both cultivars did not differ significantly. Comparison of the essential oil yields in relation to the year of study revealed greater variation in the 'Deutscher Winter' cultivar, which strongly reacted to the water deficiency in August and September 2011 by a substantial decline in the essential oil yield. In turn, at the favourable precipitation distribution in 2010, this cultivar gave higher yields of essential oil than the 'Słoneczko' cultivar (Table 5).

Content of thymol in essential oil is considered to be one of important factors of thyme quality (Stahl-Biskup and Sáez, 2002). In our study this content ranged from 49.9% to 59.2% and was comparable with that determined by Zawiślak (2007), Galambosi et al. 2002, and Marzec et al. (2010). Asllani and Toska (2003) found it at the lower level (23.1%-50.1%) in Albanian thyme, Horváth et al. (2006) in herbs from Hungary (40,5%) and Badi et al. (2004) in essential oil from thyme grown in Iran (36% - 45%). The chemical composition of thyme's essential oil depends upon several factors, such as environment of growing (Galambosi et al. 2002), development stage (Hudaib et al., 2002) and chemotypes (Thompson et al., 2003). In our study, the content of thymol in the essential oil was dependent on the harvest date and cultivars (Table 5). The highest level of thymol was recorded in herb harvested in the third decade of August, (i.e. 140 days after sowing), and succeeding harvests brought about decrease its content. The similar dependence observed Chauhan et al. (2011), and Hudaib et al. (2002). Senatore (1996) who investigated the essential oil of a wild Italian thyme, Thymus pulegioides L., the highest content of thymol found in herb gathered in May (39.1%) and much smaller in September (20.8%). This does not correspond however with the experiment of Christensen and Grevsen (2006) who found more thymol in herb cultivated thyme harvested in October in comparison with that harvested in September. In our experiment markedly more thymol conatined herb of 'Słoneczko' cultivar in comparison with 'Deuscher Winter' (Table 5). Also Dudaš and Böhme (2004), reported that 'Słoneczko' characterized the highest amounts of thymol in essential oil among compared cultivars.

CONCLUSIONS

Examined in the experiment harvest dates of the thyme (140, 160, and 180 days after sowing) had a significant effect on the herb yield and its quality. Along with delaying harvest, yield increased while quality of herb decreased. The highest yield of herb was recorded at latest date of harvest (180 days after sowing), the herb however characterized low quality, contained smaller quantity of essential oil and lesser amount of valuable thymol. The optimal harvest time of both cultivars compared ('Słoneczko' and 'Deutscher Winter') proved to be 160 days after sowing (mid of September), when fairly good yields were accompanied by reasonable quality of herb. The weather conditions during the vegetation period had a significant effect on the herb yield and a minor impact on the its quality. Higher yields of herb was recorded in 2010 year which characterized abound and uniformly distributed rainfall.

LITERATURE CITED

Adams, R.P. 2001. Identification of essential oil compounds by gas chromatography/quadrupole mass spectroscopy. Allured, Carol Stream, IL, USA.

Alizadeh, A., O. Alizadeh, S. Sharafzadeh, S. Mansoori. 2011. Effect of different ecological environments on growth and active substances of garden thyme. Adv. Envir. Bio. 5: 780-783.

Asllani, U., V. Toska. 2003. Chemical composition of Albanian thyme oil (Thymus vulgaris L.). J. Essent. Oil Res. 15(3): 165-167.

Badi, H.N., D. Yazdani, S.M Ali, N. Fatemeh. 2004. Effects of spacing and harvesting time on herbage yield and quality/quantity of oil in thyme, Thymus vulgaris L. Ind. Crops Prod. 19(3): 231-236.

Begrow, F., J. Engelbertz, B. Feistel, R. Lehnfeld, K. Bauer, E. J. Verspohl. 2010. Impact of thymol in thyme extracts on their antispasmodic action and ciliary clearance. Planta Med. 76(04): 311-318.

Berbeć, S., B. Kołodziej. 2007. The effects of herbs irrigation. Herba Pol. 53(3): 141-145.

Büechi, S., R. Vögelin, M.M. Von Eiff, M. Ramos, J. Melzer. 2005. Open trial to assess aspects of safety and efficacy of a combined herbal cough syrup with ivy and thyme. Res. Complement. Med. 12(6): 328-332.

Calin-Sanchez, A., A. Figiel, K. Lech, A. Szumny, A.A. Carbonell-Barrachina. 2013. Effects of drying methods on the composition of Thyme (Thymus vulgaris L.) essential oil. Drying Technol. 31(2): 224-235.

Carlen, C., M. Schaller, C.A. Carron, J.F. Vouillamoz, C.A. Baroffio. 2010. The new Thymus vulgaris L. hybrid cultivar (Varico 3) compared to five established cultivars from Germany, France and Switzerland. Acta Hort. 860: 161-166.

Cetin, B., S. Cakmakci, R.Cakmakci. 2011. The investigation of antimicrobial activity of thyme and oregano essential oils. Turk. J. Agric. For., 35: 145-154.

Chauhan, N.K., S. Singh, H. Lohani, Z.S. Heider. 2011. Effect of different harvesting time on growth, yield and quality of thyme under the agro-climatic conditions of Doon Valley, Uttarakhand. J. Chem. Pharm. Res. 3 (6): 982-986.

Christensen, L.P., K. Grevsen. 2006. Effect of development stage at harvest on the composition and yield of essential oils from thyme and oregano. Dev. Food Sci. 43: 261-264.

Dudaš, S., M. Böhme. 2004. Characteristics of thyme varieties in comparison. Gartenbauwissenschaftliche Tagung, 39.

Dambrauskiene, E., P. Viskelis. R. Venskutonis. 1999. Effect of nitrogen fertilizers on the yield of first year thyme and its quality. Scientific Works, 18: 107–112.

European Pharmacopoeia. 4th ed. Version 4.08. 2004, Strasbourg: EDQM: 3158-9.

Galambosi, B., Z.S. Galambosi, R. Pessala, I.Hupila, A. Aflatuni, P.K. Svoboda, M. Repcak, 2002. Yield and quality of selected herb cultivars in Finland. Acta Hort. 576: 139-149.

Haraguchi, H., T. Saito, H. Ishikawa, H. Date, S. Kataoka, Y. Tamura, K. Mizutani. 1996. Antiperoxidative components in Thymus vulgaris. Planta Med. (62): 217 – 221.

Horváth, G., L.G. Szabó, É. Héthelyi, É. Lemberkovics. 2006. Essential oil composition of three cultivated Thymus chemotypes from Hungary. J. Ess. Oil Res, 18(3): 315-317.

Hudaib, M., E. Speroni, A.M. Di Pietra, V. Cavrini. 2002. GC/MS evaluation of thyme (Thymus vulgaris L.) oil composition and variations during the vegetative cycle. J. Pharm. Biom. Annl. 29(4): 691-700.

Joulain, D., W.A. König. 1998. The Atlas of Spectral Data of Sesquiterpene Hydrocarbons. E.B. Verlag, Hamburg.

Kołodziej, B. 2009. The effect of plantation establishment method and foliar fertilization on the yields and quality of thyme. Annales UMCS, Sec. E, 64(2): 1-7.

Król, B. 2009. The effect of foliar fertilization with Tytanit and Ekolist in thyme culture. Annales UMCS, Sec. E, 64(1): 1-6.

Letchamo, W., A. Gosselin, J. Hoelzl, R. Marquard. 1999. The selection of *Thymus vulgaris* cultivars to grow in Canada. J. Ess. Oils Res. 11(3): 337-342.

Maghdi, B.H., Z.M. Maki. 2003. Review of common Thyme. J .Med. Plan. 2(7): 1-12.

Marzec M., C. Polakowski, R. Chilczuk, B. Kolodziej. 2010. Evaluation of essential oil content, its chemical composition and price of thyme (*Thymus vulgaris* L.) raw material available in Poland. Herb Pol., 56(3): 37-52.

Mewes, S., H. Krüger, F. Pank. 2008. Physiological, morphological, chemical and genomic diversities of different origins of thyme (*Thymus vulgaris* L.). Gen. Res. Crop Evol. 55.8: 1303-1311.

NIST. 2002. Mass Spectral Library. NIST/EPA/NIH, Gaithersburg, MD, USA.

Ozguven, M., S. Tansi. 1998. Drug yield and essential oil of *Thymus vulgaris* L. as in influenced by ecological and ontogenetical variation. Tur. J. Agric. Forest. 22: 537-542.

Raal, A., E. Arak, A. Orav. 2005. Comparative chemical composition of the essential oil of *Thymus vulgaris* L. from different geographical sources. Herba Pol. 1/2: 10-17

Rey, C. 1991. The effect of date and height of cut in the first year on the yield of sage and thyme. Revue Suisse de Viticulture. Arboric. Hortic. 23: 137–143.

Sacchetti, G., S. Maietti, M. Muzzoli, M. Scaglianti, S. Manfredini, M. Radice, R. Bruni. 2005. Comparative evaluation of 11 essential oils of different origin as functional antioxidants, antiradicals and antimicrobials in foods. Food Chem. 91(4): 621-632.

Seidler-Łożykowska, K., 2007. The effect of weather conditions on essentials oil content in thyme (*Thymus vulgaris* L.) and marjoram (*Origanum majorana* L.). Roczn AR Pozn., Ogrodn 383(41): 605-608.

Senatore, F. 1996. Influence of harvesting time on yield and composition of the essential oil of a thyme (*Thymus pulegioides* L.) growing wild in Campania (Southern Italy). J. Agric. Food Chem. 44: 1327-32.

Sharafzadeh, S., M. Khosh-Khui, K. Javidnia, O. Alizadeh, K. Ordookhani. 2010. Identification and comparison of essential oil components in leaf and stem of garden Thyme grown under greenhouse conditions. Adv. Environ. Biol. 4(3): 520-523.

Skowera, B., J. Puła. 2004. Pluviometric extreme conditionsin spring season in Poland in the years 1971-2000. Acta Agroph. 3, (1): 171- 178.

Stahl-Biskup, E., F. Sáez. 2002. Thyme: the genus Thymus.London: Taylor & Francis.

Thompson, J. D., J. C. Chalchat, A. Michet, Y. B. Linhart, B. Ehlers. 2003. Qualitative and quantitative variation in monoterpene co-occurrence and composition in the essential oil of *Thymus vulgaris* chemotypes. J. Chem. Ecol., 29(4): 859-880.

Zawiślak, G. 2007. Analysis of chemical composition of essential oil in the herb of thyme (*Thymus vulgaris* L.) grown in South-Eastern Poland. Herba Pol., 53(3): 241-244.

26

YIELD AND QUALITY OF BLACK CUMIN (*Nigella sativa* L.) POPULATIONS: THE EFFECT OF ECOLOGICAL CONDITIONS

Nimet KARA[1]*, Duran KATAR[2], Hasan BAYDAR[1]

[1]*Suleyman Demirel University, Faculty of Agriculture, Department of Field Crops, Isparta, TURKEY*
[2]*Eskisehir Osmangazi University, Faculty of Agriculture, Department of Field Crops, Eskisehir, TURKEY*
**Corresponding author: nimetkara@sdu.edu.tr*

ABSTRACT

The experiment was carried out in two vegetation seasons of 2013 and 2014 at the Isparta and Eskisehir ecological conditions. The main purpose of the study was to determine the effects of years and locations on seed yield, some yield-related traits, essential oil content and fixed oil ratio of black cumin populations. In the study, black cumin populations were obtained from Burdur, Tokat, Usak, Eskisehir and Antalya provinces. The experiments were arranged according to a randomized complete-block design with three replications.
The average seed yield varied between 201.0-407.1 kg ha^{-1} in 2013 and 458.9-790.3 kg ha^{-1} in 2014. In compared to locations; while yield components of black cumin populations significantly varied according to locations and years, seed yield in both years were higher in Isparta location on all populations due to higher rainfall than Eskisehir. In compared to populations; the highest seed yield was determined in Usak population and it was followed by Eskisehir population. But the essential oil wasn't detected in Usak population. The highest essential oil content was obtained from Tokat population in both locations and years, it was followed by Eskisehir population. The fixed oil ratios were higher in Tokat, Usak and Antalya populations in both locations and years.

Keywords: Black cumin, Yield, Essential oil, Fixed oil

INTRODUCTION

Black cumin (*Nigella sativa* L.) is a medicinal plant annual belonging to *Ranunculacea* family which grown naturally in Southwest Asia and the Mediterranean Region (Toncer and Kizil, 2004). Black cumin seeds contain protein, alkaloids (nigellicines and nigelledine), saponin (α- hederin) fixed and essential oil (Ozel et al., 2009). The fixed and essential oil of black cumin contain various bioactive molecules such as thymoquinone, thymol, tocopherol, trans retinol and selenium (Sultan et al., 2009).

12 species of black cumin are naturally grown in Turkey's flora. The seeds of *Nigella sativa*, *Nigella damascena* and *Nigella arvensis* are used in folk medicine and as a spice. The black cumin is cultivated commonly in Afyon, Isparta, Burdur and Konya regions in Turkey (Baytop, 1984) and only the species of black cumin (*Nigella sativa* L.) is traded (Kar et al., 2007). Black cumin is intensively used in as uncrushed in bakery products (bread, muffins, biscuits, etc.) and in some cheese (brynza, cottage cheese, etc.) in Turkey (Akgul, 1993). However, the production is limited compared to many other crops and at the level that will correspond to a

very little part of domestic consumption (Kar et al., 2007). Recently, industry facilities based on the evaluation of medicinal and aromatic plants in which included the black cumin. Industrialists emphasize that the demand to fixed oil of black cumin is much and so, the necessity of increasing local production for the supply of raw material (Yilmaz, 2008). For this reason, black cumin is a plant having potential in Turkey in the future. However, there is very little studies about adaptation and quality in our country and the studies are generally related to row spacing and fertilizing. Local population are generally used as seed production and breeding materials by breeders and farmers due to the lack of cultivars. Toncer and Kizil (2004) mentioned that seed yield in black cumin was 828 kg ha^{-1} and essential oil content was between 0.27% and 0.35%. Ozel et al. (2009) reported that seed yield in black cumin at 1406.3-2482.3 kg ha^{-1}, and essential oil content was at 0.24-0.43%. Sener et al. (1985) detected that there was 0.26% lauric, 1.06% myristic, 20.4% palmitic, 1.56% stearic, 4.75% oleic, 64.6% linoleic and 7.18% arachidonic acid in fixed oil of black cumin. Nergiz and Otles (1993) stated that there was about 32% oil in black cumin seeds.

Climate and soil conditions that are the main factors on growth and development of the plants, the physiological growth of the plants, being synthesized of the active substances, the amount and quality of obtained essential oil. Especially, the effect of ecological factors on yield and essential oil of medicinal plants is higher compared to the other plants. Because the quality in medicinal and aromatic plants is as important as yield, the ones below a certain quality cannot be cultivated even if they are very high yield. Therefore, they should only be cultivated in the convenient regions for the ecologies of these plants (Yasar, 2005).

This study was conducted in order to determine the seed yield, some yield components, essential oil content and fixed oil rate of black cumin populations in Isparta and Eskisehir ecological conditions having different altitudes and climate conditions (especially rainfall) where black cumin cultivation is performed.

MATERIALS AND METHODS

Experimental conditions

The experiment was conducted during the growing seasons of 2013 and 2014 years at the Isparta and Eskisehir ecological conditions in Turkey. In this study, black cumin populations were used obtained from Burdur, Tokat, Usak, Eskisehir and Antalya because there is currently no black cumin cultivar registered in Turkey.

The province of Isparta located in the transition region in the inner part of the western Mediterranean region in between the Mediterranean Sea and the Central Anatolia Region bear the climate features (annual average rainfall is 524 mm) of both of the regions with 1050 meters altitude. The province of Eskisehir located in Aegean, Marmara and Central Anatolia Region with 788 meters altitude has semi-arid (annual average rainfall is 349.8 mm) and cool continental climate. Winter months are frosty and summer months are hot and the nights are cool. There are big differences between day and night temperatures.

The experiment land was plowed, cultivated and then prepared for planting with a single pass of a disk-harrow. The experiments were arranged according to a randomized complete-block design with three replicates. Seeds were sown by hand in the first week of March in the ratio of 15 kg ha^{-1} in both the experimental years in Isparta location and were sown in the second week of March in both the experimental years in Eskisehir location. The distance between rows was 20 cm and with in the rows were 5 cm. Each plot area was 7.2 m^2 (6 m x 1.2 m) and consisted of 6 rows. Seeds were sowed at 1-2 cm depths using a dibbler.

Nitrogen and phosphorus fertilizers were applied at a rate of 60 kg ha^{-1} and 40 kg ha^{-1} in the form of ammonium sulphate and P$_2$O$_5$, respectively (Tuncturk et al., 2012).

The total quantity of phosphorus fertilizer was applied at the time of sowing. Total nitrogen fertilization was applied in two equal doses, at the time of sowing and plant height when 15 cm. In both years and locations, the experiments were non-irrigated at any growing stage. All the necessary cultural practices were similarly applied to the plots during vegetation periods in both locations.

Climatic data of the experimental locations

Meteorological data for the growing seasons of locations are shown in Table 1. During the vegetative periods (from March to end of August) in 2013 and in 2014 a total precipitation of 289.5 and 284.2 mm, average temperature of 20.6 and 19.7 °C and an average humidity of 61.0 and 66.1% were recorded in Isparta, a total precipitation of 122.9 and 143.6 mm, average temperature of 16.7 and 16.6 °C and an average humidity of 55.6 and 61.8% were recorded in Eskisehir.

Soil structure

Soil at a depth of 60 cm was sampled before the experiment and subjected to a physicochemical analysis. Data for the years and locations are shown in Table 2.

Yield and its components

When the seeds ripen, plants from 4 rows in the center of each plot were harvested manually. Seed yield (kg ha^{-1}) and its components including plant height (cm), the number of branches (branches plant^{-1}), the number of capsule (capsule plant^{-1}) and 1000-seed weight (g) were determined.

Essential Oil Distillation

In order to determine the seed essential oil content, 100 g powdered black cumin samples in 0.5 L water from each population were extracted by hydro-distilllation for 3 hours using Clevenger apparatus according to the standard procedure described in European Pharmacopoeia (Stainier, 1975) for determining the oil content (v/w, %).

Fixed Oil

4 g dried and powdered black cumin samples from each population were extracted with *n*-hexan for 6 hours using soxhlet apparatus (Buchi Universal Extraction System B-811, Germany) for determining the fixed oil content (%).

Statistical Analysis

All the data were analyzed according to the analysis of variance (ANOVA) using SAS Statistical Package Program, the significant differences between the group means were separated using the LSD (Least Significant Difference) test.

Table 1. Meteorological data of the experiment locations*

Climatic factors	Years	Months						Total or Average
		March	April	May	June	July	August	
ISPARTA								
Precipitation (mm)	2013	25.1	59.9	66.5	34.4	88.2	15.4	289.5
	2014	78.6	44.8	107.0	42.8	0.8	10.2	284.2
	Long Years	52.8	56.6	50.8	28.4	18.4	0.8	207.8
Average Temperature (°C)	2013	7.2	11.9	17.1	20.5	23.0	23.7	20.6
	2014	7.2	11.0	14.5	19.1	23.7	23.2	19.7
	Long Years	5.9	10.8	15.6	20.1	22.3	23.9	19.7
Relative humidity (%)	2013	60.0	58.0	52.0	50.0	44.0	41.0	61.0
	2014	63.7	60.4	62.4	52.7	45.3	45.9	66.1
	Long Years	61.3	64.2	50.3	53.0	45.8	44.5	63.8
ESKISEHIR								
Precipitation (mm)	2013	40.1	30.9	18.5	31.3	2.1	0.0	122.9
	2014	23.1	15.2	27.2	70.6	7.5	0.0	143.6
	Long Years	27.6	43.1	40.0	23.7	13.1	9.2	156.7
Average Temperature (°C)	2013	7.1	10.8	18.2	20.0	21.6	22.4	16.7
	2014	6.5	11.3	16.4	19.9	23.7	21.9	16.6
	Long Years	4.9	9.6	14.9	19.1	22.1	21.8	15.4
Relative humidity (%)	2013	59.8	63.2	51.5	53.9	51.9	53.1	55.6
	2014	69.0	63.7	63.3	64.1	57.8	52.8	61.8
	Long Years	64.2	62.7	59.5	55.2	51.9	53.6	57.9

*Data were taken from Isparta and Eskisehir Regional Meteorological Service

Table 2. Physical and chemical characteristics of soil in experiment locations

Years	Structure	Lime (%)	Available Phosphorus (P_2O_5) (kg ha^{-1})	Available Potassium (K_2O) (kg ha^{-1})	pH	Organic Matter (%)
ISPARTA						
2013	Clay-Loam	10.79	9.61	705	8.1	1.69
2014	Clay-Loam	9.62	9.00	650	7.9	1.38
ESKISEHIR						
2013	Clay-Loam	8.,57	8.95	1290	7.71	1.58
2014	Clay-Loam	7.13	6.69	1050	8.0	1.43

RESULTS AND DISCUSSION

Seed yield, plant height, the number of branches, the number of capsule and 1000-seed weight of black cumin populations in Isparta and Eskisehir locations are presented in Table 3. In this research, differences among of years (except for 1000-seed weight, essential oil content and fixed oil rate), populations, locations (except for 1000-seed weight in loc II of the first year) and population x location interaction were found to be statistically significant for both years (Table 3).

Seed yield

Seed yields of black cumin populations varied according to locations and years. Mean seed yield in the second year (576.0 kg ha^{-1}) were higher than yield in the first year (291.9 kg ha^{-1}) Table 3). These differences resulted from higher rainfall in growing period of plant in the second year according to first year in both locations (Table 1).

In compared to locations, the mean seed yield in Isparta location was higher than Eskisehir location in both years. These differences in the mean seed yield could be explained by the higher rainfall in Isparta location (Table 1 and 3).

In compared to populations, while the highest seed yield was determined in Usak population (626.3 and 940.3 kg ha^{-1}, respectively) in Isparta location in both years, in Eskisehir location the highest seed yield were obtained from Tokat population (244.4 kg ha^{-1}) in the first year and from Usak population (640.3 kg ha^{-1}) in the second year.

The average seed yield in population x location interaction varied between 201.0-407.1 kg ha^{-1} during 2013 and 458.9-790.3 kg ha^{-1} during 2014 (Table 3)

Table 3. Yield, yield components, essential oil content and fixed oil rate of black cumin populations

Populations / Locations	Plant height (cm)						Number of branches (branches plant⁻¹)					
	2013			2014			2013			2014		
	Loc I	Loc II	Average	Loc I	Loc II	Average	Loc I	Loc II	Average	Loc I	Loc II	Average
Burdur	22.7 b	16.1 d	19.4 c	33.5 bc	26.4 b	29.9 b	2.8 b	2.4 cd	2.6 b	3.6 b	4.4 ab	4.0 c
Tokat	34.1 a	22.2 a	28.1 a	47.1 a	24.5 b	35.8 a	3.7 ab	3.5 a	3.6 a	5.2 a	4.1 b	4.6 ab
Usak	28.7 ab	16.6 cd	22.6 b	30.9 c	26.0 b	28.4 b	4.4 a	2.7 bc	3.5 a	4.0 b	5.0 a	4.5 b
Eskisehir	33.1 a	19.8 b	26.4 a	44.1 ab	29.4 a	36.7 a	3.4 ab	2.0 d	2.7 b	5.3 a	4.5 ab	4.9 a
Antalya	26.3 ab	17.8 c	22.0 b	32.5 bc	25.8 b	29.1 b	3.9 ab	3.3 ab	3.6 a	4.1 b	4.8 ab	4.5 b
Mean	29.0 A	18.5 B	23.7 B	36.8 A	26.0 B	31.4 A	3.6 A	2.8 B	3.2 B	4.4	4.6	4.5 A
CV	9.95	2.82	8.72	12.65	3.71	10.73	12.05	9.56	8.57	4.88	6.21	5.60
Lsd	7.90	1.43	2.53	13.04	2.69	4.20	1.19	0.73	0.732	0.59	0.77	0.31
F value	8.10 *	66.67**	17.28**	7.26**	10.01**	6.86*	4.66*	17.76**	4.61*	34.85**	4.46*	21.73**

Populations / Locations	Number of capsule (capsule plant⁻¹)						1000 seed weight (g)					
	2013			2014			2013			2014		
	Loc I	Loc II	Average	Loc I	Loc II	Average	Loc I	Loc II	Average	Loc I	Loc II	Average
Burdur	3.6 b	2.1 c	2.8 c	4.4 b	3.36	3.88 c	2.95 a	2.63	2.79 a	2.84 a	2.80 a	2.82 a
Tokat	5.7 ab	4.0 a	4.8 a	6.7 a	3.63	5.2 ab	2.12 b	2.46	2.29 b	2.10 a	2.73 ab	2.41 b
Usak	7.1 a	2.6 bc	4.9 a	7.7 a	3.80	5.7 a	3.14 a	2.60	2.87 a	2.91 a	2.10 c	2.50 b
Eskisehir	5.6 ab	2.2 bc	3.9 b	7.2 a	3.81	5.5 a	2.49 ab	2.80	2.64 a	2.49 b	2.40 abc	2.44 b
Antalya	6.1 ab	2.7 b	4.4 ab	6.0 ab	3.70	4.8 b	2.64 ab	2.76	2.70 a	2.50 b	2.23 bc	2.37 b
Mean	5.6 A	2.7 B	4.1 B	6.4 A	3.6 B	5.0 A	2.67	2.65	2.66	2.57	2.45	2.51
CV	7.77	7.85	11.31	10.41	8.02	10.4	10.50	7.93	9.31	4.33	8.04	6.38
Lsd	2.73	0.56	0.88	1.82	ns	0.63	0.76	ns	0.30	0.30	0.54	0.19
F value	4.85 *	38.60**	4.57*	10.99**	1.11	6.98*	5.95 *	1.23	3.74 *	25.37**	7.21**	15.67**

Populations / Locations	Seed yields (kg ha⁻¹)						Essential oil content (%)					
	2013			2014			2013			2014		
	Loc I	Loc II	Average	Loc I	Loc II	Average	Loc I	Loc II	Average	Loc I	Loc II	Average
Burdur	204.3 c	197.7 b	201.0 d	494.0 c	493.5 bc	494.0cd	0.17 bc	0.14 c	0.15 c	0.15 c	0.11 b	0.13 c
Tokat	313.0 c	244.4 a	278.7 c	499.6 c	418.1 c	458.9 d	0.50 a	0.78 a	0.64 a	0.48 a	0.66 a	0.58 a
Usak	626.3 a	187.8 b	407.1 a	940.3 a	640.3 a	790.3 a	-	-	-	-	-	-
Eskisehir	468.3 b	192.2 b	330.3 b	690.3 b	558.5ab	624.4 b	0.33 ab	0.64 b	0.48 b	0.32 b	0.57 a	0.44 b
Antalya	274.6 c	210.6ab	242.6 c	485.3 c	540.2 b	512.8 c	0.14 bc	0.12 c	0.13 c	0.10 d	0.11 b	0.11 c
Mean	377.3A	206.5B	291.9B	621.9A	530.2B	576.0A	0.23 B	0.34 A	0.28	0.19 B	0.29 A	0.23
CV	11.64	7.38	11.26	8.02	5.77	7.18	3.14	7.31	9.85	6.54	11.76	10.64
Lsd	120.43	41.81	40.27	136.72	83.89	50.67	0.208	0.068	0.069	0.035	0.095	0.032
F value	44.60**	6.71*	49.08**	49.99**	21.50**	16.41**	18.46**	583.87**	13.98**	802.24**	223.5**	30.85**

Populations / Locations	Fixed oil rate (%)					
	2013			2014		
	Loc I	Loc II	Average	Loc I	Loc II	Average
Burdur	24.0 b	28.0 b	26.0 b	22.6 c	26.3 b	24.5 c
Tokat	32.0 a	32.3 a	32.1 a	30.3 a	31.0 a	30.6 a
Usak	29.3 a	33.6 a	32.5 a	28.6 b	32.3 a	30.5 a
Eskisehir	21.6 b	31.3 a	26.5 b	20.6 d	30.6 a	25.6 b
Antalya	31.3 a	32.0 a	31.6 a	30.2 a	30.6 a	30.4 a
Mean	27.6 B	31.4 A	29.5	26.5 B	30.3 A	28.4
CV	4.09	3.23	3.64	1.19	2.58	2.10
Lsd	3.10	2.78	1.31	0.86	2.15	0.73
F value	49.35**	13.00**	18.32**	621.00**	26.22**	58.74**

Loc 1: Isparta location, Loc 2: Eskisehir location
*, **: significant at P<0.05 and P<0.01 probability levels, respectively, ns: non significant
Means in the same columns followed by the same letters are not significantly different as statistically

These differences among the populations could be explained by the genetic potential of populations, and generally having higher the plant height, the number of branches, the number of capsule and 1000-seed weight in compared to others populations. Usak population is more early than the others (Table 4). Unlike known, early genotypes may be more yield because of its better use of available moisture and they don't coincide to high temperatures in the flowering period. In a study conducted on another plant (wheat) by Ludwig and Asseng (2010) were reported that yield of early genotypes may be higher according to late varieties due to their better use of available moisture in arid and semi-arid climatic conditions. D'Antuono et al. (2002) reported that the major yield limiting factor for *Nigella sativa* genotypes seemed low vegetative growth and reduce of assimilate supply during the pre-anthesis phase short duration of its vegetative cycle. The results obtained from the study were comparable with Tuncturk et al. (2012) findings, but were lower than those of Karaman (1999), Toncer and Kizil (2004), Tuncturk et al. (2005) and Ozel et al. (2009). These differences were due to probably variations in climatic conditions, agronomic practices, genotype and soil properties. In addition to, environmental conditions during pollination at the first stage of seed set, determine the number of capsule and branche (Sadeghi, 2009).

Table 4. Harvest times of populations in experiment locations

Populations	2013		2014	
	Loc I	Loc II	Loc I	Loc II
Burdur	23.07.13	2.08.2013	26.07.14	07.08.2014
Tokat	25.07.13	5.08.2013	29.07.14	10.08.2014
Usak	17.07.13	28.07.2013	20.07.14	07.08.2014
Eskisehir	25.07.13	5.08.2013	29.07.14	10.08.2014
Antalya	23.07.13	5.08.2013	26.07.14	10.08.2014

Loc 1: Isparta location, Loc 2: Eskisehir location

Yield components

The plant height, the number of branches, the number of capsule and 1000-seed weight of populations varied according to locations and years. Generally, the tallest plant height was determined in Eskisehir and Tokat populations, the highest numbers of branches and number of capsule were determined in Usak and Tokat populations and the highest 1000-seed weight was obtained from Usak and Burdur populations in both years and locations (Table 3). Considering the locations and years in terms of total precipitation in Table 1, plant height, the number of branches and the number of capsule were statistically higher in rainy locations and years (Table 3). In the study, the number of branches and the number of capsules increased with the elongation of plant height due to rainfall, and seed yield was positively affected increasing yield components. Ghamarnia and Jalili (2013) reported that black cumin is a medicinal plant sensitive to water stress. The same authors stated that seed yield and its components were decreased rapidly according to increased of water stress during growing periods under semi-arid climatic conditions. Sadeghi (2009) reported that there was a positive correlation between seed yield and aboveground biomass. Ozguven and Sekeroglu (2007) and Tuncturk et al. (2012) stated that yield components such as the number of branches and capsules affects directly seed yield in the black cumin (Ozguven and Sekeroglu, 2007; Tuncturk et al., 2012) .

Essential oil content

The essential oil content of black cumin populations in Isparta and Eskisehir locations are presented in Table 3. In the research, population, location and population x location interaction were found significant. No significant differences between the years were found (Table 3).

The essential oil content of the years was found to be very close to each other with 0.28% in the first year and 0.23% in the second year (Table 3).

In both years (2013 and 2014), the mean essential oil content (0.34% and 0.29%, respectively) in Eskisehir location was higher than Isparta location (0.23% and 0.19%, respectively).

In compared to populations, the highest essential oil content was obtained from Tokat population in both years and locations. The essential oil wasn't obtained from Usak population (Table 3).

The mean essential oil content in population x location interaction varied between 0.13-0.64% in 2013 and 0.11-0.58% in 2014.

Most researchers stated that the essential oil composition of black cumin varied depending on the genotype of the plant, growing, climatic and environmental conditions as well as distillation method (Ozguven and Tansi, 1989; Toncer and Kizil, 2004; Tuncturk et al., 2005; Ozel et al., 2009; Matthaus and Ozcan, 2011). Lei (2004) reported that the essential oil ratio of some essential oil plants was decreased in high rainfall and humidity conditions.

Our results agree with findings of the researchers who reported that essential oil varied between 0.36% and 0.49% (Ozguven and Tansi, 1989; Geren et al., 1997), 0.27% and 0.35% (Toncer and Kizil, 2004), 0.48.0% and 0.51% (Tuncturk et al., 2005), 0.01% and 0.50% (Ozel et al., 2009) and 0.28% and 0.36% (Matthaus and Ozcan, 2011).

Fixed oil rate

In term of fixed oil rate, population, location and population x location interaction were found significant. No significant differences between the years were found (Table 3).

The fixed oil rate of years was found to be very close to each other with 29.5% in the first year and 28.4% in the second year (Table 3).

In compared to locations, average fixed oil rate of Eskisehir location (31.4% and 30.3%, respectively) was higher than Isparta location (27.6% and 26.5%, respectively) in both 2013 and 2014 (Table 3).

In compared to populations, the highest fixed oil rate was obtained from Tokat and Usak populations in the both years and locations.

The mean fixed oil rate in population x location interaction varied between 26.0% and 32.5% during 2013 and 24.5% and 30.6% during 2014.

Many researchers determined that the fixed oil of black cumin vary between 30.4% and 36.4% (Sener et al., 1985; Matthaus and Ozcan, 2011). Ustun et al. (1990) reported that the oil contents of black cumin seeds collected from the Kutahya, Denizli and Konya provinces were 24.4%, 29.5%, and 29.7%, respectively. Our results are similar with the findings of the above-mentioned researchers. The differences between seed oil fixed oil rate can be probably due to genotype, growing, localities, climatic and environmental conditions (Sener et al., 1985; Ustun et al., 1990; Matthaus and Ozcan, 2011; Hendawy et al., 2012).

CONCLUSION

The results obtained from present study indicated that locations (Isparta and Eskisehir) had significant effects on seed yield, some yield components, essential oil and fixed oil contents of black cumin populations. While yield components of black cumin populations varied according to locations and years, seed yield was higher in Isparta location on all populations due to higher rainfall than Eskisehir. Of the average of locations, the highest seed yield was determined in Usak population, followed by Eskisehir population. But, it was not detected essential oil in Usak population. The highest essential oil content was obtained from Tokat population in both locations and years. Tokat, Usak and Antalya populations on fixed oil rate were in the same statistical group at the P<0.01 level on the basis of the LSD test.

Based on the results of the research: Tokat and Eskisehir populations for high seed yield, essential oil and fixed oil contents, and Usak population for high seed yield and fixed oil content could be evaluated in the breeding programme for line and cultivar development.

LITERATURE CITED

Akgul, A. 1993. Spices Science and Technology. Puplication of Food Technology Society. No:15, 451p., Ankara. (in Turkish).

Baytop, T. 1984. Therapy with Medicinal Plants in Turkey. p. 480 (Past and Present). Publications of the Istanbul University. No.3255, Istanbul. (in Turkish).

D'Antuono, LF., A. Moretti and FS. Antonio. 2002. Seed yield, yield components, oil content and essential oil content and composition of Nigella sativa L. and Nigella damascena L. Industrial Crops and Products. 15: 59-69.

Geren, H., E. Bayram and A. Ceylan. 1997. Effect of different sowing dates and phosphorus fertilizer application on the yield and quality characteristics of black cumin (Nigella sativa L.). Proceedings of the Second National Field Crops Congress. pp: 376-380.

Ghamarnia, H. and Z. Jalili. 2013. Water stress effects on different black cumin (Nigella sativa L.) components in a semi-arid region. Int. J. of Agronomy and Plant Production. 4 (4):753-762.

Hendawy, SF., SE. El-Sherbeny, MS. Hussein, KA. Khalid and GM. Ghazal. 2012. Response of two species of black cumin to foliar spray treatments. Australian J. of Basic and Applied Sciences. 6(10): 636-642.

Kar, Y., N. Sen and Y. Tekeli. 2007. Investigation of black cumin (Nigella sativa L.) seeds cultivated in region Samsun and country of Egypt in terms of antioxidant activity. Suleyman Demirel University J. of Science. 2(2): 197-203.

Karaman, A. 1999. A research on effect of different sowing dates on seed yield and quality of black cumin (Nigella damascena L.). Institute of Natural and Applied Sciences, University of Cukurova, M.Sc. Thesis, p: 41 Adana, Turkey. (in Turkish).

Lei, Y. 2004. Research on the introduction and transplanting of aromatic plants from the mediterranean region to Heshuo Xinjiang and Shanghai China. In: Future for Medicinal and Aromatic Plants (Eds.L.E. Craker et al.), Acta Horticulture China. 629: 261-271.

Ludwig, F. and S. Asseng. 2010. Potential benefits of early vigor and changes in phenology in wheat to adapt to warmer and drier climates. Agriculture Systems. 103: 127-136.

Matthaus, B. and MM. Ozcan. 2011. Fatty Acids, tocopherol, and sterol contents of some Nigella species seed oil. Czech J. Food Science. 29(2): 145-150.

Nergiz, C. and S. Otles. 1993. Chemical composition of Nigella sativa L. seeds. Food Chemistry. 48: 259-261.

Ozel, A., U. Demirel, I. Guler and K. Erden. 2009. Effect of different row spacing and seeding rate on black cumin (Nigella sativa L.) yields and some agricultural characters. Harran Uni. J. of Agriculture Faculty. 13(1): 17-25. (in Turkish).

Ozguven, M. and S. Tansi. 1989. A study on determination of optimal sowing dates of Nigella species at Cukurova conditions). In: Proc. VIIth Sym. on Plant Originated Crude Drugs, 19-21 May 1989, Istanbul.

Ozguven, M. and N. Sekeroglu. 2007. Agricultural practices for high yield and quality of black cumin (Nigella sativa L.) cultivated in Turkey. Acta Horticulturae. 756:329-337.

Sadeghi, S., A. Rahnavard and ZY. Ashrafi. 2009. Study importance of sowing date and plant density affect on black cumin yield. Botany Research International. 2(2): 94-98.

Sener, B., S. Kusmenoglu, A. Mutlugil and F. Bingol. 1985. A study with the seed of Nigella sativa. Gazi Uni. J. of Pharmacy Faculty. 2(1): 1-8.

Stainier, C. 1975. Role and functions of the European Pharmacopoeia. Ann Ist Super Sanita. 11(3-4): 211-219.

Sultan, MT., MS. Butt, FM. Anjum, A. Jamil, S. Akhtar and M. Nasir. 2009. Nutritional profile of indigenous cultivar of black cumin seeds and antioxidant potential of its fixed and essential oil. Pakistan J. Botany. 41: 32-40.

Toncer, O. and S. Kizil. 2004. Effect of seed rate on agronomic and technologic characters of Nigella sativa L. International J. of Agriculture & Biology. 3: 529-532.

Tuncturk, M., Z. Ekin and D. Turkozu. 2005. Response of black cumin (Nigella sativa L.) to different seed rates growth, yield components and essential oil. J. of Agronomy. 4(3): 216-219.

Tuncturk, R., M. Tuncturk and V. Ciftci. 2012. The effects of varying nitrogen doses on yield and some yield components of black cumin (Nigella sativa L.). Advances in Environmental Biology. 6(2): 855-858.

Ustun, G., L. Kent, N. Cekin and H. Civelekoglu. 1990. Investigation on the technological properties of Nigella sativa (black cumin) seed oil. J. American Oil Chemical Society. 67: 958-960.

Yasar, S. 2005. Determination of fixed and essential oil contents and soil characteristic of some perennial medical plants that grow naturally in the campus of Cukurova University. Department of Biology Institute of Natural and Applied Sciences University of Cukurova. MSc Thesis, 43p, Adana. (in Turkish).

Yilmaz, G. 2008. New approaches in growing of medicinal and aromatic plants. Graduate Course Notes (unpublished). Gaziosmanpasa University Agriculture Faculty, Field Crops, Tokat. (in Turkish).

27

YIELD STABILITY OF SOME TURKISH WINTER WHEAT (*Triticum aestivum* L.) GENOTYPES IN THE WESTERN TRANSITIONAL ZONE OF TURKEY

Fahri ALTAY*

Bilecik Seyh Edebali University, Bozüyük Vocational School, Bilecik, TURKEY
*Corresponding author: fahri.altay@bilecik.edu.tr

ABSTRACT

The purpose of the study was to evaluate the yield performance and stabilities of certain winter bread wheat cultivars grown widely in the winter wheat regions of Anatolian peninsula. Eight varieties were grown at 10 locations between 2007 and 2011 cropping seasons, in a field trial arranged in Randomized Complete Block Design with 4 replications. The combined analysis of variance was performed for the data obtained. The significant genotype x environment interactions were further investigated by the regression and the ecovalance analyses. It was concluded that cultivars Kate A1 and Mufit bey were found to be the most stable genotypes for all the environments whereas cultivar Gerek-79 was found to be the best performer for under poor soil and weather conditions.

Key words: bread wheat, yield performance, genotype x environment interaction, stability.

INTRODUCTION

Wheat is an important crop for human nutrition in the world with growing areas of the 217.2 million hectares, among cereals (FAO Stat., 2010). According to FAO's (Food and Agriculture Organization of the United Nations) reports, the biggest wheat producing countries in the world are European Union, China, India, United States, and Russia. The FAO also forecasted that the world wheat production in 2012 will be the second highest such 690 million metric tons (FAO, 2012). Wheat consumption in the world has been changed between 645 and 679 million metric tons for the past five years (International Grain Council, 2012).

Wheat is also an important staple crop in Turkey in terms of economy, nutrition and employment. Wheat growing area is 8.5 million hectares and total production is 19-20 million metric tons.

Wheat is grown in the all regions of Turkey, mostly under the rain fed conditions. Therefore, annual production is affected to large extent by the annual and seasonal distribution of precipitation (Turkish State Meteorological Service, 2011). Spatial variations in soil properties and cultural practices also contribute to fluctuations in wheat production.

Success of a wheat breeding program depends on the regional adaptability of the cultivars improved and adaptability of such cultivars in the target environments determined by its tolerance to biotic and abiotic stresses. The most important abiotic stress factor is the shortage of rainfall in the region. There are 3 critical periods of rainfall: Fall rains during the early vigor and tillering, early spring rains during the tiller survival and stem elongation, and late spring rains during grain filling period.

Pfeiffer and Braun (1989) explained sources of yield instability as spatial, temporal, and system dependent variations in the environmental conditions. Almost all breeding programs in the world aim to improve varieties with stable yields. The yield stability is generally grouped as static or dynamic stability. The static stability is defined as the lack of response to environmental variations while the dynamic stability is defined as the average response. Therefore, the static stability is an absolute value independent of the performances of the other cultivars in the trials, while the dynamic stability of a cultivar depends on the mean response of all the cultivars (Tollenaar and Lee, 2002).

Several methods have been developed to analyze and interpret genotype x environment interaction (Lin et al., 1986; Piepho, 1998). These methods can be univariate (based on regression or variance analysis) or multivariate. The earliest approach was the linear regression analysis (Mooers, 1921; Yates and Cochran, 1938). The regression approach was popularized in the 1960s and 1970s (Finlay and Wilkinson, 1963; Eberhart and Russell, 1966 and 1969; Tai, 1971). In this approach, regression graphs are used to predict adaptability of genotypes. Some other univariate stability parameters (based on variance analysis) are the environmental variance (Lin et al., 1986), the Shukla stability variance (Shukla, 1972), Wricke's ecovalence (Wricke, 1962) and the coefficient of variability (Francis and Kanenberg, 1978). As multivariate, the additive main effects and multiplicative interaction (AMMI) model have been extensively applied in the statistical analysis of multi environment cultivar trials (Kempton, 1984; Gauch, 1988; Crossa et al., 1990; Gauch and Zobel, 1997; Akcura et al., 2009; Ilker et al., 2011). In Tai's stability analysis (Tai, 1971), the interaction term is partitioned into two components: the linear response to environmental effects, which is measured by a statistic , and the deviation from the linear response, which is measured by another statistic . A perfectly stable variety has (α, λ) = (-1, 1) and a variety with average stability is expected to have (α, λ) = (0,1).

Gerek 79, released by the Eskişehir Agricultural Research Institute in 1979, has been used as standart variety in the region. Bezostaja 1 is a Russian variety was introduced to the region in 1970, has been also grown for several years. Some newly bred wheat varieties also have been grown in the region.

The main objective of this study was to assess the yield and yield stabilities of some newly developed varieties and compare them with the varieties such as Gerek 79 and Bezostaja 1 widely grown in the western transitional zone of Turkey.

MATERIALS AND METHODS

The genetic materials used in the study are given in Table 1.

Table 1. The genotypes tested and their origins in the study

Variety	Origin	Date of release	Place of release
Bezostaja 1	Russia (USSR)	1970	Eskişehir - Turkey
Gerek 79	Turkey	1979	Eskişehir - Turkey
Kate A1	Bulgaria	1988	Edirne - Turkey
Harmankaya 99	Turkey	1999	Eskişehir - Turkey
Altay 2000	Turkey	2000	Eskişehir - Turkey
Izgi 01	Turkey	2001	Eskişehir - Turkey
Sonmez 01	Turkey	2001	Eskişehir - Turkey
Mufitbey	Turkey	2006	Eskişehir - Turkey

Eight bread wheat varieties were tested in the field trials. Experiments were carried out in a total of 27 environments at 9 different locations from 2007 to 2011. The name of the locations and the number of the experiments conducted at each location were given in Table 2. Among these experiments, 25 trials were conducted under the rainfed conditions, while supplemental irrigation was applied to the other 2 experiments. Since locations vary among the years, each individual year x location combination was considered as a separate environment in the statistical analysis.

Sowings were performed by using a plot drill. Planting dates varied between September 20th and October 30th throughout the trials. Seeding rate was kept uniform such as 500 seeds m^{-2} in all experiments. Experimental layout was, the Randomized Complete Block Design with 4 replications in all trials. Plot sizes were 7 x 6 x 0.2 = 8.4 m^2 at planting and 5 x 6 x 0.2 = 6 m^2 at harvest. 60 kg P2O5 ha^{-1} and 30 kg N ha^{-1} were applied at planting, and additional 50 kg N ha-1 N was given in the early spring. Weeds were chemically controlled in the spring. Supplemental irrigation in two trials was applied; At Cumra the supplemental irrigation was given in the spring, while at Eskisehir the supplemental irrigation was given (60 mm water) at planting to secure emergence.

Statistical Analyses

Each individual trial was subjected to ANOVA. Since the genotype x environment interaction was expected, a combined ANOVA was performed to estimate this interaction. The Duncan's Multiple Range test was used to compare variety means (Steel et al., 1997).

Finlay and Wilkinson's joint regression model (1963) and Eberhart and Russel's method (1966) were applied and the regression coefficients (b), determination coefficients of the regression equations (R^2), and residual MS values (s_d^2) were calculated. To estimate the statistical parameters of regression for stability the proc reg in SAS 9.0 Software were used just by adding a statement to test hypothesis of b=1. Wricke's (1962) ecovalence values were also calculated.

Statistical analyses of the data were performed by using the SAS software (SAS, 2002) and applying General Linear Model procedure.

RESULTS AND DISCUSSION

The average grain yields for the environments are given in Table 2.

Table 2. Average grain yields of 8 winter wheat cultivars grown at 9 locations for five years in the Western Transitional Zone of Turkey

Location	Grain yield (ton ha^{-1})				
	2007	2008	2009	2010	2011
Afyon	—**	1.64	—	—	—
Altıntaş-Kütahya	3.13	2.88	3.69	1.67	6.06
Çumra-Konya (Sup. irr.)*	—	—	5.38	—	—
Emirdağ-Afyon	2.77	—	4.47	2.96	3.98
Eskişehir (Sup. irr.)*	—	—	—	—	3.93
Eskisehir	3.28	2.96	4.84	—	3.25
Hamidiye- Eskisehir	1.47	—	—	—	4.75
Konya	1.89	5.15	4.43	3.85	3.92
Uşak	1.67	2.86	3.05	—	6.02
Average	2.37	3.11	4.31	2.84	4.54

*: Supplementary irrigation.   _**: No trial was conducted

Table 2 shows, the highest mean yield obtained at Altıntaş- Kütahya location in 2011, with 6.06 tons ha^{-1}; while the lowest mean yield was recorded at Hamidiye-Eskişehir location in 2007, with 1.47 tons ha^{-1}.

The combined ANOVA indicated that the genotype x environment interaction was statistically significant. Therefore, the stability analysis could be performed to estimate the overall performance and adaptation of the genotypes (Table 3).

There was significant genotypic variation for grain yield among the 8 standard cultivars used in the stability analysis. Sonmez 01 had the highest mean yield with 3.80 tons ha^{-1}, while the lowest mean yield was obtained from Bezostaja 1, with 3.23 tons ha^{-1} (Table 4 and Figure 1).

Finlay and Wilkinson's joint regression model (1963), Eberhart and Russell's (1966) model, and Wricke's (1962) ecovalence (W$_i$) calculations were applied to the grain yield data obtained from the total 27 environments.

The stability parameters calculated through these different methods are given in Table 4.

Table 3. Combined analysis of variance for grain yield (ton ha^{-1}) of wheat genotypes tested at 9 locations for 5 years in the Western Transitional Region, Turkey.

Source of variation	Degrees of freedom	F value
Environments	26	152.12**
Reps (Environments)	81	2.76**
Genotypes	7	9.61**
Genotypes x environments	182	2.82**

CV%=16.4, R^2 =0.90** , ** significant at the 0.01 probability level

Table 4. The stability parameters estimated for 8 winter wheat genotypes.

Genotypes	a	b	Se	R^2	*P $\leq$	Wi	S_d^2	Mean ton.h^{-1}
Altay	-0.522	1.112	0.06	0.94	0.059	3443	0.13874	3.437
Bezostaja	0.030	0.899	0.06	0.89	0.119	3227	0.17011	3.231
Gerek	0.197	0.885	0.07	0.86	0.114	3337	0.21314	3.349
Harmankaya	-0.182	1.098	0.07	0.90	0.175	3727	0.21626	3.729
Izgi	0.222	0.913	0.07	0.88	0.193	3467	0.18543	3.472
KateA 1	0.011	1.057	0.08	0.87	0.488	3773	0.28059	3.773
Mufitbey	0.081	1.001	0.09	0.84	0.991	3647	0.32079	3.646
Sonmez	-0.109	1.098	0.06	0.93	0.099	3810	0.14273	3.801

*Probobilty of rejection the H_0 : b=1 hypothesis at the P $\leq$ 0 level. Any genotype with b values significantly different from 1 is accepted as nonstable   LSD(0.05): 0.154 ton.ha^{-1}

Table 4 shows that the regression coefficients (b) were not significantly different from 1. The b values ranged between 0.885 (Gerek 79) and 1.112 (Altay 2000). Residual mean square (MS) values (s_d^2) which are indicative of deviations from the regression, were close to 0 (0.13874) for Altay 2000, while Mufitbey had the highest s_d^2 (0.32074). The other genotypes b and s_d^2 values between these values.

According to Finlay and Wilkinson model, b values show the slope of the regression lines indicating adaptability of given genotypes to the range of environments tested in the study. Although high b values are generally indicative of high yield potential (Lin *et al.*, 1986), since those genotypes also generally had low intercept (a) values, they could be considered suitable for specific environments with a yield potential over a given level. Therefore, genotypes with b values close to 1 are preferred since it is indicative of wide adaptation (dynamic stability) provided their mean yield is over the general mean. Mufitbey was found to be the genotype with the b value close to 1 (1.001). Considering it also had a positive intercept value and a mean yield higher than the general mean, this genotype could be accepted to have a wide adaptation over the range of environments used in the study.

On the other hand, the Eberhart and Russell model compares the deviations of genotypic yields from their relative regression lines, indicated by s_d^2 values. This method is generally used to check the reliability of the Finlay and Wilkinson's regression line method. Therefore, these two methods were used together in this study. Since sd2values are desired to be as close as possible to 0, genotypes with the smallest s_d^2 values are considered to have reliable regression equations. However, it is known that s_d^2 values are not totally independent of level of yields, meaning that genotypes with higher yield levels generally tend to give higher s_d^2 values than the low yielding genotypes. Therefore, the genotypes in the study were compared by using the parameters of two methods together. The results indicated that genotypes Altay 2000

and Harmankaya 99 were more suitable to high yielding environments, since they had low intercept values (- 0.522 and – 0.182, respectively) and the highest b values (1.112 and 1.098, respectively). When the genotypes were compared for their wide adaptation (dynamic stability) parameters, the genotypes were Mufitbey and Kate A1 had b values equal to 1 (1.001 and 1.057, respectively). However, Sonmez had the highest mean yield and the lowest s_d^2, indicating its reliability, unless the environmental yield potential is too low since its intercept value was also low. There were also other genotypes with low s_d^2 values, with low yields. Sonmez with low s_d^2 despite its high yield was found to be the most reliable genotype based on the Eberhart and Russell analysis.

Wricke's ecovalence evaluation expressed very similar trend with Eberhart and Russell's s_d^2 values. Since this method indicates the contribution of individual genotypes to the overall genotype x environment interaction, it was expected to give similar results as Eberhart and Russell method. Consequently, it could be concluded that the Finlay and Wilkinson type analysis is a preferable method for assessment of specific or wide adaptation of genotypes, while the other 2 methods could be used to test the reliability of genotypes against yield fluctuations in the varying environments.

Regression coefficients (b) given in Table 4 indicated that, Mufitbey had the b value (1.01) close to 1 followed by Kate A1 (1.057), Izgi 01 (0.913), Harmankaya 99 (1.098) and Sonmez 01 (1.098). The other genotypes, giving the small b values also close to 1, were Bezostaja 1 (0.899), Altay 2000 (1.112) and Gerek 79, previously the most stable cultivar in rain fed conditions, with the lowest b value (0.885). The highest b value of Altay 2000 indicated same possible failure in low yielding environments, while low b values of Bezostaja 1 and Gerek 79 implied their relatively lower ability to respond to improved environmental conditions. The average yields and the b values of the genotypes are shown graphically in Figure 1.

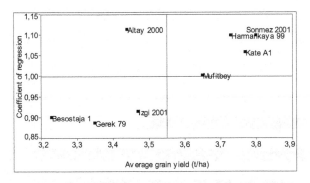

Figure 1. Grain yields and the regression coefficients of the genotypes.

The static stability defined in theory is not valid in practice. Since even in the lowest yielding environments, there would be a certain level of variation in the environmental index and a desirable genotype should be able to respond to the improved conditions to reach to the acceptable yield levels.

Figure 1 shows the genotypes with b values higher than 1 also had higher yields than the average grain yields, with the exception of Altay 2000. Another parameter used in the stability evaluations is the residual mean square (s_d^2) which is a measure of average deviations from the regression line. As suggested by Eberhart and Russell (1966), smaller values of s_d^2 indicates high level of stability. This approach has also been used by Francis and Kannenberg (1978), Motametdi (2011), and Baker (1969). However, this method has been criticized by some researchers (Pfeiffer and Braun, 1989) since the sd2 values are highly yield dependent and this may result in higher s_d^2 values for higher yield levels. Therefore, it has been suggested that the s_d^2 values should be used together with Finlay and Wilkinson's b values to test the reliability of regression equations, rather than using them alone as stability parameters (Linn et al., 1986). The s_d^2 values of the 8 genotypes tested are given in Table 4. Altay 2000 had the lowest and Mufitbey had the highest s_d^2 values, as 0.13874 and 0.32079, respectively. Average yields and sd2 values of the genotypes are shown in Figure 2.

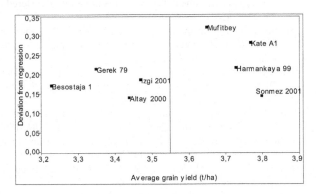

Figure 2. Grain yields and deviations from the regression of 8 wheat genotypes averaged over 27 environments.

After the comparison of the 8 genotypes by using the stability parameters estimated, Kate A1 and Mufitbey

appear to be the most stable genotypes, followed by Sonmez 01 and Izgi 01. Since, Izgi 01 with low grain yield was replaced by Harmankaya 99 in the high yielding environments. Gerek 79, Bezostaja 1 and Altay 2000 were found to be the least stable genotypes at the locations tested.

Figure 3 shows the grouping of genotypes according to Finlay and Wilkinson (1963) method. It could be seen in figure 3, Sonmez 01, Harmankaya 99, Kate A1 and Mufitbey were in the same stability group, showing good adaptation to all the environments tested in the study. Izgi 01 was moderate, Altay 2000 could have specific adaptation to the high yielding environments, Gerek 79 had specific adaptation to the low yielding environments, and Bezostaja 1 showed poor adaptation to all the environments tested.

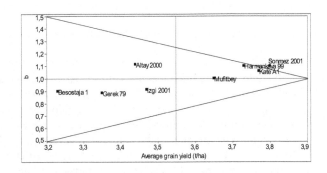

Figure 3. Regression coefficients of genotypes regressed over average grain yields.

Adaptation boundaries of the genotypes, based on expected yields for different environmental indices, calculated from regression equations are shown in Figure 4 indicated the boundaries of each genotype. Genotypes Kate A1 and Mufitbey were adaptable to all the environments used in the study. Kate A1, a cultivar introduced from Bulgaria, has been gaining acreage in the region. Mufitbey, a newly released genotype, was found to be suitable in the environments up to 5.432 tons ha^{-1} yield potential. Sonmez 01 was also found to be adaptable to environments with higher than 1.436 tons ha^{-1} index value, while Harmankaya was good in the environments with higher than 2.183 tons ha^{-1} environmental index value. On the other hand, Altay 2000 was found to have specific adaptation to high yielding environments (over 5.432 tons ha^{-1}). Izgi 01 and Gerek 79, on the contrary, were found to be suited to low yielding environments (lower than 2.183 and 1.436 tons ha^{-1} index values, respectively). Bezostaja1 did not show good adaptation to any environment in this specific set of experiments. Since it has high yield capacity and bread making quality, Bezostaja1 has been widely grown after 1970's. Later, Bezostaja1 could not compete with the new varieties especially in the high fertile transitional zone of Turkey. This study confirmed the calculated yield of Bezostaja1, by using a and b parameters for different environmental indexes, dropped back the yields of the newly developed varieties.

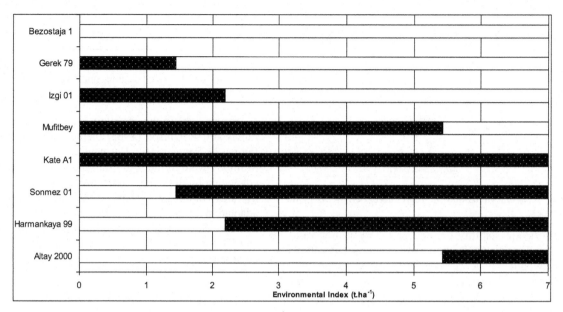

Figure 4. Adaptation boundaries of the genotypes (tons ha^{-1}).

It could be concluded that the stability parameters used were specific to the group of environments and genotypes used in this study. Therefore, their validity will be dependent on the suitability to the target region. Gerek 79 was found to be the most stable cultivar in several studies from 1980's to 1990's, when tested among a different group of genotypes. It was also concluded that newly developed cultivars appear to be superior to Gerek 79 not only in potential yield but also in yield stability.

LITERATURE CITED

Akcura, M.,Y. Kaya, S. Taner, 2009. Evaluation of durum wheat genotypes using parametric and nonparametric stability statistics. Turkish J.of Field Crops 14: 111-122

Baker, R.J., 1969. Genotype-environment interactions in yield of wheat. Canad. J. of Plant Sci. 49: 743-751.

Crossa, J., R.W. Zobel, H.G. Gauch, 1990. Additive and multiplicative interaction analysis of two international maize cultivar trials. Crop Sci. 30: 493–500.

Eberhart, S.A., W.A., Russel, 1966. Stability parameters for comparing varieties. Crop Sci. 6: 636-40.

Eberhart, S.A., W.A. Russell, 1969. Yield stability for a 10-line diallel of single-cross and double-cross maize hybrids. Crop Sci. 9, 357–361.

FAO Stat., 2010. http://faostat.fao.org (accessed 08.11.2012)

FAO, 2012. Crop prospects and food situation. (available at http://www.fao.org/docrep/015/al990e/al990e00.pdf).

Finlay, K.W., G.N. Wilkinson, 1963. The analysis of adaptation in a plant breeding programme. Austral. J. Agric. Res. 14: 742-754.

Francis, T.R., L.W. Kannenberg, 1978. Yield stability studies in short season maize. I. A descriptive method for grouping genotypes. Canad. J. of Plant Sci. 58: 1029-1034.

Gauch, H.G., 1988. Model selection and validation for yield trials with interaction. Biometrics 44, 705–715.

Gauch, H.G., R.W. Zobel, 1997. Identifying mega-environments and targeting genotypes. Crop Sci. 37, 311–326.

Ilker, E., H. Geren, R. Unsal, I. Sevim, F. A. Tonk, M. Tosun, 2011. AMMI- biplot analysis of yield performances of bread wheat cultivars grown at different locations. Turkish J. Of Field Crops 16(1): 64-68.

International Grain Council, 2012. (available at http://www.igc.int/en/downloads/gmrsummary/gmrsumme.pdf)

Kempton, R.A., 1984. The use of biplots in interpreting variety by environment interactions. J. of Agric. Sci. (Cambridge) 103: 123–135.

Lin, C.S., M.R. Binns, L.P. Lefkovitch, 1986. Stability analysis: where do we stand? Crop Sci. 26: 894–900.

Mooers, C.A., 1921. The agronomic placement of varieties. J. of the American Soc. of Agron. 13: 337–352.

Motamedi, M., 2011. Analysis for stability and location effect on grain yield of bread wheat genotypes. Advances in Environ. Biology. 5(10): 3120- 3123,

Pfeiffer, W.H, Braun H.J.,1989. Yield stability in bread wheat. In J.R. Anderson and P.B Hazel, eds. Variability in Grain Yields. Washington D.C.: John Hopkins Univ. and the Int. Food Policy Res. Inst.

Piepho, H.P., 1998. Methods for comparing the yield stability of cropping systems - a review. J. of Agron. and Crop Sci. 180: 193–213.

SAS, 2002. SAS Institute Inc., Cary, NC, USA.

Shukla G.K., 1972. Some statistical aspects of partioning genotype-environmental components of variability. Heredity, 29: 237-245.

Steel, R. G. D., Torrie, J. H., Dickey D. 1997. Principles and Procedures of Statistics: A Biometrical Approach, Third Edition, New York: McGraw-Hill, Inc.

Turkey Statistic Institute (TSI), 2010. Statistical Indicators 1923-2009, Turkish Statistical Indicators, Turkish Statistical Institute Printing Division) Ankara.

Tai, G.C.C., 1971. Genotypic stability analysis and its application to potato regional trials. Crop Sci. 11: 184–190.

Tollenaar, M, E.A. Lee, 2002. Yield potential, yield stability and stress tolerance in maize. Field Crops Res. 75: 161-169.

Turkish Grain Board, 2010. 2009 Cereal Report, 2010 Turkish Grain Board, May, Ankara, Turkey.

Turkish State Meteorological Service (TSMS), 2011.

Wricke, G., 1962. On a method of understanding the biological diversity in field research. Zeitschrift für Pflanzenzüchtung 47: 92- 146.

Yates, F., W.G. Cochran, 1938. The analysis of groups of experiments. J of Agron. and Crop Sci. 28: 556–580.

28

NITROGEN FERTILIZATION AND NARROW PLANT SPACING STIMULATES SUNFLOWER PRODUCTIVITY

Muhammad AWAIS, *Aftab WAJID, Ashfaq AHMAD, Muhammad Farrukh SALEEM, Muhammad Usman BASHIR, Umer SAEED, Jamshad HUSSAIN, M.Habib-ur-RAHMAN*

University of Agriculture, Department of Agronomy, Agro-climatology Lab., Faisalabad, PAKISTAN
Corresponding author: mauaf26@gmail.com

ABSTRACT

Plant population exerts a strong influence on growth and seed yield because of its competitive effect both on the vegetative and reproductive development. For evaluating the effect of different plant populations (83,333, 66,666 and 55,555 plants ha^{-1}) and N rates (90, 120 and 150 kg N ha^{-1}) on sunflower hybrid (Hysun-33) a field experiment was conducted during spring seasons of 2012 and 2013 using split plot design with three replications at Agronomic Research Area, University of Agriculture, Faisalabad. Results revealed that each N increment enhances the total dry matter, achene yield and its components while oil contents were undesirably affected. During both years, total dry matter, achene yield and harvest index were detected from 83, 333 plants ha^{-1}. Contradictorily, yield components (head diameter, number of achenes per head and 1000-achene weight) were highest in 55,555 plants ha^{-1} treatment. Application of 150 kg N ha^{-1} in 83,333 plants ha^{-1} plant population was best treatment to attain maximum achene yield.

Keywords: correlation, nitrogen, plant population, regression, achene, sunflower, yield.

INTRODUCTION

Pakistan has been facing a persistent shortage of vegetable oils for many years (Munir*et al.*, 2007) due to increasing population (Iqbal*et al.*, 2007) and many other critical factors. The native edible oil production does not equal to increasing demand of population. Thus country is forced to import edible oil in massive quantities involving giant expenditure in foreign exchange. Pakistan will have to seem away from the traditional oilseed crops (cottonseed, rapeseed and mustard and groundnut) to appreciably increase local production of vegetable oil (Khaliq and Cheema, 2005). Rapeseed and mustard oil cannot be used (more than 5%) for ghee formation (P.O.D.B., 2006) as it contains higher concentration of erucic acid. Similarly, cotton oil contents and its fiber are negatively correlated, as is generally grown for its fiber (Govt. of Pakistan, 2009b). The most important non-conventional oil seed crops are sunflower, soybean and safflower (Khaliq and Cheema, 2005). Among different non-conventional oilseed crops sunflower has appeared as an important crop (Badar *et al.*, 2002; Khaliq and Cheema, 2005; Ahmad *et al.*, 2009) that can reduce edible oil import (Khan *et al.*, 2003; Hu *et al.*, 2008). Sunflower can be grown nearly all over Pakistan (Khaliq and Cheema, 2005). In Pakistan area under sunflower crop was 0.28 million hectares with oilseed production of 0.378 million tons and oil production 0.14 million tons(Govt. of Pakistan, 2013).

Reduction in sunflower production was due to many factors like improper plant density, nitrogen fertilizer level and irrigation application etc. Availability of different nutrients is the main factor that controls the sunflower yield (Habib *et al.*, 2006). Fertilizer application boosts the fertility status of soil; enhance the nutrients uptake by the plant that finally increase crop yield (Adediran*et al.*, 2004). However in intensive agriculture, nitrogen is the major nutrient determining sunflower yield (Habib *et al.*, 2006; Abdel-Motagally and Osman, 2010). Nitrogen, the most deficient element in our soils, being an integral part of structural and functional protein, chlorophyll and nucleic acid plays a vital role in crop development. Increased supply of nitrogen results in faster rate of plant growth, productivity and photosynthetic capacity of leaves (Fayyaz-Ul-Hassan *et al.*, 2005). Leaf area, leaf production, light interception and photosynthesis rate were reduced by the nitrogen deficiency (Toth*et al.*, 2002; Nasim*et al.*, 2011). Increased rates of nitrogen reduced the oil contents of sunflower (Ali *etal.*, 2004; Osman and Awed, 2010; Bakht*et al.*, 2010a; Nasim*et al.*, 2011; Hussain*et al.*, 2011) but the reduction was relatively small, and it was over compensated by increase in seed yield (Scheiner*et al.*, 2002). Higherachene weight and number of achene per head were obtained from the application of 150 or 225 kg N ha^{-1} (Ali *et al.*, 2011).Plant population has a crucial significance for achieving highest yield potential of a sunflower crop. Plant population affects the radiation interception, evapotranspiration and finally water use efficiency of growing crop (Saleem*et al.,* 2008; Yasin*et al.,* 2011). Killi, (2004) reported that head diameter, total number of seeds per head, seed yield per head and 1000–seed weight were highest with highest plant population (71420 plants ha^{-1}).

The aim of the present study was to determine effect of different plant population and N rates on growth and yield of sunflower under agro environmental conditions of Faisalabad, Pakistan.

MATERIALS AND METHODS

Field experiment was conducted to study the response of sunflower to different plant populations and N rates. Investigation was carried out for two consecutive years i.e. during 2012 and 2013 at the Agronomic Research Area, University of Agriculture, Faisalabad, Pakistan. The climate of Faisalabad (31°26' N, 73°06' E) region is subtropical and arid to semi-arid at an altitude of 184 meters. The experimental field was quite uniform and its soil was sandy clay loam. The soil was analyzed for its physio-chemical properties; each year before sowing the crop (Table 1).

Table 1. Physiochemical properties of soil (0-30 cm)

Soil characteristics	2012	2013
Physical		
Sand (%)	59	63
Silt (%)	17	15
Clay (%)	24	22
Chemical		
pH	7.8	7.62
Organic matter (%)	1.07	1.09
Total soluble salt (%)	12.31	12.27
Nitrogen (%)	0.05	0.06
Available P (ppm)	6.7	6.8
Available K (ppm)	189	195

Weather summary of experimental site was presented in Fig. 1. The experiment was laid out in a randomized complete block design with split plot arrangement keeping plant population (83333, 66666 and 55555) in main plots and N rates (90, 120 and 150 kg ha^{-1}) in sub plots. The experiment was replicated thrice; measuring a net plot size of 3.6 m × 5 m. Sunflower hybrid (Hysun-33) was sown on 1st March during both the years. In each season the experimental field was wetted by heavy irrigation (locally called *rouni*) and seedbed was prepared by one deep ploughing and two cultivations, each followed by planking. Planting was done on 60 cm apart ridges in both experiments. Plant to plant distance was kept according to

plant population treatments (20 cm for 83,333, 25 cm for 66,666 and 30 cm for 55,555plants ha^{-1}). Phosphorus and potassium was used @ 60 kg ha^{-1} each in all plots. Nitrogen, phosphorus and potassium were applied in the form of Urea, DAP and SOP. One third dose of nitrogen and full of the P and K was applied at sowing. Remaining N was used in two splits; 1/3 at first irrigation and 1/3 at the flowering stage. All other cultural practices such as hoeing, earthing up and plant protection measures were kept normal for the crop. Table 2 describes the crop husbandry operations during both years. Growth sampling was conducted at a ten days interval. In each plot, four plants were harvested at the ground level. The plants that were cut were separated into leaves and stems. A sub-sample (50 g) of each fraction was taken and dried in forced draft oven at 70°C for at least 48 h up to constant weight to calculate total dry weight (g m^{-2}) at each harvest. The crop growth rate (CGR) was calculated as suggested by Hunt (1978) at each sampling date.

CGR = (W2-W1) / (T2- T1)

Where, W_1 and W_2 were the total dry weights harvested at times T_1 and T_2, respectively. Mean CGR was calculated by averaging all CGR values calculated at each destructive harvest. At final harvest, two rows with a length of 5 m for each plot were harvested. A subsample of 10 plants was obtained to measure the yield components. All the heads including subsample were threshed mechanically to estimate achene yield of entire plot and converted into kg ha^{-1}. For the measurement of achene moisture, the subsample of 500 g was weighed, dried and then weighed again, so the final yield was corrected to 10 % moisture. The sunflower head diameter was measured from the 10 randomly taken plants and average head diameter was determined. Five heads per plot were threshed, their grains counted and average grains per head was computed. Five samples each of thousand achenes were taken from each plot, weighed with an electric balance and averaged. The total dry matter was determined and for this purpose whole plot was harvested, weighed and converted into kg ha^{-1}.Harvest index (HI) was measured as the ratio of achene yield to biological yield, and expressed in percentage.

HI = (Achene yield / Biological yield) x100

Data collected were analyzed statistically by employing the Fisher's analysis of variance technique and significance of treatments means was tested using least significance difference (LSD) test at 5% probability level (steel *et el.,* 1997).

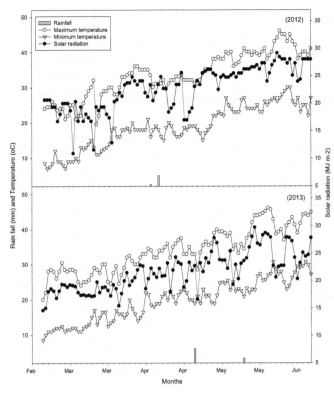

Figure 1. Mean climatic conditions of agro-ecological conditions of Faisalabad.

Table 2. Crop husbandry operations in experiment during 2012 and 2013

Operations	2012	2013
Sowing dates	01.03.2012	01.03.2013
Crop establishment	11.03.2012	11.03.2013
Fertilizer application		
P (DAP) @ 60 kg ha^{-1}	01.03.2012	01.03.2013
K (SOP) @ 60 kg ha^{-1}	01.03.2012	01.03.2013
N (Urea)	N application in three splits according to treatments	
1st Dose	01.03.2012	01.03.2013
2nd Dose	20.03.2012	20.03.2013
3rd Dose	21.04.2012	18.04.2013
Thinning	19.03.2012	19.03.2013
Earthing Up	18.04.2012	15.04.2013
Irrigations		
1	20.03.2012	20.03.2013
2	07.04.2012	05.04.2013
3	21.04.2012	18.04.2013
4	11.05.2012	06.05.2013
5	25.05.2012	21.05.2013
Sampling dates		
1	20.03.2012	20.03.2013
2	30.03.2012	30.03.2013
3	09.04.2012	09.04.2013
4	19.04.2012	19.04.2013
5	29.04.2012	29.04.2013
6	09.05.2012	09.05.2013
7	19.05.2012	19.05.2013
Harvesting	12.06.2012	08.06.2013

RESULTS AND DISCUSSION

Plant population significantly affected the crop growth rate (CGR) during both experimental years (Table 3). The sunflower achieved more CGR (15.11 and 15.32 g m^{-2} d^{-1} in 2012 and 2013, respectively) when plant population was maintained at 83,333 plants ha^{-1}. Lowest mean CGR (13.54 and 13.77 g m^{-2} d^{-1} during 2012 and 2013, respectively) was noted from 55,555 plants ha^{-1} treatment. The treatment P$_2$ (66,666 plants ha^{-1}) was statistically similar to P$_1$ (83,333 plant ha^{-1}) in both study years. The effect of N rates on CGR was also significant (Table 3). More CGR (14.85 g m^{-2} d^{-1}) was observed in plots where N was applied @ 150 kg ha^{-1} in 2012 and it was statistically at par with plots fertilizes with 120 kg N ha^{-1}.

Same trend was noted in next study year and N$_2$ treatment remained at par with N$_3$ treatment. The minimum CGR of 13.93 g m^{-2} d^{-1} was recorded in N$_1$ (90 kg ha^{-1}) during both experimental years. However, this treatment was also at par with N$_2$ (120 kg N ha^{-1}). Increased vegetative growth due to N application finally leads to enhancement in CGR. Nitrogen effects on sunflower mean CGR has been described by Nasim et al. (2011) who recorded 16.01 g m^{-2} d^{-1} mean CGR with application of 180 kg N ha^{-1}. Increase in CGR with nitrogen increment also authenticated the results of Iqbal et al. (2008) who also described positive effects of nitrogen on CGR. The CGR association with total dry matter (R^2= 1 in both years) and achene yield (R^2= 0.98 in 2012 and 0.96 in 2013) was also positive (Fig. 4).

Table 3. Influences of plant population and nitrogen on growth and yield components of sunflower hybrid

Treatments plants ha^{-1}	Crop growth rate (gm-2d-1)			Head diameter (cm)			Number of achenes per head			Thousand achene weight		
	2012	2013	Mean	2012	2013	Mean	2012	2013	Mean	2012	2013	Mean
83,333	15.11 a	15.32 a	15.22	15.2 b	15.0 b	15.1	971.0 b	942.1 b	956.5	39.3 b	38.3 b	38.78
666,666	14.51 a	14.77 a	14.64	17.4 ab	17.0 a	17.16	1030 ab	1009 a	1020	42.6 a	40.1 ab	41.32
55,555	13.54 b	13.77 b	13.66	18.8 a	18.2 a	18.48	1071 a	1024 a	1047.9	42.7 a	41.1 a	41.89
LSD 0.05	0.96	0.96		2.66	1.57		65.71	41.26		2.87	2.09	
Significance	*	*		*	*		*	*		*	*	
N rates kg ha^{-1}												
90	13.9 b	13.4 b	13.7	15.8 b	15.7 b	15.7	976.4 b	946.4 b	961.4	38.1 b	37.9 b	38
120	14.4 ab	14.7 ab	14.6	17.5 a	16.9 ab	17.2	1034 ab	996 ab	1015.7	42.4 a	40.3 ab	41.32
150	14.9 a	15.2 a	15.1	18.1 a	17.6 a	17.8	1061.8 a	1032 a	1047.3	44.1 a	41.3 a	42.69
LSD 0.05	0.7	0.81		1.26	1.42		65.74	57.17		3.95	2.72	
Significance	*	*		**	*		*	*		*	*	
Interaction	NS	NS		NS	NS		NS	NS		NS	NS	
Year mean	14.38	14.62	14.5	17.1	16.72	16.91	1024.4	991.9	1008.1	41.52	39.82	40.67

Mean sharing different letters in a column differ significantly at p≤0.05 *, ** = significant and highly significant respectively; NS = Non-Significant, LSD= least significant difference

Highly significant increase in plant height was recorded with increasing plant population during both study seasons (Table 5). P$_1$ (83,333 plants ha^{-1}) produced taller plants (187.3 cm) as compared to P$_2$ (66,666 plants ha^{-1}) and P$_3$ (55,555 plants ha^{-1}) that attained a plant height of 181.2 and 177.0 cm during season 2012. A similar trend was also seen during 2013. Decreased number of plants per unit area may increase the crop canopy ability to receive the radiation due to more light penetration than denser plants. Increased light intensity disturbs the level of auxin in plants that finally decreased plant height (Allam et al., 2003). Amjad et al. (2011) observed 191 cm tall plants with a plant population of 83,333 plants ha^{-1}. Positive effects of increasing plant population on plant height were also reported by killi (2004). Our results are not consistent with xiao et al. (2006) who described a negative effect of increasing plant population on plant height. Plant height highly significantly increased with increasing N rates (Table 5). Maximum plant height (191.1 cm and 186.4 cm) was observed with the 150 kg ha^{-1} N rate in 2012 and 2013, respectively. Minimum plant height (172.5 and 171.3 cm in 2012 and 2013, respectively) was obtained with the application of 90 kg N ha^{-1}. While the effect of N$_2$ (120 kg ha^{-1}) on plant height was statistically at par with N$_1$ and N$_3$. Nitrogen is an important component of nucleic acid, nucleotide and protein that play a vital role in metabolism.

So increase in plant height may be due to the role of N in enhancing vegetative growth (Al-thabet, 2006). These findings also in concur with Poonia (2000) and Ozer (2004) who reported that nitrogen application increased sunflower plant height.

Head size contributes considerably to achene yield as it affects number of achenes per head. The different plant populations significantly affected the head diameter (Table 3). Maximum head diameter (18.76, 18.19 cm) was noted for 55,555 plants ha^{-1} plant population during 2012 and 2013 respectively. More photosynthetic products are available for sunflower head development due to reduced plant height in plots where plant population of 55,555 plants ha^{-1} was maintained. Head diameter for 66,666 plants ha^{-1} plant population was statistically similar with the plant population of 55,555 plants ha^{-1} during both experimentation years. The head diameter was minimum (15.20 cm in 2012 and 14.99 cm in 2013) in plots where 83,333 plants ha^{-1} plant population was maintained. However, this treatment was at par with P$_2$ (66,666 plants ha$^-$) in 2012. Increase in spacing between plants enhanced head diameter (Ali and Osman, 2004). Similar findings were documented by Amjad et al. (2012) who attained 16.43 cm head size with 83,333 plants ha^{-1} plant population. Among N rates, plots fertilized with 150 kg N ha^{-1} produced maximum head diameter (18.08 cm) in 2012. The situation was similar in 2013 and treatment 150 kg N ha[-1] attained a head diameter of 17.55 cm. A regular reduction in head diameter was seen with decreasing N rates from 150 to 90 kg N ha[-1]. The treatment N_1 (90 kg N ha[-1]) produced minimum head diameter (15.77 and 15.70 cm in 2012 and 2013, respectively) and this treatment was statistically at par with N_2 (120 kg N ha[-1]) in 2013. Nitrogen rate of 150 kg ha[-1] produced a head diameter of 20.13 cm (Amjad*et al.*, 2012).

Number of achenes per head was significantly affected by plant population and N levels (Table 3). Maximum number of achenes per head was observed from 55,555 plants ha[-1] plant population (1071.6 and 1024.1) in 2012 and 2013 against minimum attained from 83,333 plants ha[-1] plant population (971.0 and 942.1) in 2012 and 2013 respectively. Plant population of 55,555 plants ha[-1] produced 9.39% and 8% more number of achenes per head as compared to 83,333 plants ha[-1] plant population during both experimental years. The treatment P_2 (66,666 plants ha[-1]) was statistically at par with P_1 and P_3 in 2012. Denser plants compete more with each other for light, water and nutrition (Amjad*et al.*, 2011) that compel sunflower to produce long plants with small achene number (Iqbal and Ashraf, 2006), head diameter (Tenebe*et al.*, 2008) and achene production (Beg *et al.*, 2007). Number of achenes per head was increased regularly with increasing N rates from 90 to 150 kg N ha[-1] (Table 3). Maximum number of achenes per head (1061.8 in 2012 and 1032.8 in 2013) was attained from N_3 (150 kg N ha[-1]) that was statistically at par with N_2 (120 kg N ha[-1]) that produced 1034.9 and 996.50 achenes per head in 2012 and 2013, respectively. The lowest number of achenes per head (976.4 and 946.4) was noted from N_1 (90 kg N ha[-1]) and this treatment was also statistically similar to N_2 during both study years. Nitrogen contribution towards improvement in source, sink and number of achenes per head was also described by Rondanini*et al.* (2006). The biomass production characters like light capture, leaf area and radiation use efficiency was reduced with N deprivation (Khan *et al.*, 1999) that finally produced less number of achenes per head (Iqbal and Ashraf, 2006). Our results collaborate with findings of Malik *et al.* (2004); Khaliq and Cheema (2005); Beg *et al.* (2007) andAmjad*et al.* (2011).

Different plant populations significantly influenced the 1000-achene weight in our study (Table 3). Thousand achene weight did not coincide with the achene yield as with increasing plant population thousand achene weight decreased gradually in order of $P_3 > P_2 > P_1$ treatment. The lowest plant population (55,555 plants ha[-1]) produced maximum 1000-achene weight (42.68 and 41.10 g) during 2012 and 2013, respectively. While minimum-1000 achene weight (39.27 and 38.29 g in 2012 and 2013, respectively) was attained from P_1 (83,333 plants ha[-1]). Reduction in 1000-achene weight with increasing plant population was also confirmed from previous studies (El-Mohandes*et al.*, 2005; Al-thabet, 2006). Differences in 1000-achene weight among different N levels were also significant (Table 3). Thousand achene weight was increased with increasing N rates and maximum 1000-achene weight (44.09 g) was obtained from 150 kg N ha[-1] in 2012. The trend was similar in 2013. The minimum 1000-achene weight of 38.07 in 2012 and 37.92 in 2013 was noted from 90 kg N ha[-1]. The optimum N supply ensures the improvement in source efficiency and sink capacity. Nasim*et al.* (2012) recorded a 1000-achene weight of 49 g with 180 kg N ha[-1]. Increase in 1000-achene weight in response to N application was also described by Anwar-ul-Haq*et al.* (2006).

Table 4 illustrated that different plant populations have highly significant effect on achene yield during two years experiment (2012 and 2013). Maximum achene yield (3164.4 kg ha[-1] in 2012 and 3030.1 kg ha[-1] in 2013) was attained from the plant population of 83,333 plants ha[-1] and it was followed by the 66,666 plants ha[-1] plant population that produced achene yield of 2890.3 and 2711.1 kg ha[-1] in 2012 and 2013, respectively. The minimum achene yield (2413.2 and 2289.6 kg ha[-1]) was produced by 55,555 plants ha[-1] plant population in 2012 and 2013 respectively. A regular increase in achene yield with increasing plant population (55,555 to 83,333 plants ha[-1]) was mainly due to increase in number of plants at harvest as described byAllam*et al.* (2003). Amjad*et al.* (2012) attained achene yield of 3662 kg ha[-1] with plant population of 83333 plants ha[-1].Highly significant and significant enhancement in achene yield was observed with increasing N rates during 2012 and 2013, respectively (Table 4). Application of 150 kg N ha[-1] gave maximum (3066 and 2859.8 kg ha[-1] in 2012 and 2013, respectively) achene yield. Minimum achene yield of 2537.5 kg ha[-1] in 2012 and 2451.6 kg ha[-1] in 2013 was resulted from 90 kg N ha[-1]. The treatment N_2 produced achene yield (2864.4 kg ha[-1] in 2012 and 2719.4 kg ha[-1] 2013) statistically similar with N_3. Significant increase in growth (TDM, CGR) and yield components (head diameter and number of achenes per head) with nitrogen increment ultimately produced significant increase in achene yield (Osman and Awed, 2010). Al-Thabet*et al.* (2006) and Awais*et al.* (2013) recorded a seed yield of 3952 kg ha[-1] and 3196.8 kg ha[-1] from the 150 kg N ha[-1] application, respectively. The regression lines drawn revealed a positive association of total dry matter with achene yields (Fig. 3).

Table 4. Influences of plant population and nitrogen on growth and yield of sunflower hybrid

Treatments	Achene yield (kg ha⁻¹)			Total dry matter (kg ha⁻¹)			Harvest index (%)			Oil contents (%)		
Plants ha⁻¹	2012	2013	Mean	2012	2013	Mean	2012	2013	Mean	2012	2013	Mean
83,333	3164 a	3030 a	3097	10197 a	9930 a	10064	31.2 a	30.5 a	30.86	40.63	39.71	40.17
666,666	2890 b	2711 b	2800	9670 a	9460 a	9565	29.9 a	28.8 ab	29.36	40.69	39.81	40.25
55,555	2413 c	2289 c	2351	8824 b	8636 b	8730	27.4 b	26.6 b	26.98	40.74	40.01	40.38
LSD 0.05	244.12	244.12		832.69	798.83		2.23	2.37		1.67	2.0	
Significance	**	**		*	*		*	*		NS	NS	
N rates kg ha⁻¹												
90	2537 b	2451 b	2494	9167 b	8760 b	8963	27.6 b	28.0	27.8	41.8 a	41.2 a	41.52
120	2864 a	2719 a	2791	9552 ab	9425 a	9488	30.2 a	28.8	29.5	41.1 ab	39.7 ab	40.39
150	3066 a	2859 a	2962	9971 a	9842 a	9907	30.7 a	29.1	29.9	39.2 b	38.6 b	38.89
LSD 0.05	210.88	262.06		612.92	664		2.44	3.58		2.12	2.04	
Significance	**	*		*	*		*	NS		*	*	
Interaction	NS	NS		NS	NS		NS	NS		NS	NS	
	2822.6	2676.9	2749.8	9564	9342.5	9453.3	29.49	28.63	29.07	40.69	39.84	40.27

Mean sharing different letters in a column differ significantly at p≤0.05 *, ** = significant and highly significant respectively; NS = Non-Significant, LSD= least significant difference

TDM accumulation was less upto 2ⁿᵈ harvest (30 DAS) then a rapid increase in TDM was observed until physiological maturity (Fig. 2). The total dry matter (TDM) was significant among different plant population and N rates (Table 4). The data revealed that TDM followed the same pattern for plant population in the whole study. P_1 (83,333 plants ha⁻¹) achieved maximum TDM of 10197.7 kg ha⁻¹ in 2012 and 9930.6 kg ha⁻¹ in 2013. However P_1 (83,333 plants ha⁻¹) was statistically at par with P_2 (66,666 plants ha⁻¹) in producing TDM (9670.4 and 9460.4 kg ha⁻¹ in 2012 and 2013, respectively). The plant population (83,333 plants ha⁻¹) gained 13.46 % and 13.03 % more TDM over 55,5555 plants ha⁻¹ plant population in 2012 and 2013, respectively. During both experimentation years, P_3 (55,5555 plants ha⁻¹) produced minimum TDM (8824.1

and 8636.5 kg ha⁻¹). Diepenbrock*et al.* (2001) attained 9850 kg ha⁻¹ TDM from 80,000 plants ha⁻¹. Similar results were recorded by Nasrollahi*et al.* (2011).In the year 2012, N rate of 150 kg ha⁻¹ produced higher TDM (9971.8 kg ha⁻¹) which was statistically similar with 120 kg N ha⁻¹ with a mean TDM of 9552.8 kg ha⁻¹ (Table 4). Lowest mean TDM (9167.6 kg ha⁻¹) was noted in N_1 treatment where N was applied @ 90 kg ha⁻¹. However N_1 and N_2 were also at par with each other in producing TDM. Almost similar situation was recorded in 2013 and maximum (9842.6 kg ha⁻¹) and minimum (8760.0 kg ha⁻¹) TDM was found in the same treatments as in 2012. However N_1 and N_2 were statistically different with each other in 2013. Similarly N_2 and N_3 were also statistically at par with each other during 2012 and 2013.

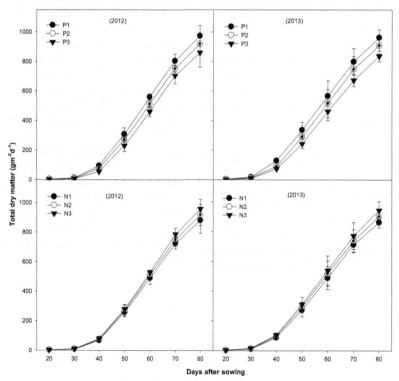

Figure 2. Total dry matter during the crop cycle as affected by various plant populations and N rates

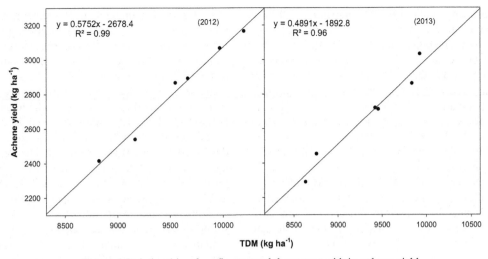

Figure 3. Relationship of sunflower total dry matter with its achene yield

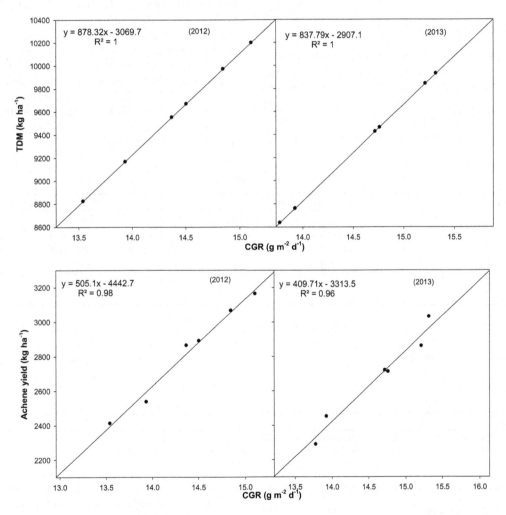

Figure 4. Relationship of sunflower crop growth rate with its total dry matter and achene yield

Harvest index was calculated to estimate efficiency of sunflower to translocate assimilates into economic yield. According to table 4 different plant populations have significant effect on harvest index in both years of study (2012 and 2013). In 2012 highest harvest index (31.21 %) was recorded when 83,333 plants ha^{-1} plant population was maintained but it did not vary significantly with the harvest index gained by the crop having a population of 66,666 plants ha^{-1} (29.92 %). The statistically lowest harvest index (27.36 %) was calculated from 55,555 plants ha[-1] plant population. In 2013 maximum (30.51 %) and minimum harvest index (26.60 %) was calculated with the same plant population as in 2012. However in 2013 P$_2$ (66,666 plants ha[-1]) achieved a harvest index of 28.79 % which was statistically similar with P$_1$ and P$_3$. Harvest index of 34.68 % was obtained at 10,0000 plants ha[-1] plant population (Nasrollahi*et al.*, 2011).Nitrogen effects on harvest index were significant in 2012 but non-significant in 2013 (Table 4). During 2012, statistically maximum (30.66 %) and minimum harvest index (27.57 %) was attained when sunflower was supplied with 150 and 90 kg N ha[-1], respectively. Same treatments gave maximum (29.12 %) and minimum harvest index (27.97 %) in 2013 but these two treatments did not differ significantly. Khaliq and Cheema (2005) calculated harvest index of 33.32 % from 200 kg N ha[-1].

A sunflower crop rich in oil contents of excellent quality is the final objective of a farmer. The effect of plant population on oil contents was non-significant during both years of experimentation (Table 4). The sunflower oil contents varied from 39.71 % to 40.74 % among different plant populations. Non-significant differences for oil contents among various plant populations were also stated by Al-Thabet (2006). Our results are not consistent to outcomes of Killi (2004) and Amjad*et al.*, (2012) who informed significant increase in oil concentration and also to Osman and Awed (2010)

who described reduction in oil contents with increasing plant population.Negative and significant effects of N increment on oil contents are present in Table 4. The sunflower oil contents follow the same pattern for different N rates in both years (2012 and 2013). The oil contents were decreased progressively as N rates increased from 90 to 150 kg ha[-1]. The highest oil contents (41.84 % and 41.21 %) were observed for N$_1$ (90 kh N ha[-1]) and lowest (39.18 % and 38.60 %) for N$_3$ (150 kg ha[-1]) during 2012 and 2013 respectively. The treatment N$_2$ (120 kg ha[-1]) was statistically similar with N$_1$ and N$_3$ in providing oil contents (41.05 % in 2012 and 39.73 % in 2013). As nitrogen is an important constituent of the proteins, so increasing nitrogen rates increased protein synthesis and reduced oil contents (Hussain*et al.*, 2011). Amjad*et al.* (2012) obtained 38 % oil from sunflower with 150 kg N ha[-1]. Significant influences of different plant populations and N rates on oil yield of sunflower were presented in table 5. Maximum oil yield (1285.3 kg ha[-1] in 2012 and 1200.6 kg ha[-1] in 2013) was achieved from the plant population of 83,333 plants ha[-1]. The minimum oil yield (982.2 and 914.7 kg ha[-1]) was produced by 55,555 plants ha[-1] plant population in 2012 and 2013 respectively.The maximum oil yield (1204.6 and 1102.7 kg ha[-1] in 2012 and 2013, respectively) was obtained with the application of 150 kg N ha[-1]. Minimum oil yield of 1060.8 kg ha[-1] in 2012 and 1009.7 kg ha[-1] in 2013 was recorded with 90 kg N ha[-1].

Table 5. Influences of Plant population and nitrogen on plant height and oil yield of sunflower

Treatments plants ha[-1]	Plant height (cm)			Oil yield (kg ha[-1])		
	2012	2013	Mean	2012	2013	Mean
83,333	187.3 a	186.3 a	186.8	1285.3 a	1200.6 a	1242.95
666,666	181.2 b	179.6 ab	180.4	1178.3 b	1077.0 b	1127.65
55,555	177.0 c	173.0 b	175.0	981.2 c	914.7 c	947.95
LSD 0.05	3.97	8.35		90.78	82.63	
Significance	*	*		*	*	
N rates kg ha[-1]						
90	172.5 b	171.3 b	171.9	1204.6 a	1102.7 a	1153.65
120	182.0 ab	181.2 ab	181.6	1180.3 a	1079.9 a	1130.1
150	191.1 a	186.4 a	188.8	1060.8 b	1009.7 b	1035.25
LSD 0.05	12.21	10.06		108.83	63.02	
Significance	**	**		*	*	
Interaction	NS	NS		NS	NS	
Year mean	181.9	179.6	180.75	1148.6	1064.1	1106.3

Mean sharing different letters in a column differ significantly at p≤0.05
*, ** = significant and highly significant respectively; NS = Non-Significant, , LSD= least significant difference

Correlation between achene yield and components of yield

Table 6 reveals correlation co-efficients of achene yield with growth and components of achene yield. Achene yield was correlated positively and highly significantly with crop growth rate, harvest index and total dry matter and during both years (2012 and 2013). Pooled data also showed similar results. However correlation of

achene yield with head diameter and number of achenes per head was negative and non-significant. Similarly correlation of 1000-achene weight with achene yield was positive and non-significant in 2012 but negative and non-significant in 2013. Pooled data also indicates a non significant association of 1000-achene weight with achene yield.

Table 6. Correlation of sunflower achene yield with some studied parameters

Parameters	Correlation co-efficient (r)		
	2012	2013	Pooled
Crop growth rate	0.99^{**}	0.98^{**}	1.00^{**}
Head diameter	-0.34^{ns}	-0.49^{ns}	-0.41^{ns}
Number of achenes per head	-0.22^{ns}	-0.22^{ns}	-0.22^{ns}
1000-achene weight	0.14^{ns}	-0.16^{ns}	0.01^{ns}
Total dry matter	0.99^{**}	0.98^{**}	1.00^{**}
Harvest index	0.98^{**}	0.98^{**}	0.99^{**}

ns= Non-significant,**,* significant at 1% and 5% probablity

CONCLUSIONS

Results in table 3,4 & 5 and figure 2,3 & 4 reveals that sunflower (Hysun-33) should be planted in highpopulationof 83,333 plants ha^{-1} (keeping row to row and plant to plant distance of 60 and 20 cm, respectively) with nitrogen level of 150 kg ha^{-1}to attain maximum achene yield. Increased plant populations (66,666 and 83,333 plants ha^{-1}) have a decreasing effect onyield components (head diameter, thousand achene weights and number of achene per head and) however this effect was over-compensated by more number of plants per meter square.

LITERATURE CITED

Abdel-Motagally, F.M.F. and E.A. Osman. 2010. Effect of nitrogen and potassium fertilization on productivity of two sunflower cultivars under east of EI-ewinate conditions. American-Eurasian J. Agric. Environ. Sci. 8: 397-401.

Adediran, J.A., L.B. Taiwo, M.O. Akande, O.J. Idowu, R.A. Sobulo and J.A. Adediran. 2004. Application of organic and inorganic fertilizer for sustainable maize and cowpea yields in Nigeria. J. Plant. Nut. 27(7): 1163-1181.

Ahmad, S., R. Ahmad, M.Y. Ashraf, M. Ashraf and E.A. Waraich. 2009. Sunflower (*Helianthus annuus*L.) response to drought stress at germination and seedling growth stages. Pak. J. Bot.41(2): 647-654.

Ali, E.A. and E.B.A. Osman. 2004. Effect of hill spacing fertigation using drip irrigation system in sandy calcareous soil on the productivity of some safflower genotypes. The 2nd Syrian-Egyptian Conf., El-Baath Univ., Syria.

Ali, H., M. Riaz, A. Zahoor and S. Ahmad. 2011. Response of sunflower hybrids to management practices under irrigated arid-environment. African J. Biotech. 10(14): 2666-2675.

Ali, H., S.A. Randhawa and M. Yousaf. 2004. Quantitative and qualitative traits of sunflower (*Helianthus annus*L.) as influenced by planting dates and nitrogen application. Int. J. Agri. Bio. 6: 410-412.

Allam, A.Y., G.R. El-Nagar and A.H. Galal. 2003. Response of two sunflower hybrids to planting dates and densities. ActaAgronomicaHungarica, 51: 25-35.

Al-Thabet, S.S. 2006.Effect of plant spacing and nitrogen levels on growth and yield of sunflower (*Helianthus Annus* L.). J. King Saud Univ. Agric. Sci. 19: 1-11.

Amjad, A., M. Afzal, I. Rasool, S. Hussain and M. Ahmad. 2011. Sunflower (*Helianthus annuus* L.) hybrids performance at different plant spacing under agro-ecological conditions of Sargodha, Pakistan. International conference on food engineering and biotechnology.IPCBEE vol. 9, IACSIT Press, Singapoore.

Amjed, A., A. Ahmad, T. Khaliq, M. Afzal and Z. Iqba. 2012. Achene yield and quality response of sunflower hybrids to nitrogen at varying planting densities. International

Conference on Agriculture, Chemical and Environmental Sciences (ICACES') Oct. 6-7, Dubai (UAE).

Anwar-ul-Haq, A., M.A. Rashid, M.A. Butt, M. Akhter, Aslam and A. Saeed. 2006. Evaluation of sunflower (*Helianthus annuus*L.) hybrids for yield and yield components in central Punjab. J. Agric. Res.44: 277-285.

Awais, M., A.Wajid, A. Ahmad and A. Bakhsh.2013. Narrow plant spacing and nitrogen application enhances sunflower (*Helianthus annuus*L.) productivity. Pak. J. Agri. Sci. 50(4): 689-697.

Badar, H., M.S. Javed, A. Ali and Z. Batool. 2002. Production and marketing constraintslimiting sunflower production in Punjab (Pakistan). Int. J. Agric. Bio. 4: 267-271.

Bakht, J., M. Shafi, M. Yousaf and Hamid Ullah Shah.2010a. Physiology, phenology and yield of sunflower (autumn) as affected by NPK fertilizer and hybrids. Pak. J. Bot. 42: 1909-1922.

Beg. A., S.S. Pourdad and S. Alipour. 2007. Row and plant spacing effects on agronomic performance of sunflower in warm and semi-cold areas of Iran. Helia, 30: 99-104.

Diepenbrock, W., M. Lang and B. Feil. 2001. Yield and quality of sunflower as affected by row orientation, row spacing and plant density. Die Bodenkultur, 52: 29-36.

El-Mohandes, S., E.A. Ali and E.B.A. Osman. 2005. Response of two sunflower hybrids to the number of NPK fertilizers splittings and plant densities in newly reclaimed soil. Assiut J. Agric. Sci. 36(5): 27-38.

Fayyaz-ul-Hassan, G. Qadir and M.A. Cheema. 2005. Growth and development of sunflower in response to seasonal variations. Pak. J. Bot.37: 859-864.

Govt. of Pakistan.2009b. Economic Survey of Pakistan. Ministry of Food, Agriculture and Livestock, Finance Division, Economic Advisor Wing, Islamabad, Pakistan, PP: 23.

Govt. of Pakistan. 2013. Economic Survey of Pakistan, 2012-2013. Finance Division, Economic Advisory Wing, Islamabad, Pakistan, PP: 23.

Habib, H., S.S. Mehdi, A. Rashid and M. A. Anjum. 2006. Genetic association and path analysis for seed yield in sunflower. Pak. J. Agric. Sci. 43: 136-139.

Hu, J., B. Yue, W. Yuan and B.A. Vick. 2008. Growing sunflower plants from seed to seed in small plots in green house. Helia, 48: 119-126.

Hunt, R. 1978. Plant growth analysis. Edward Arnold, U.K.: 26-38.

Hussain, S.S., F.A. Misger, A. Kumar and M.H. Baba. 2011. Response of nitrogen and sulphur on biological and economic yield of sunflower (*Helianthus annuus*L.). Res. J. Agri. Sci. 2: 308-310.

Iqbal, J., B. Hussain, M.F. Saleem, M.A. Munir and M. Aslam. 2008. Bioeconomics of autumn planted sunflower (*Helianthus annuus*L.) hybrids under different NPK application. Pak. J. Agri. Sci. 45: 19-24.

Iqbal, J., M.A. Malik, B. Hussain and M.A. Munir. 2007. Performance of autumn plantedsunflower hybrids under different planting patterns. Pak. J. Agric. Sci. 4: 587-591.

Iqbal, N. and M.Y. Ashraf. 2006. Does seed treatment with glycinebetaine improve germination rate and seedling growth of sunflower (*Helianthus annuus*L.) under osmotic stress?.Pak. J. Bot.38: 1641-1648.

Khaliq, A. and Z.A. Cheema. 2005. Influence of irrigation and nitrogen management on some agronomic traits and yield of hybrid Sunflower (*Helianthus annuus* L.). Int. J. Agri. Biol. 7: 915-919.

Khan, M. A., K. Ahmad and J. Ahmad. 1999. Effect of Potassium Levels on the Yield of Sunflower. Pakistan J. Biol. Sci. 2(2): 402-403.

Khan, M.S., M.S. Swati, I.H. Khalil and A. Iqbal. 2003. Heterotic studies for various characters in sunflower (*Helianthus annuus* L.). Asian J. of Plant Sci. 2: 1010-1014.

Killi, F. 2004. Influence of different nitrogen levels on productivity of oil seed and confection sunflower (*Helianthus annuus*L.) under varying plant populations. Int. J. Agri. Bio. 6: 594-598.

Malik, M.A., M.F. saleem. M. Sana and A. Rehman. 2004. Suitable level of N, P and K for harvesting the maximum economic return of sunflower. Int. J. Agri. Bio. 6: 240-242.

Munir, M.A., M.A. Malik and M.F. Saleem. 2007. Impact of integration of crop manuring and nitrogen application on growth, yield and quality of spring planted sunflower (*Helianthus annuus*L.). Pak. J. Bot. 39: 441-449.

Nasim, W., A. Ahmad, A. Wajid, J. Akhtar and D. Muhammad. 2011. Nitrogen effects on growth and development of sunflower hybrids under agro-climatic conditions of Multan. Pak. J. Bot. 43: 2083-2092.

Nasim, W., A. Ahmad, H.M. Hammad, H.J. Chaudhary and M.F.H. Munis. 2012. Effect of nitrogen on growth and yield of sunflower under semi-arid conditions of Pakistan. Pak. J. Bot. 44(2): 639-648.

Nasrollahi, H., A.H. Shirani-Rad, A. Khourgami and K. Haghiabi. 2011. Effect of plant density on yield and oil percent of sunflower early cultivars in second culture. Int. J. Sci. Advanced Tech. 1(10): 71-77.

Osman, E.B.O., M.M.M. Awed. 2010. Response of sunflower (*Helianthus annuus*L.) to phosphorus and nitrogen fertilization under different plant spacing at New Valley.Ass. Univ. Bull. Environ. Res. 13(1): 11-19.

P. O. D. B. 2006. Oilseed development strategy.Pakistan oilseed development board.Ministry of food, agriculture and livestock, Islamabad-Pakistan.

Rondanini, D.P., R. Savin, A.J. Hall. 2006. Estimation of physiological maturity in sunflower as a function of fruit water concentration. Europ. J. Agron. 30: 1-15.

Saleem, M.F., B.L. Ma, M.A. Malik, M.A. Cheema and M.A. Wahid. 2008. Yield and quality response of autumn-planted sunflower (*Helianthus annuus*L.) to sowing dates and planting patterns. Can. J. Plant Sci. 88: 101-109.

Scheiner, D.J., F.H. Gutierrez-Boem and R. Lavado. 2002. Sunflower nitrogen requirement and 15N fertilizer recovery in Western Pampas, Argentina. Eu. J. Agron. 17: 73-79.

Steel, R.G.D., J.H. Torrie and D.A. Dickey. 1997. Principles and Procedures of Statistics. A Biometrical Approach.3rd ED. McGraw Hill Book. Int. Co. New York: PP: 400-428.

Tenebe, V.A., U.R. Pal, C.A.C. Okonkwo and B.M. Auwalu. 2008. Response of rainfed sunflower (*Helianthus annuus*L.) to nitrogen rates and plant population in the semi-arid savanna region of Nigeria. J. Agron. and Crop Sci. 177: 207-2015.

Tóth, V.R., I. Mészáros, S.J. Palmer, S. Veres and I. Précsényi. 2002. Nitrogen deprivation induces changes in the leaf elongation zone of maize seedlings. BiologiaPlantarum, 45: 241-247.

Yasin, M., A. Mahmood and Z. Iqbal. 2011. Growth and yield response of autumn sunflower (*Helianthus annuus*L.) to varying planting densities under subtropical conditions. Int. J. Agric. Appl. Sci. 3(2): 86-88.

29

EFFECTS OF DIFFERENT NITROGEN LEVELS ON THE GRAIN YIELD AND SOME YIELD COMPONENTS OF QUINOA (*Chenopodium quinoa* Willd.) UNDER MEDITERRANEAN CLIMATIC CONDITIONS

Hakan GEREN

Ege University, Faculty of Agriculture, Dept. of Field Crops, Izmir, TURKEY
Corresponding author: hakan.geren@ege.edu.tr

ABSTRACT

Quinoa (*Chenopodium quinoa* Willd.), is a pseudo-cereal crop that has been cultivated in the Andean region in South America. The quinoa as a field crop has a great potential in the improvement of food for humans and animals even under the conditions of marginal lands. For getting high crop yields, nutrients in balanced amount are a basic requirement. Experiments were carried out at the Bornova experimental fields of Field Crops Dept. of Agriculture Fac., Ege Univ., Turkey during 2013 and 2014 main crop growing season in order to evaluate the effect of seven nitrogen levels (0, 50, 75, 100, 125, 150 and 175 kg ha^{-1}) on the grain yield and some yield components of cultivar Q-52 of quinoa. Results indicated that the effects of nitrogen treatments and years on all characteristics tested were significant. Nitrogen level of 150 kg ha^{-1} was proved to be the best level for nitrogen supplementation of soil for grain yield (2.95 t ha^{-1}) and crude protein content (16%) of quinoa under Mediterranean ecological conditions of Bornova.

Key words: quinoa, *Chenopodium quinoa*, nitrogen level, grain yield.

INTRODUCTION

Quinoa (*Chenopodium quinoa* Willd.) member of *Chenopodiaceae* family is a seed crop that has been cultivated for thousands of years in the Andean region for its nutritious grain and leaves (Pearsall, 1992). It is an annual broad-leaved plant, also adaptable to the conditions of marginal lands (Rea et al., 1979), allotetraploid (2n=4x=36) (Simmonds, 1971), 1–2 m tall with deep penetrating roots and can be cultivated from sea level to over 4000 m above sea level (Jacobsen, 2003). The crop, a pseudo-cereal, contains gluten-free high-quality protein, so it can play an important role in the diet of people suffering from celiac disease (Kuhn et al., 1996; Doweidar and Kamel, 2011). The protein of quinoa seed is rich in essential amino acids, particularly methionine, threonine and lysine, which are the limiting amino acids in most cereal grains (Bhargava et al., 2007; Comai et al., 2007). The Organization of the United Nations for Food and Agriculture (FAO) has declared the year 2013 as the year of quinoa (Anonymus, 2013.).

Preliminary success of quinoa cultivation in Turkey has opened the avenues to explore the production potential of quinoa comprehensively (Ince Kaya, 2010). Quinoa can be successfully grown on marginal soils showing its very low nutrient requirements (Jacobsen, 2003). However,

quinoa is also highly responsive to soil nitrogen (Erley et al., 2005; Gomaa, 2013). Basra et al. (2014) also stated that nitrogen level of 75 kg N ha^{-1} was proved to be optimum level for nitrogen supplementation of soil for quinoa growth and development to harvest maximum economic yield under ecological conditions of Egypt.

The use of modern commercial fertilizers in agricultural production results in increased crop yields in addition to the effect of better plant nutrition through commercial fertilizers signify themselves not only in increasing yields, but also in an increase in the total biomass production (Finck, 1982). Erley et al. (2005) evaluated the response of quinoa to nitrogen fertilization rates of 0, 80 and 120 kg N ha^{-1}, the results evidenced that quinoa responded strongly to nitrogen fertilization and quinoa yielded between 1.8 and 3.5 t ha^{-1}.

A nitrogen fertilization requirement of quinoa crop is still under study world widely because of variability of ecological conditions. For instance, Erley et al. (2005) reported that quinoa responded well to nitrogen and achieved yield up to 350 kg ha^{-1} at 120 kg N ha^{-1}, grain yield boosted by 94% as compared to control. However, Jacobsen et al. (1994) found increase in yield (averaging 12%) at 80 to 120 kg N ha^{-1}. Quinoa responds to nitrogen application by not only increase the crop growth and yield but also the quality of grain. Thanapornpoonpong (2004) explored the significant effect of different nitrogen rates (0, 0.8 and 1.2 g N pot[-1]) on plant height, grain yield per plant, harvest index, thousand kernel weight, protein content of seed and amino acid profile of amaranth and quinoa.

Jacobsen et al. (1994) found that quinoa grain yield increased with the increasing nitrogen fertilization rate from 40 to 160 kg N ha[-1]. Shams (2012) studied the response of quinoa to five nitrogen fertilization levels of 0, 90, 180, 270 and 360 kg ha[-1]. The author found clearly that grain and biological yields increased gradually with increasing nitrogen levels up to the highest level. Gomaa (2013) informed that the growth traits, seed yield and seed quality of quinoa plant can be improved by the application of inorganic and biofertilizers (nitrobin or phosphorin). Kakabouki et al. (2014) stated that there were significant differences in quinoa crude protein (CP) content among fertilizer treatments (2000 kg ha[-1] cow manure, 100 and 200 kg N ha[-1]) and all fertilization treatments resulted in values higher than those of the control and, the highest CP content (27%) in quinoa was observed for 200 kg N ha[-1] application.

Quinoa is a native field crop to Latin America which has the potential to grow with fewer inputs, water and tolerate a variety of biotic and abiotic stresses as compared to common cereals (Rea et al., 1979; Basra et al., 2014), and is a new introduction to Turkey. Therefore, getting to know the possibility of production of this crop is important not only in Turkey, but also in all over the world. Nevertheless, information on the productivity and quality of quinoa crop under intensive farming management in Mediterranean environment is not well documented. The objective of this experiment was to determine the optimum level of nitrogen fertilizer for getting maximum grain yield of quinoa (Chenopodium quinoa Willd.) under Mediterranean climatic conditions.

MATERIALS AND METHODS

Location of Experiment

The experiment was carried out during 2013 and 2014 growing season at Bornova experimental area (38°27.236 N, 27°13.576 E) in Agricultural Faculty of Ege University, Izmir, Turkey, at about 20 m above sea level with typical Mediterranean climate characteristics (Table 1). Experimental area is located in the Mediterranean zone of the country with quite mild winters and hot summers. Field studies were started in mid spring with low air temperature and satisfactory moisture levels were experienced in the germination and emergence period of relatively small seeds in both years, therefore, stands were excellent. The soil was a silty-clay loam (30.6% clay, 36.7% silt, and 32.7% sand) with pH 7.32, organic matter 1.16%, salt 0.074%, 0.123% total N, available phosphorus (1.4 ppm) and available potassium (350 ppm).

Table 1. Some meteorological parameters of experimental area at Bornova in 2013 and 2014.

Months	Average temperature (°C)			Total precipitation (mm)			Relative humidity (%)		
	2013	2014	LYA	2013	2014	LYA	2013	2014	LYA
April	17.3	17.0	16.1	30.2	132.2	46.4	54.0	60.6	62.9
May	22.7	20.8	21.0	43.7	15.3	25.4	54.7	57.0	59.6
June	25.7	25.0	26.0	27.1	48.5	7.5	50.7	53.0	52.9
July	28.4	27.8	28.3	0.0	1.0	2.1	42.0	49.9	51.2
X̄ - Σ	23.6	22.7	22.9	101.0	197.0	81.4	50.4	55.1	56.7

LYA: Long year average, X̄: mean, Σ: total

Field applications and experimental design

Q-52 cultivar of quinoa (Chenopodium quinoa Willd.) was used as crop material. The experiment was carried out with a randomized complete block design with three replicates; seven different nitrogen levels (0, 50, 75, 100, 125, 150 and 175 kg ha[-1]) were tested on quinoa crop. Seeds were sown by hand in rows 35 cm apart at a depth of 2-3 cm on 8th of April 2013 and 7th of April 2014 at a rate of 20 kg ha[-1] (Jacobsen, 2003). Each plot was consisted of 6 rows with 5 m length (10.5 m[-2]). Half a dose of nitrogen fertiliser (urea) was applied before sowing, and the rest of nitrogen was applied at 7-10 leaf stage as NH_4NO_3. All plots even control were fertilised using 100 kg ha[-1] P_2O_5 before sowing (Jacobsen et al., 1994).

Following establishment, plants were hand-thinned to 9-10 cm apart on rows so that the final populations were 28.6 plants m[-2]. Overhead sprinkler system was installed on the field during the both growing seasons. No insecticide was used to control insects. Hand hoeing was done once after emergence and fluazifop-p-butyl (500 g ha[-1] of a.i.) was applied at 3-5 leaf stage of crops to control narrow leafed weeds in both years.

Measurements

Plants shoots were harvested at maturity stage (~13% moisture) during July in both years, collecting mid 4 rows of plots in order to avoid border effects (net 5.6 m[-2]). 30 plants from the central rows in each replication were randomly tagged and data were recorded on these plants for the following traits (Bhargava et al., 2007): Plant height (cm): the average height from the ground level to the tip of the inflorescence on the main stem at the time of harvesting was measured. Dry weight (g plant[-1]): sample plants excluding the roots, inflorescence, branches and leaves were air dried and weighted. Harvest index (%): This was calculated by the following formula: HI: (grain yield plant[-1])/(dry weight plant[-1]). Thousand seed weight (g): A sample of 1000 seeds from the bulked seed of each plot was weighed. Grain yield (kg ha^{-1}): The seed of all the crops of each plot were bulked, after threshing by a stationary thresher, weighed and the grain yield/plot was then converted to kg per hectare. Crude protein content (%) of seed was determined using the Kjeldahl method (N%) with a conversion factor of 6.25.

Statistical analysis

All data were statistically analyzed using analysis of variance (ANOVA) with the Statistical Analysis System (SAS, 1998). Probabilities equal to or less than 0.05 were considered significant. If ANOVA indicated differences between treatment means a LSD test was performed to separate them (Stell et al., 1997).

RESULTS AND DISCUSSION

The results are summarized in Table 2. The year and nitrogen level effects were the main sources of variation in all characters tested, while the interactions (YxN) were not significant for harvest index, thousand grain weight and CP content of seed.

Table 2. Effect of different nitrogen levels on the yield and some yield components of quinoa.

N levels (ha^{-1})	2013	2014	Mean	2013	2014	Mean
	-------- Plant height (cm) --------			----- Grain yield (g plant^{-1}) -----		
0 kg	43.8	53.1	48.5	2.4	3.3	2.9
50 kg	48.4	58.9	53.6	4.1	5.0	4.5
75 kg	57.1	72.2	64.6	6.1	6.1	6.1
100 kg	66.2	75.1	70.7	8.7	8.4	8.6
125 kg	77.1	80.4	78.7	9.2	10.0	9.6
150 kg	82.8	93.0	87.9	9.2	11.2	10.2
175 kg	87.1	101.1	94.1	7.9	9.1	8.5
Mean	66.1	76.2	71.2	6.8	7.6	7.2
LSD(.05)	Y:1.8 N:3.4 YxN:4.8			Y:0.2 N:0.3 YxN:0.5		
	------- Harvest index (%) -------			----- Thousand grain weight (g) -----		
0 kg	12.3	14.4	13.3	3.38	3.33	3.36
50 kg	15.8	19.6	17.7	3.34	3.26	3.30
75 kg	34.4	36.8	35.6	3.28	3.18	3.23
100 kg	43.2	44.6	43.9	3.28	3.14	3.21
125 kg	43.6	46.2	44.9	3.21	3.11	3.16
150 kg	44.6	48.5	46.6	3.13	3.06	3.10
175 kg	38.8	41.4	40.1	3.11	3.04	3.08
Mean	33.3	35.9	34.6	3.25	3.16	3.20
LSD(.05)	Y:0.7 N:1.3 YxN:ns			Y:0.03 N:0.05 YxN:ns		
	----- Grain yield (kg ha^{-1}) -----			----- CP content of seed (%) -----		
0 kg	867	988	927	8.2	7.6	7.9
50 kg	1190	1490	1340	9.8	8.3	9.1
75 kg	1855	1936	1896	12.1	10.0	11.1
100 kg	2378	2513	2446	14.5	11.7	13.1
125 kg	2675	2939	2807	15.4	14.2	14.8
150 kg	2599	3308	2953	16.1	15.5	15.8
175 kg	2361	2677	2519	17.1	16.0	16.5
Mean	1989	2264	2127	13.3	11.9	12.6
LSD(.05)	Y:32 N:59 YxN:84			Y:0.5 N:0.9 YxN:ns		

Y: year, N: nitrogen, YxN: interaction, ns: not significant.

Plant height

The plant height was affected by YxN interaction. The highest plant height (101.1 cm) was obtained from 175 kg N ha^{-1} application in 2014, whereas the lowest was 43.8 cm for 0 kg N ha^{-1} application in 2013 (Table 2). Year effect was also significant and average quinoa height of second year (76.2 cm) was higher than the first year (66.1 cm) due to the total precipitation of the second year which was clearly higher than first year (Table 1).

In our study, the plant height of quinoa increased noticeably by increasing nitrogen fertilizer rate up to 175 kg N ha^{-1} in both seasons. Many researchers informed that the plant height of quinoa increases with the increasing nitrogen level are mainly due to the role of nitrogen in stimulating metabolic activity which contribute to the increase in metabolites amount and consequently lead to internodes elongation and increase plant height with the increasing nitrogen rate (Jacobsen et al., 1994, Erley et al., 2005; Shams, 2012). Jacobsen et al. (1994) expressed that plant height of quinoa increased with increasing N fertilization rate from 40 to 160 kg N ha^{-1}. Shams (2012) found clearly that plant height in quinoa increased gradually with increasing nitrogen levels up to 360 kg N ha^{-1}. Our findings are in accordance with those researcher's results.

Grain yield per plant

The nitrogen x year interaction was highly significant on the grain yield per plant (Table 2). The highest grain yield (11.2 g plant[-1]) was obtained from 150 kg N ha[-1] level in the second year, whereas the lowest yield (2.4 g plant[-1]) was recorded in control plots in the first year. Year effect was also significant and average grain yield per plant of second year (7.6 g) was higher than the first year (6.8 g), most probably due to the average monthly temperatures in the study site which was consistent with the 20-year average, providing better humidity and precipitation for the maturation of crops in 2014 compared to 2013 (Table 1).

Increases were significant among the nitrogen fertilizer rates till 125 kg N ha[-1], but there was not any significant difference between 125 and 150 kg N ha[-1] in 2013 season, while the increases were significant among the nitrogen rates till 150 kg N ha[-1] in 2014. In the study, the results of two years average monitored that increased nitrogen levels from 0 to 150 kg N ha[-1] increased grain yield per plant but later on decreased. Shams (2012) informed that grain yield per plant in quinoa increased gradually (1 g to 10 g plant[-1]) with increasing nitrogen levels up to 360 kg N ha[-1]. The possible reasons may be the difference in agro-ecological conditions and quinoa genotypes regarding maturation period.

Harvest index

The nitrogen x year interaction effect was not significant on harvest index. The highest average harvest index (46.6%) was obtained from 150 kg N ha[-1] treatment, whereas the lowest (13.3%) was in control (Table 2). Year effect was also significant on harvest index and average value of first year (33.3%) was lower than the second year (35.9%).

In the present study, the harvest index of quinoa increased by increasing nitrogen treatments till 150 kg N ha[-1] level but later on decreased. Erley et al. (2005) stated that harvest index of quinoa (31%) was not affected by nitrogen fertilization from 0 to 120 kg ha[-1]. Basra et al. (2014) informed that harvest index increased by increasing nitrogen treatments from 0 to 100 kg N ha[-1] level but later decreased at 120 kg N ha[-1] level. The increases in harvest index of quinoa with increasing nitrogen levels are mainly due to the role of N in stimulating metabolic activity which contributed to the increase in metabolites amount most of which is used building yield and its components (Shams, 2012).

The productive capacity of any crop plant depends, not only on its photosynthetic efficiency, but also on the effective translocation of assimilates to the seeds, which is measured by the harvest index. This partitioning between vegetative and reproductive parts can be modified by agronomic practices such as sowing date, plant density, fertilization and irrigation, etc (Bertero et al., 2004; Bhargava et al., 2007). In our study, harvest index presented tremendous variability and ranged from 12.3% to 48.5% and being affected by nitrogen levels. However,

this range is quite narrow as compared to the report of Rojas et al. (2003) who reported harvest index in quinoa in the range from 6% to 87%. Our findings were in agreement with some researchers (Bertero et al., 2004; Erley et al., 2005 and Shams, 2012), and the findings displayed that application of 150 kg N ha[-1] was the optimum level for nitrogen supplementation of soil for quinoa growth under the ecological conditions of Bornova region.

Thousand grain weight

The nitrogen x year interaction was also not significant on thousand grain weight. The highest average thousand grain weight of quinoa (3.36 g) was measured in control plot, whereas the lowest (3.08 g) was in 175 kg N ha[-1] treatment (Table 2). Year effect was also significant on this treat and average value of the first year (3.25 g) was higher than the second year (3.16 g).

In this study, increasing nitrogen levels caused a limited but significant decrease on thousand grain weight in both growing season. Basra et al. (2014) stated that thousand grain weight of quinoa (2.1 g) was not affected by nitrogen fertilization from 0 to 120 kg ha[-1]. In contrast, another study showed that thousand grain weight of 1.77 g was the highest after application of 0.8 g N per pot, with increasing nitrogen level to 1.2 g N per pot it was decreased to 1.58 g (Thanapornpoonpong, 2004). Studies in abroad also showed that nitrogen level has little effect on thousand grain weight (Gomaa, 2013). Gomaa (2013) informed that nitrogen fertilizers application with nitrobin increased the average thousand grain weight from 0 (3.3 g) to 119 (4.9 g) kg N ha[-1], however, no significant effect was determined between 119 or 238 (4.7 g) kg N ha[-1], and the highest nitrogen level (357 kg ha[-1]) was 3.3 g. Our findings partially confirmed those researcher's results.

Grain yield

There was significant nitrogen x year interactions on the grain yield of quinoa. Grain yield in quinoa crops ranged from 867 kg to 3308 kg ha[-1] in both growing season (Table 2). There were significant differences between nitrogen treatments concerning the yield. The highest grain yield (3308 kg ha[-1]) was found in the second year at 150 kg N ha[-1] level, whereas the lowest yield (867 kg ha[-1]) was in the first year at control plot. Year effect was also significant indicating better humidity conditions for the maturation of crops in 2014 compared to 2013, and, average grain yield of quinoa in the second year (2264 kg ha[-1]) was higher than the first year (1989 kg ha[-1]).

In our study, the grain yield of quinoa increased with the increasing nitrogen level from 0 to 125 kg N ha[-1] in the first year, however, the 150 kg N ha[-1] treatment caused the highest grain yield compared with the other N treatments in the second year, whereas rate of increase diminished the grain yield at the highest (175 kg N ha[-1]) rates in both years. These results could be explained by the increasing vegetative growth and decreasing inflorescence due to the nitrogen application. Some researchers (Erley et al., 2005; Bhargava et al., 2006) explained that plant height, maturation period and yield of quinoa and amaranth increased under optimum soil conditions, but at high levels of nitrogen fertilizer, grain yield was decreased caused by plant lodging. On the contrary, lodging was not observed in our experiment in both years.

According to the two years results, the grain yield was enhanced to 357% at 150 kg N ha^{-1} compared to control (0 kg N ha^{-1}) in the study. Erley et al. (2005) stated that grain yield of quinoa was affected by nitrogen fertilization from 0 to 120 kg ha^{-1} being 1790 kg to 3495 kg ha^{-1}. Jacobsen et al. (1994) informed that there was a significant grain yield increase when the amount of nitrogen fertilizer was increased from 40 to 160 kg N ha^{-1} and, the yield decreased by 24–1% when the nitrogen supply was reduced from 160 to 40 kg N ha^{-1}, while the yield decrease was 120 kg N ha^{-1} and 2–7% when the nitrogen supply was reduced to 80 and 120 kg N ha^{-1}, respectively. In addition, Shams (2012) reported that the increases in quinoa grain yield per hectare with the increase in N fertilizer application from 90 up to 360 kg N ha^{-1} over the control treatment were 518%, 769%, 936% and 1394% in average of both years. Gomaa (2013) informed that the application of 0, 119 and 238 kg N ha^{-1} with biofertilizers led to consistent increase the grain yield per hectare of quinoa as compared with untreated plants (control) over the years. Kakabouki et al. (2014) reported that nitrogen fertilization increased also the grain yield of quinoa under different tillage system.

Crude protein content of seed

The nitrogen x year interaction was not significant on protein content of quinoa seed. The highest average crude protein content (16.5%) recorded at 175 kg N ha^{-1} level, whereas the lowest protein content was 7.9% in untreated plants (control plot). Mean CP content was significantly higher in 2013 (13.3%) than in 2014 (11.9%).

The two years average indicated that nitrogen fertilization practices affected CP content of quinoa grain significantly for both years. CP content in grain was progressively increased with the increasing levels of nitrogen up to 175 kg N ha^{-1}. The highest CP content in the grain recorded by 175 kg N ha^{-1} treatment which was 16.5% higher than that in the control treatment, but statistically similar to 150 kg N ha^{-1} treatment (15.8%). The higher protein content at higher nitrogen levels was mainly due to the structural role of nitrogen in building up amino acids (Finck, 1982; Bhargava et al., 2006; Miranda et al., 2013). The progressive increase in protein contents of quinoa seed with the increasing nitrogen rates were also reported by many research workers (Jacobsen et al., 1994; Erley et al., 2005; Shams, 2012).

Erley et al. (2005) informed that average CP content of quinoa cultivars (Faro and Cochabamba) increased gradually (12.3% to 14.6%, respectively) with the increasing nitrogen levels from 0 kg N to 120 kg N ha^{-1} and, Miranda et al. (2013) reported an average CP content of 18.8% using cold resistance quinoa cultivars (Regalona

Baer and Villarrica). Kakabouki et al. (2014) also stated that increasing nitrogen level increased CP content of quinoa from 7% to 27% under different tillage system. These results in various references were slightly higher than what we were able to determine in our ecology, which might be due to differences in soil texture, crop cultivars and environmental conditions.

The major fact that determines the grain protein content is nitrogen availability, and quinoa is highly responsive to nitrogen fertilizer (Basra et al., 2014) and higher CP content, in a crop with high yield, can be obtained just by application of higher nitrogen quantities.

CONCLUSION

To come to conclusion, it should be emphasized that quinoa, a new introduction to the Mediterranean coastal part of Turkey is a promising crop material with an acceptable level of adaptability and grain yield pecularities, considering the potential harsh effect of global warming in near future. The results of our two-year study testing the effect of seven different nitrogen levels (0, 50, 75, 100, 125, 150 and 175 kg ha^{-1}) on quinoa crop showed that it was possible to produce an average of 2.95 t ha^{-1} quinoa grain yield with an average of 16% crude protein content under 150 kg N ha^{-1} treatment in the regions with Mediterranean-type climates. Future experiments on quinoa crop should be conducted at different locations with various agronomical treatments and additional fertilizer rates to be sure that results are relatively consistent over time.

LITERATURE CITED

Anonymus. 2013. Cereal's of mother: Quinoa, Journal of Tubitak Sci. & Tech., June 2013, 547:34-35.

Basra,S.M.A., S.Iqbal and I.Afzal. 2014. Evaluating the response of nitrogen application on growth, development and yield of quinoa genotypes, International Journal of Agriculture & Biology, 16(5):886-892.

Bertero,H.D., A.J.de la Vega, G.Correa, S.E.Jacobsen and A.Mujica. 2004. Genotype and genotype-by-environment interaction effects for grain yield and grain size of quinoa (*Chenopodium quinoa* Willd.) as revealed by pattern analysis of international multi-environment trials, Field Crops Research, 89:299–318.

Bhargava,A., S.Shukla and D.Ohri. 2006. *Chenopodium quinoa*-An Indian perspective, Industrial Crops and Products, 23:73–87.

Bhargava,A., S.Shukla and D.Ohri. 2007. Genetic variability and interrelationship among various morphological and quality traits in quinoa (*Chenopodium quinoa* Willd.), Field Crops Research, 101:104–116.

Comai,S., A.Bertazzo, L.Bailoni, M.Zancato, C.V.L.Costa and G.Allegri. 2007. The content of proteic and nonproteic (free and protein bound) tryptophan in quinoa and cereal flours, Food Chem. 100:1350-1355.

Doweidar,M.M. and A.S.Kamel. 2011. Using of quinoa for production of some bakery products (gluten-free), Egyptian J. Nutrition, 26(2):21-52.

Erley,G.S., H.P.Kaul, M.Kruse and W.Aufhammer. 2005. Yield and nitrogen utilization efficiency of the pseudocereals amaranth, quinoa, and buckwheat under differing nitrogen fertilization, European Journal of Agronomy. 22 (1): 95-100.

Finck,A. 1982. Fertilizer and fertilization "Introduction and practical guide to crop fertilization". Weinheim; Deerfield Beach, Florida; Basel: Verlag Chemie. ISBN 3-527-25891-4 (Weinheim).

Gomaa,E.F. 2013. Effect of nitrogen, phosphorus and biofertilizers on quinoa plant, Journal of Applied Sciences Research, 9(8):5210-5222.

Ince Kaya,Ç. 2010, Effects of various irrigation strategies using fresh and saline water applied with drip irrigation system on yield of quinoa and salt accumulation in soil in the Mediterranean region and evaluation of saltmed model, MSc. Thesis, Çukurova Univ., Inst. of Natural and Applied Sci., Dept. of Agricultural Structures and Irrigation,122p.

Jacobsen,S.E. 2003. The worldwide potential for quinoa (*Chenopodium quinoa* Willd.), Food Rev. Int. 19(1–2):167–177.

Jacobsen,S.E., I.Jørgensen and O.Stølen. 1994. Cultivation of quinoa (*Chenopodium quinoa*) under temperate climatic conditions in Denmark, J. Agrc. Sci. 122: 47-52.

Kakabouki,I., D.Bilalis, A.Karkanis, G.Zervas, E.Tsiplakou and D.Hela. 2014. Effects of fertilization and tillage system on growth and crude protein content of quinoa (*Chenopodium quinoa* Willd.): An alternative forage crop, Emir. J. Food Agric., 26(1):18-24.

Kuhn,M., S.Wagner, W.Aufhammer, J.H.Lee, E.Kübler and H.Schreiber. 1996. Einfluß von pflanzenbaulicher Maßnahmen auf die Mineralstoffgehalte von Amaranth, Buchweizen, Reismelde und Hafer. Dt Lebensm Rundschau, 92:147-152.

Miranda,M., A.Vega-Gálvez, E.A.Martínez, J.López, R.Marín, M.Aranda and F.Fuentes. 2013. Influence of contrasting environments on seed composition of two quinoa genotypes: Nutritional and functional properties, Chilean Journal of Agricultural Research, 73(2):108-116.

Pearsall,D.M. 1992. The origins of plant cultivation in South America. In: C.W.Cowan, P.J.Watson (Eds.), The Origins of Agriculture. Smithsonian Institute Press, Washington, DC, pp:173-205.

Rea,J., M.Tapia and A.Mujica. 1979. Prácticas agronómicas. In: Quinoa y Kañiwa, Cultivos Andinos, pp:83–120. Tapia,M., H.Gandarillas, S.Alandia, A.Cardozo and A.Mujica. (eds.). FAO, Rome, Italy.

Rojas,W., P.Barriga and H.Figueroa. 2003. Multivariate analysis of genetic diversity of Bolivian quinoa germplasm. Food Rev. Int., 19:9–23.

SAS Institute. 1998. INC SAS/STAT user's guide release 7.0, Cary, NC, USA.

Shams,A.S. 2012. Response of quinoa to nitrogen fertilizer rates under sandy soil conditions, Proc. 13th International Conf. Agron., Fac. of Agric., Benha Univ., Egypt, 9-10 September 2012, p:195-205.

Simmonds,N.W. 1971. The breeding system of *Chenopodium quinoa*. I. Male Sterility, Heredity, 27:73-82.

Stell,R.G.D., J.A.Torrie and D.A.Dickey. 1997. Principles and Procedures of Statistics. A.Biometrical Approach 3rd Edi. Mc Graw Hill Book. INC. NY.

Thanapornpoonpong,S. 2004. Effect of nitrogen fertilizer on nitrogen assimilation and seed quality of amaranth (*Amaranthus* spp.) and quinoa (*Chenopodium quinoa* Willd), Doctoral Dissertation, Doctor of Agricultural Sciences of the Faculty of Agricultural Sciences, Institute of Agricultural Chemistry, Georg-August-University of Göttingen.

30

YIELD COMPONENTS IN MUNG BEAN [*Vigna radiata* (L.) Wilczek]

Huseyin CANCI[*], Cengiz TOKER*

Akdeniz University, Faculty of Agriculture, Department of Field Crops, Turkey.
*Corresponding author: huseyincanci@akdeniz.edu.tr

ABSTRACT

This study was conducted to evaluate the yield components in mung bean [Vigna radiata (L.)Wilczek] using the correlation, path and factor analyses. It was found that there was considerable variation for the characters studied. Factor 1 composed of 100-grain weight, podlength, pod width, branches per plant and pods per plant. The grain weight was stronglycorrelated with pod length and pod width. Pods per plant were significantly and positivelyassociated with branches per plant. Factor 2 consisted of biological, straw and grainyields. The seed yield was highly associated with biological and straw yields. Factor 3comprised of seed per pod, pods and flowers per peduncle. The fourth factor was onlythe days to flowering. The last factor was plant height. The total factors had 74% ofthe total variance induced by the characters. It was firstly concludedthat the factor analysis together with path and correlationcoefficients could successfully be used for determining characters usable for selection in themung bean breeding programs.

Key words: Correlation analysis, Factor analysis, Mung bean, Path analysis, Selection criteria,Vigna radiata

INTRODUCTION

Genetic relationships between yield and yield related characters are prerequisite in selecting desirabletypes for the target environment. Some of the yield components are highly interrelated. On the other hand, grain yield is governed by many genetic aswell as environmental factors that are interdependent. Heritability for grain yield is lowin mung bean (Tickoo and Jain, 1988); as well as in chickpea (Toker, 1998; 2004). Path coefficientanalysis ishelpful to determine the direct contribution of yield components and their indirect contributionsover other traits on grain yield (Dewey and Lu, 1959). Path analysis has been widely used to determine direct andindirect selection criteria in food legumes (Duarte and Adams, 1972; Bahl et al., 1976; Islam and Shaikh, 1978; Toker and Cagirgan, 2003).

Cattel (1965) explained that the factor analysis has decreased a large number of correlated variables to a small number of main factors. Ithas been successfully utilized in wheat (Lee and Kaltsikes, 1973), in switch grass(Godshalk and Timothy, 1988) and in barley(Cagirgan and Yildirim, 1990) as well as in chickpea (Toker, 2004; Toker and Cagirgan, 2004). Until today, any selection criteria have not been proposed to determine characteristicsrelated to grain yield in mung bean. The objective of this study was to determine the yield component of mung bean by using the path and factor analysis.

MATERIALS AND METHODS

A total of nineteen mung bean genotypes, 17 from Nuclear Institute for Agriculture andBiology (NIAB); one genotype from market of Faisalabad, Pakistan; and one genotype fromGazipasa, Antalya, Turkey were grown in the lowland conditions (approximately 30° 44' E, 36°52' N, 51 m from sea level) of the west Mediterranean region of Turkey during 1999-2000 and2000-2001 growing seasons. Grains of genotypes were sown on May 18, 2000 and on May 7,2001 in a Randomized Complete Block Design with 3 replications and oneexperimental plot consisted of two rows of 2 m length 30 cm apart and 10 cm in the row spacing. Fertilization was applied at a rate of 20 kg nitrogen and 50 kg phosphorus perhectare prior to sowing. The experimental area was irrigated with sprinkler system with 10 daysintervals. Weeds were controlled by hand without using any chemicals.Some important phenologic, morphologic and agronomic characters were recorded.These characters were described in Descriptors for *Vigna mungo* and *V. radiata* (IBPGR,1985). *Phenological descriptors*: Days to flowering (DF) was recorded in days as number ofdays after sowing when 50% plants in the plot set the first flower. *Morphological descriptors*:Plant height (PH) was recorded in cm as average height from ground to top of two plants atmaximum growth. Branches per plant (BP) were average number of stems from two plants atflowering. Pods per plant (PP) were average number of pods from two plants at podding.Flowers per peduncle (FN) were recorded in number as average of flowers from two plants. Podsper peduncle (PN) were average number of pods from two plants. Pod length (PL) was recordedin cm as average length of pods of two plants at maximum growth. Pod width (PW) wasrecorded in cm as width of pod of two plants at maximum growth. Grains per pod (GP) wererecorded as grains of pod in two plants at maximum growth. *Agronomicaldescriptors*: Grain yield (GY) was recorded in g and then converted to kg ha-1 basis as afterthreshing seed weight each genotype. Biological yield (BY) was recorded in g after harvesting astotal dry weight each genotype. Straw yield (SY) was calculated following to the formula:[(Biological yield) – (Grain yield)] as g. 100-Seed weight (SW) was recorded in g as average oftwo times randomly 100 grains selected.Path and factor analyses were performed according to Dewey and Lu (1959) and Cattel(1965), respectively. Analyses were performed by using MINITAB statistical package programs (MINITAB, 2000).

RESULTS AND DISCUSSION

Considerable variations were found for all the 13 characteristics studied, even though limitedgenotypes have been evaluated (Table 1).It could be seen in Table 2 that grain yield was significantly andpositively correlated the biological yield (r = 0.688), pods per plant (r = 0.682), pods perpeduncle (r = 0.654), plant height (r = 0.602), days to flowering (r = 0.593), branches per plant (r= 0.585), straw yield (r = 581), grains per pod (r = 0.574), flowers per peduncle (r = 0.556) andpods width (r = 0.510). The biological yield was strongly and positively associated with strawyield (r = 0.989), plant height (r = 0.834), days to flowering (r = 0.690) and pods per plant (r =0.479). Grain weight 100^{-1} was highly and positively related with pod length (r = 0.905), podwidth (r = 0.880), plant height (r = 0.831), pods per peduncle (r = 0.692) and days to flowering (r= 0.625).Biological yield (6.034) had the highest direct and positive effect, while 100-grain weight(0.011) was the lowest contribution to grain yield (Table 3).Biological yield was followed by straw yield with negative direct effect (-5.848) and days toflowering with positive direct effect (0.797). The indirect effect of biological yield via straw yield (-5.784) was negative and high on grain yield (Table 3).

Table 1. The mean, standard error, minimum and maximum values of 13 characters in mung bean

Characters	Mean ±SE		Minimum	Maximum
Days to Flowering (days)	58.2	±0.94	20.0	76.0
Plant height (cm)	48.1	±1.44	19.5	91.0
Branches per plant	3.2	±0.07	2.0	6.0
Pods per plant	25.0	±1.13	8.0	62.5
Flowers per peduncle	4.3	±0.07	3.5	7.0
Pods per peduncle	2.9	±0.03	2.0	4.0
Pod length (cm)	9.2	±0.17	5.6	20.0
Pod width (cm)	0.48	±0.01	0.3	0.6
Grains per pod	9.9	±0.15	5.0	13.0
Biological yield (g plot-1)	665.0	±44.10	41.0	2520.0
Straw yield (g plot-1)	516.9	±39.10	22.0	2150.0
Grain yield (kg ha-1)	1209.6	±72.90	33.3	3916.6
100-grain weight (g)	5.5	±0.14	3.1	8.6

Table 2. Correlations among 13 characters in mung bean (df = 17)

Characters	PH	BP	PP	FN	PN	PL	PW	GP	BY	SY	GY	SW
DF	0.805**	0.525*	0.354	0.558*	0.831**	0.714**	0.779**	0.693**	0.690**	0.675**	0.593**	0.625**
PH		0.525*	0.587**	0.558*	0.587**	0.384	0.453*	0.555*	0.834**	0.831**	0.602**	0.222
BP			0.755**	0.840**	0.737**	0.462*	0.466*	0.731**	0.427	0.371	0.585**	0.275
PP				0.536*	0.532*	0.290	0.345	0.482*	0.479*	0.397	0.682**	0.140
FN					0.825**	0.555*	0.571**	0.800**	0.297	0.238	0.556*	0.357
PN						0.811**	0.865**	0.820**	0.409	0.343	0.654**	0.692**
PL							0.918**	0.723**	0.194	0.138	0.433	0.905**
PW								0.740**	0.312	0.258	0.510*	0.880**
GP									0.257	0.185	0.574**	0.596**
BY										0.989**	0.688**	0.090
SY											0.581**	0.057
GY												0.268

DF = Days to flowering, PH = Plant height, BP = Branches per plant, PP = Pods per plant, FN = Flowers per peduncle, PN = Pods per peduncle, PL = Pod length, PW = Pod width, GP = Grains per pod, BY = Biological yield, SY = Straw yield, GY = Grain yield, SW = 100-seed weight. Degrees of freedom is df. P < 0. 456 and 0.575 statistically significant at 0.05 and 0.01 probability levels, respectively.

Table 3. The direct and indirect contribution of characters to grain yield in mung bean

	GW	PL	PW	BP	PP	BY	SY	GP	PP	FP	DF	PH
GW	**0.011**	0.010	0.010	0.003	0.002	0.001	0.001	0.007	0.008	0.004	0.007	0.003
PL	-0.231	**-0.259**	-0.234	-0.118	-0.074	-0.049	-0.035	-0.184	-0.207	-0.141	-0.182	-0.098
PW	-0.256	-0.267	**-0.291**	-0.135	-0.100	-0.091	-0.075	-0.215	-0.251	-0.166	-0.227	-0.132
BP	-0.080	-0.133	-0.135	**-0.289**	-0.218	-0123	-0.107	-0.211	-0.212	-0.243	-0.201	-0.152
PP	0.010	0.020	0.024	0.052	**0.068**	0.033	0.027	0.033	0.036	0.037	0.030	0.024
BY	0.543	1.170	1.884	2.575	2.892	**6.034**	5.968	1.552	2.465	1.791	4.161	5.032
SY	-0.332	-0.806	-1.508	-2.168	-2.324	-5.784	**-5.848**	-1.082	-2.006	-1.392	-3.950	-4.859
GP	-0.017	-0.021	-0.022	-0.021	-0.014	-0.008	-0.005	**-0.029**	-0.024	-0.023	-0.020	-0.016
PP	0.119	0.140	0.149	0.127	0.092	0.070	0.059	0.141	**0.172**	0.142	0.143	0.101
FP	-0.016	-0.025	-0.027	-0.038	-0.024	-0.014	-0.011	-0.036	-0.038	**-0.045**	-0.031	-0.025
DF	0.498	0.569	0.621	0.555	0.355	0.550	0.539	0.553	0.663	0.547	**0.797**	0.642
PH	0.018	0.032	0.037	0.043	0.029	0.069	0.068	0.046	0.048	0.046	0.066	**0.082**

DF = Days to flowering, PH = Plant height, BP = Branches per plant, PP = Pods per plant, FN = Flowers per peduncle, PN = Pods per peduncle, PL = Pod length, PW = Pod width, GP = Grains per pod, BY = Biological yield, SY = Straw yield, GY = Grain yield, SW = 100-seed weight.

It could be seen in Table 4 that5 factors explained 74% of the total variance of the characters. Factor 1, 2, 3, 4and 5 explained 0.26%, 0.18%, 0.15%, 0.08% and 0.07% of total variance expressed. Factor 1comprised of 100- grain weight (-0.729), pod length (-0.655), pod width (-0.631), branches perplant (0.591) and pods per plant (0.486), whereas factor 2 composed of biological yield (0.675), straw yield (0.613) andgrain yield (0.612) with positive loadings. Factor 3consisted of grains per pod (0.626), pods per peduncle (0.505) and flowers per peduncle (0.471), while factor 4 encompassed days to flowering with negative loading (–0.697). The last factorconsisted of only plant height with negative loading (–0.598).

Table 4. Factor loadings and communalities of 13 characters on five principal factors in mung bean

Characters	Factors					Communality
	1	2	3	4	5	
100-grain weight	**-0.729**	0.479	0.130	-0.155	-0.029	0.80
Pod length	**-0.655**	0.472	0.236	-0.047	-0.145	0.73
Pod width	**-0.631**	0.579	0.147	-0.113	-0.009	0.76
Branches per plant	**0.591**	-0.105	0.342	-0.525	-0.072	0.75
Pods per plant	**0.486**	0.212	0.466	-0.050	-0.224	0.55
Biological yield	0.623	**0.675**	-0.228	0.052	0.231	0.95
Straw yield	0.593	**0.613**	-0.327	0.003	0.280	0.91
Grain yield	0.458	**0.612**	0.310	0.265	-0.134	0.76
Grains per pod	0.010	-0.050	**0.626**	0.124	-0.327	0.51
Pods per peduncle	0.078	0.260	**0.505**	-0.400	0.371	0.62
Flowers per peduncle	0.450	-0.373	**0.471**	-0.204	0.211	0.65
Days to Flowering	-0.038	0.066	-0.575	**-0.697**	-0.273	0.89
Plant height	0.489	0.223	-0.193	-0.006	**-0.598**	0.68
Variance	3.34	2.31	1.92	1.10	0.95	9.62
% Variance	0.26	0.18	0.15	0.08	0.07	0.74

Bold and italic numbers are the main factors.

Variation is the first requirement for selection in plant breeding. Bosand Caligari (1995) pointed out that the more genetic variation in characters is the more genetic gain.Ahmed et al. (1981) reported that pods per plant were the most important selectioncriteria to increase potential yield in mung bean. In black gram [*Vigna mungo* (L.)Hepper], it was shown that plant yield was significantly correlated with grains per pod, pods perplant, main branches per plant and plant height (Majid et al., 1982). Shamsuzzaman et al. (1983)studied for genetic, phenotypic and environmental correlations in mung bean. They found thatplant height was strongly associated with main branches per plant and pods per plant. Similarresults were obtained by Remanandan et al. (1988) in pigeonpea.Our results are in agreement with findings of Karadavut (2009). Biological yield could be accepted as the most valuable characteristicamong the traits.Biological yield had the highest direct effect on grain yield (Table 3)and biological yield could be increasedvia straw yield, branches per plant andpods per peduncle. The biological yield in chickpeawas found to be the most important selection criteria for the contributing grain yield due to the highest and the positive direct effect (Canci and Toker, 2009).The more branches per plant resulted in the more pods per plant. To utilize pods per peduncle characteristic, genotypes with high numbering flowers per peduncle should be selected. Besides, the days to flowering and plant height should also beevaluated. In a similarway,selection criteria to be usedin chickpea were evaluated (Toker andCagirgan, 2004). Toker (2004) stressed that biological yield should be evaluated in the selectionto increase the grain yield in chickpea breeding programs.

In conclusion, biological or straw yield could beused as selection criteria in mung bean. Besides, selection of genotypes with large seed, high podwidth and length could also be considered.

ACKNOWLEDGEMENTS

This study was supported by Akdeniz University The Scientific Research Projects Coordination Unit.A special thanks would like to Dr. A.M. Haq from National Institute ofAgriculture and Biology (NIAB), Faisalabad, Pakistan and Yasar Ozyigit (Asst. Prof. Dr.) fromKorkuteli Vocational School, Horticultural Program, Akdeniz University, Antalya, Turkey forkindly providing plant materials.

LITERATURE CITED

Ahmed, Z.U., M.A.Q. Shaikh, M.A. Majid and S. Begum, 1981. Correlation studies in agronomiccharacters of mung bean (*Vigna radiata*). Bangladesh Journal of Agricultural Science 8:3-36.

Bahl, P.N., R.B. Mehra and D.B. Raju, 1976. Path analysis and its implications for chickpea breeding. Z. Pflanzenzüchtung 77: 67-71.

Bos, I., and P. Caligari, 1995. Selection Methods in Plant Breeding. Published by Chapman andHall, 2-6 Boundary Row, London, SE1 8HN, UK.

Cagirgan, M.I. and M.B. Yildirim, 1990. An application of factor analysis to data from control andmacro mutant populations of 'Quantum' barley. J. Fac. of Agric. of Akdeniz University4: 125-138.

Canci, H. and C. Toker, 2009. Evaluation of Yield Criteria for Drought and Heat Resistance in Chickpea (*Cicer arietinum* L.). J. Agronomy & Crop Science 195: 47-54.

Cattel, R.B., 1965. Factor analysis: An introduction to essentials. I. The propose and underlyingmodels. Biometrics 21: 190-215.

Dewey, D.L. and K.H. Lu, 1959. A correlation and path-coefficient analysis of components ofcrested wheatgrass seed production. Agron. J. 51: 515-518.

Duarte, R.A. and M.W. Adams, 1972. A path coefficient analysis of some yield componentsinterrelations in field beans (*Phaseolus vulgaris* L.). Crop. Sci. 12: 579-582.

Godshalk, E.B. and D.H. Timothy, 1988. Factor and principal component analyses as alternativeto index selection. Theor. Appl. Genet. 76: 352-360.

IBPGR, 1985. Descriptors for *Vigna mungo* and *V. radiata* (Revised). International Board forPlant Genetic Resources, Rome, Italy.

Islam, M.Z. and M.A.Q. Shaikh, 1978. Correlation and path coefficient analysis of yield and yieldcomponents in lentil. Bangladesh Journal of Agricultural Science 5: 67-72.

Karadavut, U. 2009. Path analysis for yield and yield components in Lentil (*Lens culinaris* Medik.). Turkish J. of Field Crops, 14: 97-104.

Lee, J. and P.J. Kaltsikes, 1973. Multivariate statistical analysis of grain yield and agronomic characters in durum wheat. Theor. Appl. Genet. 43: 226-231.

Majid, M.A, S. Khanum, M.A.Q. Saikh and A.D. Bhuiya, 1982. Genetic variability and correlationstudies in black gram. Bangladesh Journal of Agriculture 7: 98-102.

MINITAB 2000. Minitab Statistical Software vers. 13.1

Remanandan, P., D.V.S.S.R. Sastry and M.H. Mengesha, 1988. ICRISAT Pigeonpea GermplasmCatalog: Evaluation and Analysis. Patencheru, A.P. 502 324, India.

Shamsuzzaman, K.M., M.R.H. Khan and M.A.Q. Saikh, 1983. Genetic variability and charactersassociation in mung bean [*Vigna radiata* (L.) Wilczek]. Bangladesh Journal ofAgricultural Res. 8: 1-5.

Tickoo, J.L. and H.K. Jain, 1988. Mungbean. In: B. Baldev, S. Ramanujam and H.K. Jain, (Eds.), Pulse Crops (Grain Legumes, Oxford and IBH Publishing Co. Pvt. Ltd. New Delhi, India, pp. 161-188.

Toker, C., 1998. Estimate of heritabilities and genotype by environment interactions for 100- grain weight, days to flowering and plant height in kabuli chickpeas (*Cicer arietinum* L.). Turk. J. Field Crops 3, 16-20.

Toker, C. and M.I. Cagirgan, 2003. Selection criteria in chickpea (*Cicer arietinum*). Acta Agric.Scan. Section B, Soil and Plant Science 53: 42-45.

Toker, C., 2004. Evaluation of selection criteria using phenotypic and factor analysis inchickpea. Acta Agric. Scan. Section B, Soil and Plant Science 54:45-48.10

Toker, C. and M.I. Cagirgan, 2004. The use of phenotypic correlations and factor analysis indetermining characters for grain yield selection in chickpea (*Cicer arietinum* L.).Hereditas 140: 226-228.

31

PRODUCTIVITY OF INTERCROPPING MAIZE (*Zea mays* L.) AND PUMPKINS (*Cucurbita maxima* Duch.) UNDER CONVENTIONAL VS. CONSERVATION FARMING SYSTEM

Nebojša MOMIROVIĆ[1], Snežana OLJAČA[1], Željko DOLIJANOVIĆ[1]*, Milena SIMIĆ[2], Mićo OLJAČA[1], Biljana JANOŠEVIĆ[1]

[1]*Belgrade University, Faculty of Agriculture, Nemanjina 6, Belgrade, SERBIA*
[2]*Maize Research Institute, Slobodana Bajica 1, Belgrade, SERBIA*
**Corresponding author: dolijan@agrif.bg.ac.rs*

ABSTRACT

The evaluation of pumpkin (Cucurbita maxima Duch.) and maize (Zea mays L.) intercropping productivity, under different farming systems: conventional farming vs. conservation farming was carried out on the chernozem type of soil at Zemun Polje, Serbia. Results obtained by the bivariate analysis of variance showed significant differences between different proportions of components in intercropped maize and pumpkins. Regarding the land equivalent ratio (LER), two rows of pumpkins and two rows of maize (proportion 2/3:1/3) were the optimum spatial arrangement in conventional farming system, while proportion 1/3:2/3 was optimal in conservation farming system. The yield of pumpkins proportionally increased with the increase of the plant population, although the intraspecies competition of pumpkins was very pronounced in intercropping with maize. The average fruit yield of pumpkins in the first year was lower in conservation farming practices in comparison with conventional farming practices. On the other hand, situation was complete opposite with pumpkin yield in second year of investigation, while significant decrease in maize yield was observed in the plots where conservation farming practices were applied. Growing pumpkins in mixture with maize probably costs a small farmer very little more effort, than the production of a sole stand of maize. At least where the productivity of mixture is dominated by one species, as with maize in maize-pumpkins intercropping, the competitive effect of the recessive species on the dominant is small.

Key words: Conventional farming; Conservation farming; Intercropping; Maize; Pumpkin.

INTRODUCTION

The cropping system is very important for weed control, considering that weeds can cause great damages to crops and decreased yield (Videnović et al. 2013). Intercropping, as a cropping system, is defined as the intensification and diversification of cropping in time and space dimensions (Francis 1986). The intensification of land and resource uses in the space dimension is an important aspect of multiple cropping in efforts to develop energy-efficient and sustainable agriculture. The considerable variation in soil and climate has resulted in large variation in yield performance of maize hybrids annually (Ilker et al. 2009). Biological potentials of intercropping, such as enhanced efficiency of incident light use if two species, occupying the same land area (different pattern of foliage display, then different rooting patterns, etc.), can be realized when the interspecific competition is smaller than intraspecific competition in the same environment (Liebman and Staver 2001; Dolijanović et al. 2013). The final advantages of intercropping are a greater resource use and significant yield benefits (Francis 1986; Gliessman 1986; Oljača

1998; Dolijanović et al. 2007; Yang et al. 2010). Deficit irrigation, by reducing irrigation water use, can aid in coping with situations where water supply is restricted. (Kuscu et al. 2013). In some situations, farmers are interested primarily in obtaining full yield of one main crop, but sowing other species into the main crop for additional benefits: more food and fodder, improved soil conservation, and better weed control (Willey 1979a; Birkás et al. 2006). The additive intercropping design is based on these principles. However, the essence of the replacement series is to obtain optimum yields of both crops, which can be achieved only in case of the most favorable relations of crops in the mixture.

Relationships between components in intercropped maize (*Zea mays* L.) and pumpkins (*Cucurbita maxima* Duch.) were studied by numerous authors (Conteras Magana and del Castillo 1991; Powers et al. 1993; Galloway and Weston 1996; Silwana and Lucas 2002). Pumpkin, because of its ability to tolerate shade and cool temperatures, and to cover ground rapidly with its creeping growth habit, is often intercropped with maize. This intercropping system could be effective in suppressing weed growth and increasing crop yields and should be incorporated into the integrated weed management and Cropping System Design.

Intercropping systems have potential difficulties and limitations that prevent their application on large areas (Biabani et al. 2008; Mudita et al. 2008; Biabani 2009). There are many reasons for this: lack of machinery necessary for such purposes, difficulties for plant protection, lack of varieties and hybrids better-adapted to such growing conditions, small scale farming and lack of marketing possibilities *etc*. In Serbia, maize is being produced on 1.2 million hectares across the country (Mitrović et al. 2012), usually as a monocrop. The best cultivars for sole crop might not be the most suitable for mixed cropping.

In some regions, lately, where stressful conditions including drought and high temperatures occur frequently, broad environmental tolerance associated with mixed population of different crops may play a significant role in yield stability.

The aim of this study was to determine the relationship between maize and pumpkins, grown in the association under different farming practices and also to identify the most efficient and productive maize - pumpkins intercrop combinations. It is expected that maize and pumpkins will have higher yields in intercropping system due to their ability to use natural resources differently and make better overall use of natural resources than grown separately.

MATERIALS AND METHODS

The evaluation of pumpkin (*Cucurbita maxima* Duch. cv. Dill's Atlantic Giant) and maize (*Zea mays* L. cv. ZP SC 709d "double ear") intercropping productivity, under different farming systems, conventional farming vs. conservation farming, was carried out in the experimental field of the Maize Research Institute (MRI) in Zemun Polje near Belgrade (44°52'N 20°20'E) during 2010 and 2011 on the chernozem type of soil. Randomized complete block design was applied with a four-replication. The maize crop was sown in 0.8 m plant spacing and 0.35 m row spacing (35714 plants ha^{-1}) in pure stands. The corresponding values for pumpkins were 1.6 m and 2 m (3125 plants ha^{-1}). There were three combinations in mixtures following the method of replacement series (de Wit 1960): 2/3 pumpkins: 1/3 maize - where two rows of pumpkins replaced two rows of maize; 1/2 pumpkins: 1/2 maize - where a single row of pumpkins replaced two rows of maize; 1/3 pumpkins: 2/3 maize - where a single row of pumpkins followed by four rows of maize. Plant spacing in mixtures was the same as in the pure stands of both crops.

The previous crop was winter wheat. After stubble disking, deep ploughing to the depth of 35 cm was done in autumn. The common cropping practices were applied within the conventional farming system. Plots under conventional farming were fertilized with 120 kg N ha^{-1}, 100 kg P$_2$O$_5$ ha^{-1} and 80 kg K$_2$O ha^{-1} in form of combined NPK fertilizer. The total amount of P$_2$O$_5$ and K$_2$O and 50% of N fertilizers were applied in autumn and the rest of N fertilizers in spring, prior to sowing.

The pre-emergence application of Prometryne (1.5 kg a.i. ha^{-1}) and Pendimethalin (1.320 kg a.i. ha^{-1}) was done. Winter vetch, as a cover crop, was sown in narrow rows after wheat stubble had been disked in plots in which conservation tillage was applied. When the cover crop was destroyed with Paraquat 0.8 kg a.i. ha^{-1} early in spring, decomposed wheat straw was applied in the amount of 5 t ha^{-1} to cover more than 60% of soil surface just before sowing of maize and pumpkins. Maize was sown in both years in the second decade of April, while four-week-old pumpkin transplants were planted on the beginning of May. Grain and fruit yields and other traits were recorded from each plot.

For the purpose of analyses, two broad approaches were used: the bivariate analysis of variance (Pearce and Gilliver 1978; Mead 1986) and Land Equivalent Ratio (LER) (Willey 1979b). Land equivalent ratio (LER) was calculated and used to evaluate the advantages in yields from intercropping: $LER = \frac{Y_{1.2}}{Y_{1.1}} + \frac{Y_{2.1}}{Y_{2.2}}$

where $Y_{1.1}$ and $Y_{2.2}$ are the crop yield for maize and pumpkin grown in monoculture, and $Y_{1.2}$ and $Y_{2.1}$ are yield of investigation crops in the mixture.

Results achieved were developed statistically with the analysis of variance (ANOVA) method. Yields per unit area of one component were plotted against that of another component as bivariate diagrams (Snaydon and Satorre 1989).

RESULTS AND DISCUSSION

In the experimental location, a total precipitation amounted to 466.5 mm and 486.0 mm, during the growing season (April-September) of 2010 and 2011, respectively, while a 30-year average amounted to 410.8 mm (Figure 1). The precipitation distribution significantly differed over year, hence in June and September of 2010, the precipitation amounted to 107.1 mm and 128.8 mm, respectively, that was two-fold higher than the long-term average. On the other hand, in 2011, August was characterized with high precipitation (145.0 mm). Temperature conditions were similar in both years of investigation (12.8 and 12.1 °C) and also in accordance with the long-term average (12.2 °C). With regard to climatic requirements of these crops, it appeared that conditions were normal for the growth and development in both years 2010 and 2011.

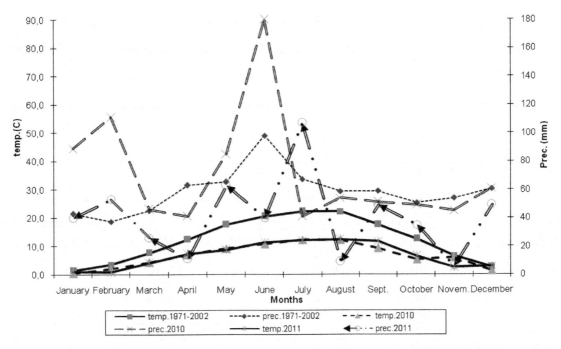

Figure 1. Meteorological data for long term (1971-2002) and experimental period (2010-2011)

Yields obtained in the sole crops of pumpkins were significantly higher than yields achieved in intercropping with maize, especially in the conservation farming system. The reduction of maize and pumpkin yields in different variants of intercropping in relation to sole crops was also established in previous studies (Silwana and Lucas 2002). Similar regularities were shown in the number and mass of pumpkin fruits per plants. Differences between treatments in number of pumpkin fruits per plants were not statistically significant (Table 1). The greatest number and mass of pumpkin fruits were obtained in the 2/3:1/3 variant in both farming systems, while the lowest values were gained in the 1/2:1/2 variant in the conventional farming system. It is interesting to emphasize that the average number of fruits per plants was lower in the sole pumpkin crops, but the fruit mass were greater than those in intercrops (Tables 1 and 2). The lowest ear number and the smallest grain mass per plant in maize were obtained in a sole maize crop, which is a logical consequence of the intensive intraspecies competition (Tables 3 and 4). Such regularity was particularly evident in the conservation farming system. Due to favorable meteorological conditions in the first year, the number of ears and grain mass per plant in maize were significantly greater in this farming system than in the second year of investigation, both in intercrops and sole crops. However, in respect to the conventional farming system, such trend was observed in the ear number but not in the grain mass per plant.

Table 1. Fruit number per plant of pumpkin in maize-pumpkin intercrop in different farming systems

Intercropping variants	Conventional farming system			Conservation farming system		
	2010	2011	Mean	2010	2011	Mean
1/3 p : 2/3 m	4.0	4.3	4.2	4.5	4.5	4.5
1/2 p : 1/2 m	4.3	3.8	4.0	4.5	5.0	4.8
2/3 p : 1/3 m	5.3	4.5	4.9	4.8	4.5	4.6
Mean.	4.5	4.2	-	4.6	4.7	-
Sole crop	4.8	4.5	4.6	4.5	4.5	4.5
S.E. (D.F.=24)	0.20	0.13	-	0.04	0.09	-

p-pumpkin; m-maize; SE–standard error

Table 2. Fruit mass (kg) of pumpkin in maize-pumpkin intercrop in different farming systems

Intercropping Variants	Conventional farming system			Conservation farming system		
	2010	2011	Mean	2010	2011	Mean
1/3 p : 2/3 m	18.2	11.8	15.0	15.0	13.1	14.0
1/2 p : 1/2 m	16.3	9.4	12.8	15.6	14.1	14.8
2/3 p : 1/3 m	19.0	18.3	18.7	18.3	16.3	17.3
Mean.	17.8	13.2	-	16.3	14.5	-
Sole crop	26.3	25.2	25.7	28.2	25.6	26.9
S.E. (D.F.=24)	1.54	2.51	-	2.16	2.02	-

p-pumpkin; m-maize; SE–standard error

Table 3. Ear number per maize plant in maize-pumpkin intercrop in different farming systems

Intercropping variants	Conventional farming system			Conservation farming system		
	2010	2011	Mean	2010	2011	Mean
1/3 p : 2/3 m	2.1	1.9	2.0	2.0	1.5	1.7
1/2 p : 1/2 m	2.2	1.9	2.0	2.1	1.7	1.9
2/3 p : 1/3 m	2.2	1.9	2.1	2.0	1.5	1.8
Mean.	2.1	1.9	-	2.0	1.6	-
Sole crop	2.0	1.8	1.9	1.9	1.3	1.6
S.E. (D.F.=24)	0.04	0.03	-	0.04	0.06	-

p-pumpkin; m-maize; SE–standard error

Table 4. Grain mass (g) per maize plant in maize-pumpkin intercrop in different farming systems

Intercropping variants	Conventional farming system			Conservation farming system		
	2010	2011	Mean	2010	2011	Mean
1/3 p : 2/3 m	467.9	480.8	474.3	475.5	411.2	443.3
1/2 p : 1/2 m	504.5	500.0	502.2	514.7	454.9	484.8
2/3 p : 1/3 m	525.5	541.1	533.3	493.8	411.2	452.5
Mean.	499.3	507.3	-	494.6	425.7	-
Sole crop	448.7	471.0	459.6	437.1	358.5	397.8
S.E. (D.F.=24)	12.28	11.00	-	11.65	13.95	-

p-pumpkin; m-maize; SE–standard error

The effects of all intercropping variants on maize grain yield were favorable in the conventional farming system. The maize grain yield was higher in the first combination in mixtures 1/3:2/3 pumpkins to maize, in both farming systems and in both years (Table 5). The lower participation of maize in the mixture was resulted in the lower maize grain yield in both farming systems. In second year, the highest (8410.5 kg ha^{-1}) and lowest yield (6401.5 kg ha^{-1}) of a sole crop of maize was obtained in the conventional and conservation farming system, respectively. According to results on pumpkin yields presented in Table 6, it is notable that the conservation farming system had some advantages in relation to the conventional farming system. The only higher yield achieved in the conventional farming system was in the 2/3:1/3 intercropping variant in which pumpkins prevailed, hence the effect of a greater number of this species, in the interaction with reduced tillage, on the yield was adverse. In this way, the highest pumpkin yield in the conventional farming system was in the 2/3:1/3 intercropping variant, particularly in 2010 (33649.5 kg ha^{-1}). The lowest pumpkin yields were obtained in the 1/3:2/3 variant, in both farming systems (10535.3 and 11032.4 kg ha^{-1}), and yield proportionally increased with the increase of the plant population of pumpkins, although the intraspecies competition of pumpkins was very pronounced in intercropping with maize.

Table 5. Maize grain yield (kg ha^{-1}) in maize-pumpkin intercrop in different farming systems

Intercropping Variants	Conventional farming system			Conservation farming system		
	2010	2011	Mean	2010	2011	Mean
1/3 p : 2/3 m	5567.8	5721.8	5644.8	5658.1	4893.0	5275.6
1/2 p : 1/2 m	4504.2	4464.3	4484.3	4595.9	4061.7	4328.8
2/3 p : 1/3 m	3127.3	3220.3	3173.8	2938.6	2447.1	2692.9
Mean	4399.8	4469.1	-	4397.5	3800.6	-
Sole crop	8011.9	8410.5	8211.2	7804.6	6401.5	7103.1
S.E. (D.F.=24)	729.73	784.81		720.66	583.34	

p-pumpkin; m-maize; SE–standard error

Table 6. Pumpkin yield (kg ha^{-1}) in maize-pumpkin intercrop in different farming systems

Intercropping variants	Conventional farming system			Conservation farming system		
	2010	2011	Mean	2010	2011	Mean
1/3 p : 2/3 m	12499.9	8570.6	10535.3	11783.5	10281.3	11032.4
1/2 p : 1/2 m	18104.7	9084.9	13594.8	18371.8	17772.6	18072.2
2/3 p : 1/3 m	33649.5	28858.5	31254.0	30108.3	25022.3	27565.3
Mean.	21418.0	15504.7	-	20087.9	17692.1	-
Sole crop	65871.9	57903.3	61887.6	66379.8	59934.5	63157.2
S.E. (D.F.=24)	8471.14	8205.35		8610.91	7764.73	

p-pumpkin; m-maize; SE–standard error

From the biological aspect, the explanation of the advantages of intercrops over sole crops is similar to the explanation of the survival of the species in the natural communities (Dolijanović et al. 2007; Dahmardeh et al. 2010). The weaker competitive pressure in intercrops, the greater advantage of intercropping. Due to this, Vandermeer (1989) introduced a new term in ecology of intercropping: competitive production principle. This principle points out to the optimal participation of species into intercrops.

The land equivalent ratio (LER) has been the best as a numerical parameter related to the interactions of species in intercrops. The effective LER, the most often applied in replacement series was used in studies of maize-pumpkin intercrops in the experimental field of the MRI, Zemun Polje (Riley 1984). Based on the LER index values presented in Table 7 and Figure 2, it can be concluded that maize-pumpkin intercrop did not expressed effectiveness over sole crops. Namely, the values of the LER index were below one in both years and farming systems.

Table 7. Land equivalent ratio in maize-pumpkin intercrop in different farming systems

Year	Intercropping variants	L_m (relative yield of maize)		L_p (relative yield of pumpkins)		LER	
		Cv	Co	Cv	Co	Cv	Co
2010	1/3 p : 2/3 m	0.69	0.72	0.19	0.18	0.88	0.90
	1/2:1/2	0.56	0.59	0.27	0.28	0.83	0.87
	2/3p:1/3 m	0.39	0.38	0.51	0.45	0.90	0.83
	Average	0.55	0.56	0.32	0.30	0.87	0.86
2011	1/3 p : 2/3 m	0.68	0.76	0.15	0.17	0.83	0.93
	1/2:1/2	0.53	0.63	0.16	0.30	0.69	0.93
	2/3p:1/3 m	0.38	0.38	0.50	0.42	0.88	0.80
	Average	0.53	0.59	0.27	0.30	0.80	0.89
	SE (D.F. = 24)	0.039	0.047	0.048	0.034	0.022	0.015

Cv-conventional farming system; Co-conservation farming system; SE–standard error

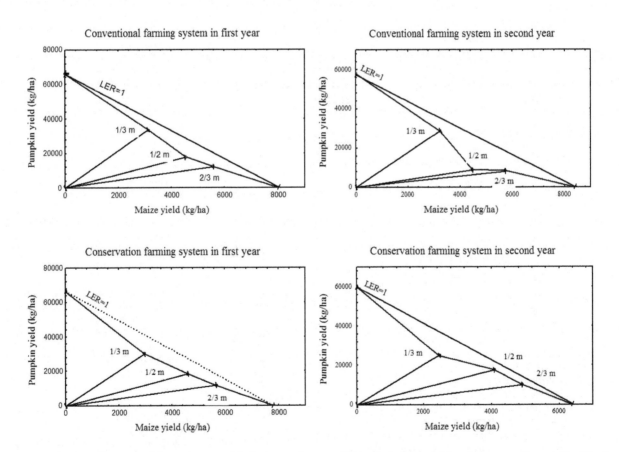

Figure 2. Bivariate diagrams of the maize-pumpkin intercropping system (the dashed line joining the monoculture yield of the two components represents conditions where LER=1; the lines radiating from the origin indicates various proportions of component crops).

The lowest values of the LER index were found in the conventional farming system in 2011, when the yield reduction in comparison to yields of sole crops amounted to 47% (maize) and 73% (pumpkins). Although intercropping does not always provide yield advantages over sole cropping, LER>1 have been reported for many different intercropping systems (Oljača et al. 2000; Ullah et al. 2007; Marer et al. 2007; Dhima et al. 2007; Hugar and Palled 2008; Biabani et al. 2008). It has already been known that maize crop have been more competitive than pumpkin crop. Such competitive ability of maize was especially effective in 1/3:2/3 mixtures in both farming systems and seasons. This variant as well as the 1/2:1/2 variant can be recommended for farming practices, particularly when maize crop is grown. On the other hand, the 2/3:1/3 variant from the aspect of yields cannot be recommended. Efficiency increasing of maize-pumpkin intercrops can be achieved by the increase of the number of plants of both crops per area unit. These results indicate a close relationship between yield and plant density. So, Budakli Carpici et al. 2010, cited that dry matter yield was influenced by plant densities. Dry matter yield increased and reached maximum at 180 000 plants ha-1 and then declined as plant density increased further.

Advantage of the conventional farming system in maize-pumpkin intercrops is obvious. The only exception was a higher yield of maize obtained in conservation farming system in second season. When more effective growing practices are applied (such as conservation farming system) under unfavorable agroecological conditions, deficiency of certain factors will favor the utilization of others. It is always necessary to determine the types of interactions in intercrops in order to find out the best pattern arrangement of maize and pumpkins in which the competition is the lowest, while yield components (quantity and quality) are the highest.

ACKNOWLEDGEMENTS

This study was supported by the Ministry of Education, Science and Technological Development of the Republic of Serbia through the Project TR-31037.

LITERATURE CITED

Biabani, A., M. Hashemi and S.J. Herbert. 2008. Agronomic performance of two intercropped soybean cultivars. International Journal of Plant Production, 2: 215-222.

Biabani, A. 2009. Agronomic performance of intercropped Wheat cultivars. Asian Journal Plant Sci., 8: 78-81.

Birkás, M., A. Dexter, T. Kalmár and L. Bottlik. 2006. Soil quality-soil condition-production stability. Cereal. Research Communication, 34: 135-138.

Budakli Carpici, E., N. Çelik and G. Bayram. 2010. Yield and quality of forage maize as influenced by plant density and nitrogen rate. Turkish Journal of Field Crops, 15: 128-132.

Conteras Magana, E.S. and F. del Castillo. 1991. Preliminary trial of the performance of squashes, snap bean, and cucumbers with mulching and row covers. Rev. Chap., 5: 73-74.

Dahmardeh, M., A. Ghanbari, B.A. Syahsar and M. Ramrodi. 2010. The role of intercropping maize (Zea mays L.) and cowpea (Vigna unquiculata L.) on yield and soil chemical properties. African Journal of Agricultural Research, 5: 631-636.

De Wit, C.T. 1960. On competition. Verslag Landbouw-Kundige Onderzoek, 66: 1–28.

Dhima, K.V., A.S. Lithourgidis, I.B. Vasilakoglou and C.A. Dordas. 2007. Competition indices of common vetch and cereal intercrops in two seeding ratio. Field Crops Research, 100: 249-256.

Dolijanović, Ž., S. Oljača, D. Kovačević and M. Simić. 2007. Effects of different maize hybrids on above-ground biomass in intercrops with soybean. Maydica, 52: 265–271.

Dolijanović, Ž., S. Oljača, D. Kovačević, M. Simić, N. Momirović and Ž. Jovanović. 2013. Dependence of the productivity of maize and soybean intercropping systems on hybrid type and plant arrangement pattern. Genetics, Belgrade, 45: 135-144.

Francis, C.A. 1986. Introduction: Distribution and importance of multiple cropping. In: *Multiple Cropping System*, (Ed. CA Francis), Macmillan Publishing Company, New York, USA, pp. 1-20.

Galloway, B.A. and L.A. Weston. 1996. Influence of cover crop and herbicide treatment on weed control and yield in no-till sweet corn and pumpkin. Weed Technology, 10: 341-346.

Gliessman, S.R. 1986. Plant interaction in multiple cropping systems. In: Multiple Cropping System, (Ed. CA Francis), Macmillan Publishing Company, New York, USA, pp. 82-96.

Hugar, H.Y. and Y.B. Palled. 2008. Studies on maize-vegetable intercropping systems. Karnataka Journal of Agricultural Sci., 21: 159-161.

Ilker, E., F. Aykut Tonk, Ö. Çaylak, M. Tosun and I. Özmen. 2009. Assessment of genotype x environment interactions for grain yield in maize hybrids using ammi and gge biplot analyses. Turkish Journal of Field Crops, 14: 123-135.

Kuscu, H., A. Karasu, M. Oz, A.O. Demir, and I. Turgut. 2013. Effect of irrigation amounts applied with drip irrigation on maize evapotranspiration, yield, water use efficiency, and net return in a sub–humid climate. Turkish Journal of Field Crops, 18: 13-19.

Liebman, M. and P.C. Staver. 2001. Crop diversification for weed management. In: Ecological Management of Agricultural Weeds (Eds. M Liebman, CL Mohler, PC Staver) Cambridge University Press, Cambridge, UK, pp. 336-351.

Marer, S.B., B.S. Lingaraju and G.B. Shashidhara. 2007. Productivity and economics of maize and pigeonpea intercropping under rainfed condition in northern transitional zone of karnataka. Karnataka Journal of Agricultural Sci., 20: 1-3.

Mead, R. 1986. Statistical methods for multiple cropping. In: Multiple Cropping System, (Ed. CA Francis), Macmillan Publishing Company, New York, USA, pp. 317-351.

Mitrović, B., D. Stanisavljević, S, Treskić, M. Stojaković, M. Ivanović, G. Bekavac and M. Rajković. 2012. Evaluation of experimental maize hybrids tested in multi-location trials using AMMI and GGE biplot analyses. Turkish Journal of Field Crops, 17: 35-40.

Mudita, I.I., C. Chiduza, S.J. Richardson-Kageler and F.S. Murungu. 2008. Performance of maize (Zea mays L.) and soya bean [Glycine max (L.) Merrill] cultivars of varying growth habit in intercrop in sub-humid environments of Zimbabwe. European Journal of Agronomy, 7: 229–236.

Oljača, S. 1998. Productivity of intercropped maize and bean on the irrigated and non irrigated conditions, PhD thesis. University of Belgrade, Faculty of Agriculture, 138. p. (in Serbian)

Oljača, S., R. Cvetković, D. Kovačević, G. Vasić and N. Momirović. 2000. Effect of plant arrangement pattern and irrigation on efficiency of maize (*Zea mays*) and bean (*Phaseolus vulgaris*) intercropping system. Journal of Agricultural Sci. Cambridge, 135: 261-270.

Pearce, S.C. and B. Gilliver. 1978. The statistical analysis of data from intercropping experiments. Journal of Agricultural Sci. Cambridge, 91: 625-632.

Powers, L.E., R. McSorley, D.A. Dunn and A. Montes. 1993. The agroecology of cucurbit-based intercropping system in the Yeguare Valley of Honduras. Agriculture Ecosystem and Environment, 48: 139-147.

Riley, J. 1984. A general form of the "Land Equivalent Ratio", Experimental Agriculture, 20: 19-29.

Silwana, T. and E.O. Lucas. 2002. The effect of planting combinations and weeding on the growth and yield of component crops of maize/bean and maize/pumpkin intercrops. Journal of Agricultural Sci. Cambridge, 138: 193-200.

Snaydon, R.W. and E.H. Satorre. 1989. Bivariate diagrams for plant competition data: modification and interpretation. Journal of Applied Ecology, 26: 1043-1057.

Ullah, A., M.A. Bhatti, Z.A. Gurmani and M. Imran. 2007. Studies on planting patterns of maize (Zea mays L.) facilitating legumes intercropping. Journal of Agricultural Research, 45: 113-118.

Vandermeer, J.H. 1989. The Ecology of Intercropping. Cambridge University Press, Cambridge. 231 p.

Videnović, Ž., Ž. Jovanović, Z. Dumanović, M. Simić, J. Srdić, V. Dragičević and I. Spasojević. 2013. Effect of long term crop rotation and fertiliser application on maize productivity. Turkish Journal of Field Crops, 18: 233-237.

Willey, R.W. 1979a. Intercropping-Its importance and research needs. II. Agronomy and research approaches. Field Crop Abstract, 32: 73-85.

Willey, R.W. 1979b. Intercropping-Its importance and research needs. I. Competition and yield advantages. Field Crop Abstract, 32: 1-10.

Yang, C.H., Q. Chai and G.B. Huang. 2010. Root distribution and yield responses of wheat/maize intercropping to alternate irrigation in the arid areas of northwest China. Plant, Soil and Environmental, 56: 253–262.

32

EFFECT OF LONG TERM CROP ROTATION AND FERTILISER APPLICATION ON MAIZE PRODUCTIVITY

Živorad VIDENOVIĆ, Života JOVANOVIĆ, Zoran DUMANOVIĆ, Milena SIMIĆ*, Jelena SRDIĆ, Vesna DRAGIČEVIĆ, Igor SPASOJEVIĆ

Maize Research Institute, Zemun Polje, Department of Cropping practices, Belgrade, SERBIA
*Corresponding author: smilena@mrizp.rs

ABSTRACT

The effect of crop rotation and the application of fertilisers on maize yield were investigated in a 12-year study (1998-2009) on the chernozem soil type at Zemun Polje, Serbia. The treatments included four cropping systems: continuous maize cropping (CS1); two crop rotation: maize - soybean (CS2) and maize - winter wheat (CS3), and three crop rotation maize - winter wheat - soybean (CS4) and the following fertilising treatments for maize: F1 - no fertiliser, F2 - 180 kg ha^{-1} NPK, F3 - 270 kg ha^{-1} NPK and F4 - 360 kg ha^{-1} NPK. The amount of applied nitrogen fertiliser in soybean was twice lower than in maize. The grain yield, on the average for all years, was the lowest (5.37 t ha^{-1}) in continuous maize cropping. In a dominant type of the cropping system in Serbia (CS3), the maize grain yield was 6.82 t ha^{-1} and in CS2, was higher (7.60 t ha^{-1}), even though the amount of nitrogen fertilisers applied, was lower by 50%. The highest average yield was obtained in CS4 (9.03 t ha^{-1}). The application of fertilisers generally significantly influenced maize yield in comparison with control. These results favoured cropping systems with legumes preceded maize due to lower investments necessary to obtain higher yields.

Key words: cropping system, fertilizing, maize, yield

INTRODUCTION

Maize growing practices include several important measures that significantly affect the yield (Videnović et al., 2011). The crop rotation is just one of them and it, contrary to others, does not require financial investments. Only long-term planning of the production and the distribution of crops in a farm are required (Jovanović et al., 2000a, 2000b, 2004; Videnović et al., 2007). However, attention is not often paid to the preceding crops suitable to certain crops and therefore needs and prices are mainly affected by their selection and the sowing scope. In Serbia, there are three major maize growing systems: continuous maize cropping (15%), two crop rotation (maize -winter wheat - 60% and maize - soybean - 15%) and three crop rotation (maize - winter wheat - soybean - 5%) (Jovanović et al., 2004; Kovačević et al., 2005, 2010). Furthermore, in smaller fields, maize was grown after some other crops such as alfalfa and other legumes, vegetables, pastures etc. (5%). Studies carried out by Stranger and Lauer (2008) showed that extended rotations involving forage crops reduce N inputs, increase maize yield and are more agronomically sustainable than current short-terms rotations.

It is known that each crop utilises nutrients from the soil to varying extent and that soybean, as a legume crop, leaves significant amounts of nitrogen in the soil for the succeeding crop (Carpenter-Boggs et al., 2000; Varvel, 2000; Jovanović et al., 2000a; Stranger et al., 2008; Riedell et al., 2009). Therefore, smaller amounts of nitrogen fertilisers are necessary in soybean cultivation. Additionally, the composition and abundance of microflora differ among various individual crops, which affect the scope of transformation of organic and mineral matters into forms available to plants. Carpenter-Boggs et al. (2000) reported that the rotation and N fertilisation significantly affected the net N mineralisation in soil samples. Adiku et al. (2009) state that the crop rotation and residue management practices can significantly affect maize performances.

The cropping system level is also very important for weed control, considering that weeds can cause great damages to corps and decreased yields (Stefanović et al. 2011). The management aimed at the increasing cropping system diversity with the application of different herbicides can lead to the development of more efficient and sustainable weed and crop management practices (Smith and Gross, 2006).

Although, research in preventive methods and cropping practices have improved weed control in row crops, the effective long-term weed management should include the crop rotation, fertiliser placement, competitive varieties etc. (Melander et al., 2005).

Based on a long-term experiment, the objective of this study was to compare the advantages and disadvantages of maize growing in continuous cropping, two crops rotation with soybean or winter wheat and the three crops rotation and to determine which growing system is the most suitable for the successful and profitable production of maize in Serbia considering the effects of the application of different rates of mineral fertilisers and their interactions with observed maize growing systems.

MATERIALS AND METHODS

Site Description and Experimental Design.

The experiment was conducted at Zemun Polje, in the vicinity of Belgrade, Serbia (44°52'N 20°20'E), during the 1998-2009 period. The soil was slightly calcareous chernozem with 47 % clay and silt and 53 % sand. The trial was set up according to the split-plot arrangement with four replications. The size of the sub-plot was 14.332 m^2. The treatments included four cropping systems (CS): CS1 - continuous maize cropping; CS2 - two crop rotation: maize - soybean; CS3 - two crop rotation: maize - winter wheat; CS4 - three crop rotation: maize - winter wheat - soybean and the following fertiliser (F) treatments for the maize crop: F1 - no fertilisers, F2 - 180 kg ha^{-1} (80 kg ha^{-1} N, 60 kg ha^{-1} P_2O_5 and 40 kg ha^{-1} K_2O), F3 - 270 kg ha^{-1} (120 kg ha^{-1} N, 90 kg ha^{-1} P_2O_5 and 60 kg ha^{-1} K_2O) and F4 - 360 kg ha^{-1} (160 kg ha^{-1} N, 120 kg ha^{-1} P_2O_5 and 80 kg ha^{-1} K_2O). The amount of applied nitrogen fertiliser in soybean was twice lower: F2 - 40 kg ha^{-1} N, F3 - 60 kg ha^{-1} N and in F4 - 80 kg ha^{-1} N.

Cropping Practices and Measurements

Total amounts of P_2O_5 and K_2O fertilisers were spread over the soil surface in the autumn. Nitrogen fertilisers for all crops were applied in spring prior to the seedbed preparation. During the 12-year period of investigation two- and three-crop rotations rotated in one plot 6 and 4 times, respectively.

Shallow stubble ploughing to the depth of 15 cm was performed after wheat and soybean harvest. Ploughing was performed in autumn to the depth of 25 cm. The soil preparation was done 7-10 days prior to sowing with a seedbed tiller. The sowing density of the late maturity (FAO 700) maize hybrid ZPSC 704 was 62,111 plants per ha; of soybean late maturity cultivar Lana 500,000 plants/ha and of winter wheat variety Pobeda 600 grains per m^2. The pre-emergence application of herbicides Atrazine 500 SC in the amount of 1 L ha^{-1} (atrazine 500 g a.i.) and Harness 2 L ha^{-1} (acetochlor 900 g a.i.) had been done in maize until 2007, when atrazine was replaced with terbuthylazine. During the growing season, inter-row cultivation was performed to suppress weeds, so they would not affect the growth and the development of the plants or would not reduce the maize yields.

All observed data were processed by the analysis of variance (ANOVA) for three factors – year, Y (12), cropping system, CS (4) and amount of fertilisers, F (4). Treatment means were compared using the Fisher's least significant difference (LSD) test (P = 0.05) (Steal and Torrie, 1980).

Meteorological Conditions

Twelve-year meteorological conditions during the growing season (April-September) varied significantly, hence years were divided into three groups based on the precipitation sums and the suitability of conditions for maize production: Y1 - the first group of years with the precipitation up to 300 mm, Y2 - the second group of years with the precipitation sum ranging from 300 to 400 mm and Y3 - the third group of years with the precipitation sum over 400 mm, (Figure 1).

Y1 - two unfavorable years: 2000 with 203.3 mm and 2003 with 271.8 mm. These precipitation sums were not sufficient for maize production. Figure 1 shows that there were periods with the extreme precipitation deficit, especially in critical developmental stages of maize.

Y2 - five years with moderately favourable conditions: 1998 with 319.5mm; 2002 with 373.6 mm; 2007 with 364.5 mm; 2008 with 305.9 mm and 2009 with 322.6 mm. Not only the precipitation sums were greater in these years, but also their distribution was more favourable, which resulted in higher yields than in Y1 years.

Y3 - five years with favourable conditions for maize production: 1999 with 637.3 mm; 2001 with 651.0 mm; 2004 with 477.7 mm; 2005 with 487.0 mm and 2006 with 445.3 mm. These years were favourable due to both, total precipitation sums and their good distribution over the growing season.

RESULTS AND DISCUSSION

Maize grain yield was significantly affected by the investigated parameters. The statistical analysis showed that grain yield of maize varied over years, cropping systems and fertilising treatments (Tables 1 and 2). The highly significant difference in the grain yield of maize was found to be among years. The lowest yield (3.67 t ha^{-1} or 32.7%), on the average, was recorded in 2007. On the other hand, the highest yield (11.41 t ha^{-1} or 100.0%) was obtained in 2005. It is interesting to mention that there were some years with precipitation sums lower than in 2007, but with somewhat higher yields, and also there were years with precipitation sums higher than in 2005, but the yield was lower. This points out that the precipitation distribution, duration of dry spells, wind frequency, extent of evapotranspiration, as well as some other factors could significantly affect maize grain yields. In addition, it was confirmed in the trial that in Y1 years a significant difference in grain yield was not obtained between cropping systems and the application of fertiliser. In contrast, in a very favourable year of 2005, the grain yield of maize was absolutely the highest in all cropping systems and fertiliser treatments.

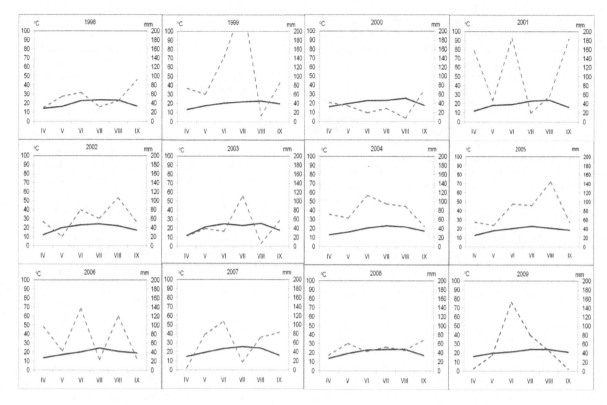

Figure 1. Walter climate diagrams of precipitation sums and average air temperatures for the April-September period during the 12-years period of investigation (1998-2009)
Climate data for Belgrade - Source: Republic Hydrometeorological Service of Serbia

Table 1. The results of the analysis of variance (ANOVA), and significance levels for maize grain yield

Sources of variance	df	F
Replication	3	1,327
Year (Y)	11	185,499*
Cropping system (CS)	3	432,329*
Y x CS	33	14,857*
Fertilisers (F)	3	208,238*
Y x F	33	18,887*
CS x F	9	12,643*
Y x CS x F	99	3,383*

* Statistically significant at P = 0.05

The studied cropping systems also significantly affected maize grain yields (Table 2, Figure 2). The yield of maize, on the average for all years and fertiliser treatments, was the lowest in CS1 (5.37 t ha^{-1}). The yields recorded in CS2, CS3 and CS4 amounted to 7.60, 6.82 and 9.03 t ha^{-1}, respectively (Table 2).

Table 2. Maize grain yield (t ha^{-1}) over the years (Y), cropping systems (CS) and levels of applied fertilizers (F), 1998-2009.

Crop systems	Fertilizer levels				Average
	F1	F2	F3	F4	
CS1	3,64	5,81	6,10	5,95	5,37
CS2	6,53	8,07	7,99	7,81	7,60
CS3	5,85	6,96	7,27	7,21	6,82
CS4	8,52	9,20	9,18	9,23	9,03
Average	6,14	7,51	7,63	7,55	7,21

LSD$_{0.05}$: Y= 0.55; CS = 0.22; C = 0.22; YxCS = 0.73; YxF = 0.49; CSxF = 0.32; YxCSxF = 0.823

In relation to maize grain yield in CS3 (6.82 t ha^{-1} or 100%), maize grain yield was 1.45 t ha^{-1} (21.3%) lower in CS1. Maize grain yield after soybean (CS2), was higher (0.78 t ha^{-1} or 11.40%) even though the 50% lower amount of nitrogen fertiliser was applied. Moreover, in CS4 grain yield was 2.21 t ha^{-1} higher than in CS3 or 32.40%. Numerous factors affected yields over tested treatments.

Reductions in yields in the continuous maize cropping were primarily recorded after dry winters, particularly if summers were also dry as it was in 1998, 2000 and 2007. Berzsenyi et al. (2000), obtained similar results. In long-term study, they found that yield-increasing effect of the crop rotation was inversely proportional to the ratio of maize or wheat in the sequence and was greatest in the Norfolk rotation and alfalfa - maize - wheat triculture that included legume crop. Varvel (2000) concluded that N, obtained from either fertiliser or legumes in continuous cropping or rotation systems, is probably one of the most, if not the most important aspect in reducing yield variability. Growing maize in extended rotations that include forage legumes may be a more sustainable practice than growing maize in either continuous cropping or two-year rotation with soybean (Riedell et al., 2009).

CR(Avg F1-F4)

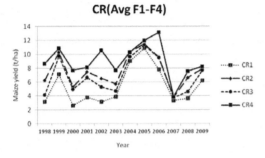

Figure 2. Effect of cropping systems on maize yield in t ha^{-1} (averagely for all fertiliser variants): CS1 - continuous maize cropping; CS2 - two crop rotation (maize - soybean); CS3 - two crop rotation (maize - winter wheat); CS4 - three crop rotation (maize - winter wheat - soybean)

The maize grain yield was significantly higher in all fertilising treatments (F2-, F3 and F4 - 7.51, 7.63, 7.55, respectively) in comparison to control (F1 - 6.14 t ha^{-1}). On the other hand, there are no significant differences among applied rates of fertilisers, due to the effects of meteorological conditions. Unfavourable meteorological conditions during the long-term studies might reduce the effects of fertilisers on maize yields in favourable years, although obtained differences were not, on the average, significant. Average maize yields were not therefore high in the treatment with the highest amount of fertilisers (F4), (Carpici et al., 2010). It could be concluded that the application of high rates of mineral fertilisers under the Zemun Polje conditions on slightly calcareous soil is not economically justified (Videnović et al., 2007).

The specificity of the obtained results was that the yield of maize grown after soybean (CS2) was higher than yield in CS3 by 0.78 t ha^{-1} (11.4%), although amounts of applied nitrogen fertilisers were 50% lower than those used for maize and winter wheat. In such a way, higher yields of maize were obtained with lower inputs. Therefore, soybean is a very suitable and desirable preceding crop for maize and thus this growing system deserves special attention. Significantly higher yields were observed at high levels of fertilisation, especially in rotations where the proportion of maize or wheat was 50% or higher (Berzsenyi et al., 2000; Ididcut and Kara, 2011).

The reduced application of N fertilisers (zero N to the grain legume and less N to the following crop), improved possibilities for using reduced tillage techniques and greater diversification of the crop rotation, which helps to reduce problems caused by weeds and pathogens and therefore pesticide applications (Nemecek et al., 2008; Videnović et al., 2011).

ACKNOWLEDGEMENTS

This study was supported by the Ministry of Education, Science and Technological Development of the Republic of Serbia through the Project TR-31037.

LITERATURE CITED

Adiku, S.G.D, J.W. Jones, F.K. Kumaga, A. Tonyigah. 2009. Effects of crop rotation and fallow residue management on maize growth, yield and soil carbon in a savannah-forest transition zone in Ghana. Journal of Agricultural Sci, 147: 313-322.

Berzsenyi, Z., B., Gyorffy, D. Lap. 2000. Effects of crop rotation and fertilisation on maize and wheat yields and yield stability in a long-term. European Journal of Agronomy, 13: 225-244.

Carpenter-Boggs, L., L.J.Jr. Pikul, F.L. Vigil, E.W. Riedell. 2000. Soil nitrogen mineralization influenced by crop rotation and nitrogen ferilization. Soil Sci. Society of America Journal, 64: 2038-2045.

Carpici, B.E., N. Celik, B. Gamze. 2010. Yield and quality of forage maize as influenced by plant density and nitrogen rate. Turkish Journal of Field Crops, 15: 128-132.

Idikut, L., S.N. Kara. 2011. The effect of previous plants and nitrogen rates on second crop corn. Turkish Journal of Field Crops, 16: 239-244.

Jovanović, Ž., G. Cvijanović, D. Kovačević. 2000a. Effects of precedings and mineral fertilisers on available content NPK nutritions in chernozem type soil. Proceedings of the Ist Eco Conference „Safe food", September 27-30, Novi Sad, Serbia, 393-398.

Jovanović, Ž., Ž. Videnović, M. Vesković, B. Kresović, D. Kovačević. 2000b. Crop Rotations Effects in Biological Cropping Systems on Maize Yield. Proceedings of the 3th International Crop Science Congress, August 17-22, Hamburg, Germany, 52.

Jovanović, Ž., M. Vesković, P. Jovin, D. Kovačević. 2004. Effect of Different Growing Systems on Maize Yield According to The Concept of Sustainable Agriculture. Proceedings of the VIIIth ESA Congress, Copenhagen, Denmark, 611-612.

Kovačević, D., S. Oljača, Ž. Dolijanović, Ž. Jovanović, V. Milić. 2005. Influence of crop rotation on the yield of some important crops. International Conference TEMPO, October 6-8, Čačak, Serbia, 422-428. (In Serbian)

Kovačević, D. 2010. Crop Science. 1. ed. Faculty of Agriculture, Belgrade, Serbia, 771p. (In Serbian)

Melander, B., A.I. Rasmussen, P. Bàrberi. 2005. Integrating physical and cultural methods of weed control— examples from European research. Weed Science, 53: 369-381.

Nemecek, T., J.S. von Richthofen, G. Dubois, P. Casta, R. Charles, H. Pahl. 2008. Environmental impacts of introducing grain legumes into European crop rotations. European Journal of Agronomy, 21: 380-393.

Riedell, E. W., L.J.Jr. Pikul, A.A. Jaradat, E.T. Schumacher. 2009. Crop rotation and nitrogen input effects on soil fertility, maize mineral nutrition, yield, and seed composition. Agronomy Journal, 101: 870-879.

Stranger, F.T., G.J. Lauer. 2008. Corn grain yield response to crop rotation and nitrogen over 35 years. Agronomy Journal, 100: 643-650.

Smith, G.R., L.K. Gross. 2006. Weed community and corn yield variability in diverse management systems. Weed Science, 54: 106-113.

Steel, R. G. D., J. H. Torrie. 1980. Principles and Procedures of Statistics, Second Edition, New York: McGraw-Hill, USA.

Stefanović, L., M. Simić, B. Šinžar. 2011. Weed control in maize agroecosystem. 1. ed. Serbian Society of Genetics. Belgrade, Serbia, 678p.

Varvel, E.G. 2000. Crop rotation and nitrogen effects on normalized grain yields in a long-term study. Agronomy Journal, 92: 938-941.

Videnović, Ž., Ž. Jovanović, G. Cvijanović, L. Stefanović, M. Simić. 2007. The contribution of science to thedevelopment of contemporary maize growing practices in Serbia. In: Science as sustainable development background. Maize Research Institute „Zemun Polje", Belgrade, Serbia, 267 – 285. (In Serbian)

Videnović Ž., M. Simić, J. Srdić, Z. Dumanović. 2011. Long term effects of different soil tillage systems on maize (*Zea mays* L.) yields. Plant, Soil and Environment, 57: 186-192.

33

EFFECT OF PHOSPHORUS DOSES AND APPLICATION TIME ON THE YIELD AND QUALITY OF HAY AND BOTANICAL COMPOSITION OF CLOVER DOMINANT MEADOW IN HIGHLANDS OF TURKEY

Ali KOC

Department of Field Crops, Faculty of Agriculture, Osmangazi University, Eskisehir, TURKEY
Corresponding author: alikoc@ogu.edu.tr

ABSTRACT

This study was conducted at the Atatürk University's farm in Erzurum over 2 years (2004-2005) to evaluate the effects of phosphorus doses and application season on the dry matter production, botanical composition and crude protein, ADF and NDF content of an alsike clover-dominated natural hay meadow. The experiment was planned in a Randomized Complete Block Design with split-plot arrangement of application seasons (autumn and spring) was considered as a main plot and five phosphorus doses (0, 11, 22, 33 and 44 kg P ha^{-1}) as subplots. P fertilization significantly affected the dry matter yield while the application season had no effect on dry matter yield. P application increased the legumes and decreased the other families' percentage in the botanical composition, whereas it did not affect the grasses percentage. Due to the nature of alsike cover, the proportion of the legumes decreased and the proportion of grasses increased significantly in the second year. P application had no effect on the crude protein and ADF content of hay, while NDF content was affected by both doses and the season of P application. Depending on changes in botanical composition between years, the chemical content of the hay changed between years. The results indicate that P fertilizer has positive effects on dry matter production and no serious effect on the forage quality of meadows and there are no significant differences between autumn and spring application. Therefore, an 11 kg ha^{-1} application of P in the spring can be suggested for sustainable dry matter production in legumes-dominated natural hay meadows in highland areas.

Keywords: Botanical composition, dry matter, hay meadow, hay quality, phosphorus fertilizer

INTRODUCTION

Animal husbandry plays a significant universal role in highlands areas since short and cool growing seasons restrict diversity and production in cropping systems. In highland areas, animal raisers have to store great amounts of roughage to meet animal needs for the long feedlot period caused by harsh climatic conditions. Meadows have a significant share in stored hay for winter period in the Eastern Anatolian highlands in Turkey, although meadows have lost productivity due to mismanagement practices such as early grazing, excessive irrigation and unsuitable fertilization applications (Koc and Gokkus, 1998). Many studies have been conducted on improving productivity on the meadows in the region (Altin, 1975; Gokkus, 1990; Gokkus and Koc, 1995) and quite significant results were obtained from fertilizer studies (Gokkus, 1989; Koc et al., 2005).

The response of meadows to the type and quantity of fertilizer changes depend on botanical composition, precipitation and soil properties (Benedycka et al., 1992; Altin et al., 2005). Generally, legume abundant stands give good response to phosphorus fertilizer application

because legumes obtain their nitrogen requirements via symbiotic pathway (Miller and Reetz, 1995) and on the other hand, under sufficient nitrogen supply in root zone, phosphorus uptake by plants increases (Fageria, 2001). Therefore, phosphorus application increases both hay yield and legumes component within the botanical composition of the stand (Papnastasis and Papachristou, 2000; Hatipoglu et al., 2001; Aydin and Uzun, 2005).

Fertilization may improve not only dry matter yield but also affect the chemical content of produced hay. In general, phosphorus application causes an increase in crude protein content due to either enhancing nitrogen uptake by plants (Benedycka et al., 1992) or increasing legume component of the sward (Aydin and Uzun, 2005). Inconsistent results have been reported on the effect of fertilization on ADF and NDF content of hay. While some studies reported a positive effect (Ball et al., 2001; Rayburn, 2004), others reported negative or no effect (Pieper et al., 1974; Dasci and Comakli, 2011; Budakli-Carpici, 2011).

Solubility of phosphate minerals increases with temperature and deficiency are common during the cool season (Connor et al., 2011). Therefore, it is essential to have enough plant available P in the root zone in the beginning of growing in the spring to achieve sustainable production. In general, the application season of P fertilizer was suggested by many researchers in the autumn (Altin, 1975; Vallentine, 1989; Hatipoglu et al., 2001; Aydin and Uzun, 2005), whereas, some researchers suggest autumn or early spring application (Barker and Collins, 2003; Altin et al., 2005) and others suggests early spring application of P fertilizers especially in regions with a very cold winter (Kemp et al., 2005).

The application of P fertilizer in early spring to hay meadows can be preferred since it saves labour and costs in the autumn. Conversely, surface runoff increases in the spring during snow melting due to frozen soil and dislodges P fertilizer and causes pollination in streams in addition to nutrient loss. Therefore, the objective of this experiment was to measure the effect of autumn or early spring P fertilizer applications on hay production, hay quality and botanical composition in clover dominate natural hay meadows in regions with a cold winter.

MATERIALS AND METHODS

The field experiment was conducted on natural hay meadows of Ataturk University in Erzurum located in the Eastern Anatolia region of Turkey ($39^\circ 55^1$ N and $41^\circ 61^1$ E at an altitude of 1800 m), from October 2003 to July 2005. The average temperature and total precipitation at the study site were 5.7 $^\circ$C and 425 mm, respectively. During the experimental years, total annual precipitation was 441 and 480 mm, and average temperatures were 4.4 and 5.1 $^\circ$C, respectively. The distribution of precipitation was uneven, with the vast majority of annual precipitation occurring from October to the beginning of July. The water table, which is recharged by subsurface flow, and rises almost to the surface in the beginning of spring and drops down to approximately 1 m at the end of the growing season, is a significant water resource for the plants in the experimental area. The total precipitation was higher than the long term average in both experimental years and the average temperature during the experimental period was lower than the long term average.

Some major soil characteristics determined by the method described by Soil Survey Laboratory Staff (1992) were as follows; the soil texture was loam, organic matter was 3.3%, lime was 3.9%, EC was 2.9 mhos cm^{-1}, pH was 7.6 in soil saturation extract, the corresponding available K and Olsen P content were 650 and 20.5 kg ha^{-1}, respectively. The soils of the experimental site were poor in phosphorus, although it was rich in potassium.

The experiment was established on a natural meadow which was dominated by alsike clover (*Trifolium hybridum*) with some cool season grasses such as *Alopecurus pratensis, Poa pratensis, Hordeum violaceum*

and other forbs such as *Ranunculus kotschyi,* and *Cerastium sp.* Alsike clover is a short-lived perennial, although it is usually considered to be a biennial plant (Townsend, 1985).

The experiment consisted of four replications of a Randomized Complete Block Design with a split-plot arrangement of application season (autumn and spring) as whole plot and five phosphorus doses (0, 11, 22, 33 and 44 kg P ha^{-1}) as subplots. Phosphorus fertilizer was applied as triple superphosphate in the beginning of October or as soon as the snow melted in the spring. Each plot was 5 X 3 m in size, with a 0.5 m buffer inside each edge and a 2 m buffer outside.

When the dominant plant species were at the flowering stage, plant samples were taken by clipping four 0.5 by 0.5 m areas within each plot down to the soil surface. After harvesting, plants in each quadrate were separated by hand and classified as grasses, legumes and the others. Thereafter, the samples were oven dried at 70°C to a constant weight and weighted in order to determine dry matter yield and botanical composition (Jones, 1981).

After weighting the oven dried samples were mixed and ground to pass through a 1 mm sieve for chemical analysis. Total N content of samples was determined using the Kjeldahl method and multiplied by 6.25 to give the crude protein content (Jones, 1981). ADF and NDF content were measured using an ANKOM 200 fiber analyser (ANKOM Technology, Fairport, NY) following the procedure described by Van Soest et al. (1991).

All data was subjected to analysis of variance based on general linear models in order to determine factorial arrangement of treatments and repeated to determine the factorial arrangement of treatments using the SPSS statistical package (SPSS 1999). Means were separated using Duncan's Multiple Range Test.

RESULTS

There were no significant differences between the autumn and spring application of P, although P doses had a significant effect on dry matter production in the alsike clover dominant hay meadow (Table 1). Based on the 2 years average, dry matter production in P_0 plots was 4.69 t ha^{-1}, while it increased 5.34 t ha^{-1} in P_{11} treatment. There were no significant increases in dry matter production after P_{11} doses in line with increased P doses. Dry matter production was significantly higher in the second year than in the first year and the response to P doses were different between years. In the first year, the highest dry matter yield was recorded in P_{44} plots, although the result of this treatment was statistically similar to that of P_{11} plots, while the highest dry matter yield was recorded in P_{22} plots in the second experimental year (Figure 1). This different response to P doses between years was responsible for year X P doses interaction.

Table 1. The effect of doses and application season of phosphorus fertilizer on dry matter yield and botanical composition of a legume-dominated natural hay meadow.

	Dry Matter Yield (t ha^{-1})	Botanical Composition		
		Grasses	Legumes	Other Families
P_0	4.69 B	43. 22	35.40 B	21.38 A
P_{11}	5.34 A	38.53	40.13 AB	21.65 A
P_{22}	5.57 A	41.13	40.11 AB	18.77 AB
P_{33}	5.29 A	41.77	43.78 A	14.46 B
P_{44}	5.42 A	40.96	42.22 A	16.82 B
Average	**5.26**	**40.95**	**40.33**	**18.62**
Spring	5.27	42.28	41.59	16.25 B
Autumn	5.25	39.96	39.06	20.98 A
Average	**5.26**	**40.95**	**40.33**	**18.62**
2004	3.38 B	17.04 B	64.79 B	18.29
2005	7.14 A	65.20 A	15.86 A	18.94
Average	**5.26**	**40.95**	**40.33**	**18.62**
P	**	ns	*	**
S	ns	ns	ns	**
Y	**	**	**	ns
P x S	ns	ns	ns	ns
P x Y	**	ns	ns	ns
S x Y	ns	ns	**	**
P x S x Y	ns	ns	ns	ns

P: Phosphorus, S: Season, Y: Year, ns: non-significant, *: $p<0.05$, **: $p<0.01$. Means in the same column with different letters are significant.

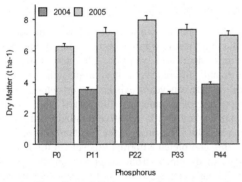

Figure 1. The effect of P doses on dry matter production in relation to years.

The application season of P fertilizers had no significant effect on grasses and legumes percentage, although it had a significant effect on the other families' percentage in the botanical composition. Autumn application caused significant increases in the percentage of the other families in the botanical composition. P application caused increases in the fraction of legumes and decreases in the fraction of the other families in the botanical composition (Table 1). The grasses percentage in the botanical composition was higher in the second year (65.20%) compared to the first year (17.04%). No significant interaction effects among treatments occurred for the percentage of grasses in the botanical composition. Legumes content in produced hay was lower in the second year than the first year. There was no significant effect of P application season on legumes proportion in the first year, although the legumes percentage was slightly higher

in the plots that received P in the spring in the second year. This different response of legume content in produced hay to the P application season between years caused P application season X year interaction (Figure 2). There were no considerable differences in the other families' content of produced hay between years, although year X season interaction was significant because spring P application caused significantly decreases in the percentage of the other families in the botanical composition in the second year, while there were no differences in the first year between P application seasons (Figure 3).

Crude protein content did not change depending on P application, although spring application of P caused slightly increases in crude protein content (Table 2). The hay harvested in the first experimental year had a higher crude protein content than that of the second year. No significant interaction effects among treatments occurred for crude protein content of the hay. NDF content of hay slightly increased with P application, although the increases were not consistent with P doses. The effect of years and P application season were not significant. While the higher NDF content was recorded in the plots received 22 or 33 kg P ha^{-1} in the first year, it was recorded in the plots received 11 or 44 kg P ha^{-1} in the second year. Therefore, year X P doses interaction was significant for NDF content (Figure 4). ADF content of hay was affected by neither doses nor application times of P fertilizer, although the year effect was significant. The higher ADF content was recorded in the second year of harvested hay.

However, no significant interaction effect among treatments was recorded for ADF content.

DISCUSSION

The data from the present study indicated that hay production in a clover-dominated natural meadow increased using phosphorus application, although the application of season had no significant effect on hay production. There was no consistent response to P doses between years because better results for P doses were recorded in the plots that received 11 kg P ha^{-1} in the first year, while it was recorded in the plots received 22 kg P ha^{-1} in the second year (Figure 1). The experimental area soils were poor in plant available P content and the contribution of legumes to botanical composition was higher, therefore, the response to P application regarding dry matter production was an expected result because P application always give a dedicated yield increases in plants under P deficient soils (Benedycka et al., 1992; Altin et al., 2005, Henkin et al., 2010; Venterink, 2011; Sigua et al., 2011), especially if legumes abundance is higher in sward composition then this response would be clearer due to better response of legumes to P fertilizer (Marschner and Romheld, 1983; Henkin et al., 2010). On the other hand, harsh and long winters are a general characteristic of the experimental area and always cause lower soil temperature in the beginning of growing season. This situation increases severity of P deficiency because lower microbial activity and plants shows good response to applied P fertilizer (Connor et al., 2011).

Table 2. The effect of doses and application season of phosphorus fertilizer on crude protein, NDF and ADF content of a legume-dominated natural hay meadow forage.

	CP (%)	NDF (%)	ADF (%)
P$_0$	11.91	50.21 C	39.64
P$_{11}$	12.32	52.11 AB	40.17
P$_{22}$	12.05	51.11 BC	40.07
P$_{33}$	12.72	53.01 A	40.07
P$_{44}$	12.61	53.25 A	40.40
Average	**12.32**	**51.94**	**40.07**
Spring	12.54 A	52.32	40.31
Autumn	12.10 B	51.55	39.83
Average	**12.32**	**51.94**	**40.07**
2004	13.40 A	51.72	38.29 B
2005	11.24 B	52.16	41.85 A
Average	**12.32**	**51.94**	**40.07**
P	ns	**	ns
S	*	ns	ns
Y	**	ns	**
P x S	ns	ns	ns
P x Y	ns	*	ns
S x Y	ns	ns	ns
P x S x Y	ns	ns	ns

P: Phosphorus, S: Season, Y: Year, ns: non-significant, *: $p<0.05$, **: $p<0.01$. Means in the same column with different letters are significant.

While higher dry matter production obtained in the plots received 11 kg P ha^{-1} in the first year, it was recorded in the plots that received 22 kg P ha^{-1} in the second year (Figure 1). This different response to P doses

between years could be related to increases in nitrogen availability in the root zone due to release of nitrogen from dead nodules because alsike clover in the botanical composition decreased sharply in the second year since the plant completed a normal lifecycle in the first year. It is well known that nitrogen availability increases P uptake by plants (Venterink, 2011), therefore, the response of dry matter yield to P doses increased in the second year.

The proportion of grasses in dry matter yield showed no response to P application, although the proportion of legumes increased and the proportion of the other families decreased with the increasing P availability (Table 1). P fertilizer stimulates the growth of legumes (Russell et al., 1965; Aydin and Uzun, 2005), therefore, higher legumes contribution to dry matter production were observed in the P applied plots in the experiment. Similar results were also reported by other researchers (Henkin et al., 2010; Venterink, 2011; Sigua et al., 2011). The decreases in the proportion of the other families mainly originated from increases in legume proportion in the sward. The over-compensatory effect of legumes on the other families rather than grasses may originate from nitrogen obtaining differences between the family groups because legumes obtain their nitrogen by via the symbiotic pathway (Miller and Reetz, 1995). Conversely, grasses use water and nutrients in the root zone efficiently by an intensive root system rather than the other root system (Koc et al., 2008). Therefore, the over-compensatory effect of legumes may mainly address the plants belong to other families.

While the proportion of grasses in dry matter production increased in the second year, the proportion of legumes decreased sharply, although there was no significant difference in the proportions of the other families between years. Alsike clover, which is the main component of legumes in the sward, is a short lived perennial plant that is sensitive to shade and temperature (Davies, 2001). As most of the plants of this species most probably completed a normal lifecycle in the first year and started to disappear from vegetation in the second year and the gaps originated by disappearance of alsike clover might be recruited grasses. Conversely, decomposed nodules and roots of alsike clover must encourage grasses growth due to improving the soil nitrogen status. Consequently, the contribution of grasses to dry matter production increased. Similarly, as the nitrogen availability increased in the sward, a suppressive effect of grasses increase was reported by other researchers (Aerts et al., 2003; Beltman et al., 2007). Most probably, after depleting nitrogen gained form decomposition of roots and dead nodules of alsike clover, new alsike clover seedlings emerged from seed stocks because grasses become sparse since the tiller density and size of grasses decreased and plant dominance changed in advanced time. However, after 2 or 3 years of alsike clover dominance, 2 or 3 year grasses dominance existed in the experimental area (personal observation). The other families' percentage in the botanical composition decreased slightly, while the legumes percentage increased partly in the spring P applied plots in the second year (Figure 2 and

3). These differences may be addressed to changing nutrient availability depending on the applying season. Similar results have also been reported by other researchers (Aerts et al., 2003; Beltman et al., 2007). Although a partly different response of families to the P applying season were recorded in this study, there were no seriously deviation from a general trend in the changes of botanical composition dependent on P fertilizer.

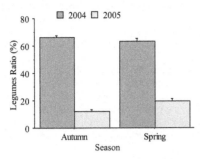

Figure 2. The effect of P application season on the proportion of legumes in botanical composition in relation to years

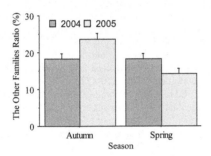

Figure 3. The effect of P application season on the proportion of the other families in botanical composition in relation to years

Although P application caused a slightly increase in the crude protein content, this increase was not significant statistically, whereas the hay harvested in the first year contained more crude protein than that harvested in the second year. This difference between years mainly originated from decreases in the proportion of legumes in dry matter production in the second year (Table 1) because legumes always have higher protein content than grasses (van Gessel, 1970; Tan and Menteşe, 2003). Similarly, the crude protein content was slightly higher in the hay harvested spring P applied plots due to higher legumes contribution to botanical composition.

The NDF content of hay increased with P application. Similarly, the changes in the NDF content of hay depending on P doses differed between years (Figure 4). These differences may originate from changes in plant composition in dry matter or translocation models of plants depending on changing nutrient ability because P plays a significant role in several physiological and biochemical plant activities (Mehrvarz and Chaich, 2008). While the hay harvested from different plots had similar ADF content, the NDF content changed depending on P doses and application times. These results implied that the plants store more hemicelluloses as P availability

increases because NDF consists of ADF plus hemicelluloses (Collins and Fritz, 2003). Neither doses nor application time of P fertilizer had any effect on the ADF content of hay, although the difference between years was significant. The ADF content of hay was higher in the second year than the first year. This increase mainly originated form higher grasses contribution to produced hay in the second year because grasses always have a higher cellulosic component than legumes (van Gessel, 1970; Tan and Mentese, 2003).

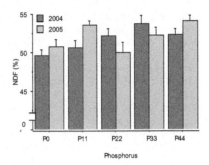

Figure 4. The effect of P doses on NDF content in relation to years.

CONCLUSION

The results of this study indicated that phosphorus fertilizer is important for improving dry matter production and botanical composition without considerable reduction in hay quality. The suitable P dose was 11 kg ha^{-1} while the legumes was dominate, whereas it was 22 kg P ha^{-1} when the grasses become dominate in the sward. The sward where grasses became dominate produced sustainable dry matter at the dose of 11 kg P ha^{-1} and therefore, an 11 kg P ha^{-1} dose can be suggested for sustainable dry matter production on the hay meadow on the high altitude areas without considering soil P status because low soil temperature in the spring restricted P availability for plants. On the other hand, autumn application of P fertilizer has generally been suggested by specialists in Turkey, although there were no significant differences recorded for the efficiency between spring and autumn application of P fertilizer in this study. Considering the study results, spring application of P fertilizer can be suggested for hay meadows or natural rangelands, or at least the areas where surface runoff occurs in the spring.

LITERATURE CITED

Aerts, R., H. de Coluwe and B. Beltman, 2003. Is the relation between nutrient supply and biodiversity co-determined by the type of nutrient limitation? Oikos 101: 489-498.

Altin, M., 1975. An Investigation on Effects of Nitrogen, Phosphorus and Potassium Fertilization on Yield Crude Protein and Ash and Plant Composition of Natural Meadow and Rangeland. Atatürk University Publ. No: 326, Faculty of Agriculture Publ. No: 159.

Altin, M., A. Gokkus and A. Koc, 2005. Range and Meadow Improvement. Ministry of Agricultural and Rural Areas, Ankara.

Aydin, I. and F. Uzun, 2005. Nitrogen and phosphorus fertilization of rangeland affect yield, forage quality and the botanical composition. Europ. J. Agron. 23: 8–14.

Ball, D.M., M. Collins, G.D. Lacefield, N.P. Martin, D.A Mertens,. K.E Olson, D.H. Putnam, D.J. Undersander and M.W. Wolf. 2001. Understanding Forage Quality. American Farm Bureau Federation Publication 1- 01, Park Ridge, IL.

Barker, D.J. and M. Collins, 2003. Forage fertilization and nutrient management. In: Barnes RF, Miller DA, Nelson CJ, (ed.) Forages: An Introduction to Grassland Agriculture. 263-293, 6th ed. Iowa State Univ. Press, Ames, IO.

Beltman, B., J.H. Willems and S. Gusewell, 2007. Flood events overrule fertilizer effects on biomass production and species richness in riverine grasslands. J. Veget. Sci. 18: 625-634.

Benedycka, Z., S. Benedycki and S. Grzegorczyk, 1992. Phosphorus utilization in the dependence on nitrogen fertilization by greensward. Fourth International Imphos Conference, Phosphorus, Life and Environment, Gand, Belgium.

Budakli-Carpici, E., 2011. Changes in leaf area index, light interception, quality and dry matter yield of an abandoned rangeland as affected by the different levels of nitrogen and phosphorus fertilization. Turk. J. Field Crops. 16: 117-120.

Collins, M. and J.O. Fritz, 2003. Forge quality. In: Barnes RF, Miller DA, Nelson CJ, editors. Forages: An Introduction to Grassland Agriculture. 363-390. 6th ed. Iowa State Univ. Press, Ames, IO.

Connor, D.J., R.S. Loomis and K.G. Cassman, 2011. Crop Ecology: Productivity and Management in Agricultural Systems. 546, Cambridge Press. New York.

Dasci, M. and B. Comakli, 2011. Effects of fertilization on forage yield and quality in range sites with different topographic structure. Turk. J. Field Crops. 16: 15-22.

Davies, A., 2001. Competition between grasses and legumes in established pastures. In: Tow PG, Lazenby A, editors. Competition and Succession Pastures. Institute of Grassland and Environmental Research, Plas Gogerddan, 63-83, Aberystwyth, Ceredigion, UK.

Fageria, V.D., 2001. Nutrient interactions in crop plants. J. Plant Nutrition. 24: 1269-1290.

Gokkus, A., 1989. Effect of fertilizing, irrigation and grazing on hay and crude protein yields of meadows at Erzurum Plain. Turk. J. Agric. and For. 13: 1002–1020.

Gokkus, A., 1990. Effect of fertilizing, irrigation and grazing on chemical and botanical composition of meadows at Erzurum plain. Atatürk University J. Faculty of Agric. 21(2): 7-24.

Gokkus, A. and A. Koc, 1995. Hay yield, botanical composition and useful hay content of meadows in relation to fertilizer and herbicide application. Turk. J. Agric. and For. 19: 23-29.

Hatipoglu, R., M. Avcı, N. Kılıcalp, T. Tukel, K. Kokten and S. Cınar, 2001. Research on the effects of phosphorus and nitrogen fertilization on the yield and quality of hay as well as the botanical composition of a pasture in the Çukurova region. Proc. Turkish Field Crops Congress, 17-21 September 2001; Tekirdağ, pp. 1–6.

Henkin, Z., N.G. Seligman and I. Noy-Meir, 2010. Long-term productivity of mediterranean herbaceous vegetation after a single phosphorus application. J. Veget. Sci. 21: 979-991.

Jones, D.I.H., 1981. Chemical composition and nutritive value. In swart measurement handbook In: Handson J, Baker RD, Davies A, Laidlows AS, Leawer JD, editors. 243-265, The British Grassland Soc.

Kemp, P.D., L.M. Condron and C. Matthew, 2005. Pastures and soil fertility. In: White J, Hodgson J, editors. New Zealand Pasture and Crop Science. 67-82, Oxford Univ. Press, South Melbourne.

Koc, A. and A. Gokkus, 1998. Suggestions from previous studies for beter range and meadow management in eastern Anatolia region. Eastern Anatolia Agricultural Congress, 14-18 September 1998; Erzurum. Vol: I pp. 419-428.

Koc, A., M. Dasci and H.I. Erkovan, 2005. Effect of fertilizer and cutting stage on hay yield and invasive plant density of meadow. Proc. Turkish Field Crops Congress, 5-9 September 2005; Antalya. pp. 863–866.

Koc, A., H.I. Erkovan and Y. Serin, 2008. Changes in vegetation and soil properties under semi-nomadic animal raising areas in highlands of Turkey. Current World Environment 3: 15-20.

Marschner, H. and V. Romheld, 1983. In vivo measurement of root-induced pH changes at the soil–root interface: effect of plant species and nitrogen sources. Zeitschrift fur Pflanzenernahr Bodenkunde 111: 241-251.

Mehrvarz, S. and M.R. Chaichi, 2008. Effect of phosphate solubilizing microorganisms and phosphorus chemical fertilizer on forage and grain quality of barley (Hordeum vulgare L.). American-Eurasian J. Agric. and Environ. Sci. 3: 855-860.

Miller, D.A. and H.F. Reetz, 1995. Forage fertilization. Forages. Iowa State University Press, USA.

Papanastasis, V.P. and T.G. Papachristou, 2000. Agronomic aspect of forage legumes: management and forage quality. Legumes for Mediterranean Forage Crops, Pastures and Alternative Uses. Cahier Options Mediterraneennes, Ciheam, Zaragoza. pp. 113-126.

Pieper, R.D., R.J. Kelsey and A.B. Nelson, 1974 Nutritive quality of nitrogen fertilized and unfertilized blue grama. J. Range Manage. 27(6): 470-472.

Rayburn, E.D., 2004. Forage management, Understanding Forage Analysis Important to Livestock Producers. West Virginia Univ. Extension Service. http://www.wvu.edu/agexten/forglvst/analysis.pdf.

Russell, J.S., E.M. Brouse, H.F. Rhoades and D.F. Burzlaff, 1965. Response of sub-irrigated meadow vegetation to application of nitrogen and phosphorus fertilizer. J. Range Manage. 18: 242-247.

Sigua, G.C., S.W. Coleman, J. Albano and M. Williams, 2011. Spatial distribution of soil phosphorus and herbage mass in beef cattle pastures: effects of slope aspect and position. Nutr Cycl Agroecosyst, 89: 59-70.

Soil Survey Laboratory Staff, 1992. Soil Survey Laboratory Methods Manual. USDA-SCS. Soil Survey Investigations Report No: 42.

SPSS Inc, 1999. SPSS for Windows: Base 10.0 Applications Guide. Chicago, Illinois.

Tan, M. and O. Mentese, 2003. Effect of anatomic structure and chemical composition on forage quality. Atatürk University J.Faculty of Agric. 34: 97-103.

Townsend, C.E., 1985. Miscellaneous perennial clovers. In: Gibson RB, Gillett JW, Knight WE, Rupert EO, Smith RR, Van Keuren RW editors. Clover Science and Technology. p: 563-578, American Society of Agronomy. Madison, Wisconsin, USA

Valentine, J.F., 1989. Range Development and Improvements. 3rd ed. Academic Press Inc., San Diego, California, p. 524.

van Gessel, T.P., 1970. Factors influencing the mineral content of herbage. Agri Digest 20: 35-41.

Van Soest, P.J., J.B. Robertson and B.A. Lewis, 1991. Methods for dietary fiber, neutral detergent fiber and non-starch polysaccharides in relation to animal nutrition. J. Dairy Sci. 74: 3583-3597.

Venterink, H.O., 2011. Does phosphorus limitation promote species-rich plant communities? Plant Soil. 345: 1-9.

34

INFLUENCE OF CORM PROVENANCE AND SOWING DATES ON STIGMA YIELD AND YIELD COMPONENTS IN SAFFRON (*Crocus sativus* L.)

*Reza Amirnia [1], Mahdi Bayat *[2], Asadollah Gholamian[3]*

[1]*Department of Agronomy, Faculty of Agriculture, Urmia University, Urmia, Iran*
[2]*Department of Agriculture, Mashhad Branch, Islamic Azad University, Mashhad, Iran*
[3]*Department of Biology, Mashhad Branch, Islamic Azad University, Mashhad, Iran*
* Corresponding author: saadat_rooish@yahoo.com*

ABSTRACT

In order to study effect of different sowing dates on stigma yield and yield components in saffron. Field trials were arranged in split plot design was sowing dates (05-May, 05-Jun, 05-Jul, 05-Aug, 05-Sep and 05-Oct) as main plot and four ecotypes as sub plots (Mashhad, Torbat-jam, Gonabad, Birjand) in Mashhad at three distinct years (2009-11). The results from analysis of variance indicated significant differences between ecotypes and sowing dates in individual years, at last it is distincted that two Mashhad and Torbat-jam ecotypes had the highest stigma yield (98.6 and 92.5 mg/m^2) in Mashhad climate. Also the best sowing dates were 05-Jun to 05-July in Mashhad. It revealed that climate had significant effects on stigma yield and yield components, so that delaying in cultivation leads to decrease all agronomical traits of saffron. In other hand, results from metereological data indicated that average of temperature and average of sunny hours in comparison with average of percipitation and average of relative humidity, showed maximum effects on stigma yield and yield components, positively. Finally, results from phenotypic correlation indicated that stigma yield showed significant and positive correlation with all studied traits and results from step by step regression indicated that daughter corm number and corm fresh weight had positive and maximum direct effects to improve stigma yield. Therefore, it can be concluded that increasing cultivation density and selecting bigger corms are useful factor to increasing stigma yield.

Keywords: Direct and indirect selection, Saffron ecotypes, Step by step regression

INTRODUCTION

Saffron is one of the oldest and most expensive spices in the world (Winterhalter and Straubinger, 2000). It obtains from the dried stigmas of *Crocus sativus* L., a member of the large family Iridaceae, and has been cultivated in several countries such as Iran, India, Greece, Morocco, Spain and Italy (Ghorbani, 2006). Iran is the greatest producer saffron in the world and 75 % of the production is coming from Iran (Amirnia et al., 2012). Nowadays, saffron has attracted much attention to food science due to the demand for natural food colors and additives which could replace the synthetic colorants, flavor enhancers and aromas (Kafi, 2006) and also used in dyes and perfumes (Basker and Negbi, 1983). The novel use of saffron in recent years has been associated in cancer cure. There has been increased interest in the biological effects and potential medical applications of saffron, particularly those based on their cytotoxic, anticarcinogenic and antitumour properties (Abdullaev, 2002; Abdullaev and Espinosa-Aguirre, 2004; Fernandez, 2004; Magesh et al., 2006; Chryssanthi et al., 2009; Dalezis et al., 2009). Under this perspective, there is an increasing interest by producers and consumers in high quality product with clear geographical origin (Anastasaki, et al., 2010).

From an agronomic point of view, saffron is well adapted to different environmental conditions ranging from dry sub-tropical to continental climates (Azizbekova and Milyaeva, 1999; Mollafilabi, 2004; Sampathu et al., 1984). Anastasaki et al., (2010) and Luana Maggi et al., (2011) reported that price of saffron strongly depends on its country of production as many factors interfere to saffron's quality such as edaphoclimate conditions (soil, climate, rainfall), harvest time and postharvest treatment. During the last 40 years, the intensive manual labor required for daily flower picking and the separation of stigmas has significantly reduced the cultivated area of saffron in many European countries that traditionally produced the crop (Fernandez, 2004; Gresta et al., 2009).

For all the previously mentioned reasons, this study was undertaken to evaluate the influence of corm provenance and sowing dates on dry stigma yield and quantitative characteristics of saffron in order to explore the best saffron ecotype and sowing date for establishing a new farm in Mashhad condition.

MATERIALS AND METHODS

To study effects of sowing dates on yield and yield components in saffron an experiment performed in the experimental fields in Mashhad-Iran. Mashhad region locate in 36'15" latitude, 59'28" longitude and 985 meters above sea level and soil in experimental field was silt-clay and pH=7.8. In this study different ecotypes of saffron evaluated in three distinct years (2008-09, 2009-10 and 2010-11) so that at the end of each year, we through out pervious corms and cultivated new corms in the same farm. In each arable year, the experiment was conducted as split-plot in BCRD which different sowing dates (05-May, 05-Jun, 05-Jul, 05-Aug, 05-Sep, 05-Oct) used as main plots and different saffron ecotypes (Mashhad, Torbat-jam, Gonabad and Birjand) as sub plot. Four saffron samples from different regions of Iran, traditional saffron production areas, were studied in this work. Samples 1, 2 and 3 were obtained from the zone of Torbat-Jam, Mashhad and Gonabad in Razavi Khorasan province of Iran. Sample 4 was obtained from the area of Birjand in the south Khorasan province of Iran. To

prevent Fusarium and Penicillium infestations, corms were dipped in a prochloraz solution (0.1%) and dried under forced ventilation for 5–7 h to remove the surface water.

In each year after preparing the field in April, 75 kg ha^{-1} pure nitrogen, 75 kg ha^{-1} pure phosphorous, 50 kg ha^{-1} pure potassium were used. The cultivation practices used were those commonly used for this crop, and an organic fertilizer (mature manure) was applied. Each plot contains 8 culture lines with 3 meter length and 25 cm distances from another line (the plot area was 6 m^2). Corm distances on lines were 8 cm and 15 cm depth (density was 50 corms/m^2). To avoid marginal effects and minimizing errors, plots situated beside each other by 50 cm distances. To enhance accuracy, margins considered as two lines at beginning and at the end of plots as well as 50 cm at two another sides of plots. All agronomical operation (ecotypes and infield preparation, traits measuring and experimental design) in three years were similar, and at the end of each year, the infield was plowed and corms were harvested for measuring. Rainfed conditions met the water requirements at the start of growth (October to November), and plants were drip irrigated from December to April. Meteorological data in show in below Table 1.

Table 1. Meteorological statistics of Mashhad field station during 2009-2011

Month	Temperature (°C)			Relative humidity (%)			Precipitation (mm)			Number of sunny hours		
	2009	2010	2011	2009	2010	2011	2009	2010	2011	2009	2010	2011
April	21.84	17.71	20.12	41.24	64.69	54.76	12.40	38.70	35.10	286.90	226.00	220.20
May	26.17	23.94	25.94	34.82	43.98	30.44	10.20	5.10	3.50	340.50	337.40	367.30
June	27.70	27.02	29.35	26.95	30.29	23.81	0.00	1.30	1.00	367.40	365.90	377.60
July	27.45	27.95	26.68	20.87	33.23	22.19	0.00	2.90	0.00	380.00	361.10	376.60
August	23.85	22.16	22.67	25.31	40.26	24.92	0.00	21.30	2.40	330.60	331.40	329.60
September	18.91	17.26	20.68	40.87	37.58	38.07	0.00	0.00	0.50	262.60	295.10	285.10
October	8.19	12.09	12.28	57.38	50.55	46.18	10.30	5.80	8.60	177.50	206.30	233.50
November	6.83	5.50	9.47	67.48	70.63	42.03	12.50	22.20	0.00	130.30	136.80	236.90
Desember	3.42	7.04	3.58	67.67	62.08	54.32	29.20	4.20	13.10	160.10	148.10	183.70
January	6.60	4.89	4.50	68.50	67.88	64.67	28.80	22.60	48.00	132.20	137.60	145.10
February	10.85	11.82	5.37	60.23	69.28	67.81	40.20	50.50	22.40	196.90	127.80	116.60
March	11.50	13.74	14.78	71.27	58.48	45.08	102.10	27.60	10.10	161.30	176.60	245.10
Average	16.11	15.92	16.29	48.55	52.41	42.86	20.48	16.85	12.06	243.86	237.51	259.78
Sum	2202.3	2201.1	2206.4	2591.6	2638.9	2525.3	2254.7	2212.2	2155.7	4935.3	4860.1	5128.3

Data were collected on the following 13 characters in each pot: Number of Flower (NF), Fresh Weight of Flower (mg) (FWF), Dry Weight of Flower (mg) (DWF), Fresh Weight of Stigma (mg) (FWS), Dry Weight of Stigma (mg) (DWP), Number of daughter Corm (NDC), Fresh Weight of daughter Corm (mg) (FWDC), Dry Weight of daughter Corm (mg) (DWDC), Length of Leaf (cm) (LL), Dry Weight of Leaf (mg) (DWL), Biomass (BIO), Harvest Index (HI) and Stigma Yield per m2 (mg m^{-2}) (SY). Then the averages of the traits were used to statistical calculations in SAS ver. 9.12.

RESULTS AND DISCUSSION

Year Effects

Results from analysis of variance (Table 2) indicated FWF, DWF, FWS, DWS and DWL did not show

significant differences at different years, so it can be stated environmental changes did not affected these traits, while NF, NDC, FDWC, DWDC, LL, BIO, HI and SY affected. These results are supported by previous findings (Mollafilabi, 2004; Unal and Cavusoglu, 2005; Gresta, et al., 2008). The results from means comparisons of these traits (Table 3) were maximum and minimum in 2011 and 2009, respectively. For instance, SY was 98.8 mg m^{-2} in 2011, while 80.8 mg m^{-2} in 2009. Also, NF ‹NDC ‹ FWDC ‹DWDC and LL traits, as major factor to improve stigma yield in next year were enhanced 4.8, 0.4, 0.9, 0.5 and 2.4 mg m^{-2} in 2011 in comparison to 2009, respectively. Analysis of metereological data in 2009-11 (Table 1) indicated in 2011 the parameters include average of temperature, average of sunny hours were upper, while average of relative humidity and average of precipitation were lower. According to these results it can be stated that meteorological parameters include average of temperature, average of sunny hours had maximum and positive effects on yield and yield components in saffron. Koocheki et al., (2006) stated that flowering and growth seasons of saffron are in autumn and winter seasons that have maximum precipitation, so saffron has not need for irrigation, necessary. On the other word, Saffron has low water requirements and is typically cultivated under rainfed conditions because it is well adapted to the rainfall pattern of diverse Mediterranean areas (Alizadeh, 2006). Zhang (1994) after study the precipitation and temperature effects on winter wheat stated temprature changes are more important on wheat seed yield in comparison with precipitation.

Table 2. Variance analysis of studied traits in saffron genotypes

SOV	df	NF	FWF	DWF	FWS	DWS	NDC	FWDC	DWDC	LL	DWL	BIO	HI	SY
Year	2	432.9 **	129.4	2.6	49.1	1.6	4.4 **	17.2 **	5.4 **	115.4 *	0.1	6017265.3 **	0.002 **	5973.2 *
E$_1$	6	8.5	51.8	1.2	13.0	0.4	0.1	0.1	0.1	12.7	0.1	98980.9	0.000	784.4
Month	5	1440.5 *ᵃ	1057.8 *ᵃ	122.5 *ᵃ	121.6 *	12.8 **	9.3 **	37.1 **	24.8 **	401.2 *ᵃ	1.6 **	38381266.5 *ᵃ	0.001 **	67660.8 *ᵃ
Year×Month	10	46.5 **	112.8	1.7	5.2	0.2	0.5 *ᵃ	1.3 **	0.7 **	11.8 *	0.1 *	670373.2 **	0.000	1089.8
E$_2$	30	7.7	64.8	0.8	2.7	0.5	0.1	0.1	0.2	5	0.0	208471.5	0.000	533.8
Ecotype	3	44.1 **	2497.1 *ᵃ	42.3 **	55 **	1.8 **	14.1 **	8.5 **	3.5 **	177.6 *ᵃ	2.0 **	10799168.9 *ᵃ	0.0001 *ᵃ	3346.6 **
Year×Ecotype	6	32.4 **	431.4 **	3.5 **	9.7 **	0.3 *ᵃ	0.4 **	1.4 **	0.3	10.2 0	0.3 **	460823.7	0.0001 *ᵃ	644.7 *
Ecotype×Month	15	41.7 **	52.2	2.6 **	2.9	0.1	0.1	0.9 **	0.5 *	12.4 **	0.1 **	393396.5	0.0001 *ᵃ	1311.4 **
Year×Ecotype×Montl	30	26.9 **	43.1	1.6 **	1.7	0.1	0.2 **	0.9 **	0.6 **	18.1 **	0.1 **	722014.1 **	0.0001 *ᵃ	889.5 **
E$_3$	108	6.9	34.1	0.7	1.9	0.1	0.1	0.1	0.3	3.7	0.0	269103.7	0.0000	219.9
CV%	---	15.6	3.7	3.9	6.9	6.9	13.2	6.5	14.6	10.5	5.9	8.0	7.8	16.6

ᵃ - Abbreviations are described in materials and methods

Table 3. Mean comparison of studied traits in saffron genotypes

		NF	FWF	DWF	FWS	DWS	NDC	FWDC	DWDC	LL	DWL	BIO	HI	SY
Year														
	2009	14.7 c	350.8 a	42.9 a	28.5 a	5.2 a	2.6 a	5.5 c	3.2 c	18.5 ab	148.5 a	6178 c	0.09 a	80.8 b
	2010	16.3 b	353.5 a	43.3 a	28.6 a	5.2 a	2.2 b	6.1 b	3.5 b	16.8 c	151.1 a	6576.3 b	0.08 b	84.6 ab
	2011	19.5 a	352.1 a	43.2 a	27.1 a	4.9 a	2.2 b	6.4 a	3.7 a	19.2 a	149.7 a	6740.1 a	0.07 b	98.8 a
Month														
	05-May	17.8 bc	353.3 bc	44.2 a	27 b	5.1 b	2.2 bc	5.7 b	3.2 b	16.5 c	151.1 a	6223.3 b	0.084 ab	90.8 b
	05-Jun	24.1 a	355.9 ab	44.8 a	29.9 a	5.9 a	3 a	7.2 a	4.5 a	22.3 a	160.6 a	7761.1 a	0.077 b	142.2 a
	05-Jul	24 a	359.7 a	44.8 a	30.8 a	5.8 a	2.8 a	7.2 a	4.5 a	22 a	161.5 a	7770.2 a	0.075 b	138.9 a
	05-Aug	14.5 c	350.5 bc	43.2b	27.6 b	4.7 bc	2.3 b	5.7 b	3.1 b	18.1 b	149.2 b	6111.0 bc	0.078 ab	68.6 c
	05-Sep	12.0 d	349 cd	41.6 c	26.8 b	4.6 c	1.9 c	5.4 c	2.9 b	16.4 c	142.5 c	5824.6 c	0.079 ab	55 cd
	05-Oct	8.7 e	344.3 d	40.3 d	26.4 b	4.5 c	1.7 d	4.7 d	2.6 c	13.9 d	133.8 d	5298.7 d	0.087 a	39 d
Ecotype														
	Gonabad	16 b	345.1 b	42.3 b	27 d	4.9 d	1.7 d	5.6 d	3.2 b	15.8 c	139.9 c	6030.3 b	0.083 a	82.2 b
	Mashhad	17.7 a	358.4 a	44 a	29.3 a	5.3 a	2.9 a	6.4 a	3.7 a	19.8 a	159.2 a	6942 a	0.078 b	98.6 a
	Torbat-Jam	17.5 a	357.4 a	43.8 a	28.4 b	5.2 b	2.5 b	6.2 b	3.6 a	19.4 a	156.9 a	6813.3 a	0.077 b	92.5 a
	Birjand	16.1 b	347.5 b	42.5 b	27.7 c	5.0 c	2.1 c	5.7 c	3.3 b	17.7 b	143.1 b	6207 b	0.082 a	83 b

ᵃ - Abbreviations are described in materials and methods

As last, it can state to saffron plant is a winter crop and the most growth of this plant occurred in autumn and winter months, on the other hand the mount of precipitation (rain and snow) usually is more but the evaporation is less in these months. So no water shortage would be seen and saffron will respond to increase temperature and sunny hours positively to improve yield components. Also researchers such as Molina et al., 2005a, 2005b; Gresta, et al., 2008; Lundmark, et al., 2009; Luana Maggi et at., 2011 pointed out that environments and climatic conditions (e.g. temperature, soil water content) effect severely on quantitative and qualitative traits of saffron.

Sowing Dates

The results from analysis of variance (Table 2) indicated different sowing dates affected all studied traits and they can have an important role in saffron production. Means comparisons of studied traits in different months (Table 3) indicated to decline in studied traits following to delaying in sowing date. These scholars in their studied observed that almost all of traits decreased with delayed sowing time. At last, according to the results of this study, it is revealed that sowing dates 05-Jun. and 05-Jul. were the best months to culture in Mashhad due to positive effects on yield components and stigma yield, while 05-Oct. was the worst month for culturing due to negative effects on yield components and stigma yield. For instance, delaying from 05-Jun. and 05-Jul to 05-Oct. lead to reduce NF from 24 to 9 numbers in per square meter, respectively. Also delaying in culture lead to reduce NDC, FWDC and DWDC traits. These traits as the main yield components had major roles in increase of stigma yield in next year. On the other hand, it is revealed that the amount of SY at 05-Jun and 05-Jul sowing dates and 05-Oct. sowing date were 142.2 and 138.9 and 39 mg/m^2, respectively. Gresta et al., (2009) observed soil temperature and moisture in sowing date (Jun. and Jul. months) were important factors in flowering and stigma yield of saffron, so lower soil temperature (in sowing date) lead to more stigma yield. At last, these investigators stated that soon sowing date is better than late. On the other hand, Molina et al., (2004) reported that the most favorable temperature to differentiate flowers in saffron is 23-25 °C for growing corms at summer; these conditions observed from June to July in Mashhad.

Overall, it revealed delaying in sowing date would reduce SY and FN as well as vegetative yield (leaves and daughter corm production). The reason of these unfavorable effects in delaying sowing dates is that saffron corms is in dormancy period from Aug. to Oct. seemingly, but they are very active physiologicaly, so translocating of mother corms to new farm in these months lead to readuce flowers and leaves buds severly, and subsequently it will reduce stigma yield specially at the first cultivation year, (Benschop, 1993). Mollafilabi (2004) and Kafi (2006) suggested we can culture saffron corms from early Jun. to mid Sep. month. But it should avoid cultivating in mide summer (Aug. and Sep. months) due to hot weather and lowering humidity; because it would damage corms due to losing its humidity during transition.

Ecotype

Results from analysis of variance (Table 2) indicated significant differences between ecotypes in all studied traits. Although saffron is sterile and reproduced in asexual method (corm) but if saffron corms cultivated in a ragion for several decades, undergo morphological and genetical changes. Results from means comparisons (Table 3) revealed there were maximum differences between ecotypes due to FWS, DWS, NDC, FWDC, LL and DWL traits. These traits are as selective traits for stigma yield at current year and important yield components in next year. Therefore Mashhad and Birjand ecotypes were the best and worst from those traits respectively, so it can predicted that these ecotypes would have maximum and minimum stigma yield at current year, and even next years respectively. Finally, in Mashhad climate, it can be stated ecotypes Mashhad and Torbat-jam that produced 98.6 and 92.5 mg/m^2 dry stigma are the most adapted ecotypes, while two ecotypes Gonabad and Birjand that produced 82.2 and 83 mg/m^2 dry stigma are the least adapted ecotypes in Mashhad climate. So as it would be predicted, Mashhad ecotype adapted to Mashhad climate- as testifier could achieve maximum yield and then Torbat-jam ecotype, because of meteropological similarity between two regions Mashhad and Torbat-jam. On the other hands, declining yield in ecotypes Gonabad and Birjand in Mashhad climate is caused by climate variations between two regions. Ehsanzadeh et al., (2004) reported extraordinary differences between saffron ecotypes. At last, they stated that among ten studied ecoptyes in Shahrkord region, three ecotypes Birjand, Qaen and Shahrkord had maximum yield and quality because of climatic similarity between three regions. Freeman (1973) reported saffron is adapted to different

environments that affect quality and quantity characteristics, but subtropical weather is the best conditions. Siracusa et al., (2010) stated that flower number and stigma yield were significantly affected by environment and corm provenance.

The main point in this study is that difference between ecotypes becease of I) meteropological changes amonge Mashhad and original regions which corms gathered, II) different qualities between saffron corms. Therefore, it can be stated that saffron is a sensitive plant to environmental conditions, so it is necessary to pay attention to corm quality and meterological similarity between origin and destination climate for establishing new farms and at last it is essential to evaluate different ecotypes to select the best one for each region. Laidiaw (2005) in a study on pastural plant introduced some factors include light, temperature, humidity and nutritions as important parameters to plant productions, so it is necessary to attention to climate condition for selecting appropriate ecotype. Dhar et al., (1988) stated that stigma yield is dependent to factors include soil characters, cultivation methods, density, corms weight, geographical situation, climate, agronomical management and harvesting period. It has been shown that the size of the mother corm has a significant effect on vegetative development and the production of daughter corms (Negbi, 1999; De Mastro and Ruta, 1993; De Juan et al., 2003).

Intractions between Ecotypes and Soweing Dates

Results from analysis of variance (Table 2) indicated significant differences between ecotypes and sowing dates intractions for studied traits. This result indicated that studied ecotypes in different sowing dates were not similar. The results from means comparisons between ecotypes and sowing dates intractions (Table 4) indicated Mashhad ecotype at 05-Jun and Birjand ecotype at 05-Oct were the best and the worst ecotpyes for majority of traits espectialy for stigma yield. Comstock and Moll (1963) stated that regard to interactions between genotype and environment, selection of variants upon only one environmental condition is not appropriate; hence it is better to evaluate genotypes in range of places, years and environmental changes, these help us to estimate adaption and stability of genotypes yield achieving a constant criteria to select the best and most efficient variants (Johnson et al., 1995).

Table 4. Mean comparison of interactions between studied traits in saffron genotypes

Ecotype	Month	NF	FWF	DWF	FWS	DWS	NDC	FWDC	DWDC	LL	DWL	BIO	HI	SY
Mashhad	05-May	15.3 f-h	359.9 a-c	45.1 a-d	28.4 c-g	5.4 d-f	2.8 b-d	6.6 e-f	3.7 c-f	18.5 d-f	161.6 a-e	6944.8 d-g	0.078 c-e	82.5 e-g
	05-Jun	28.4 a	361.0 a-c	45.9 a	31.2 a-b	6.1 a-b	3.4 a	7.5 b	4.8 a-b	22.8 a-b	168.4 a-b	8246.4 a-b	0.075 e	173.7 a
	05-Jul	25.2 a-c	366.2 a	45.4 -c	33.4 a	6.2 a	3.5 a	8.0 a	5.1 a	24.6 a	165.9 a-c	8500.5 a	0.074 e	158.3 a-b
	05-Aug	16.4 e-g	355.2 b-f	43.9 c-g	28.2 d-g	4.8 g-j	2.8 b-d	6.3 f-g	3.3 d-g	18.6 d-f	157.9 c-f	6515.3 e-h	0.075 e	79.6 f-h
	05-Sep	13.7 f-j	356.2 a-e	43.1 f-i	27.5 e-i	4.7 g-k	2.5 d-g	5.3 i-j	2.8 f-h	18.4 d-g	152.6 d-i	5891 h-k	0.081 b-e	63.9 g-k
	05-Oct	7.3 k	352.0 c-g	40.6 k-m	27.1 e-i	4.7 g-k	2.4 d-g	4.8 k-l	2.5 g-h	15.7 f-j	148.7 f-j	5553.8 h-k	0.084 b-e	33.6 l
Torbat	05-May	21.2 b-d	355.7 b-e	44.3 b-f	26.7 f-i	5.1 e-g	2.4 d-g	5.9 g-h	3.3 d-g	16.9 e-i	155.4 d-h	6466.1 f-i	0.080 b-e	107.7 d-e
	05-Jun	21.7 b-d	360.0 a-c	44.7 a-e	30.7 b-c	6.0 a-c	3.3 a-b	7.4 b-c	4.8 a-b	25.5 a	162.7 a-d	8084.3 abc	0.076 e	131.7 c-d
	05-Jul	24.3 a-c	364.2 a-b	45.6 ab	31.1 a-b	5.8 b-d	3.1 a-c	7.3 b-c	4.4 a-c	23.2 a-b	169.8 a	7883.9 abc	0.075 e	140.6 b-c
	05-Aug	14.9 f-i	357.1 a-d	44.1 b-g	28.0 d-h	4.8 g-k	2.5 d-f	5.6 h-i	3.1 e-h	17.8 d-g	154.3 d-h	6254.1 g-j	0.078 d-e	70.8 f-i
	05-Sep	12.9 g-j	355.4 b-e	42.4 h-j	27.0 e-i	4.6 h-k	2.1 f-i	5.9 g-h	3.2 e-h	17.8 d-g	152.2 e-i	6277.3 g-j	0.074 e	59.4 g-l
	05-Oct	10.0 j-k	352.1 c-g	41.7 h-j	26.9 f-i	4.6 h-k	1.6 i-l	5.2 i-k	2.9 f-h	15.3 f-j	146.7 g-k	5914 h-k	0.078 c-e	44.5 j-l
Gonabad	05-May	18.0 d-f	351.2 c-g	43.4 e-h	26.5 g-i	5.0 e-h	2.0 f-j	5.1 i-k	2.8 f-h	17.0 e-h	145.8 h-k	5793.9 h-k	0.087 a-d	90.7 e-f
	05-Jun	20.8 c-e	355.3 b-f	44.9 a-e	29.3 b-e	5.8 b-d	2.7 c-d	7.0 b-d	4.2 b-c	22.3 a-c	156.7 c-g	7391.2 b-e	0.078 c-e	119.6 c-d
	05-Jul	22.6 b-d	356.6 a-e	44.4 a-f	29.7 b-d	5.6 d	2.6 c-e	7.0 c-e	4.4 a-c	21.0 b-d	151.8 e-j	7493.8 b-d	0.075 e	125.1 c-d
	05-Aug	14.9 f-i	344.7 g-i	42.3 h-j	27.4 e-i	4.8 g-k	2.0 f-i	5.3 i-j	3.0 f-h	18.3 d-g	142.2 i-k	5876.7 h-k	0.081 bcde	70.3 f-i
	05-Sep	10.9 h-k	341.9 g-i	40.0 l-m	26.8 f-i	4.6 h-k	1.9 g-k	5.3 i-j	2.8 f-h	15.0 g-k	136.3 j-k	5621.8 h-k	0.082 b-e	49.7 i-l
	05-Oct	9.4 j-k	335.1 i	39.6 l-m	26.3 g-i	4.5 j-k	1.5 j-l	4.6 l	2.5 g-h	12.7 j-k	125.8 m	5064.8 k-l	0.090 a-b	42.7 k-l
Birjand	05-May	16.4 e-g	346.3 e-h	43.7 d-h	26.2 g-i	5.0 f-i	1.4 k-l	5.0 j-l	2.8 f-h	13.5 i-k	141.7 j-k	5688.4 h-k	0.089 a-c	82.2 e-g
	05-Jun	25.4 a-b	347.3 d-h	43.6 e-h	28.5 c-g	5.6 c-d	2.5 d-g	6.8 e-d	4.2 b-d	18.5 d-f	154.4 d-h	7322.3 c-f	0.077 d-e	143.6 b-c
	05-Jul	24.0 a-c	351.7 c-g	43.8 d-h	29.1 b-f	5.4 d-e	2.1 d-g	6.6 e-f	4.0 b-e	19.2 c-e	158.6 b-f	7202.4 c-f	0.076 d-e	131.7 c-d
	05-Aug	11.7 h-k	345.1 f-i	42.7 g-i	26.7 g-i	4.6 i-k	1.7 h-l	5.6 h-i	2.9 f-h	17.7 d-g	142.2 i-k	5798.2 h-k	0.079 b-e	53.6 h-l
	05-Sep	10.7 i-k	342.4 g-i	40.9 j-l	25.7 h-i	4.4 j-k	1.3 l	5.3 i-j	2.9 f-h	14.2 h-k	128.8 l-m	5508.2 j-l	0.080 b-e	47.1 i-l
	05-Oct	8.0 k	337.8 h-i	39.2 m	25.5 l	4.4 k	1.2 l	4.1 m	2.3 h	11.9 k	114.0 n	4662 l	0.097 a	35.2 l

[a] - Abbreviations are described in materials and methods

Correlation Coefficient

Results from phenotypic correlation coefficient (Table 5) indicated that there were positive and significant correlations between SY and all studied traits that the maximum correlation belonged to FWF and DWF ($r = 0.77^{**}$ and 0.75^{**} respectively) and minimum correlation belonged to NF and HI (both $r= 0.64^{**}$). Also Katar et al., (2012) in their study stated that there were high positive and significant correlations between yield and yield components.

Table 5. Correlation coefficient between studied traits in saffron genotypes

	NF	FWF	DWF	FWS	DWS	NDC	FWDC	DWDC	LL	DWL	BIO	HI
FWF	0.45 **	FWF										
DWF	0.70 **	0.81 **	DWF									
FWS	0.54 **	0.60 **	0.63 **	FWS								
DWS	0.71 **	0.59 **	0.75 **	0.91 **	DWS							
NDC	0.53 **	0.66 **	0.67 **	0.67 **	0.69 **	NDC						
FWDC	0.53 **	0.66 **	0.67 **	0.67 **	0.69 **	0.99 **	FWDC					
DWDC	0.71 **	0.60 **	0.70 **	0.66 **	0.73 **	0.54 **	0.55 **	DWDC				
LL	0.70 **	0.61 **	0.71 **	0.66 **	0.72 **	0.55 **	0.54 **	0.98 **	LL			
DWL	0.72 **	0.58 **	0.69 **	0.70 **	0.79 **	0.59 **	0.58 **	0.96 **	0.96 **	DWL		
BIO	0.71 **	0.59 **	0.70 **	0.71 **	0.78 **	0.58 **	0.59 **	0.95 **	0.95 **	0.98 **	BIO	
HI	0.62 **	0.56 **	0.62 **	0.62 **	0.67 **	0.74 **	0.74 **	0.71 **	0.71 **	0.70 **	0.70 **	HI
SY	0.64 **	0.75 **	0.77 **	0.65 **	0.68 **	0.68 **	0.68 **	0.66 **	0.66 **	0.65 **	0.65 **	0.64 **

[a] - Abbreviations are described in materials and methods

Step by Step Regression

To determine the most effective traits on stigma yield, step by step regression method was used (Table 6) and stigma yield as dependent variable and other traits as independent variables were spoted. On the other hand, to achieve the real understanding of traits effects on stigma yield, we used step by step regression for each year individually, and for three years overall. So results from step by step regression analysis indicated FWS and DWF accounted for 71% of total variance of stigma yield in arable 2009; the regression coefficient was SY = -143.2 + 5.1 DWF + 2.5 FWS. Results from step by step regression analysis indicated FWDC and DWF accounted for 80% of total variance of stigma yield in arable 2010; the regression coefficient was SY = -2.2 + 14.9 DWF + 2.8 FWDC. Results from step by step regression analysis indicated FWF accounted for 76% of total variance of stigma yield in arable 2011; the regression coefficient was SY = -226.2 + 1.0 FWF. Finally, results from step by step regression for all years indicated FWF, NDC and NF accounted for 68% of total variance of stigma yield. Low amount of Standard deviation for NF, FWF and NDC (0.17, 0.14 and 0.04) as well as low amount of Durbin Watson Index (1.7) indicated well propriety and confidency of model; the regression coefficient was SY = -135.1 + 0.09 NDC + 0.75 FWF + 0.68 NF.

Table 6. Regression coefficients of saffron yield using stepwise method

Year	Model	Unstandardized Coefficients B	Std.Error	Unstandardized Coefficients Beta	t	R2	Durbin Watson
2009	Intercept	-143.2	47.2		-3.0 **		
	DWF	5.1	1.5	0.55	3.5 **		
	FWS	2.5	1.1	0.37	2.3 *		
						0.71	1.2
2010	Intercept	-2.2	36.9		-0.06		
	DWF	14.9	2.9	0.63	5.23 **		
	FWDC	2.8	0.9	0.36	6.02 **		
						0.80	1.5
2011	Intercept	-226.2	43.94		-5.15 **		
	FWF	1.07	0.13	0.88	8.56 **		
						0.76	1.9
Total	Intercept	-135.1	47.78		-2.83 **		
	NDC	0.09	0.04	0.21	2.19 *		
	FWF	0.75	0.14	0.47	5.21 **		
	NF	0.68	0.17	0.32	3.93 **		
						0.68	1.7

[a] - Abbreviations are described in materials and methods

Overall, from results of step by step regression, it can be stated FWDC and NDC are the most effective and important factors for increasing stigma yield in saffron. According to reproduction of saffron by corms, increasing NDC and FWDC not only can improve yield components (include NF, FWS FWF DWF) but also improve dry saffron yield via more production of daughter corms and high NDC at the next years extremely.

CONCLUSION

According to present study, it can be concluded beside that saffron is a sterile crop and reproduce only by asexual method, it was affected climate severely. So it is necessary to select appropriate ecotypes to each region. We indicated two ecotypes Mashhad and Torbat-jam as more fitted ecotype to Mashhad climate. Then it revealed delaying in cultivation not only reduced yield and component yield at the first year, but also affected saffron yield in next years. Both sowing dates 05-Jun. and 05-Jul. also were selected as the best sowing dates in Mashhad region as well as the traits NDC and FWDC were selected as the main triats for increasing dry saffron yield; so high-weight corms (up to 8 g) are appropriate and necessary for producing dry saffron especially at the first year. Anyway according to this study, to establish new saffron farms, it is better to pay attention to corm quality and climate accordance between the region have produced mother corms and the region will be established new farm. Finaly we have concluded that the major component of well-establishing new farm in new climate is to select most suitable ecotypes based on proper analysis.

LITERATURE CITED

Abdullaev, F.I., 2002. Cancer chemopreventive and tumoricidal properties of Saffron (Crocus sativus L.). Exp. Biol. Med. 227 (May), 20–25.

Abdullaev, F.I., J.J. Espinosa-Aguirre, 2004. Biomedical properties of saffron and its potential use in cancer therapy and chemoprevention trials. Cancer Detect. Prev. 28, 426–432.

Alizadeh, A., 2006. Irrigation. In: Kafi, M., Koocheki, A., Rashed, M.H., Nassiri, M. (Eds.), Saffron (Crocus sativus) Production and Processing. Science Publishers, Enfield, pp. 79–90.

Amirnia, R., M. Tajbakhsh, M. Ghiyasi, Y. Rezaee Danesh, M. Bayat, M. Izadkhah Shishavan, 2012. Saffron Golden Crop of Iran. Tibbi ve Aromatik Bitkiler Sempozyumu. 13-15 Eylul 2012, Tokat, Turkey.

Anastasaki, E., C. Kanakis, C. Pappas, L. Maggi, C. P. del Campo, M. Carmona, G. L. Alonso, M. G. Polissiou, 2010. Differentiation of saffron from four countries by mid-infrared spectroscopy and multivariate analysis. Eur Food Res Technol. 230:571–577.

Azizbekova, N.S.H., E.L. Milyaeva, 1999. Saffron in cultivation in Azerbaijan. In: Negbi, M. (Ed.), Saffron: Crocus sativus L. Harwood Academic Publishers, Australia, pp. 63–71.

Basker, D., M. Negbi, 1983. Uses of saffron. Econ. Bot. 37, 228–236.

Benschop, M., 1993. Crocus. In. The physiology of flower bulbs. Hertogh, A. de. Nard, M. Leed. (eds.) Amesterdam, Elsevier pp. 257-272.

Chryssanthi, D.G., P. Dedes, F.N. Lamari, 2009. Crocetin, the active metabolite of crocins, inhibits growth of breast cancer cells and alters the gene expression pattern of metalloproteinases and their inhibitors in the cell line MDA-MB-231. In: 3rd International Symposium on Saffron Forthcoming Challenges in Cultivation Research and Economics, Krokos, Kozani, Greece, p. 58.

Comstock, R.E., R.H. Moll, 1963. Genotyp-environment interaction , Page 146-196. In: Statistical genetics and plant breeding , Nat. Acad. Sci. Pabl. No. 982 Washington, D.C.

Dalezis, P., E. Papageorgiou, E. Geromichalou, G. Geromichalus, 2009. Antitumor activity of crocin, crocetin and safranal on murine P388 leukemia bearing mice. In: 3rd International Symposium on Saffron Forthcoming Challenges in Cultivation Research and Economics, Krokos Kozani, Greece, p. 58.

De Juan, A., A. Moya, S. Lapez, O. Botella, H. Lapez, R. Mu~noz, 2003. Influence of the corm size and the density of plantation in the yield and the quality of the production of corms of Crocus sativus L. ITEA 99, 169–180.

De Mastro, G., C. Ruta, 1993. Relation between corm size and saffron (Crocus sativus L.) flowering. Acta Hortic. 334, 512–517.

Dhar, A. K., R. Spru, K. Rekha. 1988. Studies on saffron in Kashmir. 1. Variation in natural population and its cytological behavior Crop Improvement. 15 (1): 48-52.

Ehsanzadeh, P., A. A. Yadollahi, A. M. M. Maiboodi. 2004. Productivity, Growth and quality attributes of 10 Iranian saffron accessions under climatic condition of Chahar Mahal Bakhtiari in Iran. ISHS Acta Horticulturae, 650, 1 International Symposium of Saffron Biology and Biotechnology, 31 May, Albacete, Spain. P. 291.

Fernandez, J.A., 2004. Biology, biotechnology and biomedicine of saffron. Rec. Res. Dev. Plant Sci. 2, 127–159.

Freeman, G. H. 1973. Statistical method for the analysis of genotype environment interaction. Heredity. 31: 339-354.

Ghorbani, M., 2006. The economics of saffron in Iran. II. International symposium on saffron biology and technology. ISHS Acta Horticulturae, 739, 321–332.

Gresta, F., G. Avola, G.M. Lombardo, L. Siracusa, G. Ruberto, 2009. Analysis of flowering, stigmas yield and qualitative traits of saffron (Crocus sativus L.) as affected by environmental conditions. Scientia Horticulturae. Volume 119, Issue 3, Pages 320-324.

Gresta, F., G. M. Lombardo, L. Siracusa, G. Ruberto. 2008. Effect of mother corm dimension and sowing time on stigmas yield, daughter corms and qualitative aspects of saffron (Crocus sativus L.) in aMediterranean environment. J. Sci. Food Agric. 88, 1144–1150.

Johnson, H.W., H.F. Robinson, R.E. Comstock, 1995. Estimate of genetic and environment variability in soybean. Agron. J. 47: 314-318.

Kafi, M., 2006. Saffron ecophysiology. In: Kafi, M., Koocheki, A., Rashed, M.H., Nassiri, M. (Eds.), Saffron (Crocus sativus) Production and Processing. Science Publishers, Enfield, pp. 39–58.

Katar, D., Y. Arslan, I. Subaşi, 2012. Genotypic Variations on Yield, Yield Components and Oil Quality in Some Camelina (Camelina Sativa L. Crantz) Genotypes. Turkish Journal Of Field Crops,17(2): 105-110.

Koocheki, A., M. Nassiri, M.A. Behdani, 2006. Agronomic attributes of saffron yield at agroecosystems scale in Iran. Proceedings of the 2nd International Symposium on Saffron Biology and Technology. Mashhad, Iran, 28-30 October 2006, p. 33-40.

Laidiaw, A.S., 2005. The effect of extremes in soil moisture content on perennial ryegrass grow The International Grassland congress, IRELAND UNITED KINGDOM.

Luana Maggi, a., a. Manuel Carmona, D. Simon, b. Kelly, c. Niusa Marigheto, L. Gonzalo, a. Alonso, 2011. Geographical origin differentiation of saffron spice (Crocus sativus L. stigmas) Preliminary investigation using chemical and multi-

element (H, C, N) stable isotope analysis. Food Chemistry, 128: 543–548.

Lundmark, M., H. Vaughan, L. Lapointe, 2009. Low temperatures maximise growth of Crocus vernus (L.) Hill via changes in carbon partitioning and corm development. J. Exp. Bot. 60, 2203–2213.

Magesh, V., J.P.V. Singh, K. Selvendiran, M.G. Ekambara, D. Sakthisekaran, 2006. Antitumor activity of crocetin in accordance to tumor incidence, antioxidant status, drug metabolizing enzymes and histopathological studies. Mol. Cell. Biochem. 287, 127–135.

Molina R.V., M. Valero, Y. Navarro, A. Garcia-luis, J.L. Guardiola, 2004. The effect of time of corm lifting and duration of incubation at inductive temperature on flowering in the saffron plant (Crocus sativus L.). Scientia Horticulturae. 103: 79.91.

Molina, R.V., M. Valero, Y. Navarro, A. Garci´a-Luis, J.L. Guardiola, 2005a. Low temperature storage of corms extends the flowering season of saffron (Crocus sativus L.). J. Hort. Sci. Biotechnol. 80, 319–326.

Molina, R.V., M. Valero, Y. Navarro, J.L. Guardiola, A. Garci´a-Luis, 2005b. Temperature effects on flower formation in saffron (Crocus sativus L.). Sci. Hort.-Amsterdam 103, 361–379.

Mollafilabi, A., 2004. Experimental findings of production and echo physiological aspects of saffron (Crocus sativus L.). Acta Horticulturae (ISHS). 650: 195-200.

Negbi, M., 1999. Saffron cultivation: past, present and future prospects. In: Negbi, M. (Ed.), Saffron. Crocus sativus L. Harwood Academic Publishers, Australia, pp. 1–18.

Sampathu, S.R., S. Shivashankar, Y.S. Lewis, 1984. Saffron (Crocus sativus L.): cultivation, processing, chemistry and standardization. Critical Reviews in Food Science and Nutrition 20, 123–157.

Siracusa, L., F. Gresta, G. Avola, G.M. Lombardo, G. Ruberto. 2010. Influence of corm provenance and environmental condition on yield and apocarotenoid profiles in saffron (Crocus sativus L.). Journal of Food Composition and Analysis. 23:394–400.

Unal, M., A. Cavusoglu. 2005. The effect of various nitrogen fertilizers on saffron (Crocus sativus L.) yield. Akdeniz Üniv. Ziraaf Fak. Dergisi, 18(2): 257-260.

Winterhalter, P., M. Straubinger. 2000. Saffron renewed interest in an ancient spice. Food Rev. Int. 16: 39-59.

Zhang, Y., 1994. Numerical Experiments for The Impacts of Temperature and Precipitation on The Growth and Development of Winter Wheat, Journal of Environment Science: 194-200, pages 5.

35

EFFECTS OF ROW SPACING AND SEEDING RATE ON HAY AND SEED YIELD OF EASTERN ANATOLIAN FORAGE PEA (*Pisum sativum* ssp. *arvense* L.) ECOTYPE

Mustafa TAN[1], Kader KURSUN KIRCI[2], Zeynep DUMLU GUL[1]*

[1] *Ataturk University Faculty of Agriculture Department of Field Crops, Erzurum, TURKEY*
[2] *Ministry of Food, Agriculture and Livestock Provincial Directorate, Malatya, TURKEY*
* Corresponding author: zdumlu@atauni.edu.tr

ABSTRACT

The Eastern Anatolian forage pea ecotype generally has a thin stem, small leaves and small seeds. The ecotype is different from the improved breeds and thus cultural techniques for this ecotype are somehow different. Current study was conducted to investigate the effects of row spacing and seeding rate on yield and some other characteristics of the Eastern Anatolian forage pea (*Pisum sativum* ssp. *arvense* L.) ecotype. Experiments were carried out in randomised complete blocks design with three replications under irrigated conditions of Erzurum during the years 2010 and 2011. Three different row spacing (20, 40 and 60 cm) and four different seeding rates (60, 90, 120 and 150 kg ha^{-1}) were used in experiments. Hay yield, hay crude protein, ADF and NDF ratios, seed yield, straw yield, biological yield and harvest index parameters were investigated in this study. Results can be summarised as follows: Row spacing and seeding rate had significant effects on dry matter and seed yield of the forage pea. Based on the results of the present study and under Erzurum conditions, it was recommended that the Eastern Anatolian forage pea ecotype should be cultivated with 40 cm row spacing and 120 kg ha^{-1} seeding rate for hay yield and with 40 cm row spacing and 90 kg ha^{-1} seeding rate for seed yield.

Key Words: Forage pea, ecotypes, seed yield, plant traits

INTRODUCTION

Grain feed cultured in the Eastern Anatolia Region are limited only with barley and common vetch. Short vegetation periods and low summer temperatures of Bayburt, Erzurum, Agri, Kars and Ardahan plateaus hinders the cultivation of important grain forage crops like corn and soybean. Therefore, beside roughage production, there is a significant deficit also in grain forage production. There is a need in the region for alternative grain forage crops to be used beside barley and able to provide support to meet the forage demand of the region.

The forage pea (*Pisum sativum* ssp. *arvense* L.) has promising potential to be widespread in the region to meet the forage demand. It has been commonly cultivated in Ardahan, Kars and Bayburt like Eastern Anatolian Provinces for many years for both hay and grain. It is highly resistant to cold climate conditions and adapted to the ecology of East Anatolia. The forage pea is a nutritious forage source for livestock breeders. The crude protein ratio is approximately 18.05% during the flowering period (Tan et al., 2012), 8.94% after seeds were taken (Deniz, 1967) and 26.5% in grains (Acikgoz, 2001). Forage peas with their rich protein contents are

also used as forage additives in Western European Countries.

Current local forage pea ecotypes in the regions are used both for hay and seed production. However, there were no studies carried out about the cultivation techniques of these ecotypes. The 1000-kernel weights (60-80 g) of these ecotypes, also known as Eastern Anatolian populations, are significantly lower than improved cultivars (Tan et al., 2012). The sowing densities of Eastern Anatolian ecotypes are also partially different from the improved cultivars. Therefore, there is a need for basic studies to be carried out on seeding rate and row spacing of local population since such cultivation techniques vary based on plant characteristics and ecological conditions of the region.

Cultivation techniques for hay yield and seed yield may exhibit differences from each other. Uzun and Acikgoz (1998) reported that dense sowing decreased the lodging rate, harvest index, 1000-kernel weight and seed yield of the plants. Chayferus and Okuyucu (1987) reported increasing grain and crude protein yields, although decreasing 1000-kernel weights with decreasing row spacing in peas. Previous researches recommend row spacing for peas as 15-60 cm based on the purpose of production and ecological conditions (Senel, 1958; Johnston et al., 2002; Bozoglu et al., 2004; Inanc, 2007). Seeding rate may sometimes reach 20 kg ha^{-1} (Potts, 1980).

MATERIALS AND METHODS

Experiments were conducted with the local forage pea ecotype (*Pisum sativum* ssp. *arvense* L.), commonly cultivated in Ardahan Province, over the experimental fields of Ataturk University Faculty of Agriculture under irrigated conditions during the years 2010 and 2011. This ecotype has a thin stem and small leaves; flowers are violet and seeds are spherical with greenish-brown colour; 1000 kernel weight is approximately 60-80 g (Tan et al., 2012). Field experiments were designed in a factorial arrangement of a randomised complete blocks design with 3 replications. Three different row spacing (20, 40 and 60 cm) and four different seeding rates (60, 90, 120 and 150 kg ha^{-1}) were used in 12 combinations in each replication. Plot size was arranged as 2.4 x 5 m. There were 12 plant rows at 20 cm spacing, 6 rows at 40 cm spacing and 4 rows at 60 cm spacing. Sowing was performed within the last week of April of both years with a hand drill. Pre-sowing fertilisation was performed during seed-bed preparation as to have 50 kg N ha^{-1} and 80 kg P$_2$O$_5$ ha^{-1} standard fertiliser rates (Tan and Serin, 2013). The

flooding irrigation method was used to meet the water requirements of the plants during the summer months.

Harvest for hay yield was performed in July during the formation of lower pods and harvest for seed yield was performed in August when 75% pods were ripened. Half of each plot was harvested for dry matter and the other half for seed yield. Some characteristics related to dry matter yield and seed yield were investigated in this study. The two-year results were statistically evaluated using MSTAT-C procedures and mean separations were made on the basis of the Least Significant Differences (LSD).

Experimental soils have clay-loam texture, salt-free and with slight alkalinity (pH: 7.82). Soils have medium level available phosphorus (88 kg P$_2$O$_5$ ha^{-1}) and low organic matter content (1.79%), and soils are rich in potassium (1980 kg K$_2$O ha^{-1}).

Long-term and experimental period monthly average temperatures and total precipitations of the years 2010 and 2011 for Erzurum Province are provided in Table 1. The monthly average temperatures of both years were close to long-term averages, although the monthly total precipitations were above the long-term averages. This was more distinctive in April and May of the year 2011.

Table 1. Temperature and rainfall values of the research periods in 2010 and 2011 and long-term average (1975-2009) in Erzurum.

Months	Average temperature (°C)			Total rainfall (mm)		
	2010	2011	LTA*	2010	2011	LTA*
April	5.6	5.6	5.4	54.2	147.7	58.4
May	10.4	9.6	10.5	63.6	105.2	70.0
June	15.9	14.6	14.9	50.5	55.3	41.6
July	19.5	19.6	19.3	55.5	26.6	26.2
August	20.3	19.4	19.4	9.0	21.8	15.1
Mean/total	14.3	13.8	13.9	232.8	356.6	211.3

*LTA: Long Term Average (1975-2009)

RESULTS AND DISCUSSION

Different row spacing and seeding rates had significant effects on dry matter yield of the forage pea (Table 2). While the highest dry matter yield (4237 kg ha^{-1}) was obtained from 40 cm row spacing, yields of larger and narrower row spacing were found to be lower. Increasing seeding rates also increased dry matter yields of forage pea. However, such an increase was not found to be significant above the seeding rate of 90 kg ha^{-1}. The effects of row spacing x seeding rate interaction on dry matter yield were also found to be significant. Such a case indicates the interactions between row spacing and seeding rate. The highest dry matter yields (4560 and 5021 kg ha^{-1}) were observed at 40 cm row spacing with 120 and 150 kg ha^{-1} seeding rates. Dry matter yields of the first year were higher than the yield of the second year. Year x row spacing interaction was also significant. The row spacing of 40 cm had the highest dry matter yields in both years (Figure 1).

The effects of different row spacing and seeding rates on crude protein ratios were found to be insignificant. Crude protein ratios of different row spacing treatments varied between 16.28-16.38% and ratios of different seeding rates varied between 16.00-16.62%. The effects of years on crude protein ratios were not also significant. While the crude protein ratio of the first year with better plant growth and higher dry matter yield was 15.02%, the value of the second year was found to be 17.61%.

With regard to ADF ratios of the forage pea, only the effects of seeding rate and row spacing x seeding rate interaction were found to be significant (P<0.05). ADF ratios of 60, 90, 120 and 150 kg ha^{-1} seeding rates were respectively observed as 23.71, 22.59, 24.22 and 22.12%. Row spacing was not effective on ADF ratios, although it was found to be effective in interaction with seeding rate. Considering both treatments, while the highest ADF ratio (25.50%) was obtained from 20 cm row spacing with 60 kg ha^{-1} seeding rate, the lowest value (21.28%) was obtained from 60 cm row spacing with the highest seeding rate (Table 1). Conversely, treatments did not have significant effects on NDF ratios, which is another indicator of fibrous parts. NDF ratios of row spacing treatments varied between 32.61-33.11% and ratios of seeding rate treatments varied between 31.89-33.45% (Table 2).

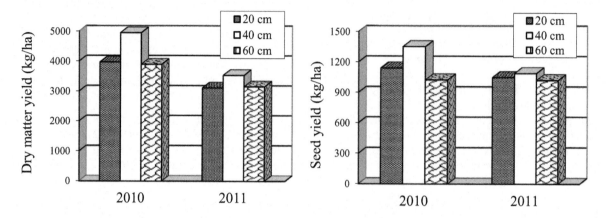

Figure 1. Row spacing x year interaction on dry matter and seed yields

Table 2. Dry matter yield, crude protein, ADF and NDF ratios of forage pea as affected by row spacing and seeding rate (average of 2 years).[1]

Row spacing (cm)	Seeding rates (kg ha⁻¹)	Dry matter yield (kg ha⁻¹)	Crude protein ratio (%)	ADF ratio (%)	NDF ratio (%)
20	60	2728	15.49	25.50	33.10
	90	3740	16.17	21.43	32.18
	120	3593	16.67	25.08	34.20
	150	4081	16.81	22.97	33.00
Mean		3536 B	16.28	23.74	33.11
40	60	3154	16.60	22.85	33.65
	90	4214	16.34	23.58	31.62
	120	4560	16.53	24.32	32.25
	150	5021	16.06	22.11	34.14
Mean		4237 A	16.38	23.21	32.92
60	60	3255	15.91	22.79	33.15
	90	3611	16.42	22.77	31.87
	120	3661	16.67	23.28	32.21
	150	3554	16.15	21.28	33.21
Mean		3520 B	16.29	22.53	32.61
Seed Rates	60	3046 B	16.00	23.71 AB	33.30
	90	3855 A	16.31	22.59 BC	31.89
	120	3938 A	16.62	24.22 A	32.89
	150	4219 A	16.34	22.12 C	33.45
Year	2010	4267 A	15.02 B	23.37	32.91
	2011	3262 B	17.61 A	22.96	32.85
F-test (LSD Value)					
	Row spacing	** (343)	ns	ns	ns
	Seed rate	** (396)	ns	* (1.16)	ns
	R spacing x S. rate	** (687)	ns	* (2.02)	ns
	Year	* (210)	** (0.85)	ns	ns
	R. spacing x Year	* (364)	ns	ns	ns
	S. rate x Year	ns	ns	ns	ns
	RS x SR x Y	ns	ns	ns	ns

[1]Values followed by different letters in a column represent significant differences.
ns: no significance, *: P<0.05, **: P<0.01

The effects of row spacing, seeding rate, row spacing x seeding rate, row spacing x year interactions on seed yield of forage pea ecotype were found to be significant (Table 3). The highest seed yield (1220 kg ha^{-1}) was observed in 40 cm row spacing, although seed yield of 20 cm row spacing (1094 kg ha^{-1}) was also significantly high. The lowest seed yield (1025 kg ha^{-1}) was obtained from the largest row spacing treatment. Seeding rates of 90, 120 and 150 kg ha^{-1} also had high seed yields, although the differences between them were not significant. The seeding rate of 60 kg ha^{-1} had low seed yield (922 kg ha^{-1}). Considering row spacing and seeding rate together, seed yields of 20 cm x 150 kg ha^{-1} and 40 cm x 90-150 kg ha^{-1} treatments seemed to be high. With regard to row spacing x year interaction, 40 cm row spacing had higher yields in both years (Figure 1).

While the effects of row spacing on straw yield were not significant, seeding rate, year and row spacing x year interaction had significant effects on straw yields. As it was in dry matter and seed yields, the increasing seeding rates generally increased straw yields. However, such an increase was not significant above the seeding rate of 90 kg ha^{-1}. A similar case was also observed in biological yield. While the lowest biological yield (3080 kg ha^{-1}) was obtained from the lowest seeding rate, increasing seeding rates also increased biological yields, although again, such an increase was not significant for seeding rates above 90 kg ha^{-1}. Biological yield was effected by the changes in row spacing (P<0.05). A biological yield of 40 cm row spacing was higher than the yield of other treatments and this yield was also significantly higher than the yield of 60 cm row spacing. Both straw and biological yields were found to be higher in the first year (2010) with better plant growth.

The harvest index was not affected from single row spacing and seeding rate treatments, although it was significantly affected by the interaction of these two treatments. The highest harvest index values were observed in 20 cm x 150 kg ha^{-1}, 40 cm x 90-150 kg ha^{-1} and 60 cm x 60-120 kg ha^{-1} treatments (Table 3).

Table 3. Seed, straw and biological yields and harvest index of forage pea as affected by row spacing and seeding rate (average of 2 years).[1]

Row spacing (cm)	Seeding rates (kg ha^{-1})	Seed yield (kg ha^{-1})	Straw yield (kg ha^{-1})	Biological yield (kg ha^{-1})	Harvest index (%)
	60	969	2377	3346	28.9
	90	1044	2860	3904	26.5
20	120	1134	2772	3906	29.4
	150	1229	2646	3876	31.3
Mean		1094 AB	2664	3758 AB	29.0
	60	886	2130	3016	29.9
	90	1251	2464	3715	34.7
40	120	1338	3029	4367	31.6
	150	1404	3214	4618	30.5
Mean		1220 A	2709	3929 B	31.7
	60	922	1956	2877	34.0
	90	1079	2494	3573	31.0
60	120	1134	2681	3814	30.4
	150	951	2759	3710	27.1
Mean		1025 B	2472	3494 B	30.6
	60	926 B	2155 B	3080 B	30.9
Seed Rates	90	1124 A	2606 AB	3731 A	30.7
	120	1202 A	2872 A	4029 A	30.4
	150	1195 A	2873 A	4068 A	29.6
Year	2010	1172	2866 A	4038 A	28.9 B
	2011	1051	2364 B	3415 B	31.9 A
F-test (LSD Value)					
	Row spacing	** (139)	ns	* (354)	ns
	Seed rate	** (161)	** (495)	* (545)	ns
	R spacing x S. rate	* (209)	ns	ns	* (4.2)
	Year	ns	** (350)	** (386)	ns
	R. spacing x Year	* (148)	* (455)	* (501)	* (4.5)
	S. rate x Year	ns	ns	ns	ns
	RS x SR x Y	ns	ns	ns	ns

[1]Values followed by different letters in a column represent significant differences.
ns: no significance, *: P<0.05, **: P<0.01

The number of plant per unit area is the most significant factor effective on yield and it mostly depends on row spacing and seeding rate. Yield decrease is evident at seeding rates above or below the optimum rates (Martin et al., 1994; Johnston et al., 2002). Increasing the seeding rates generally increases the hay yield of peas (Turk et al., 2011). In a study carried out in Erzurum, the highest dry matter yield of common vetch, which has a very similar seed size and plant height with Eastern Anatolian forage pea ecotypes, was obtained from 24 cm row spacing with 120 kg ha^{-1} seeding rate (Serin et al., 1996). The effects of sowing density on hay quality were not distinctive. The effects of seeding rate were found to be significant only on ADF ratios. The ADF ratio was higher in lower seeding rates. This may be due to increasing fibrous stem and branches parallel to decrease the number of plants per unit area. Small effects of sowing density on forage quality were also reported by Juskiw et al. (2000). Yavuz et al. (2011) reported insignificant effects of row spacing and seeding rate on the ADF and NDF ratios of vetch.

In the present study, seed, straw and biological yields were higher at 40 cm row spacing and 90-120 kg ha^{-1} seeding rate. For higher seed yields in peas, Inanc (2007) recommended a row spacing of 35 cm and Bozoglu et al. (2004) recommended 40 cm. With regard to seeding rates, Uzun and Acikgoz (1998) had the highest seed yield with a seeding rate of 100 seeds/m^2. Turk et al. (2011) reported a decreasing number of seeds per pod and seed yield with increasing seeding rates. Yavuz et al. (2011) investigated the effects of seeding rates on vetch yields and observed increasing seed and biological yield until a seeding rate of 100 kg ha^{-1}. Conversely, Girgel (2006) indicated increasing straw yields with dense seeding rates.

In the present study, it was proven that row spacing and seeding rate were significant parameters to be considered in the cultivation of the Eastern Anatolian forage pea ecotype. While the proper row spacing was not found to be different from the previously recommended ones, the seeding rate was found to be relatively lower than the values reported in previous pea studies. This was mostly due to some morphological characteristics such as the small seed size, thin stem and small leaves of Eastern Anatolian forage pea ecotype. Considering the row spacing x seeding rate interaction, 40 cm row spacing with 120 kg ha^{-1} seeding rate was recommended for hay yield and 40 cm row spacing with 90 kg ha^{-1} seeding rate was recommended for seed yield.

LITERATURE CITED

Acikgoz, E. 2001. Forage Crops. Uludag Univ. Publ. No: 182, Bursa, Turkey, 584 p.

Bozoglu, H., E. Pesken, A. Gulumser. 2004. Effect of row spacing and potassium humate application on yield and some traits of peas. Ankara Univ. Faculty of Agric., Journal of Agric. Sci., 10 (1): 53-58.

Chayferous, C., F. Okuyucu. 1987. Studies on effects of row spacings on yield and other some characters of two different forage pea (*Pisum arvense* L.). Journal of Ege Univ. Faculty of Agric., 25: 3, Izmir.

Deniz, O. 1967. Crude Protein, Digestible Nutrients, Calcium and Phosphorus Content of Forage Pea. Livestock and Grassland Research Institute, Ayyildiz Pub. Inc., Ankara, 91 p.

Girgel, U. 2006. A Research on the Yield Components of Different Sowing Densities on Bolero Pea (*Pisum sativum* L.) Cultivars under the Kahramanmaras Conditions. Sutcu Imam Univ. Institute of Natural and Applied Sciences, Msc Thesis, Kahramanmaras.

Inanc, S. 2007. The Effects of Different Row Spacings on the Yield and Yield Components in Pea (*Pisum sativum* ssp. *arvense*). Yuzuncu Yil Univ. Institute of Natural and Applied Sciences, Msc Thesis, Van.

Johnston, A.M., G.W. Clayton, G.P. Lafond, K.N. Harker, T.J. Hogg, E.N. Johnson, W.E. May, J.T. McConnell. 2002. Field pea seeding management. Canadian J. Plant Sci., 82: 639-644.

Juskiw, P.E., J.H. Helma, D.F. Salmona. 2000. Forage yield and quality for monocrops and mixtures of small grain cereals. Crop Sci., 40(1): 138-147.

Martin, I.J., L. Tenoria, L. Ayerbe. 1994. Yield, growth and water use of conventional and semi leafless peas in semiarid environments. Crop Sci., 34: 1576-1583.

Potts, J.M. 1980. The influence of sowing date, harvest date and seed rate on the yield of forage peas. Grass and Forage Science, 35: 41-45.

Senel, M. 1958. Variety screening for vinter vetches (1934-1954). Experimental Research Station of Adana, Ministry of Agriculture, Technical Bul: No: 5, Adana, Turkey.

Tan, M., Y. Serin. 2008. Forage Legumes. Ataturk Univ. Faculty of Agric. Publ. No: 190, Erzurum, Turkey.

Tan, M., A. Koc, Z. Dumlu Gul. 2012. Morphological characteristics and seed yield of East Anatolian local forage pea (*Pisum sativum* ssp. *arvense* L.) ecotypes. Turkish Journal of Crop Sci., 17(1): 24-30.

Turk, M., S. Albayrak, O. Yuksel. 2011. Effect of seeding rate on the forage yields and quality in pea cultivars of differing leaf types. Turkish Journal of Field Crops, 16(2): 137-14.

Uzun, A., E. Acikgoz. 1998. Effect of sowing season and seeding rate on the morphological traits and yields in pea cultivars of differing leaf types, J. Agronomy and Crop Sci., 181: 215-222.

Yavuz, T., M. Surmen, N. Cankaya. 2011. Effect of row spacing and seeding rate on yield and yield components of common vetch (*Vicia sativa* L.). Journal of Food, Agriculture & Environment, 9(1): 369-371.

36

CHICKPEA PERFORMANCE COMPARED TO PEA, BARLEY AND OAT IN CENTRAL EUROPE: GROWTH ANALYSIS AND YIELD

R.W. Neugschwandtner[1]*, S. Wichmann[1], D.M. Gimplinger[1], H. Wagentristl[2], H.-P. Kaul[1]

*[1]BOKU - University of Natural Resources and Life Sciences, Vienna,
Department of Crop Sciences, Division of Agronomy, Tulln, Austria
[2] BOKU - University of Natural Resources and Life Sciences, Department of Crop Sciences,
Experimental Farm Groß-Enzersdorf, Groß-Enzersdorf, Austria
Corresponding author: reinhard.neugschwandtner@boku.ac.at

ABSTRACT

An increase of grain legume production is essential for meeting feed protein need in Europe. Warming climates offer the opportunity for adapting crops with a more warm-season growth habit such as chickpea (*Cicer arietinum* L.) in cool, northern latitude areas. Therefore, yield and growth analysis of chickpea were assessed in a two-year field experiment in Central Europe (Raasdorf, Austria) and compared to pea, barley and oat, which are well adapted crops in that region. Chickpea had a lower above-ground biomass and grain yield compared to pea, barley and oat in 2006, whereas only pea was more productive than chickpea in the dry year of 2007. The relative good performance of chickpea regarding crop growth rate and relative growth rate compared to pea, barley and oat under severe drought in 2007 indicated that chickpea may be an interesting crop in the Central European production area in the face of possible climate change.

Keywords: Biomass production, Central Europe , chickpea, *Cicer arietinum*, growth rate

INTRODUCTION

The European Union has a deficit of protein sources for livestock relying therefore to a large extent on soybean meal imports. Increasing grain legume production in Europe could provide an alternative. Furthermore, grain legumes contribute to the diversification and long-term productivity of sustainable agricultural systems as they can satisfy a bulk of their N demand from atmospheric nitrogen through symbiosis with nitrogen fixing soil bacteria (*Rhizobium* spp.) thus minimizing the demand for N fertilizer inputs within crop rotations (van Kessel and Hartley, 2000). Positive yield effects on subsequent non-legume crops result from the soil-N sparing of the legumes and the transfer of biologically fixed N via crop residues (Chalk, 1998; Kaul, 2004).

Warming climates are prolonging growing seasons in northern latitudes and thus causing an impact on agricultural systems (Menzel et al., 2006). Under these conditions, promising opportunities may arise for adapting crops with a more warm-season growth habit such as chickpea (*Cicer arietinum* L.) in comparatively cool, northern latitude areas (Gan et al., 2009). Currently, pea (*Pisum sativum* L.) and soybean (*Glycine max* (L.) Merr.) are the most important grain legumes in Eastern Austria. Alternative crops like chickpea could become of interest for this region due to the forecasted change in climatic conditions with presumably longer periods of drought stress. Chickpea genotypes can effectively cope with drought conditions due to several morphological and physiological advantages of the crop (Serraj et al., 2004; Toker et al., 2007; Cutforth et al., 2009; Zaman-Allah et al., 2011). A former experiment showed that chickpea could compete on an average grain yield level of 3 t ha^{-1} with pea and soybean in Central Europe (Wichmann et al., 2005).

Chickpea is on the fourth position among grain legumes in the world regarding grain production after soybean, bean (*Phaseolus vulgaris* L.) and pea. It is mainly produced in areas classified as arid or semiarid environments (Canci and Toker, 2009ab). Chickpea is of high importance in human diets in many areas of the world. Additionally, chickpea grains can be used as energy and protein-rich feed in animal diets and chickpea straw as alternative forage for ruminants (Bampidis and Christodoulou, 2011). Although chickpea is not a common crop in Central Europe, it could provide an alternative for food and feed protein production in Central Europe in the face of possible climate change.

The objective of this study was to evaluate chickpea suitability in Central Europe, focussing on grain yield and growth analysis as compared to the legume pea and the non-legumes barley (*Hordeum vulgare* L.) and oat (*Avena sativa* L.), which are common crops in that region.

MATERIALS AND METHODS

Experimental factors

Chickpea genotypes were tested under different treatments of nitrogen fertilization in comparison to common varieties of pea and the non-legumes barley and oat. The experiment was set up in a randomized complete block design with two replications. The *Cicer arietinum* variety "Kompolti" and commercial seeds of a *C. arietinum* genotype of unknown origin (named by us after the trade company "Hirschhofer") (both are Kabuli type chickpeas) were planted. The seeds had been multiplied on-farm by our own. *Pisum sativum* cv. "Attika" and "Rosalie", *Hordeum vulgare* cv. "Xanadu" and *Avena sativa* cv. "Jumbo" were used as standards of comparison. The nitrogen fertilizer calcium ammonium nitrate (CAN) (27% N, 10% Ca) and the depot fertilizer Basacote® Plus 6M (16% N, 3.5% P, 10% K, 1.2% Mg, 5% S and micronutrients) were applied at two fertilization levels (10 and 20 g N g^{-2}) supplemented by an unfertilized control. Fertilizers were applied right after sowing.

Environmental conditions

The experiment was carried out in Raasdorf (48° 14' N, 16° 33' E) in Eastern Austria on the experimental farm Gross-Enzersdorf of BOKU University. The soil is classified as a chernosem of alluvial origin and rich in calcareous sediments (pH 7.6). The texture is silty loam; the content of organic substance is at 2.2-2.3%.

The mean annual temperature is 10.6°C, the mean annual precipitation is 538 mm (1980-2009). Table 1 shows the long-term average monthly temperature and precipitation (1980-2009) from February to July and the deviations during the 2006 and 2007 growing season. The temperature was considerable higher in 2007 than in 2006 (except for the month July). Monthly precipitation was highly above average in March and April in 2006. Contrary to that, the growing season 2007 was characterized by severe spring drought without rainfall from end of March to beginning of May.

Table 1. Long-term average monthly temperature and precipitation (1980-2009) and deviations during the 2006 and 2007 growing seasons

	Temperature (°C)			Precipitation (mm)		
	Mean (1980-2009)	**2006 (±)**	**2007 (±)**	**Mean (1980-2009)**	**2006 (±)**	**2007 (±)**
February	1.7	-1.9	3.8	26.4	-7.7	17.7
March	5.8	-2.1	2.3	38.5	7.7	28.0
April	10.7	1.3	2.1	35.3	30.3	-34.4
May	15.6	-0.5	1.6	56.1	16.7	-9.8
June	18.5	0.6	2.8	72.3	-9.9	-3.9
July	20.8	2.8	1.9	59.1	-52.3	-6.2

Experimental treatments and measurements

Seeds were sown with an Oyjard plot drill at a row distance of 12 cm on plots of 30 m². Seeds of chickpea were inoculated with *Mesorhizobium ciceri* (Jost GmbH), seeds of pea with *Rhizobium leguminosarum* (Radicin No4, Jost GmbH) according to product specifications before sowing. Table 2 gives detailed information on the experimental conditions and crop management practices. The development of above-ground dry matter was determined by harvesting (0.24 m² per plot) at intervals of about 14 days until end of June and drying (100°C, 24 h). For grain and straw yield assessment, the final harvest was performed at full ripeness of the plants on 0.96 m² per plot.

Table 2. Field experiment and crop management practices

		2006	2007
Sowing date		14 April	11 April
Row distance		12 cm	
Sowing rate (seeds m⁻²)	Chickpea	90	
	Pea	90	
	Barley	350	
	Oat	350	
Weeding		Mechanical hand weeding	
Harvest date 1		5 May	14 May
Harvest date 2		22 May	31 May
Harvest date 3		9 June	14 June
Harvest date 4		27 June	26 June
Harvest date 5	Chickpea	1 August	23 July
	Pea	20 July	9 July
	Barley	18 July	23 July
	Oat	24 July	23 July

Growth analysis

Crop growth rate (CGR) and relative growth rate (RGR) were calculated for each period between subsequent harvest dates according to Hunt (1982) as follows:

$$CGR = (W_2 - W_1)/(t_2 - t_1)$$

$$RGR = (\ln W_2 - \ln W_1)/(t_2 - t_1)$$

where W_2 and W_1 indicate the final and initial above-ground plant dry weight and t_2 and t_1 indicate the end and the start day of each period. Growth analysis is one way to verify ecological adaptation of crops to new environments (Namvar et al., 2011).

Statistics

Statistical analyses were performed using software SAS version 9.2. Analysis of variance (PROC GLM) with subsequent multiple comparisons of means were performed. Means were separated by least significant differences (LSD), when the F-test indicated factorial effects on the significance level of $p<0.05$. Genotype differences of chickpea and pea were not significant, so data were pooled for analysis.

RESULTS AND DISCUSSION

Above-ground biomass, final yield and harvest index

Grain yield of chickpea and the other crops was not affected in both years by N fertilization. Straw yield and consequently above-ground biomass (AGB) were increased by N fertilization in the more humid year 2006, whereas no effects of fertilization were found in the dry year 2007. No interactions of crop×fertilization were observed for the yield parameters (data not shown). Obviously the experimental site had a very fertile soil that supplied enough plant available N even in the unfertilized control plots. Farzaneh et al. (2009) found similar results with chickpea and barley in pot experiments.

A significant interaction of crop×year was observed for above-ground biomass (AGB), grain yield, straw yield and harvest index (Fig. 1). The AGB of chickpea was significantly lower than that of the other crops in 2006 whereas it was slightly lower than the AGB of pea and slightly higher than the AGB of barley and oat in 2007 (Fig. 1a). Drought occurring at the end of the growing period of the plants in 2006 might not have affected plant growth and yield substantially. Generally, the water restrictions limited AGB productions of all plants in the dry year of 2007. Precipitation in the late growing period in 2007 could not compensate for the drought damages that occurred earlier that season. Kurdali et al. (2002) reported that drought highly reduced dry matter production as well as nodulation and N_2 fixation of chickpea. Reduced N_2 fixation may additionally impair yields as reported for soybean (Salvagiotti et al., 2008). However, negative effects of drought on chickpea's AGB productions were less pronounced compared to other crops.

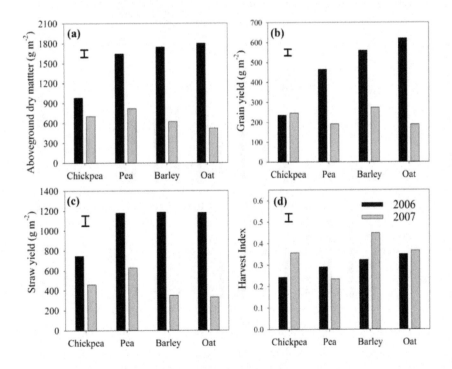

Figure 1. (a) Above-ground biomass, (b) grain yield, (c) straw yield and (d) harvest index depending on crop and year, error bars are LSD

Chickpea had the lowest grain yield among the four crops in 2006, while it had performed better in terms of grain yield than pea and oat in the dry year of 2007 (Fig. 1b). Barley had the highest grain yield among the studied crops in 2007 since it is one of the important drought resistant crops (Toker et al., 2009). Similar to our observations, Angadi et al. (2008) reported that chickpea grown in semiarid conditions in the Canadian prairie produced under mid- to late-season drought stress a higher grain yield compared to the crops canola (*Brassica napus* L. and *B. rapa* L.), mustard (*Brassica juncea* L.), pea and spring wheat (*Triticum aestivum* L.). Chickpea is better adapted to moderate to severe water stress due to its ability to maintain a positive turgor over a wide range of water potentials (Cutforth et al., 2009), a deep and prolific root system (Serraj et al., 2004) and a conservative pattern of water use (Zaman-Allah et al., 2011). Chickpea grain yields (2006: 235 g m^{-2}, 2007: 246 g m^{-2}) in Central Europe seem to be satisfying in face of grain yield reported in experiments conducted in Turkey, one of the main chickpea producing countries, which were at 186 g m^{-2} (Özalkan et al., 2010).

The straw yield of chickpea was lower than that of the other crops in 2006 but it was higher than those of barley and oat in 2007 (Fig. 1c). Chickpea had the lowest harvest index in 2006, while its harvest index was in a similar range to that of oat, higher compared to pea and lower compared to barley in 2007 (Fig. 1d). Under the dry conditions of 2007, chickpea´s harvest index was considerably higher than in 2006. Chickpea has an indeterminate growth habit (Yildirim et al., 2013), thus under favourable growing conditions the plant continues

the vegetative growth and delays pod setting, seed filling and maturity (Liu et al., 2003; Gan et al., 2009). Sinha et al. (1982) reported that sufficient water availability after flowering is used by chickpea for a prolonged vegetative growth resulting thereby in further increase of total above-ground biomass but not necessarily in a grain yield gain. Consequently, chickpea can achieve a good harvest index under drought conditions. Angadi et al. (2008) reported that the harvest index of chickpea was not so severely impaired by drought conditions in the Canadian Prairie as compared to that of pea, wheat and oilseed crops. Our results are also in accordance with Fulkai (1995) who reported an increased harvest index under water stress at two of three experimental sites. A crop's ability to maintain a high harvest index is important for its adaption to semiarid conditions (Ludlow and Muchow, 1990). An increased biomass partitioning into seed contributes especially under dry conditions to the relatively large seed yield of the lower biomass producing pulse crops, like chickpea, compared with other crops (Angadi et al., 2008).

Growth analysis

The crop growth rates (CGR) of chickpea, pea, barley and oat are shown in Fig. 2. Crop growth rate is a function of canopy gross photosynthesis and crop respiration (Evans, 1993). These processes are influenced by environmental conditions such as temperature, solar radiation, water and nutrient supply (Connor et al., 2011). Fertilization positively influenced the crop growth rates of all tested crops until middle of June, i. e. until harvest date 3, which was set in accordance with flowering of pea and chickpea (data not shown).

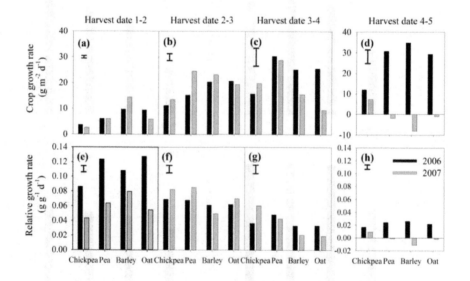

Figure 2. Crop growth rate (a-d) and relative growth rate (e-h) depending on crop and year, error bars are LSD

The CGR was lowest at early growth. There was a significant crop×year interaction in all observation periods (Figs. 2a-d). The crop growth rate of chickpea was consistently lower than that of the other crops in 2006 whereas chickpea's CGR remained lower until middle of June (harvest date 3) in 2007 (Fig. 2a, b), subsequently exceeding the CGR of barley and oat but staying below that of pea (Fig. 2c). After end of June (harvest date 4-5), the CGR of chickpea was the highest one among the four crops (Fig. 2d). Obviously, chickpea performed better compared to the other crops in the dry conditions of 2007. Although rainfall (below the long-term average) occurred in June and July 2007, water limitations due to severe spring drought with absent rainfall for some weeks may have strongly impaired CGR of pea, barley and oat at the end of the growth period (harvest date 4-5) while the slower development of chickpea allowed to profit from this late water supply. Koutroubas et al. (2009) reported that different weather conditions were among the main causes of seasonal variation in rainfed chickpea growth and productivity under Mediterranean conditions. Our results showed generally lower variations of chickpea's CGR between the two years indicating that the crop growth of chickpea may have been less affected by environmental conditions compared to that of pea, barley and oat.

The relative growth rate (RGR) was the highest at early growth and decreased with time (Figs. 2e-h). This is due to the increase of the share of non-assimilatory tissues with time (Nogueira et al., 1994). The RGR was only affected by fertilization in 2006, when RGR of all fertilized treatments was higher than that of the unfertilized control (data not shown). The RGR of chickpea was lower than that of the other crops in the first growth period in 2006, whereas it remained in the other periods on a similar level as with pea, barley and oat. In 2007, chickpea's RGR was the lowest again in the first period (Fig. 2e). In the second period the RGR of the legumes was higher than that of the cereals (Fig. 2f). Starting with sampling date 3 the RGR of chickpea was the highest among the tested crops being the only one with positive values after end of June (date 4) (Fig. 2g, h) until harvest. Consequently, chickpea had the highest biomass increase per unit of biomass and per unit of time among the tested crops after mid of June 2007.

CONCLUSION

Yield and growth analysis assessment of chickpea compared with pea and cereals was performed in a two-year experiment in Central Europe. Chickpea had a lower above-ground biomass and grain yield compared to pea, barley and oat in the humid year 2006, whereas only pea was more productive than chickpea in the dry year of 2007. This and the good performance of chickpea regarding the crop growth rate and the relative growth rate during late crop development under severe drought in 2007 indicate that chickpea may be an interesting crop for the Central European production area in the face of possible climate change.

LITERATURE CITED

Angadi, S.V., McConkey, B.G., Cutforth, H.W., Miller, P.R., Ulrich, D., Selles, F., Volkmar, K.M., Entz, M.H. and S.A. Brandt. 2008. Adaptation of alternative pulse and oilseed crops to the semiarid Canadian Prairie: Seed yield and water use efficiency. Can J Plant Sci 88:425-438.

Bampidis, V.A. and V. Christodoulou. 2011. Chickpeas (*Cicer arietinum* L.) in animal nutrition: A review. Anim Feed Sci Technol 168:1-20.

Canci, H. and C. Toker. 2009a. Evaluation of annual wild *Cicer* species for drought and heat resistance under field conditions. Genet Resour Crop Ev 56:1-6.

Canci, H. and C. Toker. 2009b. Evaluation of yield criteria for drought and heat resistance in chickpea (*Cicer arietinum* L.). J Agron Crop Sci 195:47-54.

Chalk, P.M. 1998. Dynamics of biologically fixed N in legume-cereal rotation: a review. Aust J Agric Res 49:303-316.

Connor, D.J., Loomis, R.S. and K.G Cassman. 2011. Crop Ecology: Productivity and Management in Agricultural Systems. 2. ed. Cambridge, UK: Cambridge University Press.

Cutforth, H.W., Angadi, S.V., McConkey, B.G., Entz, M.H., Ulrich, D. Volkmar, K.M., Miller, P.R. and S.A. Brandt. 2009. Comparing plant water relations for wheat with alternative pulse and oilseed crops grown in the semiarid Canadian prairie. Can J Plant Sci 89:823-835.

Evans, J. 1982. Symbiosis, nitrogen and dry matter distribution in chickpea (*Cicer arietinum* L.). Exp Agr 18:339-351.

Farzaneh, M., Wichmann, S., Vierheilig, H. and H.-P. Kaul. 2009. The effects of arbuscular mycorrhiza and nitrogen nutrition on growth of chickpea and barley. Pflanzenbauwiss 13:15-22.

Fulkai, T.S. 1995. Growth and yield response of barley and chickpea to water stress under three environments in southeast Queensland. I. Light interception, crop growth and grain yield. Aust J Agric Res 46:17-33.

Gan, Y.T., Warkentin, T.D., McDonald, C.L., Zentner, R.P. and A. Vandenberg. 2009. Seed yield and yield stability of chickpea in response to cropping systems and soil fertility in northern latitudes. Agron J 101:1113-1122.

Hunt, R. 1982. Plant Growth Curves. Functional Approach to Plant Growth Analyses. London: Edward Arnold.

Liu, P.H., Gan, Y.T., Warkentin, T. and C.L. McDonald. 2003. Morphological plasticity of chickpea in a semiarid environment. Crop Sci 43:426-429.

Ludlow, M.M. and R.C. Muchow. 1990. A critical evaluation of the traits for improving crop yields in water limited environments. Adv Agron 43:107-153.

Kaul, H.-P. 2004. Pre-crop effects of grain legumes and linseed on soil mineral N and productivity of subsequent winter rape and winter wheat crops. Bodenkultur 55:95-102.

Koutroubas, S.D., Papageorgiou, M. and S. Fotiadis. 2009. Growth and nitrogen dynamics of spring chickpea genotypes in a mediterranean-type climate. J Agr Sci 147:445-458.

Kurdali, F., Al-Ain, F. and M. Al-Shamma. 2002. Nodulation, dry matter production, and N_2 fixation by fababean and chickpea as affected by soil moisture and potassium fertilizer. J Plant Nutr 25:355-368.

Menzel, A., Sparks, T.H. and N. Estrella. 2006. European phenological response to climate change matches the warming pattern. Glob Change Biol 12:1969-1976.

Namvar, A., Sharifi, R.S. and T. Khandan. 2011. Growth analysis and yield of chickpea (*Cicer arietinum* L.) in relation to organic and inorganic nitrogen fertilization. Ekologija 57:97-108.

Nogueira, S.S.S., Nagai, V., Braga, N.R., Do, M., Novo, C.S.S. and M.B.P. Camargo. 1994. Growth analysis of chickpea (*Cicer arietinum* L.). Sci Agr 51:430-435.

Özalkan, C., Sepetoğlu, H.T., Daur, I. and O.F. Şen. 2010. Relationship between some plant growth parameters and grain yield of chickpea (*Cicer arietinum* L.) during different growth stages. Turk J Field Crops 15:79-83.

Salvagiotti, F., Cassman, K.G., Specht, J.E., Walters, D.T., Weiss, A. and A. Dobermann. 2008. Nitrogen uptake, fixation and response to fertilizer N in soybeans: A review. Field Crops Res 108:1-13.

Serraj, R., Krishnamurthy, L., Kashiwagi, J., Kumar, J., Chandra, S. and J.H. Crouch. 2004. Variation in root traits of chickpea (*Cicer arietinum* L.) grown under terminal drought. Field Crops Res 88:115-127.

Sinha, S.K., Khanna-Chopra, R., Aggarwal, P.K., Chaturvedi, S. and K.R. Koundal. 1982. Effect of drought on shoot growth: significance of metabolism to growth and yield. In: Drought resistance in crops with emphasis on rice. ed. International Rice Research Institute, 153-170, International Rice Research Institute, Los Baños, Laguna, Philippines.

Toker, C., Lluch, C., Tejera, N.A., Serraj, R. and K.H.M. Siddique. 2007. Abiotic stresses. In: Yadav, S.S., Redden, R., Chen, W. and Sharma, B. (eds) Chickpea breeding and management. CAB Int., Wallingford, UK, pp. 474-496.

Toker, C., Gorham, J. and M.I. Çagirgan. 2009. Certain ion accumulations in barley mutants exposed drought and salinity. Turk J Field Crops 14:162-169.

van Kessel, C. and C. Hartley. 2000. Agricultural management of grain legumes: has it led to an increase in nitrogen fixation? Field Crops Res 65:165-181.

Wichmann, S., Wagentristl, H. and H.-P. Kaul. 2005. Der Anbau von Kichererbsen in Vergleich zu Körnererbsen und Sojabohnen, Mitt Ges Pflanzenbauwiss 17:28-29.

Zaman-Allah, M., Jenkinson, D.M. and V. Vadez. 2011. A conservative pattern of water use, rather than deep or profuse rooting, is critical for the terminal drought tolerance of chickpea. J Exp Bot 62:4239-4252.

Yildirim, T., Canci, H., Inci, N.E., Baloglu, F.O.C., Ikten, C. and C. Toker. 2013. Inheritance of female sterility in induced *Cicer* species. Turk J Field Crops 18:78-81.

37

THE EFFECTS OF DIFFERENT NITROGEN DOSES ON TUBER YIELD AND SOME AGRONOMICAL TRAITS OF EARLY POTATOES

Leyla GULLUOGLU[1], H. Halis ARIOGLU[2], Halil BAKAL[2]

[1]*Cukurova University, Vocational School of Ceyhan, Adana, TURKEY*
[2]*Cukurova University, Faculty of Agriculture, Department of Field Crops, Adana, TURKEY*
[*]*Corresponding author: gulluoglu@cu.edu.tr*

ABSTRACT

This study was conducted to determine the effects of different nitrogen rates on tuber yield and some agronomical traits of early potatoes grown in 2011 and 2012. The field trial was conducted at the experimental field of Cukurova University, in the Randomized Complete Block Designs, with three replications, with Marfona a medium early table potato. Zero, 40, 80, 120, 160, 200, 240, 280, 320 and 360 kg ha^{-1} pure nitrogen doses were applied. According to the mean values of two years, the total tuber yield was 24.0 ton ha^{-1} in the control plot. The highest yield per hectare (53.31 ton ha^{-1}) was obtained from the 240 kg ha^{-1} N applied plots. Beyond 240 kg ha^{-1} peak, as the dose was increased, the tuber yield was decreased. In addition, the marketable tuber yield ratio was 95.60 % with 0 N dose. When the N dose increased to 200 kg ha^{-1}, the marketable tuber yield has also increased up to 97.41. After 200 kg ha^{-1} dose, the marketable tuber yield has started to decrease.

Keywords: Early potato production, Nitrogen fertilizer, Potato, Tuber yield

INTRODUCTION

Potato an annual tuber crop with high adaptation to various climatic zones, is successfully cultivated almost in every part of the world as nutritional source (Arioglu, 2014).

Potato, which contains valuable nutrients such as carbohydrates in the form of starch, proteins, vitamins and iron (Fe) in its tubers, is consumed by humans directly as table potatoes and also as processed foods such as chips and French fries. In addition, its varieties that contain high levels of starch are utilized as raw materials in the production of flour, starch and alcohol, as well as animal feeding (Arioglu, 2014).

Due to its high nutritional value, potato could be considered one of the most important nutrients that could help to solve the starvation problem in the under-developed countries. Considering the starvation and malnutrition problems of millions of human beings, the Food and Agriculture Organization of the United Nations declared 2008 as the "International Year of the Potato" to promote awareness (Arioglu and Gulluoglu, 2014).

In the coastal regions of Turkey, where the Mediterranean type climate prevails with warm winters, potato is produced as early as possible with high tuber yield (Arioglu and Caliskan, 1999; Arioglu et al., 2002). The total potato production of Turkey in 2013 was 4.8 million tons; 300,000 tons of which were cultivated as the early potatoes (Anonymous, 2013).

The potato plant takes up substantial amounts of plant nutritional elements from the soil during the growing period. It stores approximately 1/2 to 1/3 of the plant nutritional elements in its vegetative parts. The remaining plant nutritional elements are stored in the tuber (Mikkelsen, 2006; Arioglu, 2014).

It was reported that when 3 tons of potato tubers are harvested per hectare, 150 kg of N, 60 kg of P_2O_5, 350 kg of K_2O, 90 kg of CaO and 30 kg of MgO per hectare are taken from the soil by the plants (Beukema and Van Der Zaag, 1979).

Potato plants require nitrogen fertilizers substantially. Studies conducted in the Erzurum province of Turkey by Karadogan (1996) and Ozturk et al. (2007) indicated that as the doses of nitrogen were increased, tuber yields were also increased. Since potatoes are produced in sandy soils and require substantial amounts of irrigation, the soil is subject to serious nitrogen leaching. Depending on the purpose of cultivation, it is vital that the nitrogen fertilizers should be applied to fields to compensate losses in the potato cultivation. Substantial amounts of nitrogen fertilizers applied at the early stages stimulate the development of vegetative parts with delay in tuber initiation ((Mikkelsen, 2006). So, this type effect of nitrogen fertilizers should be considered in the early potato production (Ozer ve Arıoglu, 1994; Arioglu, 2014).

In order to regulate tuber initiation and to prevent nitrogen leaching, the application of nitrogen fertilizers at two different times was suggested, such as the application of ½ of the fertilizer at the time of planting, and half of the remaining ½ at the time of the hilling (5 to 6 weeks after the plants emerge from the soil surface) and the rest when the tubers reach the size of an egg, increases tuber yield (Rosen and Bierman, 2008; Davis et al., 2014). The leaching of nitrogen should be considered in sandy soils so the amount of nitrogen fertilizers applied should be increased and the fertilizer should be applied by the sprinkler irrigation at several times (Yilmaz, 1992; Hochmuth and Hanlon, 2014).

The amount of nitrogen fertilizers to be used in potato cultivation should be based on the target crop quantity and the type of the soil, since potato plants require a certain amount of nitrogen intake from the soil for certain tuber yield levels. When aiming for certain levels of tuber yield from potato crop, the properties of the variety and the suitability of the environmental factors should also be considered (Rosen, 2013; Arioglu, 2014). The results of research conducted for main crop in the various regions suggested the application of 100 to 210 kg ha^{-1} nitrogen fertilizers as the proper amount in order to reach high quality and quantity tuber yield per unite area in Turkey (Taskiran and Esendal, 1988; Kasap, 1994; Tuncturk et al., 2004). Previous research showed that the quantity of nitrogen fertilizers that should be applied in potato cultivation changes from region to region. But we do not yet have the results of early potatoes in the region. Therefore the purpose of this study was to investigate the optimal nitrogen fertilizer rate and its timing to obtain economic yield in the early potato production in the Cukurova region.

MATERIALS AND METHODS

Research Materials

The study was conducted at the Research Fields of the Cukurova University, Faculty of Agriculture (Southern Turkey, 36°59^1 N, 35°18^1 E; 23 elevation). The design of the field trial was the Randomized Complete Block with 3 replications, in 2011 and 2012 years. Marfona, a medium early potato variety, was used as a plant material. Potassium Sulfate (50% K$_2$O), Triple Superphosphate (42-44% P$_2$O$_5$), Ammonium Sulfate (21%

N) and Ammonium Nitrate (33% N) were used as chemical fertilizers.

Soil Properties of the Research Area

The soil of the experimental field had a loamy structure with pH levels of 7.28-7.29. The salt content of the soil was 0.052 % to 0.060 %. The usable P$_2$O$_5$ content was 14.17 % at the upper layers with the content decreasing in the lower layers. In addition, the nitrogen content of the soil was 0.122 % at the upper layers and 0.056 % at the lower layers. The lime content was 33.02 % at the upper layers with the increased levels in lower layers.

Climate of the Research Area

In the Adana province of Turkey, winters are mild and rainy, whereas summers are dry and warm, which is a typical of a Mediterranean climate. The average monthly temperatures during the research period were 16.7 to 28.9 °C in 2011, whereas it was in the 18.8 to 29.4 °C interval in 2012. The maximum temperature for Adana was 39.3 °C in 2011 and was 40.6 °C in 2012. The total precipitation was 184.7 mm and 110.4 mm in 2011 and 2012, respectively (Turkish State Meteorological Service Adana Regional Directorship, 2011 and 2012).

Planting and Management

The field was cultivated deeply by the moldboard following the harvest of the previous crop in the fall, and then the soil was prepared by using diskharrow. The plantings were done by hand in the second week of January in two years, with 70 cm between row distance and 30 cm in-row distance. The plot size was set as 14 m^2 (2.8 m x 5.0 m). Before planting, 550 g Triple Superphosphate (42-44% P$_2$O$_5$) and 450 g Potassium Sulfate (50% K$_2$O) fertilizers per plot were applied to the furrows by hand and afterwards seed tubers were placed in the furrows according to the given row spaces.

During the growing period, Ridomil was applied three times to prevent late blight (*Phytophtora İnfestans*) and overhead sprinkler irrigation was applied two times. During the growing period, other standard cultural practices were applied at proper time intervals. The planned Nitrogen application doses and times are given in Table 1.

Table 1. The Application of Nitrogen Doses and Times

Treatment Number	Nitrogen Doses (Kg ha⁻¹)	Nitrogen Application Time and Doses		
		Planting	Tuber initiation*	Tuber bulking**
1	0	0	0	0
2	40	40	0	0
3	80	40	40	0
4	120	60	60	0
5	160	80	80	0
6	200	100	100	0
7	240	80	80	80
8	280	100	100	80
9	320	110	110	100
10	360	120	120	120

* 5th of April 2011; **25th of April 2012

Measurement of characteristics

The tubers were harvested in the first week of June. The plants in the middle two rows of each plot were harvested by hand. The number of tubers plant⁻¹, tuber yield plant⁻¹, marketable tuber yield ratio (tubers in 25 mm), dry matter content and tuber yield hectare⁻¹ were determined following the harvest (Onaran and Arioglu, 1999). The standard analysis of variance was used to analyze the traits measurement. The Least Significant Differences (LSD) test was used to comparing the treatments at 0.05 level (Steel and Torrie, 1980).

RESULTS AND DISCUSSIONS

Tuber Yield and Marketable Tuber Yield

The total tuber yield per hectare and the percentage of marketable tuber yield values are presented in Table 2.

Table 2. The Effect of Different Nitrogen Doses on Total Tuber Yield (ton ha⁻¹) and the Percentage of Marketable Tuber Yield (%) in the Early Potato Production at Adana-Turkey.

N Dose (kg ha⁻¹)	2011		2012		Mean	
	TTY* (ton ha⁻¹)	PMTY** (%)	TTY* (ton ha⁻¹)	PMTY** (%)	TTY* (ton ha⁻¹)	PMTY** (%)
0	20.76	95.6	27.53	95.6	24.06	95.6
40	31.98	96.8	34.38	95.9	33.18	96.3
80	36.68	97.1	48.66	96.0	42.67	96.6
120	38.80	97.1	51.98	96.6	45.39	96.8
160	41.24	97.5	54.81	97.0	48.03	97.2
200	43.53	97.7	54.44	97.1	48.98	97.4
240	45.93	97.6	**60.68**	96.4	**53.31**	97.0
280	**46.63**	96.7	54.71	96.4	50.67	96.6
320	44.50	96.4	53.14	96.4	48.82	96.4
360	42.70	96.3	52.74	96.3	47.72	96.3
Average	39.28	-	49.31	-	44.29	-
LSD(5%)	6812	0.94	4162	0.68	12180	0.56

* TTY: Total tuber yield; ** PMTY: The percentage of marketable tuber yield

Table 2 shows that the lowest tuber yields were observed when non nitrogen fertilizer was applied during the planting and the growing period, in both years. As the amount of nitrogen fertilizers applied was increased, tuber yield hectare⁻¹ also increased substantially. The highest tuber yields hectare⁻¹ were obtained from the plots where 280 kg ha⁻¹ (with 46.63 ton ha⁻¹ yield) nitrogen was applied in 2011, and 240 kg ha⁻¹ (with 60.68 ton ha⁻¹ yield) nitrogen was applied in 2012. This increase has continued up to 280 kg ha⁻¹ nitrogen application in 2011 and 240 kg ha⁻¹ nitrogen application in 2012. Besides high nitrogen levels resulted in decreases in tuber yield. In addition, the tuber yield levels in 2012 were higher when compared to those

in 2011. This could be due to the negatively affected vegetative growth of the plant as a result of low temperatures in 2011. Also, the seed tubers utilized in 2012 were comparatively bigger than those used in 2011. The highest tuber yield, based on the average values for both years, was observed in the plots where 240 kg ha⁻¹ of nitrogen fertilizer were applied, resulting in a total yield of 53.31 ton ha⁻¹. Based on the average values for both years, it was observed that as the nitrogen fertilizers applied was increased, the tuber yield was also increased. This increase has continued up to 240 kg ha⁻¹ nitrogen application followed by a decrease in yield beyond that amount (Fig. 1).

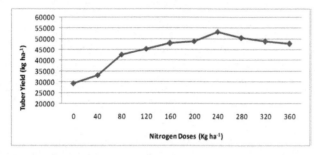

Figure 1. The Effect of Nitrogen Doses on Total Tuber Yield of Early Potato Growing at Adana-Turkey

This decrease could be explained as follows. The additional nitrogen stimulated the development of the vegetative growth of the plants. However, the vegetative parts of the plants did not utilize light energy sufficiently. But, the vegetative parts continued to respire, leading to a

substantial amount of dry mater that resulted from photosynthesis to be consumed as a result of respiration. Consequently, the tuber yields were decreased. These findings are supported by Ozer ve Arıoglu (1994), Karadogan (1996) and Ozturk et al. (2007).

In potato growing, the marketable tuber yield is an important trait. The highest marketable tuber yield percentage (97.4%) was obtained in the plots where 200 kg ha[-1] nitrogen fertilizer was applied. In 2011, this figure was 97.7%, whereas in 2012, it was 97.1% (Table 2). Similar results were reported by other researchers (Taskiran and Esendal, 1988; Kasap, 1994; Tuncturk et al., 2004; Rosen and Bierman, 2008; Rosen, 2013).

Tuber Number and Tuber yield per Plant

The number of tubers and tuber yield of per plant are presented in Table 3.

Table 3. The Effect of Different Nitrogen Doses on the Number of Tubers and the Yield per Plant in the Early Potato Production at Adana-Turkey.

N doses (Kg ha[-1])	Year				Mean	
	2011		2012			
	TN*	TY**	TN*	TY**	TN*	TY**
0	4.7	358.3	7.3	590.7	6.0	474.5
40	5.9	559.7	8.5	778.0	7.2	668.8
80	6.6	642.0	10.0	1125.0	8.3	883.3
120	6.8	679.3	10.7	1240.0	8.8	959,8
160	7.1	722.0	10.8	1270.0	9.0	996.2
200	7.4	761.7	10.8	1306.0	9.1	1034.0
240	7.8	804.0	10.9	1329.0	9.4	1066.0
280	7.6	816.0	11.2	1337.0	9.4	1077.0
320	7.4	779.0	11.5	1396.0	9.4	1037.0
360	7.0	747.3	11.5	1279.0	9.2	1013.0
Average	6.83	686.9	10.32	1165.1	8.58	920.96
LSD(%5)	1.15	119.0	1.98	189.8	1.11	108.1

* TN: Tuber number per plant (number plant[-1])
** TY: Tuber yield per plant (g plant[-1])

Table 3 shows that tuber number per plant in 2011 was in the range of 4.7 (control) to 7.8 (240 kg ha[-1]) number plant[-1], and 7.3 (control) to 11.5 (320 kg ha[-1]) number plant[-1] in 2012. The number of tubers per plant in 2011 increased up to 240 kg ha[-1] nitrogen fertilizer application, followed by decreases beyond that point. On the other hand, in 2012, as the amount of nitrogen fertilizer increased, tubers per plant increased as well. The reason for discrepancy between years could be the plants being affected by low temperatures in 2011 and the large seed tubers developed more main stem in 2012 (Arıoglu, 2014). Based on the average of two years, the highest number of tubers per plant was obtained by 240 kg ha[-1] nitrogen fertilizer was applied (9.4 number plant[-1]). As the amount of nitrogen fertilizers applied increased, the number of tubers per plant also increased, as well (Ozer and Arıoglu, 1994).

Means of plant yields in 2011 were 358.3 to 816.0 g plant[-1]. In 2011, the highest yield was obtained in the plots, where 280 kg ha[-1] nitrogen fertilizer was applied. On the other hand, the tuber yield means for 2012 was ranged between 590.7 and 1396.0 g plant[-1], compared to

the means in 2011. In 2012, the highest yield was obtained in the plot, where 320 kg ha[-1] nitrogen fertilizer was applied. The mean of 2012 were higher than those obtained in 2011. This difference could be due to low temperatures occurred in 2011. Beside large seed tubers might develop high vegetative growth, leading to increased yields. Based on the means of both years, the highest tuber yield per plant was obtained from 280 kg ha[-1] nitrogen fertilizer. The tuber yields in the control plots were 474.5 g plant[-1], whereas it was 1077 g plant[-1] in plots, where 280 kg ha[-1] nitrogen fertilizer was applied. Beyond certain rates of nitrogen fertilizer applications, the tuber yield per plant started to decrease due to the excessive amounts of vegetative parts are not affecting photosynthesis positively in April and May in the Cukurova Region. These findings were in agreement with the findings of Ozer and Arıoglu, (994), Karadogan (1996), and Ozturk et al. (2007).

Dry Matter Content

Dry matter content is presented in Table 4. Table 4 shows that as amount of the nitrogen increased, the dry matter content in the tubers decreased in two years. The values for 2011 were between 15.63% (360 kg ha^{-1}) and 18.63% (40 kg ha^{-1}).

Table 4. The Effect of Different Nitrogen Doses on the Dry Matter Content (%) of Tubers in the Early Potato Production at Adana-Turkey.

N rate (kg ha^{-1})	Year		Mean
	2011	2012	
0	17.52	16.52	17.02
40	18.63	17.77	18.20
80	18.20	17.63	17.92
120	17.90	17.38	17.64
160	16.85	16.24	16.55
200	16.80	16.15	16.48
240	16.15	15.88	16.02
280	15.90	15.23	15.57
320	15.88	15.23	15.56
360	15.63	15.27	15.45
LSD (5%)	0.82	0.79	0.87

The highest dry matter content (18.63%) was observed from the treatment 40 kg ha^{-1} nitrogen application at planting time. As the amount of nitrogen fertilizer increased up to 360 kg ha^{-1}, the dry matter content was decreased down to 15.63%. Similar results were also obtained in 2012. The dry matter content decreased from 17.77% levels to 15.23% levels. When the results for both years are compared, it could be observed that the dry matter content was higher in 2011. The reason for this difference might be large tubers. As the tuber size gets larger, the dry matter content in the tubers starts to decreases.

The means for both years indicated that as the nitrogen rate increased, the dry matter content decreased. In the plots, where 360 kg ha^{-1} nitrogen fertilizers were applied, the dry matter content in the tubers was 15.45%, whereas it was 18.20% in the plots, where 40 kg ha^{-1} nitrogen fertilizer was applied (table 3). As the amount of nitrogen fertilizers applied increased, the tubers get bigger and consequently, the dry matter content per tuber decreased, while water content increased. Therefore, in processing type potato cultivation, substantial amounts of nitrogen fertilizers should not be applied. These findings are supported by the findings of Mikkelsen (2006).

CONCLUSIONS AND SUGGESTIONS

The results and their discussions show that the application type and amount of nitrogen fertilizers at different doses and periods resulted in substantial increases in yield. As the rates of nitrogen fertilizer applications increased, tuber yields were also increased, leading to some increases in net income. The optimal and most economical nitrogen fertilizer application level was 240 kg ha^{-1}, resulting in the highest net income and net income increase levels. Based on the results of the study, it could be suggested that 1/3 of the fertilizer could be applied during the planting, 1/3 applied during the tuber initiation, and the remaining 1/3 during the period where the tubers reach the size of an egg.

LITERATURE CITED

Anonymous, 2013. İstatistik Bolumu İnternet Sitesi, Http://www.fao.org

Arioglu, H.H., M.E. Caliskan. 1999. Studies on the early potato growing possibilities in the Coastal Regions of Mediterranean. Second Potato Congress of Turkey, 28-30 June, Erzurum, pp.220-226. (in Turkish)

Arioglu, H.H., H. Incikli, B. Zaimoglu, L. Gulluoglu. 2002. Studies on the early potato growing in the Cukurova Region. Third National Potato Congress, 23-27 September 2002, Izmir, pp 117-123. (in Turkish)

Arioglu, H.H. 2014. Growing and Breeding of Starch and Sugar Crops. Text Book of Cukurova University, Publication number 188-A-57, Adana, Turkey. (in Turkish)

Arioglu, H., L. Gulluoglu. 2014. Potato production in Turkey and its problems. Agromedya, 2 (10) 76-82. (in Turkish)

Beukema, H.P., D.E. Van Der Zaag. 1979. Potato Improvment. IAC, s.224, The Netherlands

Davis, J. D., R.D. Davidson, S.Y.C. Essah. 2014. Fertilizing Potatoes. Colarado State University Extension, 5pp.

Hochmuth, G., Ed. Hanlon. 2014. A summary of N, P, and K Resarch With Potato in Florida. University of Florida UF/IFAS Extension, Gainesville,FL.,29pp

Karadogan, T. 1996. Application and rates of nitrogen and phosphorus on yield components and quality of potato. Journal of Ataturk University Faculty of Agriculture, 27 (1):50-60. (in Turkish)

Kasap, Y., 1994. The effects of different nitrogen levels and yield and dry matter contents in potato. 1st Congress of Field Crops – Izmir, Volume 1: 260-262. (in Turkish)

Mikkelsen, R. 2006. Best Management Practices for Profitable Fertilization of Potatoes. Better Crops, 90(2):12-13

Onaran, H., H.H. Arioglu. 1999. Determination of suitable plant density for different tuber sizes in table potato growing in the Niğde province. 2nd National Potato Congress, 28-30 June, Erzurum, Turkey, pp 284-297. (in Turkish)

Ozer, C., H.H. Arioglu. 1994. Determination of the effect of nitrogen fertilizer on tuber development and harvest date of early potato growing in the Cukurova Region. 1st Congress of Field Crops, 25-29 April 1994 – Izmir, Volume 1, pp 185-188.. (in Turkish)

Ozturk, E., K. Kara, T. Polat. 2007. The effects of form and application times of nitrogen on yield and tuber size of potato. Journal of Tekirdag Agriculture Faculty 4(2) 127-135. (in Turkish)

Rosen,C.J., P.M. Bierman. 2008. Best Management Practices for Nitrogen Use: Irrigated Potatoes. University of Minnesota Extesion, 7pp.

Rosen, C.J. 2013. Potato Fertilization on Irrigated Soils: Crops. University of Minnesota Extesion, 6pp.

Taskiran, A., E. Esendal. 1988. The effects of planting times and nitrogen doses on tuber yield and some characteristics of potato growing under farmer conditions. Journal of 19 Mayıs University, Faculty of Agriculture, 3(2):25-45. (in Turkish)

Tuncturk, M., R. Tuncturk, B.Yildirim, T. Eryigit. 2004. The effect of different nitrogen doses and in-row spacings on yield and quality of potato. Journal of Yuzuncu Yil University, Faculty of Agriculture, 14(2):95-104. (in Turkish)

Yilmaz, A. 1992. A study on the effect of nitrogen applied at different times and doses on tuber yield, tuber size and some quality characteristics of potato growing in the Misli Plain. Doctoral dissertation Selcuk University, Faculty of Agriculture. (in Turkish)

38

EFFECT OF DIFFERENT PLANT DENSITIES ON THE YIELD AND SOME SILAGE QUALITY CHARACTERISTICS OF GIANT KING GRASS (*Pennisetum hybridum*) UNDER MEDITERRANEAN CLIMATIC CONDITIONS

Hakan GEREN[*], *Yasar Tuncer KAVUT*

[*]*Ege University, Faculty of Agriculture, Department of Field Crops, Izmir, TURKEY*
Corresponding author: hakan.geren@ege.edu.tr

ABSTRACT

The aims of this study were to determine the adaptability of giant king grass (*Pennisetum hybridum*) under Mediterranean climate conditions, and also to determine whether proper plant densities could improve growth and yield. Study site was located at Bornova, Turkey (38°27.236 N, 27°13.576 E and 20 m). Treatment consisted of four densities of plant population (D_1:57,143; D_2:28,571; D_3:19,048 and D_4:14,286 plant ha^{-1}) with three replicates per treatment in a randomized block design. Setts were planted in mid June of 2010 and allowed to grow for 4 full growing seasons. Average result of four years indicated that there were significant effects of plant densities on the dry matter yield and other yield characteristics of giant king grass but not on silage pH. It was recommended that the production of giant king grass using D_2 (70x50 cm) was the most successful planting density regarding the dry matter (32.6 t ha^{-1}) and crude protein yield to the regions with Mediterranean-type climates under irrigation.

Key words: Giant king grass, plant density, dry matter yield, silage quality.

INTRODUCTION

The most unusual handicap of Turkish animal husbandry sector is providing cheap and high quality roughage production for the farmers. Since the roughages are 70% of general expenditures in animal production, there is an urgent need to improve forage lands in the country (Acar et al., 2015). Remembering the Mediterranean conditions of west and south part of Turkey which represents many favorable factors for crop growth, it must be emphasized that any attempt to benefit from new forage crop production alternatives are of vital importance. Corn (*Zea mays*) silage plays an important role as a winter feed in the livestock industries of many countries. The main reasons for the popularity of corn for silage purposes are the high yield obtained in a single harvest, the ease with which it can be ensiled and its high energy value as a feed source (Çarpıcı et al., 2010). However, its major shortcoming is undoubtedly its production costs (seed, soil preparing, etc). Farmers have tried to reduce production costs by better use of grazing and silage making. Giant king grass (*Pennisetum hybridum*) cultivation under field conditions in this Mediterranean environment may be one of those alternatives to produce large amount of high quality roughage (feed) instead of corn.

Giant king grass (GKG) as an interspecific hybrid (6n) has been formed between Napier grass (*Pennisetum purpureum*) (2n=4x=28 chromosomes) and pearl millet (*Pennisetum glaucum*) (2n=2x=14 chromosomes) the resulting hybrid is sterile due to the triploid condition (2n=3x=21 chromosomes) and restored by chromosome duplication with the use of colchicine (Hanna et al., 1984). *P.hybridum* is also referred to as "Maralfalfa", "Elephant grass" or sometimes "*P.violaceum*".

GKG as a field crop is a perennial grass and native to Africa, and has been the most promising and high yielding fodder giving dry matter yields that surpass most other tropical grass (Mannetje, 1992). It mostly used for "cut and carry" system over the tropical and sub-tropical area of the world (Santos et al., 2013). Giant king grass can adapt to a wide range of soil types from sandy to clayey. It can also grow in soils in the pH range of highly acidic to alkaline. The maximum growth is attained on well-drained loamy soils with high organic matter content, but it is susceptible to water logging (Singh et al., 2013). GKG has a perennial life cycle (C4) and is propagated vegetatively. It has a profuse root system, penetrating deep into the soil, and an abundance of fibrous roots spreading into the top soil horizons. The rhizomes are short and creeping, and, nodes develop fine roots and culms. The plant forms clumps and grows upwards, profusely branched with thick cane-like stems. The culms are erect and tall, varying in height from 2 to 6 m.

The propagation of giant king grass is by stem cuttings (setts), rooted stems or splitting of rhizomes (Singh et al.,

2013). The most common way of planting is using setts. The mature stems in the basal 2/3 zone with well-developed leaf buds are cut into sections of three nodes and planted upright by burying two basal nodes into the soil.

Numerous field experiments carried out on *Pennisetum* genus indicated that yield and forage quality of the crop depends on cultivar, plant density, cutting interval and harvesting purpose. Optimum plant density is an important factor in maximizing yields of tall grasses such as *Pennisetum* or *Miscanthus* (Danalatos et al., 2007; Zewdu, 2008; Ansah et al., 2010). Thus, the optimum plant density or plant population for any given situation results in mature plants that are sufficiently crowded to efficiently use resources such as water, nutrients, and sunlight, yet not so crowded that some plants die or are unproductive (Lyon, 2009). At this population, production from the entire field is optimized, although any individual plant might produce less than would have occurred with unlimited space.

For forage production in *Pennisetum purpureum*, intra-row and inter-row spacing of 0.5 m × 1 m in Ghana (Ansah et al., 2010), 1 m × 1 m in Kenya (Nyambati et al., 2010) and 50 cm × 50 cm in Japan (Ishii et al., 2005) have been suggested. Mukhtar et al. (2003) found that where plants have to go through overwintering, high population density impacted adversely the number of stubble tillers emerging from underground stems. Wijitphan et al. (2009) pointed out that there was a significant effect of plant spacing on total DM yield among 50x40, 50x60, 50x80 and 50x100 cm plant spacing of *Pennisetum purpureum*, and the highest total DM yield with 70.84 t ha⁻¹ was obtained from 50x40 cm planting spaces under Thailand condition. A field trial in Greece (Danalatos et al., 2007) showed that no significant response of *Miscanthus giganteus* to nitrogen levels (50 and 100 kg N ha⁻¹) but, a significant effect of plant density was found, with the denser populated plants (10,000–20,000 plant ha⁻¹) performing growth rates, reaching maximum dry biomass yields in excess of 38 t ha⁻¹ compared to 6,700 plant ha⁻¹.

Zewdu (2008) tested different treatments with 1.5, 1 and 0.5 m spacing between rows and 0.75, 0.5 and 0.25 m spacing between plants, which consisted of 8,889, 13,333, 26,668, 13,333, 20,000, 40,000, 26,668, 40,000 and 80,000 plants ha⁻¹ on *Pennisetum purpureum* in Ethiopia reported that the number of tillers per plant, dry matter yield were significantly affected by plant density but not crude protein content and, higher DM yields were obtained from 80,000 (7.80 t ha⁻¹) and 40,000 (39.9 t ha⁻¹) plant density in 2004 and 2005 respectively.

Giant king grass has the potential to grow with fewer inputs, water and tolerate a variety of biotic and abiotic stresses as compared to corn, and the fact that it is perennial which means that replanting every year is not required (Kukkonen, 2009), and is a new introduction to Turkey. Therefore, getting to know the possibility of production of this crop is important not only in Turkey, but also in all over the world. Nevertheless, information on the productivity and quality of giant king grass under intensive farming management in Mediterranean environment is not well documented. The objective of this research was to evaluate the influence of different planting densities on the yield, some yield and silage quality components of giant king grass under irrigated conditions of Mediterranean climate.

MATERIALS AND METHODS

Location of Experiment

The experiment was carried out during four growing seasons (2010-2013) at Bornova experimental fields (38°27.236 N, 27°13.576 E) of Agricultural Faculty of Ege University, Izmir, Turkey, at about 20 m above sea level with typical Mediterranean climate characteristics (Table 1). The soil was a silty-clay loam (30.6% clay, 36.7% silt, and 32.7% sand) with pH 7.32, organic matter 1.16%, salt 0.074%, 0.123% total N, available phosphorus (1.4 ppm) and available potassium (350 ppm).

Table 1. Some meteorological parameters of experimental area at Bornova in 2010-2013.

Months	Average temperature (°C)					Total precipitation (mm)				
	2010	2011	2012	2013	LYA	2010	2011	2012	2013	LYA
January	10.6	9.0	6.8	9.4	9.0	142.3	100.9	127.7	252.5	112.2
February	12.6	10.3	7.6	11.2	9.2	301.3	107.3	128.2	187.0	99.7
March	13.3	12.0	11.3	14.0	11.8	16.1	18.8	34.7	56.8	82.9
April	17.4	14.5	17.5	17.3	16.1	20.4	65.3	105.0	30.2	46.4
May	21.8	20.1	20.5	22.7	21.0	27.1	29.0	86.6	43.7	25.4
June	25.5	25.4	27.3	25.7	26.0	76.3	0.6	19.9	27.1	7.5
July	28.8	28.9	30.1	28.4	28.3	0.0	0.0	0.0	0.0	2.1
August	30.2	28.1	29.2	28.7	27.9	-	0.0	0.0	20.2	1.7
September	24.8	25.6	24.3	24.0	23.9	12.3	8.6	0.0	5.1	19.9
October	18.8	17.1	21.7	17.2	19.1	232.5	90.3	22.1	94.1	43.2
November	18.1	11.1	16.4	15.0	13.8	32.4	0.0	56.9	128.9	109.7
December	13.3	10.7	10.7	8.5	10.5	155.7	140.5	218.2	9.1	137.9
✗ - Σ	19.6	17.7	18.6	18.5	18.1	1016.4	561.3	799.3	854.7	688.6

LYA:Long year average, ✗: mean, Σ:total

Field applications and experimental design

"Paraíso" cultivar of giant king grass (*Pennisetum hybridum*) was used as crop material. The experiment was carried out with a randomized complete block design with three replication; four plant spacings where 70 cm among the rows and 25, 50, 75 and 100 cm within the rows (D$_1$:57,143, D$_2$:28,571, D$_3$:19,048 and D$_4$:14,286 plant ha^{-1}, respectively) were tested. Each plot was consisted of 4 rows with 5 m length (14 m^2).

The grasses were grown from stem cuttings (setts) having 4 nodes and 40-60 cm length taken from basal part of *P.hybridum* rootstock on 15th June, 2010. The setts were buried into a well-prepared seedbed with one sett per hill. Before planting, the additional leaves of the setts were trimmed and dipped at 4000 ppm indole-3-butyric acid (IBA) concentrations of hormone for 3-4 seconds up to 10-15 cm height (Hartmann et al., 2002).

The recommended dose of 90 kg N ha^{-1} was applied for all plots in two equal doses at establishment year (Ruviaro et al., 2008). Half a dose of nitrogen fertiliser (urea) was applied before planting, and the rest of nitrogen was applied when the crops were 50-60 cm plant height as NH$_4$NO$_3$. All plots were fertilised using 100 kg ha^{-1} P$_2$O$_5$ before planting. In the following years (2011, 2012 and 2013) of the field trial, the same amount of N fertilization was applied in mid April (plus again phosphorus), as ammonium sulfate, and the other half at the beginning of stem elongation, as ammonium nitrate.

Drip irrigation system was installed on the field during the establishment and growing seasons. Weed control was performed by manual hoeing only during the establishment year. No evident crop diseases or insects were detected. Harvests were made only once a year on 8 November 2010, 31 October 2011, 5 November 2012 and 30 October 2013, respectively.

Measurements, silage making and chemical analysis

In each plot, plant height was measured from in the mid two rows and area of 1-3 m (30 plants) at the day of cutting. Tiller density was also determined by counting the plant in the same area in each plot at the same day of cutting. Forage on a central area of 3 m^2 in each plot was cut at 15-20 cm above ground level and fresh weight recorded. Harvested fresh forage (2 kg) was dried to a constant weight at 65°C during 72 h for calculating dry matter (DM).

In each plot, about 5 kg of fresh whole crop material were wilted overnight (14–16 h) prior to ensiling. The wilted herbage was then chopped, using a static precision-chop forage harvester to give a chop length of 5–10 mm. The final dry matter content was approximately 30%. 250 g chopped samples were vacuum-packed into polythene bags (dimensions 250-200 mm) (Johnson et al., 2005) with addition of 0.5% salt. The vacuum bag silos were kept in storage without light for 60 days for anaerobic fermentation.

pH value of matured silage samples was also determined (Alcicek and Ozkan, 1996). Matured silage

samples of each component were dried at 65°C for 48 h. The dried samples were reassembled and ground in a mill passed through a 1 mm screen. The crude protein (CP) was calculated by multiplying the Kjeldahl N concentration by 6.25. The neutral detergent fibre (NDF) and acid detergent fibre (ADF) concentrations were measured to Ankom Technology to determine the Relative Feed Value (RFV) which was estimated according to the following equations adapted from Trotter and Johnson (1992): DM intake (DMI)=(120/NDF%), Digestible DM (DDM)=88.9-(0.779xADF%), RFV=DDM% x DMI% x 0.775.

Statistical analysis

All data were statistically analyzed using analysis of variance (ANOVA) with the Statistical Analysis System (SAS, 1998). Probabilities equal to or less than 0.05 were considered significant. If ANOVA indicated differences between treatment means a LSD test was performed to separate them (Stell et al., 1997).

RESULTS AND DISCUSSION

Experimental area is located in the Mediterranean zone of the country with quite mild & warm winters and hot & dry summers (Table 1). Field studies were started in mid June with high air temperature and satisfactory moisture levels supported by drip irrigation and owing to IBA, therefore, stands were excellent. No winter injury on the crops was detected during the experimental years. The results are summarized in Table 2. The year and plant density effects were the main sources of variation in all tested characters, while the interactions (YxD) were not significant except plant height.

Number of tiller

There were statistically significant differences among plant densities regarding average number of tiller per square meter (Table 2). Minimum plant density being D$_4$ (14,286 crop ha^{-1}) had the highest average number of tiller (237 tiller m^2), whereas maximum plant density being D$_1$ (57,413 crop ha^{-1}) was the lowest (199 tiller m^2). Year effect was also significant and average number of tiller of first year (44 tiller m^2) was quite lower than the following years (2011:262, 2012:287 and 2013:291 tiller m^2, respectively), but there was no significant difference between 2012 and 2013.

In the present study, number of tillers of giant king grass in the last 3 years was approximately 6-7 times higher than the first year tillers. Four years average result indicated that decreasing plant density from D$_1$ to D$_4$ increased the number of tiller per square meter; however there was no significant difference between D$_2$ (70x50 cm) and D$_3$ (70x75 cm). This might be due to the fact that plants with relatively wider spacing (D$_3$ and D$_4$) compared to D$_1$ or D$_2$ produced many fine-stemmed tillers and showed vigorous growth and development because of reduced competition for space, moisture and nutrients during the growing period as reported by many researchers (Mukhtar et al., 2003; Zewdu, 2008; Wijitphan et al., 2009).

Table 2. Effect of different plant densities on the yield and some yield components of GKG.

Density	2010	2011	2012	2013	Mean	2010	2011	2012	2013	Mean
	------- Tiller number (m^{-2}) -------					---------- Plant height (cm) ----------				
D_1	35	232	259	270	199	212	327	409	388	334
D_2	40	259	292	294	221	229	352	410	394	346
D_3	45	271	295	296	227	257	374	420	403	364
D_4	54	287	302	303	237	278	382	426	404	372
Mean	44	262	287	291	221	244	359	416	397	354
LSD(.05)	Y:10 D:11 YxD:ns CV(%):15.3					Y:11 D:14 YxD:17 CV(%):12.9				
	------- Dry matter yield (t ha^{-1}) -------					--------------- Silage pH ---------------				
D_1	15.6	32.2	35.5	36.9	30.1	3.86	3.77	3.65	3.65	3.73
D_2	18.4	35.1	37.4	39.4	32.6	3.80	3.59	3.63	3.64	3.67
D_3	17.7	33.1	35.5	37.8	31.0	3.78	3.57	3.61	3.62	3.65
D_4	17.7	31.6	33.4	35.2	29.5	3.60	3.54	3.60	3.61	3.59
Mean	17.3	33.0	35.5	37.3	30.8	3.76	3.62	3.62	3.63	3.66
LSD(.05)	Y:1.3 D:1.4 YxD:ns CV(%):15.4					Y:ns D:ns YxD:ns CV(%):4.1				
	------- Crude protein content (%) -------					------- Relative Forage Value -------				
D_1	7.7	7.0	6.5	6.3	6.9	125	96	91	89	100
D_2	7.2	6.4	6.3	6.2	6.5	118	94	89	88	97
D_3	6.5	5.7	5.8	6.1	6.0	111	89	87	86	93
D_4	6.4	5.6	5.7	6.0	5.9	105	86	85	85	90
Mean	6.9	6.2	6.1	6.2	6.3	115	91	88	87	95
LSD(.05)	Y:0.4 D:0.3 YxD:ns CV(%):6.7					Y:3 D:4 YxD:ns CV(%):4.5				

D_1:57,143, D_2:28,571, D_3:19,048 and D_4:14,286 plant ha^{-1}, Y: year, D: density, YxD: interaction, ns: not significant.

These results generally displaying the high growth rates, showed that giant king grass was capable of growing in the experimental site. Singh et al. (2013) emphasized that giant king grass grows vigorously, providing the extensive stem structure for leaf formation needed for photosynthesis, starting from the first year, and in this respect the crop differs from many other vigorously growing grasses such as *Miscanthus*, which needs the first year for establishment.

Wijitphan et al. (2009) informed that number of tillers per hill of *Pennisetum purpureum* was affected by plant configuration, and plant spacing of 50x40 cm with 16 tiller hill^{-1} had the highest compared to other spacing of 50x60 cm (14 tiller hill^{-1}), 50x80 cm (12 tiller hill^{-1}) and 50x100 cm (10 tiller hill^{-1}), respectively. Another study with *Pennisetum purpureum* by Mukhtar et al. (2003) informed that with the increasing plant density from 4, 8 to 16 plant m^{-2}, number of tillers increased from 50, 55 and 76 per plant in the establishment year, and 114, 110 and 147 per plant in the third year, respectively. Zewdu (2008) reported that the number of tillers per plant was significantly affected by plant density, and wider rows were more productive than narrow rows, and number of tillers of *P.purpureum* in the second years was approximately 7 times higher than the first year tillers. Our findings are in accordance with those researcher's results.

Plant height

The plant height of GKG was affected by YxD interaction. The highest plant height (426 cm) was obtained from D_4 in 2012, whereas the lowest was 212 cm for D_1 in

2010 (Table 2). Year effect was also significant and average giant king grass height of third year (416 cm) was higher than the other years (2010:244 cm, 2011:359 cm and 2013:397 cm, respectively).

In the study, four years average result displayed that the plant height of giant king grass increased noticeably by decreasing plant density most probably due to the thick-stemmed plants developed at wider rows and less competition compared to narrow spacing. Zewdu (2008) informed that there was no significant effect on plant height of *P.purpureum* due to the plant density during the establishment year, however, plant height was significantly affected by plant density in the following year, and plant height in lower plant densities were taller than higher densities. Mukhtar et al. (2003) reported that there was a highly significant difference between normal and dwarf type of *P.purpureum* varieties, and plant height of *P.purpureum* crops in low-density plots were higher than higher-density plots in Japan. Danalatos et al. (2007) indicated that plant height was not affected by plant density (6,700-10,000-20,000 plant ha^{-1}) of *Miscanthus giganteus* in Greece. Živanović et al. (2014) informed that planting density (G_1:2 and G_2:3 rhizomes m-^{2}) and year affected the *M.giganteus* plant height. G1 was taller than G2, and in the first year of the G_1 density were higher by 10%, and during the second year by 4%, but nitrogen (0-60-100 kg ha^{-1}) had a higher influence on plant height compared to the planting density in Serbia. Our findings for *P.hybridum* were in agreement with the indications of above mentioned researchers under typical Mediterranean environmental conditions.

Dry matter (DM) yield

The DM yield was not affected by interaction. There were statistically significant differences among plant densities regarding DM yield per hectare (Table 2). The highest average DM yield of giant king grass was obtained from D_2 (70x50 cm) being 32.6 t ha^{-1}, whereas the lowest average yield obtained from D_4 (70x100 cm) being 29.5 t ha^{-1}, but there was no significant difference between D_1 (70x25 cm) and D_3 (70x75 cm). Year effect was also found to be significant and the first year (17.3 t ha^{-1}) was quite lower than the following years (2011:33.0 t ha^{-1}, 2012:35.5 t ha^{-1} and 2013:37.3 t ha^{-1}, respectively).

Many factors influence the optimum plant population density for a crop: availability of water, nutrients and sunlight; length of growing season; potential plant size; and the plant's capacity to change its form in response to the varying environmental conditions (Lyon, 2009; Singh et al., 2013). According to the four years result, the maximum DM yields occurred at D_2, being quite higher than other densities. This result suggests that number of tiller per unit area and plant height and stem thickness can cause a strong competition among the crops and depress the yield components of individual plant in lower or wider densities. Nevertheless, since the numbers of tiller per unit area in sparsely populated stands (D_3-D_4) were quite higher than the other densities which were densely populated stands (D_1-D_2), it was also suggested that DM yields were the highest at D_2. Even thought, number of tiller per unit area and plant height was increased with the decreasing rate of plant density.

It was also concluded that data related to DM yield and yield components of D_2 also indicated that giant king grass in this planting density was highly adaptable to the experimental area. Highly significant differences among the years were an indication of the yield variation of plant densities depending on establishment year and the changes of climatic parameters of years.

Zewdu (2008) reported that there was a significant difference in DM yield among the different plant densities in *P.purpureum* in Ethiopia, and DM yield increased as plant density increased, and higher DM yields were obtained from 80,000 plant ha^{-1} (7.80 t ha^{-1}) and 40,000 plant ha^{-1} (39.9 t ha^{-1}) in the first and second crop seasons, respectively. Mukhtar et al. (2003) informed that total annual DM yield increased as planting density increased in both years in Japan, and higher plant density (16 plant m^{-2}) was superior than other densities (8 or 4 plant m^{-2}, respectively). Wijitphan et al. (2009) stated that there were significant effects of spaces of planting on total DM yields, and the highest DM yield of 70.84 t ha^{-1} was obtained from a 50x40 cm planting treatment. This was significantly higher than that from other planting treatments in Northeast Thailand, and the 50x100 cm space DM yield was significantly lower than other spaces of planting and the yield was 55.8 t ha^{-1}. However, narrower row spacing may facilitate stand establishment and increase forage production in the early life of the crop. Our findings are in accordance with those researcher's results.

DM yield variation in giant king grass has been attributed to many factors such as climatic conditions, soil water availability, nutrients availability, plant density, harvest time and method, etc. Some cutting experiments with giant king grass also revealed in tropic countries that the choice of cutting interval or frequency is crucial to their performance and were found to be the main factor affecting growth, yield and persistence of swards (Tegami Neto and Mello, 2007; Zewdu, 2008; Wijitphan et al., 2009).

Except establishment year, generally annual DM yields of giant king grass in the experimental area were almost twice that of corn which was the most and largely used silage crop in the region (Geren and Kavut, 2009; Çarpıcı et al., 2010). Masuda et al. (1991) found growth rate of *P.purpureum* superior to that of corn, and the final DM yield was twice that of corn. Considering the enormous amount of research effort and resources spent towards making corn more productive to where it is now, chances of large initial biomass gain in giant king grass with concentrated research effort are promising.

Silage pH

Plant density did not show any significant effect on the silage pH of giant king grass silage in the study (Table 2). The values of silage pH ranged from 3.54 to 3.86 depending on the plant densities and years. Year effect was not also significant on pH values and average pH of 3.66.

The most important physicochemical parameter for the evaluation of silage quality is a pH below 4, which was observed for all the silages tested. All indicators were characteristic of good silage conservation whatever the treatments were. The silage quality was especially confirmed by the proportion of fermentation products at the end of the storage period (Ferrari Junior et al., 2009). This might be due to the fact that all the crops used in the study were from the same genetic material and all the treatments were harvested at the same growth stage. Many research (Bernardino et al., 2005; Ruviaro et al., 2008; Ferrari Junior et al., 2009; Shen et al., 2012) reports revealed that plant height at cutting and variety differences are the major factors that affect the chemical composition and digestibility of *P.purpureum*. Woodard et al. (1991) reported that mean pH values of *P.purpureum* silages ranged from 3.8 to 4.0 made from plants harvested at the different frequencies, and the ease with which *P.purpureum* was preserved as silage was attributed to adequate levels of water soluble carbohydrate concentrations and inherently low buffering capacities in standing forages.

Crude protein (CP) content of silage

There was significant effect on CP content of silage of GKG due to the plant density during the study (Table 2). The highest average CP content (6.9%) recorded at D_1 (70x25 cm), whereas the lowest CP content was 5.9% at D_4 (70x100 cm). Mean CP content was significantly higher in 2010 (6.9%) than the following years which were no significant differences among them.

The four years average value in the study indicated that plant density practices affected CP content of giant king grass significantly, and densely populated stands gave higher CP content compared to sparsely populated stands during the study. The higher CP content at higher plant densities was mainly due to the structural role of thinner and taller tillers of the crop. Some research workers (Zewdu, 2008; Wijitphan et al., 2009) reported that CP content of *P.purpureum* was not influenced by different plant densities. Meanwhile, Wijitphan et al. (2009) informed that CP content (13.9%) of *P.purpureum* was higher in densely populated stands (50x60 or 50x80 cm) compared to sparsely populated stand (50x100 cm, 13.2%).

Numerous research reports revealed that plant height at cutting or plant age is the major factor that affects CP content and digestibility of giant king grass (Woodard et al., 1991; Nyambati et al., 2010). In our study, CP contents of giant king grass silage were relatively low compare to corn silage because of the cutting made only once at the end of the growing season in each year. However, this CP content should be improved by adding some protein-rich crop material (cowpea, alfalfa, etc) or concentrates (soybean cake, crushed vetch, etc) (Bernardino et al., 2005; Ferrari Junior et al., 2009; Pires et al., 2009). Accordingly, the addition of leguminous hays in the rates of 25% or 50% to giant king grass at the time of ensiling resulted in good fermentation and raised the CP content of the silage from 6% to 12% and reduced fibres content (Geren, 2014).

These results illustrated that giant king grass emerged as one of the best candidates in the perennial forage crops group that thrives in temperate continental climate areas, and, as C4 photosynthetic pathway species, it is characterized by high photosynthesis efficiency.

Relative forage value (RFV)

The RFV was not affected by year-plant density interaction. Higher plant densities being D_1 and D_2 had the higher average RFVs (100 and 97, respectively), whereas RFV of lower plant densities being D_4 and D_3 were the lower (90 and 93, respectively), while the differences between them were not significant. Year effect was also significant on RFV, and average RFV of the first year (115) was higher than the following years (2011:91, 2012:88 and 2013:87, respectively).

As it is well known, the RFV is an index used to predict the intake and energy value of the forages and it is derived from the DMI and DDM (Ball et al., 1996). The RFV makes marketing on the basis of feeding value possible. This practice aids the seller in pricing hay or silage and provides valuable information to buyer about how to use forage most effectively. Forages with a RFV are described as that over 151 (prime), between 150-125 (premium), 124-103 (good), 102-87 (fair), 86-75 (poor) and fewer than 75 (reject), respectively (Rohweder et al., 1978). RFV, though not a reflection of the nutrition of forage, is also important in estimating the value of forage, and all treatments had RFV ranging from 125 (good) to 85 (poor) in our experiment which densely populated stands gave higher RFV compared to sparsely populated stands.

There is very little or no information available regarding the plant density on the effect of RFV of giant king grass. Kukkonen (2009) stated that young giant king grass first harvested at a height of 90 to 120 cm tall had a CP level of 19.5% of DM, with 34% ADF, 56% NDF. These figures are comparable to alfalfa which is the top forage crop and commands the highest prices. But later, harvest of third growth the crop at a height of 120 to 220 cm on November had CP of 10.5%, 37% ADF, 66% NDF. Shen et al. (2012) reported that feeding value of king grass silage affected by different harvest time and wilting process with mean of 60% NDF and 30% ADF. Previous studies (Bernardino et al., 2005; Tegami Neto and Mello, 2007; Geren, 2014) about improving RFV of giant king grass silage displayed the cutting management (plant height or age) or some additives to the crop at the time of ensiling had greater effect on RFV.

CONCLUSION

It should be emphasized that giant king grass, a new introduction to the Mediterranean coastal part of Turkey, is a promising perennial forage crop material with an high level of adaptability and forage yield and quality pecularities.

The results of our four-year study testing the effect of four plant density (57,143; 28,571; 19,048 and 14,286 plant ha[-1], respectively) on the crop showed that based on dry matter and crude protein yield, the planting of giant king grass using 28,571 (70x50cm) plants ha[-1] should be recommended in the regions with Mediterranean-type climates and in similar agro-ecologies of the country.

Future experiments on giant king grass crop should be conducted at different locations with various agronomical treatments and especially cutting intervals, to be sure that results are relatively consistent over time.

ACKNOWLEDGEMENTS

We express our gratitude to Prof. Dr. Riza AVCIOGLU Ege University Faculty of Agriculture Dept. of Field Crops, Izmir/Turkey for his kind and valuable help in evaluating the manuscript.

LITERATURE CITED

Acar,Z., C.O.Sabanci, M.Tan, C.Sancak, M.Kizilşimşek, U.Bilgili, I.Ayan, A.Karagöz, H.Mut, Ö.Ö.Aşçi, U.Başaran, B.Kir, S.Temel, G.B.Yavuzer, R.Kirbaş and M.A.Pelen. 2015. Changes in forage crop production and new quests, 8th Technical Congress, The Chamber of Turkish Agricultural Engineers, 12 January 2015, Vol:1:508-547.

Alcicek,A. and K.Ozkan. 1996. Zur quantitativen Bestimmung von Milch-, Essig- und Buttersäuren in Silage mit Hilfe eines Destillationsverfahren, J.of Agr.Fac.of Ege Univ, 33(2-3):191-198.

Ansah,T., E.L.K.Osafo and H.H.Hansen. 2010. Herbage yield and chemical composition of four varieties of Napier (*Pennisetum purpureum*) grass harvested at three different days after planting. Agricultural and Biology Journal of North America 1, 923–929.

Ball,D.M., C.S.Hovelend and G.D.Lacefield. 1996. Forage quality in Southern Forages, Potash & Phosphate Institute, Norcross, Georgia, p:124-132.

Bernardino,F.S., R.Garcia, F.C.Rocha, A.L.de Souza and O.G.Pereira. 2005. Production and characteristics of effluent and bromatological composition of elephantgrass with different levels of coffee hulls addition, Revista Brasileira de Zootecnia., 34(6):2185-2191.

Çarpıcı,E.B., N.Çelik and G.Bayram. 2010. Yield and quality of forage maize as influenced by plant density and nitrogen rate, Turk J Field Crops, 15(2):128-132.

Danalatos,N.G., S.V.Archontoulis and I.Mitsios. 2007. Potential growth and biomass productivity of *Miscanthus x giganteus* as affected by plant density and N-fertilization in central Greece, Biomass and Bioenergy 31:145–152.

Ferrari Junior,E., V.T.Paulino, R.A.Possenti and T.L.Lucenas. 2009. Additives in silage of paraisograss (*Pennisetum hybridum* cv. Paraiso), Arch. Zootec. 58(222):185-194.

Geren,H. and Y.T.Kavut. 2009. An investigation on comparison of sorghum (*Sorghum sp.*) species with corn (*Zea mays* L.) grown under second crop production, The Journal of Ege University, Faculty of Agriculture, 46(1):9-16.

Geren,H. 2014. An investigation on some quality characteristics of ensilaged giant kinggrass (*Pennisetum hybridum*) with different levels of leguminous forages, The Journal of Ege University, Faculty of Agriculture, 51(2): 209-217.

Hanna,W.W., T.P.Gaines, B.Gonzales and W.G.Monson. 1984. Effects of ploid on yield and quality of pearl millet x Napier grass hybrids. Agron. J. 76:669-971

Hartmann,H.T., D.E.Kester, F.T.Davies,Jr. and R.L.Geneve. 2002. Plant Propagation: Principles and Practice, Prentice Hall, 7th ed.

Ishii,Y., N.Yamaguchi and S.Idota. 2005. Dry matter production and *in vitro* dry matter digestibility of tillers among napier grass (*Pennisetum purpureum* Schumach) varieties, Japanese Journal of Grassland Science 51, 153–163.

Johnson,H.E., R.J.Merry, D.R.Davies, D.B.Kell, M.K.Theodorou and G.W.Griffith. 2005. Vacuum packing: a model system for laboratory-scale silage fermentations, Journal of Applied Microbiology, 98:106–113.

Kukkonen,C. 2009. Giant king grass, an energy crop for cellulosic biofuels & electric power plants, VIASPACE Inc. Irvine, California USA

Lyon,D.J. 2009. How do plant populations affect yield? Agricultural Research Division of IANR, University of Nebraska – Lincoln.

Mannetje,L.'t. 1992. *Pennisetum purpureum* Schumach. In:'t Mannetje,L. and Jones,R.M. (eds) Plant Resources of South-East Asia, No:4, Forages, pp:191-192. (Pudoc Scientific Publishers, Wageningen, the Netherlands).

Masuda,Y., F.Kubota, W.Agata and K.Ito. 1991. Analytical study on high productivity of napier grass (*Pennisetum purpureum* Schumach) 1. Comparison of the characteristics of dry matter production between napier grass and corn plants. Journal of Japanese Grassland Science 37, 150–156.

Mukhtar,M., Y.Ishii, S.Tudsri, S.Idota and T.Sonoda. 2003. Dry matter productivity and overwintering ability of dwarf and normal napier grass as affected by planting density and cutting frequency. Plant Production Science 6, 65–73.

Nyambati,E.M., F.N.Muyekho, E.Onginjo and C.M.Lusweti. 2010. Production, characterization and nutritional quality of napier grass [*Pennisetum purpureum* (Schum.)] cultivars in Western Kenya. African Journal of Plant Science, 4:496–502.

Pires,A.J.V., G.G.P.Carvalho, R.Garcia, J.N.D.Carvalho Junior, L.S.O.Ribeiro and D.M.T.Chagas. 2009. Elephant grass ensiled with coffee hulls, cocoa meal and cassava meal. Revista Brasileira de Zootecnia 38:34–39.

Rohweder,D.A., R.F.Barnes and N.Jorgensen. 1978. Proposed hay grading standards based on laboratory analyses for evaluating quality, J. Anim. Sci., 47(3):747–759.

Ruviaro,C., A.B.Lazzeri, H.A.S.Thomaz and Z.B.Oliveira. 2008. Effects of nitrogen fertilization management on irrigated Elephant grass cv. Paraíso productivity, Irriga, Botucatu 13(1):26-35.

Santos,R.J.C.d., M.d.A.Lira, A.Guim, M.V.F.d.Santos, J.C.B.D.Junior and A.C.d.L.d.Mello. 2013. Elephant grass clones for silage production, Sci. Agric. 70(1):6-11.

SAS Institute. 1998. INC SAS/STAT user's guide release 7.0, Cary, NC, USA.

Shen,C., X.Shang, X.Chen, Z.Dong and J.Zhang. 2012. Growth, chemical components and ensiling characteristics of king grass at different cuttings, African Journal of Biotechnology, 11(64):12749-12755.

Singh,B.P., H.P.Singh and E.Obeng. 2013. Biofuel Crops: Production, Physiology and Genetics; 13th section: Elephant grass, (ed. B.P. Singh), CAB International, p:271-291.

Stell,R.G.D., J.A.Torrie and D.A.Dickey. 1997. Principles and Procedures of Statistics. A. Biometrical Approach 3rd Edi. Mc Graw Hill Book. INC. NY.

Tegami Neto,Â. and S.Mello. 2007. Avaliação da produtividade e qualidade do capim paraíso (*Pennisetum hybridum*), em função de diferentes doses de nitrogênio em cobertura e freqüência de corte, Nucleus, 4(1-2):9-12.

Trotter,DJ and K.D.Johnson. 1992. Forage-testing: why, how, and where, Purdue Univ. Cooperative Extension Service Paper:337.

Wijitphan,S., P.Lorwilai and C.Arkaseang. 2009. Effects of plant spacing on yields and nutritive values of napier grass (*Pennisetum purpureum* Schum.) under intensive management of nitrogen fertilizer and irrigation, Pakistan Journal of Nutrition, 8(8):1240-1243.

Woodard,K.R., G.M.Prine and D.B.Bates. 1991. Silage characteristics of elephant grass as affected by harvest frequency and genotype, Agronomy Journal 83(3):547-551.

Zewdu,T. 2008. Effect of plant density on morphological characteristics, yield and chemical composition of napier grass (*Pennisetum purpureum* (L.) Schumach), East African Journal of Sciences, 2(1):55-61.

Živanović,L., J.Ikanović, V.Popović, D.Simić, L.Kolarić, V.Maklenović, R.Bojović and P.Stevanović, 2014, Effect of planting density and supplemental nitrogen nutrition on the productivity of *Miscanthus*, Romanian Agricultural Research, No:31, 291-298.

Permissions

All chapters in this book were first published in TJFC, by Society of Field Crops Science; hereby published with permission under the Creative Commons Attribution License or equivalent. Every chapter published in this book has been scrutinized by our experts. Their significance has been extensively debated. The topics covered herein carry significant findings which will fuel the growth of the discipline. They may even be implemented as practical applications or may be referred to as a beginning point for another development.

The contributors of this book come from diverse backgrounds, making this book a truly international effort. This book will bring forth new frontiers with its revolutionizing research information and detailed analysis of the nascent developments around the world.

We would like to thank all the contributing authors for lending their expertise to make the book truly unique. They have played a crucial role in the development of this book. Without their invaluable contributions this book wouldn't have been possible. They have made vital efforts to compile up to date information on the varied aspects of this subject to make this book a valuable addition to the collection of many professionals and students.

This book was conceptualized with the vision of imparting up-to-date information and advanced data in this field. To ensure the same, a matchless editorial board was set up. Every individual on the board went through rigorous rounds of assessment to prove their worth. After which they invested a large part of their time researching and compiling the most relevant data for our readers.

The editorial board has been involved in producing this book since its inception. They have spent rigorous hours researching and exploring the diverse topics which have resulted in the successful publishing of this book. They have passed on their knowledge of decades through this book. To expedite this challenging task, the publisher supported the team at every step. A small team of assistant editors was also appointed to further simplify the editing procedure and attain best results for the readers.

Apart from the editorial board, the designing team has also invested a significant amount of their time in understanding the subject and creating the most relevant covers. They scrutinized every image to scout for the most suitable representation of the subject and create an appropriate cover for the book.

The publishing team has been an ardent support to the editorial, designing and production team. Their endless efforts to recruit the best for this project, has resulted in the accomplishment of this book. They are a veteran in the field of academics and their pool of knowledge is as vast as their experience in printing. Their expertise and guidance has proved useful at every step. Their uncompromising quality standards have made this book an exceptional effort. Their encouragement from time to time has been an inspiration for everyone.

The publisher and the editorial board hope that this book will prove to be a valuable piece of knowledge for researchers, students, practitioners and scholars across the globe.

List of Contributors

Arif SANLI
Süleyman Demirel University, Faculty of Agriculture, Field Crops Department, Isparta, TURKEY

Tahsin KARADOGAN
Süleyman Demirel University, Faculty of Agriculture, Field Crops Department, Isparta, TURKEY

Muhammet TONGUC
Süleyman Demirel University, Faculty of Agriculture, Agricultural Biotechnology Department, Isparta, TURKEY

Sevgi CALISKAN
Nigde University, Faculty of Agricultural Sciences and Technologies, Nigde, TURKEY

Cahit ERDOGAN
Mustafa Kemal University, Faculty of Agriculture, Field Crops Department, Hatay, TURKEY

Mehmet ARSLAN
Mustafa Kemal University, Faculty of Agriculture, Field Crops Department, Hatay, TURKEY

Mehmet Emin CALISKAN
Nigde University, Faculty of Agricultural Sciences and Technologies, Nigde, TURKEY

Velimir MLADENOV
University of Novi Sad, Faculty of Agriculture, Novi Sad, REPUBLIC SERBIA

Borislav BANJAC
University of Novi Sad, Faculty of Agriculture, Novi Sad, REPUBLIC SERBIA

Mirjana MILOŠEVIĆ
Institute of Field and Vegetable Crops, Novi Sad, REPUBLIC SERBIA

Hayrettin KUSCU
Uludag University, Mustafakemalpasa Vocational School, Department of Crop and Animal Production, Mustafakemalpasa, Bursa, TURKEY

Abdullah KARASU
Uludag University, Mustafakemalpasa Vocational School, Department of Crop and Animal Production, Mustafakemalpasa, Bursa, TURKEY

Mehmet OZ
Uludag University, Mustafakemalpasa Vocational School, Department of Crop and Animal Production, Mustafakemalpasa, Bursa, TURKEY

Ali Osman DEMIR
Uludag University, Faculty of Agriculture, Department of Biosystems Engineering, Gorukle, Bursa, TURKEY

İlhan TURGUT
Uludag University, Faculty of Agriculture, Department of Field Crops, Gorukle, Bursa, TURKEY

R.W. Neugschwandtner
BOKU - University of Natural Resources and Life Sciences, Vienna, Department of Crop Sciences, Division of Agronomy, Tulln, AUSTRIA

H. Wagentristl
BOKU - University of Natural Resources and Life Sciences, Vienna, Department of Crop Sciences, Experimental Farm Groß-Enzersdorf, Groß-Enzersdorf, AUSTRIA

H.-P. Kaul
BOKU - University of Natural Resources and Life Sciences, Vienna, Department of Crop Sciences, Division of Agronomy, Tulln, AUSTRIA

Stephen OYEDEJI
Department of Plant Biology, University of Ilorin, Ilorin, Nigeria

Augustine Onwuegbukiwe ISICHEI
Department of Botany, Obafemi Awolowo University, Ile-Ife, Nigeria

Adekunle OGUNFIDODO
Department of Mathematics, Obafemi Awolowo University, Ile-Ife, Nigeria

Hasan MARAL
Karamanoglu Mehmetbey University, Ermenek Vocational School, Karaman, TURKEY

Ziya DUMLUPINAR
Kahramanmaras Sutcu Imam University, Agricultural Faculty, Department of Agricultural Biotechnology, Kahramanmaras, TURKEY

Tevrican DOKUYUCU
Kahramanmaras Sutcu Imam University, Agricultural Faculty, Department of Agronomy, Kahramanmaras, TURKEY

Aydin AKKAYA
Kahramanmaras Sutcu Imam University, Agricultural Faculty, Department of Agronomy, Kahramanmaras, TURKEY

Halil YOLCU
Kelkit Aydın Dogan Vocational Training School, Gumushane University, Gumushane, TURKEY

Adem GUNES
Department of Soil Science, Faculty of Agriculture, Ataturk University, Erzurum, TURKEY

M. Kerim GULLAP
Narman Vocational Training School, Ataturk University, Narman, Erzurum, TURKEY

Ramazan CAKMAKCI
Department of Agronomy, Faculty of Agriculture, Ataturk University, TURKEY

Hakan GEREN
Ege University, Faculty of Agriculture, Department of Field Crops, Izmir, TURKEY

Alpaslan KUSVURAN
Kizilirmak Vocational High School, Cankiri Karatekin University, Cankiri, TURKEY

Mahmut KAPLAN
Department of Field Crops, Faculty of Agriculture, Erciyes University, Kayseri, TURKEY

Recep Irfan NAZLI
Department of Field Crops, Faculty of Agriculture, Cukurova University, Adana, TURKEY

Mehmet Salih SAYAR
Crop and Animal Production Department, Bismil Vocational Training High School, Dicle University, Bismil, Diyarbakir, TURKEY

Adem Emin ANLARSAL
Department of Field Crops, Faculty of Agriculture Cukurova University, Adana, TURKEY

Mehmet BASBAG
Department of Field Crops, Faculty of Agriculture, Dicle University, Diyarbakir, TURKEY

Deniz ISTIPLILER
Ege University, Faculty of Agriculture, Department of Field Crops, İzmir, TURKEY

Emre ILKER
Ege University, Faculty of Agriculture, Department of Field Crops, İzmir, TURKEY

Fatma AYKUT TONK
Ege University, Faculty of Agriculture, Department of Field Crops, İzmir, TURKEY

Gizem CIVI
Ege University, Faculty of Agriculture, Department of Field Crops, İzmir, TURKEY

Muzaffer TOSUN
Ege University, Faculty of Agriculture, Department of Field Crops, İzmir, TURKEY

H. Ibrahim ERKOVAN
AtaturkUniversity, Faculty of Agriculture, Department of Field Crops, Erzurum, TURKEY

M. Kerim GULLAP
AtaturkUniversity, Faculty of Agriculture, Department of Field Crops, Erzurum, TURKEY

Kamil HALILOGLU
AtaturkUniversity, Faculty of Agriculture, Department of Field Crops, Erzurum, TURKEY

Ali KOC
OsmangaziUniversity, Faculty of Agriculture, Department of Field Crops, Eskisehir TURKEY

Jinbao YAO
Provincial Key Laboratory of Agrobiology, Jiangsu Academy of Agricultural Sciences, Nanjing, CHINA

Xueming YANG
Provincial Key Laboratory of Agrobiology, Jiangsu Academy of Agricultural Sciences, Nanjing, CHINA

Miaoping ZHOU
Provincial Key Laboratory of Agrobiology, Jiangsu Academy of Agricultural Sciences, Nanjing, CHINA

Dan YANG
Provincial Key Laboratory of Agrobiology, Jiangsu Academy of Agricultural Sciences, Nanjing, CHINA

MA Hongxiang
Provincial Key Laboratory of Agrobiology, Jiangsu Academy of Agricultural Sciences, Nanjing, CHINA

Agnieszka FALIGOWSKA
Poznan University of Life Sciences, Department of Agronomy, POLAND

Marek SELWET
Poznan University of Life Sciences, Department of General and Environmental Microbiology, POLAND

Katarzyna PANASIEWICZ
Poznan University of Life Sciences, Department of Agronomy, POLAND

Grażyna SZYMANSKA
Poznan University of Life Sciences, Department of Agronomy, POLAND

Ayse Betul AVCI
Ege University, Odemis Vocational School, Izmir, TURKEY

Radosav JEVDOVIC
Institute for Medicinal Plant Research "Dr Josif Pancic", Belgrade, REPUBLIC OF SERBIA

Goran TODOROVIC
Maize research Institute, Zemun Polje, Belgrade-Zemun, REPUBLIC OF SERBIA

Miroslav KOSTIC
Institute for Medicinal Plant Research "Dr Josif Pancic", Belgrade, REPUBLIC OF SERBIA

Rade PROTIC
Institute for Science Application in Agriculture, Belgrade, REPUBLIC OF SERBIA

Slavoljub LEKIC
Faculty of Agriculture, University Belgrade, REPUBLIC OF SERBIA

Tomislav ZIVANOVIC
Faculty of Agriculture, University Belgrade, REPUBLIC OF SERBIA

Mile SECANSKI
Maize research Institute, Zemun Polje, Belgrade-Zemun, REPUBLIC OF SERBIA

Javed Iqbal
Institute of Geographical Information Systems, School of Civil & Environmental Engineering, National
University of Sciences and Technology, Islamabad, Pakistan

John J. Read
U.S. Department of Agriculture, Agricultural Research Services2, Genetics and Precision Agriculture
Research Unit, Mississippi, USA

Frank D. Whisler
Department of Plant and Soil Sciences, Mississippi State University, Mississippi, USA

Mehmet Salih SAYAR
Crop and Animal Production Department, Bismil Vocational Training School, Dicle University, Bismil, Diyarbakir, TURKEY

Yavuz HAN
GAP International Agricultural Research and Training Center, Diyarbakir, TURKEY

Halil YOLCU
Kelkit Aydın Doğan Vocational Training School, Gumushane University, Kelkit, Gumushane, TURKEY

Hatice Yücel
Eastern Mediterranean Agricultural Research Institute, Dogankent, Adana, TURKEY

Selahattin CINAR
Department of Field Crops, Faculty of Agriculture, Gaziosmanpasa University, Tokat, TURKEY

Mustafa AVCI
East Mediterranean AgricultureResearch Institue, Adana, TURKEY

Serap KIZIL AYDEMIR
Variety Registration and Seed Certification, Ankara, TURKEY

Ali Akbar GHANBARI
Seed and Plant Improvement Institute (SPII), Karaj, IRAN

Seyyed Hassan MOUSAVI
Seed and Plant Improvement Institute (SPII), Karaj, IRAN

Ahmad MOUSAPOUR GORJI
Seed and Plant Improvement Institute (SPII), Karaj, IRAN

Idupulapati RAO
Centro Internacional de Agricultura Tropical (CIAT), A. A. 6713, Cali, COLOMBIA

Yasar Tuncer KAVUT
Ege University, Faculty of Agriculture, Department of Field Crops, Izmir, TURKEY

Riza AVCIOGLU
Ege University, Faculty of Agriculture, Department of Field Crops, Izmir, TURKEY

Satang HUSSANUN
Department of Plant Science and Agricultural Resources, Faculty of Agriculture, Khon Kaen University, Khon Kaen, THAILAND

Bhalang SURIHARN
Department of Plant Science and Agricultural Resources, Faculty of Agriculture, Khon Kaen University, Khon Kaen, THAILAND
Plant Breeding Research Center for Sustainable Agriculture, Khon Kaen University, Khon Kaen, THAILAND

Kamol LERTRAT
Department of Plant Science and Agricultural Resources, Faculty of Agriculture, Khon Kaen University, Khon Kaen, THAILAND
Plant Breeding Research Center for Sustainable Agriculture, Khon Kaen University, Khon Kaen, THAILAND

Emine BUDAKLI CARPICI
Uludag University, Faculty of Agriculture, Department of Field Crops, Bursa, TURKEY

Necmettin CELIK
Uludag University, Faculty of Agriculture, Department of Field Crops, Bursa, TURKEY

Beata KRÓL
University of Life Sciences in Lublin, Department of Industrial and Medicinal Plants, POLAND,

Anna KIEŁTYKA-DADASIEWICZ
State Higher Vocational School in Krosno, POLAND

Nimet KARA
Suleyman Demirel University, Faculty of Agriculture, Department of Field Crops, Isparta, TURKEY

Duran KATAR
Eskisehir Osmangazi University, Faculty of Agriculture, Department of Field Crops, Eskisehir, TURKEY

Hasan BAYDAR
Suleyman Demirel University, Faculty of Agriculture, Department of Field Crops, Isparta, TURKEY

Fahri ALTAY
Bilecik Seyh Edebali University, Bozüyük Vocational School, Bilecik, TURKEY

Muhammad AWAIS
University of Agriculture, Department of Agronomy, Agro-climatology Lab., Faisalabad, PAKISTAN

Aftab WAJID
University of Agriculture, Department of Agronomy, Agro-climatology Lab., Faisalabad, PAKISTAN

Ashfaq AHMAD
University of Agriculture, Department of Agronomy, Agro-climatology Lab., Faisalabad, PAKISTAN

Muhammad Farrukh SALEEM
University of Agriculture, Department of Agronomy, Agro-climatology Lab., Faisalabad, PAKISTAN

Muhammad Usman BASHIR
University of Agriculture, Department of Agronomy, Agro-climatology Lab., Faisalabad, PAKISTAN

Umer SAEED
University of Agriculture, Department of Agronomy, Agro-climatology Lab., Faisalabad, PAKISTAN

Jamshad HUSSAIN
University of Agriculture, Department of Agronomy, Agro-climatology Lab., Faisalabad, PAKISTAN

M. Habib-ur-RAHMAN
University of Agriculture, Department of Agronomy, Agro-climatology Lab., Faisalabad, PAKISTAN

Hakan GEREN
Ege University, Faculty of Agriculture, Dept. of Field Crops, Izmir, TURKEY

Huseyin CANCI
Akdeniz University, Faculty of Agriculture, Department of Field Crops, Turkey

Cengiz TOKER
Akdeniz University, Faculty of Agriculture, Department of Field Crops, Turkey

Nebojša MOMIROVIĆ
Belgrade University, Faculty of Agriculture, Nemanjina 6, Belgrade, SERBIA

Sneţana OLJAĈA
Belgrade University, Faculty of Agriculture, Nemanjina 6, Belgrade, SERBIA

Željko DOLIJANOVIĆ
Belgrade University, Faculty of Agriculture, Nemanjina 6, Belgrade, SERBIA

Milena SIMIĆ
Maize Research Institute, Slobodana Bajica 1, Belgrade, SERBIA

Mićo OLJAĈA
Belgrade University, Faculty of Agriculture, Nemanjina 6, Belgrade, SERBIA

Biljana JANOŠEVIĆ
Belgrade University, Faculty of Agriculture, Nemanjina 6, Belgrade, SERBIA

Živorad VIDENOVIĆ
Maize Research Institute, Zemun Polje, Department of Cropping practices, Belgrade, SERBIA

Života JOVANOVIĆ
Maize Research Institute, Zemun Polje, Department of Cropping practices, Belgrade, SERBIA

Zoran DUMANOVIĆ
Maize Research Institute, Zemun Polje, Department of Cropping practices, Belgrade, SERBIA

Milena SIMIĆ
Maize Research Institute, Zemun Polje, Department of Cropping practices, Belgrade, SERBIA

Jelena SRDIĆ
Maize Research Institute, Zemun Polje, Department of Cropping practices, Belgrade, SERBIA

Vesna DRAGIGEVIĆ
Maize Research Institute, Zemun Polje, Department of Cropping practices, Belgrade, SERBIA

Igor SPASOJEVIĆ
Maize Research Institute, Zemun Polje, Department of Cropping practices, Belgrade, SERBIA

Ali KOC
Department of Field Crops, Faculty of Agriculture, Osmangazi University, Eskisehir, TURKEY

Reza Amirnia
Department of Agronomy, Faculty of Agriculture, Urmia University, Urmia, Iran

Mahdi Bayat
Department of Agriculture, Mashhad Branch, Islamic Azad University, Mashhad, Iran

Asadollah Gholamian
Department of Biology, Mashhad Branch, Islamic Azad University, Mashhad, Iran

Mustafa TAN
Ataturk University Faculty of Agriculture Department of Field Crops, Erzurum, TURKEY

Kader KURSUN KIRCI
Ministry of Food, Agriculture and Livestock Provincial Directorate, Malatya, TURKEY

Zeynep DUMLU GUL
Ataturk University Faculty of Agriculture Department of Field Crops, Erzurum, TURKEY

R. W. Neugschwandtner
BOKU - University of Natural Resources and Life Sciences, Vienna, Department of Crop Sciences, Division of

S. Wichmann
BOKU - University of Natural Resources and Life Sciences, Vienna, Department of Crop Sciences, Division of

D. M. Gimplinger
BOKU - University of Natural Resources and Life Sciences, Vienna, Department of Crop Sciences, Division of

H. Wagentristl
BOKU - University of Natural Resources and Life Sciences, Department of Crop Sciences, Experimental Farm Groß-Enzersdorf, Groß-Enzersdorf, Austria

H.-P. Kaul
BOKU - University of Natural Resources and Life Sciences, Vienna, Department of Crop Sciences, Division of Agronomy, Tulln, Austria

Leyla GULLUOGLU
Cukurova University, Vocational School of Ceyhan, Adana, TURKEY

H. Halis ARIOGLU
Cukurova University, Faculty of Agriculture, Department of Field Crops, Adana, TURKEY

Halil BAKAL
Cukurova University, Faculty of Agriculture, Department of Field Crops, Adana, TURKEY

Hakan GEREN
Ege University, Faculty of Agriculture, Department of Field Crops, Izmir, TURKEY

Yasar Tuncer KAVUT
Ege University, Faculty of Agriculture, Department of Field Crops, Izmir, TURKEY